D0948802

Pavlov's Physiology Factory

PAVLOV'S
Physiology Factory

Experiment, Interpretation,
Laboratory Enterprise

Daniel P. Todes

The Johns Hopkins University Press

BALTIMORE AND LONDON

The Johns Hopkins University Press
2715 North Charles Street
Baltimore, Maryland 21218-4363
www.press.jhu.edu

Library of Congress Cataloging-in-Publication Data
Todes, Daniel Philip.
    Pavlov's physiology factory / Daniel P. Todes
        p.   cm.
    Includes index.
      ISBN 0-8018-6690-1 (alk. paper)
      1. Pavlov, Ivan Petrovich, 1849–1936.   2. Physiologists—Russia
(Federation)—Biography.   3. Medical laboratories—Russia
(Federation)—History—19th century.   I. Title.
QP26.P35 T634 2002
571.1′9′092—dc21

00-012972

A catalog record for this book is available from the British Library.

For my beloved Eleonora

CONTENTS

PREFACE

This book resulted from a roadblock in my work on a biography of Ivan Pavlov. By late 1994 I was reasonably satisfied with my draft chapters on the first forty years of Pavlov's life, including my account of his research in the 1870s and 1880s on the physiology of the heart and the digestive system. The science was interesting, and I treated it in a conventional biographical manner by exploring the various contexts in which Pavlov worked, by studying his publications and those of relevant contemporaries, and by examining the archival materials (manuscripts, memoirs, correspondence, and a few laboratory notebooks) I had gathered.

This approach proved inapplicable, however, when I came to 1891 and the investigations on digestion that culminated in Pavlov's *Lectures on the Work of the Main Digestive Glands* (1897) and his Nobel Prize (1904). Before 1891 Pavlov had been a lone investigator, but in that year he became a laboratory chief—the director of a large and productive research group in his Physiology Division of the Imperial Institute of Experimental Medicine in St. Petersburg. Even a glance at Pavlov's *Lectures*—with its constant references to the experiments of his numerous coworkers—revealed that the chief's synthetic work was understandable only as the product of his laboratory system.

An examination of the few laboratory notebooks I could find only underscored this point. The notebooks from the years before 1891 contained notes, data, and other records of a single investigator puzzling over his subject. Those from after 1891 had an entirely different character: almost completely lacking experimental data (which, I later learned, were kept by Pavlov's coworkers in notebooks arranged by experimental animal), these were the records of a manager and thinker reflecting on the results obtained by his subordinates and planning their future work.

My biographical interests, then, led me to a very different task: to reconstruct the social and cognitive dynamics of Pavlov's laboratory enterprise and to place his own work, as manager, experimentalist, and thinker, within this dynamic. How did Pavlov's laboratory enterprise generate knowledge claims and other products? What was the relationship between the nature of this laboratory, Pavlov's scientific work, and his professional trajectory?

So famous has Pavlov (and his salivating dogs) become as a cultural symbol of the experimental, objectivist approach to mind that relatively few people today are aware that he first achieved prominence for his earlier contributions to digestive physiology, let alone that these contributions featured an insistence on the central physiological role of an idiosyncratic psyche. This book addresses that earlier Pavlov. I explore his transformation from lone investigator to laboratory chief, his creation of a laboratory system that would prove a prototype for the much larger system over which he presided in future years, and the investigations of digestive physiology that brought him international fame and launched him upon the later research on conditional reflexes for which he is best known today.

On Terminology

Readers with expertise in physiology, medicine, and biochemistry will, at many junctures, note that I relate without comment what now seem to be clearly tendentious interpretations by the scientists I discuss and that I do not "update" scientific terminology. For example, one might be tempted to identify the nineteenth-century term *ferment* with today's *enzyme,* the earlier *zymogen* with today's *proenzyme,* the earlier *humoral mechanism* with today's *hormone,* and the earlier *catarrh* with today's *inflammation of the gastric membrane.* I have instead followed the historical convention of using actors' terms and categories—and, whenever possible, I allow these terms and categories to define themselves through usage and context. This approach, it seems to me, allows us to better understand the reasoning of scientists and physicians who lived in a different time and place. Because terminology is embedded in a broader discourse, even apparently simple terminological changes often involved complex conceptual changes. For example, the transition in the late nineteenth century from *proteid* to *protein* seems to have been relatively straightforward, while the cluster of meanings surrounding the words *ferment* and *enzyme* did not map nearly so neatly and often differed from scientist to scientist. These transitions, like the various and changing meanings ascribed to *catarrh, dyspepsia,* and even the word *scientific,* are themselves the subject of historical inquiry.

Acknowledgments

It is most gratifying to have this opportunity to thank the many institutions and individuals who have helped me write this book.

The National Endowment for the Humanities, Fulbright-Hays, the International Research and Exchanges Board, and the John Simon Guggenheim Memorial Foundation generously funded my continuing research on a biography of Ivan Pavlov and so provided the necessary resources for this product of that work. I am also grateful to my Department of History of Science, Medicine, and Technology at the Johns Hopkins University—especially to its chairperson, Gert Brieger; its longtime administrative manager, Dolores Sawicki; and Linda Bright and Christine Ruggere of its historical collection—for constant support and assistance. The good efforts of directors and staff at numerous archives have also proven indispensable. I am particularly grateful to Vladimir Sobolev and the talented and hospitable staff at the St. Petersburg branch of the Archive of the Russian Academy of Sciences, to Iurii Vinogradov for sharing his great knowledge of Pavlov and his archival legacy, to Natalia Zagrina and the fine staff at the Dom-Muzei I. P. Pavlova in Ryazan, to Anita Lundmark at the Karolinska Institute in Stockholm, and to Denise Ogilvie at the Pasteur Institute in Paris.

This book bears the imprint of many colleagues and friends who have been generous with their time and energy, allowing me to draw upon their knowledge and insights. Five colleagues read the entire manuscript with great care. Mark Adams first encouraged me to write a biography of Pavlov, and then to write the present book along the way; he regularly discussed with me the issues arising in my research and read several versions of the manuscript with his usual intelligence and perspicacity. Two wonderful colleagues at the Johns Hopkins University, Larry Schramm and Harry Marks, also commented extensively on the manuscript to my great benefit—Larry with the sensitivities of a physiologist and lover of history, Harry with the intellectual engagement and insight that he brings to all historical subjects. Larry also kindly arranged for me to discuss Pavlov's physiology with two groups of supportive scientists at Johns Hopkins. Nikolai Krementsov brought me the benefit of both his earlier incarnation as a Pavlovian physiologist and his expertise on the history of Russian science, and Larry Holmes made important suggestions based on his unparalleled knowledge of the history of physiology.

I am also grateful to the following historians and physiologists for their critical and supportive reactions to chapter drafts and oral presentations and for helpful conversations and responses to specific queries: Lloyd Ackert, Paul Andrews, Keith Barbera, Horace Davenport, Graham Docking, Barry Dworkin, Laura Engelstein, Mary Fissell, Robert Frank Jr., Joseph Fruton, Gerald Geison, Sander Gliboff, David Grundy, Robert Kargon, Cecil Kidd, Anna Krylova, Timothy Moran, Alan Ross, James Schafer, Gary Schwartz, Chandak Sengoopta, Jay Shulkin, Tilli Tansey, Andrea Varro, John Harley Warner, Ursula Wesselmann, John West, and Charles Yeo. Linda Strange, copy editor for The Johns Hopkins University Press, sharpened the manuscript substantially by her careful commentary and queries.

Material from the Nobel Archives was kindly provided by the Nobel Committee for Physiology or Medicine. My thanks to The University of Chicago Press for permission to republish large sections of my article "Pavlov's Physiology Factory," *Isis* 88, no. 2 (1997): 205–46, which comprise almost the entirety of Chapter 3 of this book. Portions of Chapter 7 originally appeared in my article "From the Machine to the Ghost within: Pavlov's Transition from Digestive Physiology to Conditional Reflexes," *American Psychologist* 52, no. 9 (1997): 947–55. Copyright © 1997 by the American Psychological Association. Reprinted with permission.

I took everybody's comments and advice to heart, but the final decisions and interpretations are my own, and I of course bear full responsibility for them and for any errors that remain.

I am most profoundly indebted to my wife, Eleonora Filippova, who contributed immensely to the research on this book, served as my first reader and critic, and sustained me with her confidence, perspective, and love. My daughter, Sarah, was a delightful companion during trips to Russia, listened patiently to endless tales about scientists, dogs, and secretory curves, and showed great forbearance concerning "daddy's obsession with Pavlov." My dear friend Marc Levine was a constant source of wit and wisdom. Special thanks to Pavlov's granddaughters and great-granddaughter, Ludmila Balmasova, Maria Sokolova, and Marina Balmasova, for their kindness and support.

INTRODUCTION

What is a scientific laboratory? It is a small world, a small
corner of reality. And in this small corner man labors with his
mind at the task of . . . knowing this reality in order correctly
to predict what will happen, . . . even to direct this reality according to his will, to command it, if this is within our technical means.

—IVAN PAVLOV, "On the Mind in General" (1918)

In four successive years (1901–4) Ivan Pavlov was nominated for the Nobel Prize, and each time the award committee confronted the same question: to what extent were
the products of Pavlov's laboratory truly Pavlov's?

That question arose for good reason. The nominee had
himself pronounced his most substantial work, *Lectures on
the Work of the Main Digestive Glands* (1897), "the deed of
the entire laboratory" and credited his coworkers by name
for conducting the experiments on which it was based. Furthermore, he referred readers seeking evidence for his most
important arguments to the publications of his coworkers,
where many of those arguments first appeared. Did Pavlov's
major works, then, represent his own original contributions
to science or were they merely "a type of compilation of the
experimental dissertations upon which they are based"?[1]

Guided by an image of the heroic lone investigator, the
Nobel Prize Committee here confronted a different form of
scientific production. As the nineteenth century wore on
and the workshop yielded pride of place to the factory in

goods production, so, too, were leading laboratory scientists increasingly likely to be the managers of large-scale enterprises. Justus von Liebig and Felix Hoppe-Seyler in chemistry, Carl Ludwig and Michael Foster in physiology, Robert Koch and Louis Pasteur in bacteriology, and Paul Ehrlich in immunology all presided over distinctively social enterprises involving substantial capital investment, a specially designed workplace, a relatively large workforce, a developed division of labor, and a productive process that involved managerial decisions. Clearly, their achievements owed something not only to their scientific (and rhetorical) skills but also to their qualities as masters of large-scale production.[2]

Scientists at the turn of the century fully appreciated the significance of this newly emerging system of production, and they filled their journals with articles describing the physical plant, design, workforce, technologies, and managerial style of various laboratories.[3] Writing in 1896, William Henry Welch observed that the emergence of the "well-equipped and properly organised modern laboratory" had "completely revolutionised during the past half-century the material conditions under which scientific work is prosecuted." For Welch, as for many others, this development only enhanced the heroic stature of such earlier figures as Claude Bernard, who had contributed mightily to science in premodern conditions. "Bernard, that prince of experimenters, worked in a damp, small cellar, one of these wretched Parisian substitutes for a laboratory which he called the 'tombs of scientific investigators,'" Welch reminded his readers. "There can be no greater proof of the genius of Bernard than the fact that he was able to make his marvellous discoveries under such obstacles and with such meagre appliances."[4]

Nobody was more keenly aware of the advantages of the modern laboratory than another of Bernard's admirers, Ivan Pavlov. Having labored in a scientific workshop for some fifteen years—and having glimpsed the alternative during brief stays in Rudolf Heidenhain's and Carl Ludwig's laboratories—in 1891 Pavlov became master of Russia's first large-scale physiological enterprise. The laboratory system he established, however, was no carbon copy of those he had seen in Western Europe. Like scientific workshops, large-scale laboratories differed greatly one from the other. Each laboratory chief brought with him differing conceptions of his science and differing approaches to the management of complex tasks, and each responded to the challenges and opportunities of a different set of institutional circumstances.[5]

In the last decades of the nineteenth century, a variety of large-scale laboratories addressed the science of life, physiology. Carl Ludwig in Leipzig managed the research of a large group of transient, mostly unskilled coworkers—

usually aspiring physicians—who addressed a wide variety of subjects that seem linked only by the chief's perception that they were important and, perhaps, by his drive to apply physicochemical models to life processes. Carl Voit in Munich managed an interdisciplinary enterprise that coordinated the work of a more stable group of investigators—each with special expertise in anatomy, physiology, or chemistry—on highly focused studies of metabolic processes. Michael Foster in Cambridge managed only very loosely the undergraduates, graduates, and permanent researchers whom he inspired to adopt his evolutionary and broadly biological approach to a wide range of topics, but who remained free to follow their own evolving interests.[6] Ludwig and Voit were accomplished experimentalists who could always be found in the laboratory; Foster, like William Henry Welch, was not. The similarities and differences among these laboratories could be enumerated almost ad infinitum. How was work organized and intellectual credit divided? What drove the research agenda? Who controlled interpretive decisions, and how? What was the relative importance and relationship of research and teaching? What products did they generate, and for whom?

Pavlov's laboratory system shared some features with those of his fellow physiologists, but it was the distinctive result of institutional resources and his scientific-managerial vision. First and foremost, his laboratory was organized for efficient production and, as in any factory, the labors of his coworkers were tightly coordinated for a common task. In Pavlov's physiology factory, that task was prescribed by his scientific vision as applied to digestive physiology. The managerial system by which he coordinated laboratory work featured an authoritarian structure and cooperative ethos that allowed Pavlov to use coworkers as extensions of his sensory reach while enabling him constantly to monitor the work process, to control the interpretive choices in experiments, to incorporate results into his developing ideas, and to convert them into marketable products.

The nineteenth-century factory was not an assembly line run on F. W. Taylor's principles of scientific management, let alone the finely mechanized and computerized operation that one can find today. In both manufacturing and science, the new large-scale enterprises coexisted alongside traditional workshops, and factories themselves incorporated elements of the workshop tradition. Despite the increasing division of labor and reliance on technologies, artisanal skills retained their importance. As Raphael Samuel has explained, skilled workers proved critical to the operation of early factory machinery, which was often "too crude and indiscriminate for the tasks it was appointed to perform . . . The difficulty often lay with the raw materials which were too delicate, or too variable, for machinery's harsh beat. No two skins were ever precisely the same in the leatherworking trades, no two grains in furniture." Machinery, therefore, "was rarely self-acting, but required skilled hands to guide and to complete its work."[7]

Standardized methodologies and new technologies acquired increasing importance in late nineteenth-century science, but scientists, too, dealt with extremely delicate and variable materials. Pavlov and his coworkers, for example, worked largely with intact, complex animals and were well aware that no two dogs—and no two experiments—were precisely the same. Craft skills, puzzle-solving abilities, and imagination, then, proved no less important in Pavlov's laboratory—and in large-scale scientific enterprises in general—than in the scientific workshop.

The word *workshop* has long evoked a much more pleasing image than the word *factory,* especially in reference to knowledge-generating enterprises. In this spirit, turn-of-the-century scientists frequently referred to themselves as *craftsmen,* and one series of articles in *Nature* about the new large-scale laboratory enterprises of the day bore the appealing title "Famous Scientific Workshops."[8] Even today, many academic departments—in a tradition dating from the late nineteenth century—refer to their weekly sessions with a visiting speaker as a *workshop.* This word evokes a pleasing image (however historically inaccurate) of communality, the absence of hierarchy, and the happy unity of head and hands.

The word *factory,* on the other hand, calls forth images of alienated labor and a standardized, exploitative, even inhuman process that violates the spirit of intellectual inquiry. Factories may produce things efficiently, but they lack the contemplative, cooperative, egalitarian image that scholars and scientists commonly associate with the nature of their work. Some decades ago, one group of socialist historians named their journal *History Workshop;* it is difficult to imagine any historians—left, right, or center—adopting the title *History Factory.* We all, of course, might label another department, laboratory, or school a "real factory." But never our own. Pavlov himself, a committed truth-seeker who celebrated the communal ethos of science, would certainly have objected to my description of his laboratory as a factory.

I do not, however, use this word pejoratively. My intention, rather, is to emphasize the differences between the production process in a small enterprise and in a large one, to describe the system by which Pavlov directed the labors of his coworkers to his own investigative ends, and to explore—as did the Nobel Prize Committee for its own, quite different reasons—the structure and functioning of his laboratory.

The first part of this book, "The Factory," explores Pavlov's laboratory system as the union of a particular man and a particular set of institutional circumstances. It examines the resources, challenges, and incentives that the Imperial Institute of Experimental Medicine brought to this union (Chapter 1), the origins and content of the animating principles that Pavlov brought to his post there (Chapter 2), and the social-cognitive dynamics of the laboratory that resulted from the merger of the two (Chapter 3).

In Part II, "Producing Physiology," I turn to the central work process in the laboratory: experimentation and interpretation. Pavlov was indeed, as he once put it proudly, "an experimenter from head to toe," but the meaning of that phrase is hardly transparent. In recent years, the "black box" of experimentation has been opened by scientists and historians alike, revealing it to be a complex, highly contingent process that resists easy definition. We now know from a number of case studies that, as in other forms of scientific inquiry, the experimenter's confrontation with an infinitely complex reality is influenced by a broad range of factors including the institutional, disciplinary, and social context in which he or she works; the availability and choices of technique and model organism; responses to unexpected results and opportunities; and interpretive decisions in the face of complex and often contradictory data. In their reactions to these and other elements of what one scholar has termed "the mangle of practice," experimenters display various styles. In a group enterprise, the chief's style as an experimentalist is necessarily entangled in the social relations of his or her laboratory.[9]

To approach experimentation as a form of work—and science as a form of production—is not to neglect the importance of ideas, but rather to emphasize it. Ideas are inherent to any productive process, whether in a workshop or a factory, in manufacturing or in science. They may animate all participants in the work process or the manager alone; they may be codified in a manual or simply embedded in technologies and the organization of labor—but they are always an important force of production.

This point was made forcefully by both Karl Marx and Max Weber. In a famous comment about workshop production, Marx observed that "a spider conducts operations that resemble those of a weaver, and a bee puts to shame many an architect in the construction of her cells. But what distinguishes the worst architect from the best of bees is this, that the architect raises his structure in imagination before he erects it in reality." Factory production, he noted, involved the distribution of this "intelligence" among a differentiated workforce and machines.[10] Weber objected forcefully to what he perceived as the fashionable view of his day that imagination had no place in modern science because the laboratory had become a factory. Proponents of that view, he observed, lacked "all clarity about what goes on in a factory or in a laboratory. In both some idea has to occur to someone's mind." "The idea is not a substitute for work," he observed, "and work, in turn, cannot substitute for or compel an idea, just as little as enthusiasm can."[11] This, I think, is self-evident to any person—scientist or historian, craftsperson, laborer, or factory manager—who reflects on his or her own labors.

Pavlov's particular scientific style makes attention to his underlying ideas about the organism and physiology especially important, for he was always a resolutely vision-driven experimentalist. What he termed *physiological thinking* rested on an unchanging view of the organism and how best to explore it, and, contrary to his carefully cultivated image, his style of physiological investigation relied heavily on his interpretive choices amid complex and often contradictory data.

Thus Part II explores the process by which experiment and interpretation gave rise to the laboratory's most important knowledge claims. I examine the laboratory career of Pavlov's favorite dog and the key concepts that resulted (Chapter 4), the emergence of Pavlov's central explanatory metaphor and its relationship to the changing interpretation of data (Chapter 5), and Pavlov's major synthetic treatise, both as a final stage in data processing and as a rhetorically powerful literary product that communicated an appealing vision of the relationship between the laboratory and the clinic (Chapter 6). Finally, in Chapter 7, I discuss the transformation of the laboratory's view of the psyche in the years 1897–1904 and analyze the related social-cognitive processes that fueled the transition from digestive physiology to investigations of conditional reflexes.

Parts I and II, then, concern the origins, structure, and functioning of Pavlov's physiology factory and the nature of its experimental practices. Part III, "Laboratory Products," addresses several questions: what did Pavlov's physiology factory actually produce, for whom, and to what effect? Here the distinctive qualities of factory production come clearly to the fore, and we see the important role of the sheer number and diversity of laboratory products. In the years 1891–1904, the laboratory produced not only an avalanche of knowledge claims and publications aimed at various audiences but also medical advice, techniques and technologies, digestive fluids for investigative and clinical use, expertise in laboratory design, alumni, and—as the spokesman for all these products—Pavlov himself, who became a powerful symbol of a modern clinically relevant, laboratory-based physiology.

The chapters in Part III explore this process from three different perspectives. Chapter 8 focuses on a single product, the laboratory's most important direct contribution to medical practice: "the natural gastric juice of the dog," which the laboratory drew from experimental dogs and sold as a remedy for dyspepsia. Chapter 9 examines Pavlov's rise to international renown as the spokesman for and embodiment of the laboratory's product line. Here, too, I explore the developments in turn-of-the-century physiology and physiological chemistry that, just as Pavlov's reputation attained new heights, subjected some of his most important scientific conclusions to serious criticism. Finally, Chapter 10 examines the assessment of Pavlov's achievements by a single audience—the Karolinska Institute's Nobel Prize Committee for Physiology or Medicine.

That Committee finally decided that the products of Pavlov's laboratory were Pavlov's indeed, and in 1904 it made him the first physiologist to win the prize. In pondering the question of intellectual property, the Committee had confronted the different features of Pavlov's scientific profile. It was precisely the integration of these features that lay at the core of his laboratory system and his astonishing range of achievements, and that had already made him, for many of his colleagues, the symbol of a new era in laboratory physiology. This book, then, concerns this multifaceted Pavlov—the thinker, experimentalist, manager, and spokesman; in a phrase, the visionary factory physiologist.

A merchant or a big industrialist without "business imagination," that is, without ideas or ideal intuitions . . . will never
be truly creative in organization. Inspiration in the field of
science by no means plays any greater role, as academic con-
ceit fancies, than it does in the field of mastering problems of
practical life by a modern entrepreneur. On the other hand,
and this also is often misconstrued, inspiration plays no less a
role in science than it does in the realm of art.

> —MAX WEBER, "Science as a Vocation" (1919)

The principle of a factory is that each labourer, working separately, is controlled by some associating principle which directs his producing powers to effect a common result, which
it is the object of all collectively to obtain.

> —*Palgrave's Dictionary of Political Economy* (1925)

So famous is Ivan Pavlov, so striking his undeniable tal-
ents, and so important his place in twentieth-century science and culture, that it is difficult to imagine him as anything other than a phenomenon. Winner of a Nobel Prize
at age fifty-five and celebrated during his lifetime as the
"Prince of World Physiology," this passionate, frosty-
bearded Russian (with his salivating dogs) is an internationally recognized symbol of the power not only of experimental biology but of science itself.

That familiar Pavlov is difficult to recognize in the forty-year-old man who, in 1889, having failed in two attempts to land an assistant professorship at a Russian university, squeezed his experiments into the time remaining from his duties as assistant in a small and filthy laboratory. Precious time, he lamented, was slipping away unproductively. Financially strapped and with no clear prospects, he suffered from a myriad of bodily symptoms, which, drawing upon his medical training, he forlornly diagnosed as a fatal case of "tabes." By his fortieth birthday, Pavlov had a completed doctoral thesis and a number of published articles to his credit. A specialist on the nervous control of the heart and digestive glands, he was a talented and well-trained investigator with grand ideas about his science, but had he indeed perished that year few would remember him today.

Two years later, in 1891, Prince Alexander Ol'denburgskii appointed Pavlov chief of the Physiology Division in his newly created Imperial Institute of Experimental Medicine. The result was not merely that Pavlov finally obtained his own laboratory and the opportunity to conduct his experiments full-time. This union of the man and the institution fundamentally transformed his scientific profile: the workshop physiologist became a factory physiologist. Pavlov's physiology factory was born, and its chief began a rapid ascent to world renown.

The struggling physiologist with his ideas and skills, the Prince's Institute with its resources and incentives, the unlikely process that brought them together, and the laboratory that resulted from their union were all deeply steeped in their place and time—in Russian social, cultural, and political life from Tsar Alexander II's Great Reforms of the 1860s to the country's industrial revolution in the late 1880s; in the exciting era in science and medicine associated with the names of Darwin, Bernard, Ludwig, Pasteur, and Koch; in the play of networks, professional structures, and the state bureaucracy. These structured the individual journeys of our two protagonists—the physiologist and the Institute—and shaped the results of their eventual union. Chance, too, played its part.

Chapter 1

THE PRINCE AND HIS PALACE

Human dignity and human pride demand palaces to apply and manifest the power of the human intellect.

— IVAN PAVLOV, "The Contemporary Unification in Experiment of the Main Aspects of Medicine, as Exemplified by Digestion" (1899)

On November 15, 1885, a rabid setter named Pluto bit Alexander Dem'ianenkov, an officer in St. Petersburg's Corps of Guards. This event was in itself unextraordinary, for tsarist Russia's many dubious medical distinctions included its world leadership, together with Hungary and Italy, in reported cases of rabies.[1] Pluto's bite, however, proved momentous because it prompted the officer's commander, Prince Ol'denburgskii, a colorful member of the extended royal family and a philanthropist *engagé*, to convert a series of events and circumstances in Paris, Berlin, and St. Petersburg into what he hoped would be a grand bacteriological institute—a "palace" of scientific medicine similar to those being founded by Pasteur, Koch, and Lister in the West.[2]

At the moment that Pluto sank his teeth into Alexander Dem'ianenkov, Ivan Pavlov (1849–1936) was on a study leave in Europe, where he was working in the physiological laboratories of Carl Ludwig and Rudolf Heidenhain. Thirty-six years old and struggling to support a wife and a one-year-old son, Pavlov was no doubt already worrying about his bleak job prospects at home in St. Petersburg. Having acquired his medical degree as a prerequisite for a career in physiology and his doctorate for a thesis on the nerves of the heart, he would hardly have had reason to see his personal salvation in Prince Ol'denburgskii's efforts.

Social networks, the politics of institution-building, and a set of unforeseeable contingencies soon combined, however, to make the Prince and the struggling physiologist central to each other's ambitions. In December 1890, five years after Pluto's bite, Prince Ol'denburgskii celebrated the founding of his creation—though it was not the bacteriological institute he had originally envisioned—and Ivan Pavlov had found in it an institutional home that would facilitate his transformation into Russia's leading physiologist.

The Philanthropist Prince

Prince Alexander Petrovich Ol'denburgskii (1844–1932) was heir not only to his family's considerable fortune but also to its philanthropic traditions. These traditions had developed together with those of the ruling Romanovs, with whom the Ol'denburgskiis were joined by friendship and marriage.

A branch of the venerable Holstein-Hottorp family, the Ol'denburgskiis began in the early eighteenth century to intermarry with the Romanovs. One issue of this familial relation was our Prince's father, Petr Georgievich Ol'denburgskii (1812–81), the son of Petr Friedrich Georg Ol'denburgskii and Ekaterina Pavlovna Romanova, the daughter of Empress Maria Fedorovna and Tsar Paul I.

Petr Georgievich received his early education in Oldenburg before being summoned back to Russia by his uncle, Tsar Nicholas I, who assigned him to service as colonel of the palace guard, the famous Preobrazhenskii regiment. In 1845 he was awarded the title *Vashe Imperatorskoe Vysochestvo* (Your Imperial Excellency), signifying his membership in the extended royal family. He partook fully in the life of the tsarist court and enjoyed a lifelong friendship with Alexander II, the "Tsar Liberator" who reigned from 1855 to 1881 and instituted a number of modernizing reforms, most notably the emancipation of the serfs in 1861.[3]

By the time of Alexander II's accession to the throne, the Romanov family, especially its female members, had cultivated a rich philanthropic tradition.[4] Empress Maria Fedorovna (1759–1828), wife of Tsar Paul I, devoted considerable energy to the education of women—for example, as head of the Society for the Upbringing of Well-Born Young Ladies (founded in 1796) and, from 1797, as head of St. Petersburg's homes for young women without means (*vospitatel'nye doma*). The wives of the two tsars who reigned in the first half of the nineteenth century were well known for their charitable works: Empress Elizaveta Alekseevna (1779–1826), wife of Tsar Alexander I, donated the lion's share of her fortune to charity; and Empress Alexandra Fedorovna (1798–1860), wife of Tsar Nicholas I, succeeded Maria Fedorovna as trustee of the empire's philanthropic institutions.

Romanov philanthropy took a medical turn during the Crimean War (1853–56), particularly through the efforts of the Grand Duchess Elena Pavlovna (1806–73).[5] Before the war, the Grand Duchess had managed two institutes concerned with women's education and safe childbirth; during the war she organized the Krestovozdvizhenskoi Obshchiny Sester Miloserdiia (The Cross of the Exalted Commune of Sisters of Mercy), the forerunner of Russia's Red Cross.[6] She also founded an institute to provide young physicians with additional training and investigatory experience. After the Grand Duchess's death, this important and prestigious institution in St. Petersburg medicine was renamed in her honor, becoming the Clinical Institute of Grand Duchess Elena Pavlovna.

Many other female members of the royal family shared the Grand Duchess's medical interests. Empress Maria Alexandrovna (1824–80), wife of Tsar Alexander II, organized the Red Cross and a series of philanthropic societies. Grand Duchess Elizaveta Fedorovna (1864–1918) established and governed several charitable organizations in St. Petersburg and Moscow, and during the Russo-Japanese War (1904–5) she donated considerable funds for the purchase of medical trains, aid to widows and orphans, and so forth. At least one Grand Duke, Pavel Alexandrovich (1860–1919), took an interest in medical charity as honorary president of the Russian Society for the Preservation of Public Health. This Romanov tradition continued through the First World War up to the overthrow of the monarchy. Empress Alexandra Fedorovna, wife of the last tsar, Nicholas II, patronized sanitary divisions at the front, established and supervised field hospitals, and even completed the necessary courses to work as a nurse.

An enlightened member of the gentry, committed to the modernizing mission of his friend and sovereign Alexander II, Prince Petr Georgievich Ol'denburgskii partook fully in this philanthropic tradition and its medical orientation, bringing to it considerable organizational skills. Aside from service in the military and various governmental bodies, he contributed money and served as founder, president, or trustee of an extensive network of educational and medical institutions: the Imperial Juridical Academy, the Aleksandrovsk Lycée, the Free Economic Society, the Main Council of Women's Institutions of Learning, several hospitals (including one for the mentally ill), and several

Communes of the Sisters of Mercy. These activities—in which the Prince co-operated with his wife and other members of the royal family—involved constant contact with leading figures in Russian statecraft, education, and medicine and created a rich network of expertise and mobilizable connections.

Most important for our story are the St. Troitskii and St. George's Communes of the Sisters of Mercy, each of which, as E. A. Annenkova and Iu. P. Golikov have put it, constituted "an entire village" dedicated to the welfare and medical care of the indigent. Founded in 1844 by Prince Petr Ol'denburgskii's wife, Theresa, and two of Tsar Alexander I's daughters, the St. Troitskii Commune developed by the 1880s into an extensive facility for the care and medical training of indigent women. Under the trusteeship of Prince and Princess Ol'denburgskii, this commune expanded to include a church, several hospitals with outpatient clinics, a dispensary of free medicines, a shelter for "fallen women," a gymnasium, and courses that trained women as teachers or as assistants in hospitals and infirmaries. The St. George's Commune, founded in 1870 by Maria Fedorovna (1847–1928, wife of the future Tsar Alexander III), was devoted to the free care of the indigent and wounded. The Empress's personal physician, Sergei Petrovich Botkin, oversaw the St. George's Commune's medical facilities and convinced the Ol'denburgskiis, who had also participated in its creation, to finance its courses for medical paraprofessionals (*fel'd-shery*).[7]

This, then, was the tradition—with its moral vision, institutions, and social networks—into which Prince Alexander Petrovich Ol'denburgskii was born. Like his father, Alexander combined a military career with that of a philanthropist *engagé*. He distinguished himself in combat during the Russo-Turkish war of 1877–78, and in 1885 was appointed commander of the twenty-two thousand or so soldiers in the palace guard. When his father died in 1881, the Prince assumed a number of his institutional positions—for example, as trustee of the Imperial Academy of Jurisprudence, the St. Troitskii Commune, and the Charity Home for the Mentally Ill. The Prince also undertook his own initiatives. One of his favorites was the creation of St. Petersburg's Narodnyi Dom (People's House), which offered the lower classes a clinic for alcoholics and such affordable alcohol-free entertainment as theater, opera, a library, musical instruction, and inexpensive restaurants.

The Prince's wife, Princess Evgeniia Maksimilianovna (1845–1925), the granddaughter of Tsar Nicholas I, was an equally energetic activist and philanthropist who supported and oversaw numerous scholarly, medical, and artistic institutions. She served, for example, as president of the St. Troitskii Commune and honorary trustee of the St. George's Commune, as trustee of the St. Petersburg School of the St. Petersburg Women's Patriotic Society and several shelters for women and children, and as president of the Imperial Mineralogical Society and the Society for the Support of the Arts.

The couple's philanthropic energies were also evident in their management of the estate in Ramon (in central Russia, near Voronezh) that Tsar Alexander II had presented to the Princess (his niece). Here the Princess created one of only two hospitals in the entire province, a well-equipped facility staffed by a modern physician, Pavel Khizhin, who had studied surgery with the acclaimed Moscow University Professor V. A. Basov.[8] According to one account, "Tens of thousands of suffering people streamed there seeking advice and aid; hundreds of important operations were performed with complete success, thanks to the modern setting; . . . thousands were cured of the wasting fever and of that terrible disease [syphilis] that undermines the health of the population and even influences its distant progeny." The Ol'denburgskiis also created there a school and an inexpensive cafeteria for the local peasantry. "In this cafeteria a peasant could use his earnings to purchase good food and, perhaps, even to drink a small cup of vodka—but would not receive a second one."[9]

The estate's sugar factory also came to embody this spirit of enlightened gentry philanthropy. As one Russian later recalled: "I can still remember very well the Ramon sugar candies or *Monpensier*. We ate them as children in the eighties and nineties. Instead of the unappetizing molasses candies we had been used to up to that time, it was now possible to buy for 25 kopeks (12½ cents) a 1 lb. tin of good candy made with pure sugar. The Prince wanted to give the people inexpensive sweets of fine quality. The Prince's candy-making project failed, probably because of his inexperience in business matters, and his factory was placed in the hands of promoters. However, the example set by the Prince led other private concerns to manufacture candy cheaply and well."[10]

The philanthropic activities of the Prince and Princess (Figure 1) were widely recognized and celebrated—for example, on the occasion of their twenty-fifth wedding anniversary in 1893—and the Prince's contributions were recognized in 1904 through conferral of the title "honorary citizen of St. Petersburg," which carried with it the right to vote in the city Duma.[11]

Prince Ol'denburgskii's effectiveness as a patron rested not only on his considerable financial resources and energy but also on his court connections, his erratic but forceful personality, and his talent for extracting money from influential personages in the government, including the Tsar himself. Finance Minister Sergei Witte characterized the Prince as "a lively man possessed of such a quality of character that when he pesters people, including those stand-

FIGURE 1. *The Imperial family. Prince Ol'denburgskii stands* second from the left in the back row; *Tsar Alexander III sits* third from the right in the second row.

ing above Prince A. P. Ol'denburgskii himself, they agree to the payment of hundreds of thousands of rubles from the state purse, if only to rid themselves of him."[12] The Prince was "a good person" engaged in "useful activity," but was also "abnormal"—emotionally erratic, quick to anger, and capable of "the most impossible acts." In Witte's view, no other member of the royal family had so completely "inherited the qualities of [the ostensibly unbalanced] Tsar Paul."[13] Another leading figure in court circles observed similarly that the Prince was "very capable, but extremely unbalanced [and] explosive, and lacking sufficient inhibitory centers."[14] A third contemporary later recalled that the Prince's "hot temper led him to do many strange things," which rendered him especially effective in tasks for which "a certain degree of coercion was necessary, for everyone was afraid of him."[15] The Prince could also be quite charming, eliciting the following description by one foreign visitor of 1886: "With a still-youthful appearance, he was large, with an open and sympathetic appearance—amiable, but commanding respect."[16]

Like his father, Prince Ol'denburgskii was an enthusiast and a decidedly hands-on patron. Aside from donating sizable personal funds and securing state financing, he devoted considerable energy to the daily management of his creations and to publicizing their achievements. His enthusiasms were varied and mirrored those of his time. For example, during the vogue of spiritualism among the Russian intelligentsia in the 1890s, he transported a purported medium (by train) from the sugar refinery in Ramon to St. Petersburg for a séance.

Most important for us here, Prince Ol'denburgskii combined his family's rich tradition of medical philanthropy and the networks he had inherited with great enthusiasm for the possibilities of scientific medicine. Alert to Western European developments, he championed numerous innovative medical treatments rooted in the science of the day, ranging from the rabies vaccine, tuberculin, and light therapy developed by established medical authorities to the vibrating table that the Prince himself designed to cure various nervous ailments.[17]

Laying the Foundations

When Pluto bit Alexander Dem'ianenkov, Prince Ol'denburgskii reacted to this attack on his officer not simply as a misfortune but as an event made especially meaningful by the medical developments of the day—and so as an opportunity. Louis Pasteur had just issued his sensational announcement of a rabies vaccine, riveting the eyes of the medical world upon him.[18] The Prince dispatched his wounded officer to Paris, together with the military physician N. A. Kruglevskii, who was to study Pasteur's techniques and acquire a sample of the vaccine.[19] Pasteur received his Russian patient warmly, but was initially reluctant to share the vaccine itself, so Kruglevskii returned empty-handed from Paris in early January 1886. (Dem'ianenkov apparently recovered from his wounds.) Meanwhile, Ol'denburgskii directed Kh. I. Gel'man, the veterinarian attached to the Preobrazhenskii regiment, to follow Pasteur's published procedures. Having made an emulsion from Pluto's brain, Gel'man began passing the "poison" through a series of rabbits.[20] On Kruglevskii's return, Ol'denburgskii financed the construction of a small anti-rabies station in his division's military hospital. By mid-February the facilities were completed and Gel'man's work on preparation of a vaccine proceeded in earnest.[21]

In early 1886, the Medical Council of Russia's Ministry of Internal Affairs formally requested that Pasteur train Russian physicians in his anti-rabies techniques. In an impolitic reply leaked to Russia's leading medical journal *Vrach* (The Physician), Pasteur suggested that Russian rabies patients instead make haste to Paris ("from Siberia?" interjected *Vrach* incredulously). Pasteur continued, "If I could give advice I would propose that the Russian government provide financial aid to the already-established [Parisian] institute for preventive injections against rabies. I would then consider myself fortunate to receive in this institute physicians from your expansive country, who always elicit in me great sympathy."[22]

Vrach reacted editorially with controlled patriotic and professional outrage: "We understand entirely if well-off Russians make their donations to the construction of the Pasteur Institute. For the government, however, a more natural and kindred concern is the construction of its own institute, all the more so since the foundations have already been laid through the personal resources of Prince A. P. Ol'denburgskii. Aside from the abnormality of concentrating all investigations of any scientific question in one place, it suffices to note that the costs of sending the bitten to Paris will probably in a short time surpass the cost of an independent institute."[23]

Some offended Russians canceled plans to visit Pasteur's rabies station, but scores of Russians bitten by dogs and wolves soon formed the largest contingent among its foreign patients.[24] These included twenty-one people from Smolensk. Pasteur's successful treatment of them caused a sensation, and a grateful Tsar Alexander III pledged 100,000 francs to the projected Pasteur Institute.

Prince Ol'denburgskii himself took the Tsar's contribution to Paris (along with his own gift of a malachite vase). He returned to Russia with a rabbit that had been inoculated against rabies and with Pasteur's promise of further assistance. Shortly thereafter Pasteur dispatched two coworkers, Adrien Loir and Leon Perdrix, to St. Petersburg.[25] He had by this time decided to encourage the founding of anti-rabies facilities in other countries, and Ol'denburgskii's was but the first of several such "Pasteur Institutes" to appear outside Paris in 1886.[26]

In the years 1886–90, Western European developments combined with the expanding operations of the St. Petersburg rabies station to allow Prince Ol'denburgskii to make the case for a much-expanded investigative institute. The subscription drive to honor Pasteur with a new medical institute in Paris proved astoundingly successful, and the formal founding of the Pasteur Institute in 1888 was followed shortly by the announcement of plans for similar institutions in Germany and Great Britain. The St. Petersburg rabies station was soon engaged not only in treating patients but also in scientific investigations—conducted largely by Gel'man and the eminent syphilologist Eduard Shperk on various infectious diseases in animals—substantiating the earlier observation in *Vrach* that Ol'denburgskii's facility provided an appropriate nucleus for a similar institute in Russia.[27]

Inspired by Pasteur's example, the Prince began in late 1888 to lay the groundwork for a Russian bacteriological institute by inviting Ilya Mechnikov, the zoologist-pathologist already renowned for his phagocytic theory of inflammation, to become its director. Mechnikov had earlier co-founded the Odessa Bacteriological Station, only to see it temporarily closed the following year when locals protested against its reputed role in an epizootic of anthrax. This episode, and other difficulties at Odessa's Novorossiisk University, where Mechnikov had been a faculty member, convinced him that the obstacles to scientific work in Russia were virtually insurmountable. He had been scouting for a position in Western Europe and had good reason to believe he would find one at the Pasteur Institute.[28] Not wanting to risk the indignity of having Mechnikov refuse his direct request, Prince Ol'denburgskii chose the expatriate's friend and former collaborator, Nikolai Gamaleia, as an intermediary. In August 1888, Gamaleia informed Mechnikov that Ol'denburgskii wanted him to head "a new Bacteriological institute with the richest resources" and that the Prince would guarantee him "absolute independence" and agree to any of Mechnikov's conditions for taking the post. Mechnikov, however, demurred and soon accepted a position at the Pasteur Institute.[29]

Undeterred, Ol'denburgskii dispatched V. A. Kraiushkin and Gel'man to study relevant institutions in the West and requested permission from Tsar Alexander III to establish an institute similar to those projected in Paris and Berlin. In November 1888 the Tsar granted his request, but stipulated that the new institute would have to subsist "without the expenditure of state funds."[30] Interestingly, however, Ol'denburgskii soon addressed a separate request to the Ministry of the Imperial Court and Domains, and in April 1889 he received a credit of 200,000 rubles to be spent over twenty years on construction of a "bacteriological station."[31]

The construction of Prince Ol'denburgskii's institute clearly reflected its Parisian inspiration. As Paul Weindling has explained, the Pasteur Institute and Koch's Institute for Infectious Diseases offered quite different physical and organizational models.[32] Like Pasteur, Ol'denburgskii chose a spacious location on the outskirts of the city, purchasing over thirty-seven thousand square meters of land (Pasteur's Institute was housed on a mere eleven thousand) on St. Petersburg's outlying Aptekarskii Island (Pharmacist's Island), which took its name from the enormous garden of medicinal grasses founded there in the early eighteenth century.[33] The extensive compound that arose on Lopukhinskaya Street included a large building to house the Institute's laboratories and others for employee apartments, a machine shop, experimental animals, diseased animals, and patients (Figure 2).

The staffing of the new Institute, however, remained a problem, and Ol'denburgskii's approach to it reflected his attempt to build relations with the

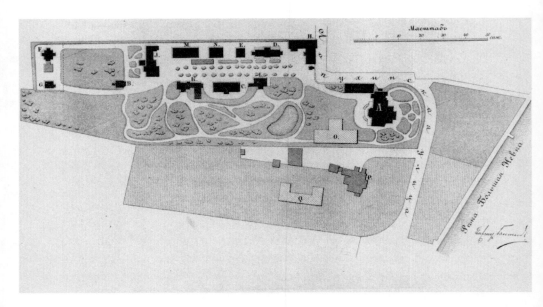

FIGURE 2. *Ground plan of the Imperial Institute of Experimental Medicine, 1892.*
The Institute was originally to be housed entirely on the northern tract of land. Prince
Ol'denburgskii purchased the southern tract after his negotiations with Nencki, and
the three southernmost buildings (O, P, and Q) represent the beginnings of a much-
expanded Institute with a large workforce and separate laboratory buildings for its
various divisions. A, main building with attached laboratories; B, greenhouse;
C, holding facilities for animals (equipped with cages, pool, bird cages); D, kennels;
E, monkey cages; F, barracks for diseased animals; G, barracks for patients;
H, mechanical works; J, director's living quarters; K, cafeteria and apartments for
praktikanty; L, apartment for administrative manager; M, warehouse; N, apartments
for attendants; O, chemistry laboratory (under construction); P, club and cafeteria for
Institute personnel, apartments; Q, furnished apartments for individual praktikanty
and their families (under construction). From A. P. Salomon, "Imperatorskii Institut
Eksperimental'noi Meditsiny v S-Peterburge," Arkhiv Biologicheskikh Nauk 1
(1892): 21

court, the key state ministry, and a skeptical medical community. He chose
as the Institute's director V. K. Anrep, a distinguished physiologist with excellent court and ministerial connections. Anrep had left a professorship at
Khar'kov University for a position at the Clinical Institute of Grand Duchess
Elena Pavlovna, and he occupied the medical community's preeminent bu-
reaucratic post as scholarly secretary of the Ministry of Internal Affairs' Medical Council. The editor of *Vrach* expressed his delight at the Prince's director-
designate: the institute's success, he noted, depended on a wise selection of
personnel and "one could not wish for a better choice [than Anrep]."[34]

The organizing committee that Ol'denburgskii created in 1888 to assist An-rep was largely bacteriological in orientation: M. I. Afanas'ev, an accomplished European-trained investigator of infectious diseases who was director of the Clinical Institute of Grand Duchess Elena Pavlovna, a professor at St. Petersburg's Military-Medical Academy, and a clinician at the city's main military hospital; A. V. Pel', a biological chemist and member of the state's Medical Council; and Gel'man, Kraiushkin, and Shperk from the rabies station. Not surprisingly, then, as late as November 1890, *Vrach* referred to the Prince's project as "the Bacteriological Institute."[35]

The other member of the organizing committee was a relative nonentity who was cobbling together a precarious existence as a *privatdozent* (an un-salaried teacher who received payment from students' fees) and laboratory assistant at the Military-Medical Academy. Ivan Pavlov lacked bacteriological expertise, professional *gravitas,* and scientific eminence. He was, to be sure, a promising middle-aged scientist who had won a coveted two-year fellowship for scientific studies in Europe (1884–86). Yet this hardly distinguished him from other talented investigators of his generation, as he would discover painfully in 1889, when he placed second in competitions for assistant professorships in physiology at Tomsk and St. Petersburg universities.

Ol'denburgskii probably included Pavlov on the organizing committee as a link to (and perhaps at the suggestion of) Russia's most powerful medical man, Sergei Botkin (1832–89). Through decades of activity within the court, state ministries, and medical institutions, and as an esteemed professor at the Military-Medical Academy, Botkin had become a hegemonic presence in St. Petersburg medicine. The Empress's personal physician, chair of the state's high-profile commission on public health, and honorary trustee at every St. Petersburg city hospital and clinic, Botkin had, by the late 1880s, installed a protégé as chief physician in each of St. Petersburg's hospitals.

A vigorous proponent of scientific medicine, Botkin had created in 1876 a small clinical laboratory on the grounds of the Medical-Surgical Academy (which became the Military-Medical Academy in 1881). Two years later, too busy to supervise work there, he had asked one of the physicians working in the laboratory, Ia. Ia. Stol'nikov, to recommend somebody for the job. Stol'nikov had suggested his good friend Pavlov, who was just completing his medical degree at the Academy. With only a two-year break for research in Europe, Pavlov had managed Botkin's laboratory ever since, guiding the doctoral research of a good many physicians, including Botkin's son. Though sometimes privately expressing scorn for his powerful patron, Pavlov turned frequently to him for advice and help. Relations were sufficiently close that when

Pavlov's wife, Serafima, took ill in 1883 she enjoyed the services of the Empress's physician. At Botkin's suggestion, Pavlov had also taught courses in physiology to medical paraprofessionals at the St. George's Commune, a project on which Botkin and Prince Ol'denburgskii collaborated and over which Princess Ol'denburgskaia presided.[36] The nature of Pavlov's relationship with the royal couple is unknown, but he was, at the very least, a known quantity. Furthermore, Pavlov maintained good relations with Botkin's former student V. A. Manassein, the editor of *Vrach*. The Botkin connection, then, made Pavlov a most sensible choice for Ol'denburgskii's organizing committee, especially as this unpaid advisory position did not carry with it the presumption of an academic position in the future Institute.

The Tuberculin Gambit

In fall 1890, with the physical construction of the Institute largely completed, Ol'denburgskii moved to phase two: winning it a permanent place on the state payroll. Just as he had earlier capitalized on Pasteur's rabies vaccine, he now moved quickly to exploit the announcement of another timely miracle of medical science: Robert Koch's treatment for tuberculosis—tuberculin. The Prince's plan was simple: invite the Tsar to stroll through the impressive grounds and facilities and use a demonstration of Koch's cure to dramatize the great blessings that the new Institute, if properly funded, could bestow upon Russia.[37]

This stratagem proved successful and nearly catastrophic. Prince Ol'denburgskii apparently pressured his medical advisors to obtain quickly the positive clinical results needed to impress the Tsar. On November 8, 1890, *Vrach* reported that the Prince had dispatched Anrep to Berlin to familiarize himself with Koch's latest discovery.[38] On November 11, immediately after his return, Anrep regaled a Sunday evening audience at the Institute with his impressions of the first patients treated with tuberculin in Berlin. "It turns out," wrote a medical reporter, "that a truly positive result is obtained at the present time only when the new substance is used on patients with lupus and various tubercular illnesses of the bones, joints, and glands. In such cases one cannot doubt the brilliant results of the treatment." Anrep proceeded in the audience's presence to give three female lupus sufferers spinal injections of tuberculin. Ten hours later, two of the three women already manifested the characteristic reaction to tuberculin described by Koch (the third required a second injection). The Institute promised a prompt report on the final results.[39] Four days later, on November 15, *Vrach* reported expectantly that "the bacteriological institute of Prince A. P. Ol'denburgskii has already begun experiments on Koch's serum." On November 22, however, there followed an announcement of an entirely different character: "We have heard that the director of the Bacteriological Institute, V. K. Anrep, has resigned his post. Undoubtedly a terrible loss for the Institute."[40]

Three days later, on November 25, Tsar Alexander III paid a visit to the Institute. Much impressed, he accepted it as a "gift." "A fervent sympathy for the suffering," the Tsar wrote shortly thereafter to Prince Ol'denburgskii, "has inspired in You the idea of building in Petersburg an institution for scientific investigations of the most important questions arising in contemporary medicine about new means to treat many serious ailments that were previously considered untreatable." The Prince had clearly spared neither effort nor expense; his Institute reflected "the spiritual qualities You have inherited from Your father" and was destined to occupy "a prominent place among institutions devoted to the protection of the people's health." As a "sign of My special goodwill toward You," the Tsar accepted the Institute as imperial property and decreed that Prince Ol'denburgskii should serve as its trustee, "in the conviction that, with the assistance of our best national scientific forces, You will assure it a future corresponding to My intentions and Your desires."[41] The Institute of Experimental Medicine thus became The *Imperial* Institute of Experimental Medicine—and was assured of state funding.

The Tsar's reference to "our best national scientific forces" captured an important element of the context in which Prince Ol'denburgskii would build the Imperial Institute. As Richard Wortman has demonstrated, Alexander III fashioned a distinctive, personalized "scenario of power" based on conservative nationalism. The economic modernization of Russia continued—indeed, accelerated greatly—at the same time as the Tsar emphasized the special traits of the Russian people and the spiritual connection between himself and his people, a connection captured and sustained by the Tsar's self-presentation as exemplar of traditional Russian virtues. Alexander III emphasized that the Empire's non-Russian lands were integral to Russia's national destiny, and so properly Russified Poles, for example, might be included among "national scientific forces." Officially sanctioned anti-Semitism made the status of Jewish subjects much more problematic (Jews were expelled from Moscow in 1891, terrorized by pogroms, and subject to numerous official strictures).[42] Dependent on state patronage and himself a member of the royal family, Prince Ol'denburgskii was of course sensitive to the values of his sovereign. In any case, as we shall see, he would be constantly reminded of them while creating his Imperial Institute.

Returning to the success of the Tsar's visit to the Institute and Anrep's sudden resignation as its director-designate, it seems likely that the same event played an important role in both: the "favorable results" reported on the treatment of nonpulmonary tuberculosis with tuberculin.[43] One day after the Tsar's visit, *Vrach* received a "bulletin" from Prince Ol'denburgskii detailing the positive effects of tuberculin on the Institute's three patients. This report, with which the Tsar had no doubt been regaled during his visit on the previous day, was signed by Shperk, Khizhin, and two physicians who had previously worked with Pavlov in Botkin's lab.[44] One of them, Pavlov's friend D. A. Kamenskii, later recalled this episode.

> Prince Ol'denburgskii generally wanted "his" institute to be foremost in the world and was delighted that the first investigations of tuberculin would be conducted at "his" institute. The great Koch proposed that under certain conditions tuberculin could become a true means for treating tuberculosis, especially of the skin. The Institute took up the verification of this proposal. Patients suffering from nonpulmonary tuberculosis were transferred to the Institute from Kalinkin hospital. When the question arose as to which physicians to invite to supervise the tubercular patients during their treatment, Ivan Petrovich [Pavlov] suggested me and V. V. Kudrevetskii. E. F. Shperk, the senior physician at Kalinkin hospital and a well-known specialist-physician, also participated in this study.
>
> One must recall that after injections with tuberculin the patients' temperature rose very sharply, to 40 degrees [Celsius; about 104 degrees Fahrenheit]; their faces glowed, there was observed a significant quickening of respiration and pulse—the patients were in a very serious state. This acute reaction lasted a day or two and then the face became red, swollen. After a day or two these symptoms disappeared and it seemed to us that tuberculin actually was a good specific remedy for lupus. But E. F. Shperk turned out to be more competent than were we; he photographed these patients when they were brought to the Institute. When we photographed them again one month later everybody saw that tuberculin had produced no benefits, and the treatment of lupus with tuberculin was terminated.[45]

By that time, however, the deed had been done.[46] We should also note that, however prescient Shperk appears in Kamenskii's account, he did sign—perhaps in deference to the Prince—the published "bulletin" about the therapeutic benefits of tuberculin.[47]

Anrep, however, did not—and a conflict over this episode may well have precipitated his demonstrative resignation as director on the very eve of the

Tsar's visit. Afanas'ev and Pel' followed suit, leaving a much-depleted and generally undistinguished committee—Shperk, Pavlov, Gel'man, and Kraiushkin—to set the Institute's course.[48] These resignations severely damaged the Institute's reputation within the medical community, making it all the more difficult to recruit "the best national scientific forces." *Vrach* never commented explicitly on the tuberculin incident, but one week after Anrep's resignation it observed editorially that recent studies of the substance were "distinguished by a haste that is incomprehensible to clinicians who are not enthusiasts."[49]

Discredited within the medical community, Ol'denburgskii became increasingly dependent on the remaining loyalists on the Institute's organizing committee. Pavlov's standing with the Prince rose quickly. By December 1890, hoping to appoint a director before the Institute's upcoming founding ceremony and unable to lure a more distinguished candidate from a skeptical medical community, Ol'denburgskii offered Pavlov the position. Guarded about the Institute's prospects, and having just been appointed assistant professor of pharmacology at the Military-Medical Academy, Pavlov declined.[50] He had his eye instead on the superbly equipped laboratory in the Institute's Physiology Division (which, unlike the directorship, he could acquire without relinquishing his more secure position at the Academy). Anrep had apparently intended the Physiology Division for V. Ia. Danilevskii, Khar'kov University's distinguished professor of physiology; but Danilevskii's candidacy evaporated with Anrep's resignation. The position became Pavlov's for the asking.[51] Ol'denburgskii considered Koch's son-in-law and collaborator Eduard Pfuhl as a potential director, but *Vrach* sternly warned him off: "One must think that this rumor is mistaken, since there can no doubt be found one of our own Russian candidates who has, moreover, a bigger name in science than Mr. Pfuhl."[52]

On December 8, 1890, the Imperial Institute of Experimental Medicine—directorless, its organization, guidelines, and staffing still unsettled, but luxuriously equipped and firmly ensconced on the state payroll—was formally born (Figure 3). The gathering that heard the Tsar's edict pronouncing the Institute royal property included representatives of the great triumvirate of western medical science—Pasteur, Koch, and Lister. Their presence underscored the symbolism of the occasion: the tsarist state had ensured that in Russia, as in its more modern neighbors to the west, science would have the "palaces" necessary to redeem its promise to medicine.[53]

Prince Ol'denburgskii had played court politics masterfully, medical politics much less well. He now had his Imperial Institute, but it remained to be seen what kind of experimental medicine might flourish there.

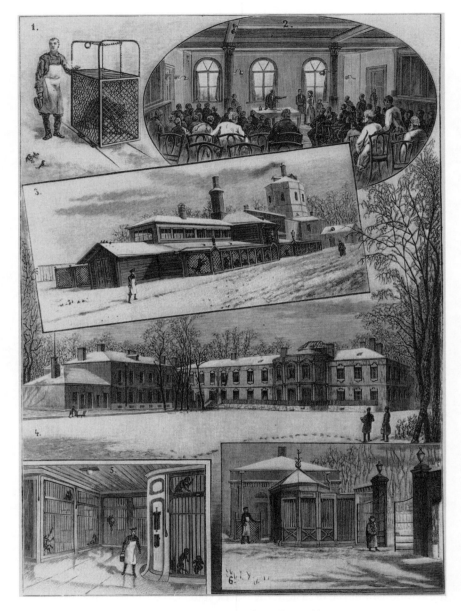

FIGURE 3. *A montage of Institute scenes, 1891, from the weekly magazine* Niva. *From* Niva, *1891, no. 7: 156–57*

Structuring Scientific Medicine

Shortly after the founding ceremony, Ol'denburgskii and his organizing committee submitted the budget and temporary statutes for an institute devoted to "the all-sided investigation of the causes of illnesses, mainly of an infectious character, and equally for the practical application of means of struggle with illnesses and their consequences." Here the Prince preached the gospel of scientific medicine. "Science knows only two means of struggle with illnesses—prevention of illness and treating the sick," he explained. Prevention was both scientifically and economically preferable, yet it remained undeveloped because science had not yet generated "precise and well-founded data on the essence of many illnesses." By their great successes, Pasteur and Koch had demonstrated the great potential of a scientific approach, and so had generated an "urgent demand for the very broadest formulation of the task of studying the causes of diseases and the means to struggle rationally against them."[54]

Scientific medicine, however, was an inherently expensive endeavor. Experimental investigations "must be conducted in conditions that satisfy absolutely the demands of science; the slightest deviation from these demands, whatever the reason, places the enterprise on shaky ground and frequently deprives any results of their significance." By financing the Institute, the Russian state joined its western counterparts in providing the extensive funding necessary to meet the exacting demands of the experimental method.[55]

As a state institution in a disease-wracked nation, the Institute would supplement its scientific investigations with a commitment to immediate, practical tasks. "As opposed to clinical medicine, the task of which consists in studying at the sickbed the action of medicinal substances and methods of treatment, experimental medicine has as its subject the study by means of direct experiments of morbid principles, the essence of the alterations they cause in the tissues and functions of organisms, and, no less, of the means to struggle against illnesses. But if this theoretical task determines the role of the Institute in science, its state designation is conditioned by the practical application on some scale of those means being investigated for the prevention and treatment of illnesses." These practical obligations would include clinical testing, serum production, and participation in the struggle against great epidemics.[56]

Ol'denburgskii's proposal for the organization and governance of the Institute distinguished his creation from many other Russian institutions by investing substantial power in various levels of the Institute itself (as opposed to the state bureaucracy). The governing trustee (*popechitel'*) served at the Tsar's pleasure and provided the Institute's "main leadership." Prince Ol'denburgskii (who served as trustee until the Bolshevik takeover) was to report annually on the Institute's activities both to the Tsar and to the responsible state body, the Ministry of Internal Affairs. He had the right to dispatch members of the Institute throughout the Empire and abroad and to grant vacations.[57] Responsibility for "direct governance" of the Institute's affairs fell to its director, who oversaw its scientific and practical activities and headed its two deliberative bodies, the Council and the Management Committee, both of which were composed of division chiefs and reported directly to the Prince.[58]

The chiefs of the Institute's scientific divisions were to be granted considerable authority; they would set the course of its research, hire an assistant, and determine which outsiders would be permitted to work there. Division chiefs would submit annual reports to Prince Ol'denburgskii. The Institute's planners originally envisioned a great deal of joint work among the divisions according to a general plan.

> Most important, of course, is the elaboration of scientific-practical questions that demand the joint work of all the divisions. The success of such works demands rational organization, with the subordination of separate investigations to a certain general plan. This is the particular, fundamental task of the Institute, giving sense to its existence as a scholarly corporation; without such a task the Institute would revert into a group of laboratories united accidentally under one general management.
>
> But one must also not forget that science, being a completely free affair demanding personal creativity and inspiration, cannot move forward without the independent initiative of scientists. Thus it is also necessary to permit in the Institute the private work undertaken by separate members of the Institute at their own discretion.[59]

The notion of planned Institute-wide scientific work was clearly important to Ol'denburgskii. In his introductory remarks to the first Council meeting in September 1891, he reminded members that the Institute was based on the view that "only through harmonious general work and mutual aid among members in separate specialties is it possible fruitfully to develop science and to intensively resolve its increasingly complex tasks."[60] (The Institute, however, quickly evolved in precisely the opposite direction.)[61]

The identity of the Institute's divisions and chiefs was, however, still unresolved. The Institute's organizing committee recommended eight scientific and two practical sections. Acknowledging that it was still "difficult to determine" what practical sections to include, the committee noted only that one would certainly be devoted to Pasteur's rabies vaccine and another to clinical studies of Koch's tuberculin.[62]

When the Ministry of Internal Affairs and the Ministry of Finance reviewed Ol'denburgskii's proposal in early 1891, only one item proved controversial: the budget—specifically, the premium salaries that Ol'denburgskii and his committee proposed for the Institute's director and division heads. The Prince insisted that this largesse was necessary not only to ensure recruitment of the "best national scientific forces" but also because of the very nature of experimental investigations:[63] "The conditions of a professor's activity are completely different from those of the laboratory activity of a scientist-experimenter. The professor knows well in advance, and knows exactly, how many and which hours of the week he must dedicate to the reading of lectures. Aside from these hours, he disposes freely of his time. The scientist-experimenter can never know in advance how much time in a given day he will devote to his activities. Once begun, an experiment can last much longer than the designated time; in the course of the work itself there may arise new tasks demanding immediate resolution and often the plan of work for an entire week can change suddenly."[64] Ol'denburgskii and the organizing committee proposed, therefore, that all division heads be paid 5,000 rubles a year and that the director receive in addition an "apartment in nature" (free lodging).

The Ministry of Internal Affairs agreed, but the Ministry of Finance objected that this proposed salary exceeded even that paid to members of the exalted Academy of Sciences and was 60 percent more than the salaries of full professors in the universities and the Military-Medical Academy. The director's salary was considerably more than that of the director of the Military-Medical Academy and university rectors, whose duties were "very responsible and difficult, not at all comparable to the more peaceful activity of heading a scholarly Institute."[65] Division heads at the Institute, argued the Ministry of Finance, should be paid the same as assistant professors—2,000 rubles a year.

The final compromise surely made Institute scientists the envy of their colleagues. The director received 5,000 rubles plus his "apartment in nature," the heads of scientific divisions received 4,000, the heads of practical divisions 2,400, and assistant heads of all divisions 1,500.[66] So, the Institute director was paid about 10 percent more than his counterparts at the universities and the Military-Medical Academy (none of whom received free lodging), and division heads received 33 percent more than full professors at those same institutions. (Pavlov's salary would considerably exceed that of full professors at Tomsk and St. Petersburg universities, where he had recently been denied an *assistant* professorship.) In addition, Prince Ol'denburgskii had at his disposal a yearly slush fund of 4,000 rubles to supplement the salaries of certain division chiefs as he saw fit.

The Institute's academic personnel also gained the considerable advantages of a high place on the Table of Ranks, the system established by Tsar Peter the Great to reward state servants. The Institute's director was awarded rank 4 (the fourth-highest rank); full members, rank 5; heads of practical divisions, rank 6; and assistants to chiefs of scientific divisions, rank 7. Thus the director enjoyed the same rank as a regional governor or the director of a central department in the state apparatus; division heads ranked with vice-governors and vice-directors. As a rank 4 state servant, the Institute's director immediately acquired the considerable advantages of hereditary nobility. Because the Institute's statutes allowed for division heads, through able and continuous service, to rise two additional ranks above that with which they began (rank 5), membership in the hereditary nobility became a reasonable prospect for them as well.[67]

The Institute also gained the right to hire those whose social standing (nationality, low rank, gender) otherwise prevented them from assuming desirable positions in state institutions. For example, a special law of 1891 allowed women to assume a number of jobs at the Institute, including that of division head (although without the right to a pension).[68]

The final Institute budget approved by the State Soviet in April 1891 amounted to 131,000 rubles annually—a truly princely sum that dwarfed the combined yearly budget of all St. Petersburg's sanitary institutions and, as *Vrach* observed archly, surpassed the annual outlay of an entire medical faculty.[69] Roughly 40 percent of the Institute's budget was designated for the salaries of scientific personnel; another 40 percent for buildings, equipment, and supplies (in 1892 Ol'denburgskii planned to spend 12,000 rubles on laboratory buildings and equipment, 6,000 to buy experimental animals, 5,000 to feed the animals, 10,000 for oil and firewood, and so forth); and about 15 percent for the expenses and salaries of the nonacademic administrative apparatus.[70] State funding began on January 1, 1892.

However lavish in sum, this budget provided relatively few resources for the daily operations of the scientific divisions. Even in 1892, when the scientific divisions were just getting off the ground, the Institute's fixed costs left few discretionary funds available. The state recognized that the Institute would inevitably require funding beyond its annual budget and so allowed it to seek these "special sums" from three sources: (1) grants from state, community, and private institutions; (2) personal contributions; and (3) income derived from the Institute's property, scientific publications, and sale of medicinal preparations.[71] As Institute resources became chronically and increasingly strained over subsequent years, these special sums became essential to its operations, lending an increasingly practical character to its activities.

The budgets for the individual divisions were not fixed by statute but set annually by the Institute's Council. Yet the heads of these divisions—and so the Council itself—had not been recruited at the time the Institute's general budget was approved. As we shall see, when personnel were chosen and the scientific divisions actually began their operations, the Institute evolved quickly in an unforeseen direction.

New Faces and Their Consequences

In the first half of 1891 Prince Ol'denburgskii awaited formal approval of the Institute's budget and continued to seek a director and scientific heads. Of the scientific divisions, only Physiology—promised, but not yet officially granted, to Pavlov—was operational.[72] The old rabies station was moved to the Institute grounds, where Shperk, Gel'man, and others continued their work on rabies, syphilis, and other diseases.

In June 1891, one month after approval of the Institute budget, Ol'denburgskii officially began to appoint the chiefs and assistants of the six scientific divisions (pharmacology and medical botany were dropped from the original eight approved by the ministries) and one practical division (a general "inoculation division" replaced the projected rabies and tuberculin divisions). By August all the division chiefs had been chosen.

Scientific Divisions

SYPHILOLOGY
Head: E. F. Shperk
Assistant: E. A. Ganike

PHYSIOLOGY
Head: I. P. Pavlov
Assistant: V. N. Massen

PATHOLOGY-ANATOMY
Head: N. V. Uskov
Assistant: N. K. Shul'ts

CHEMISTRY
Head: M. V. Nencki
Assistants: N. O. Ziber-Shumova,
S. K. Dzerzhgovskii

GENERAL MICROBIOLOGY
Head: S. N. Vinogradskii
Assistant: N. V. D'iakonov

EPIZOOTOLOGY
Head: Kh. I. Gel'man
Assistant: A. A. Vladimirov

Practical Division

INOCULATIONS
Head: V. A. Kraiushkin
Assistant: V. G. Ushakov

In October 1891, after a long and fruitless search—and almost one year after the formal founding of the Institute—the Prince finally settled the directorship upon the most distinguished member left on his organizing committee, Eduard Shperk.[73]

Theoretical and practical work in bacteriology occupied over one-half of the Institute's divisions, but this was not the bacteriological institute that the Prince had originally envisioned. The departure of Anrep, Afanas'ev, and Pel', the failure to recruit a leading microbiologist such as Mechnikov, and the political necessity to recruit "national scientific forces" had forced the Prince to create a somewhat broader investigative institution than those in Paris and Berlin.[74]

The identity of the division heads also reflected the Institute's low standing within the Russian medical community and Ol'denburgskii's failure to recruit the leading lights of Russian medical science. Of those who had been residing in Russia, only Shperk possessed what the Russian medical press commonly called a "European reputation." One division chief confided to a colleague that his counterparts "left much to be desired."[75] One physician who solicited the advice of a senior colleague about a proffered, relatively minor position at the Institute later recalled that the response was typical of medical opinion in St. Petersburg: "You know, it does not take much to build a house and even to paint it in fine white oil paint! One still must choose good personnel and organize things. We will see what comes of it!"[76] Pavlov, too, was wary. He confided to a friend that he would probably remain at the Institute only long enough to establish himself financially.[77]

Two of these division chiefs, Vinogradskii and Nencki, had been aggressively recruited by the Prince in an effort to strengthen his Institute by attracting "national scientific forces" from abroad. S. N. Vinogradskii (1856–1953) was a graduate of St. Petersburg University who held only a masters degree in botany, but he had worked for years in Zurich, and his investigations of nitrification and microbiology had begun to earn him a "European reputation." In 1890 both Ol'denburgskii and Pasteur invited him to head a division at their new institutes. Vinogradskii met with Pasteur in 1891 and was deeply impressed with the atmosphere and facilities in Paris, but after much soul-searching he decided to return to Russia. "It came to me," he later confided, "that if I enter this scientific milieu I will always, like Mechnikov, remain within it. The desire not to expatriate myself determined my decision."[78] Vinogradskii proved to be a good choice. He enhanced the Institute's stature through his growing scientific reputation (he was elected corresponding member of the Academy of Sciences in 1894), his practical work on cattle plague, and his editorship of the Institute's journal, *Arkhiv Biologicheskikh Nauk* (Archive of the Biological Sciences). In 1903, Vinogradskii founded the Russian Microbiological Society, and he served briefly (1902–5) as the Institute's director.[79] Furthermore, during the recruitment process Vinogradskii made only modest demands on his future employer. In his correspondence with Shperk in summer 1891 he requested only a single assistant, some basic equipment, and four separate rooms for microscopy, chemical analysis, sterilization of equipment, and his assistant's workplace.[80]

Not so Ol'denburgskii's prize catch, M. V. Nencki (1847–1901). Born near Kielce in what is now Poland, Nencki had participated in the failed Polish uprising against Russian rule in 1863 and so had been forced into emigration. At the time of Ol'denburgskii's overtures, he was chairman of the Department of Physiological Chemistry at the University of Bern and a distinguished scientist with interests that ranged broadly from the liver to the chemical structure of the blood to the etiology of various diseases.[81] He arrived in St. Petersburg with great clout and big plans.[82] His privileged position is clear both from the 3,000 rubles he received annually as a salary supplement from the Prince's slush fund and from his bringing with him from Switzerland not one, but two assistants. Three more foreign coworkers arrived by November 1891, followed closely by a stream of others attracted to Nencki's laboratory.[83]

Most important, Nencki was accustomed to a spacious laboratory with numerous coworkers. He rejected the relatively modest facilities originally offered for his Chemistry Division in the single building erected for the Institute's laboratories and insisted on a grand laboratory building designed, constructed, and equipped to his specifications. A contemporary later recalled, "It had originally been proposed that scientific investigations would be conducted by the division heads themselves together with their assistants (one per laboratory), and that there might be 1–2 laboratory coworkers. The laboratories were not designed for large cadres. But upon his arrival at the Institute Nencki announced that he was accustomed to a large laboratory with many people working in it. A separate building was constructed for the Chemistry division, and the entire work of the Institute acquired a different character."[84] Nencki's demand for a laboratory "designed for large cadres" became one of two key factors that soon transformed the Institute into a site for large-scale scientific investigations.

In summer 1891, after his discussions with Nencki, Prince Ol'denburgskii bought an additional tract of land, doubling the size of the Institute's grounds. There in 1892 a two-story chemistry building was erected, and construction began on a new building for the pathology-anatomy laboratory. Another new two-story structure housed expanded facilities for the Inoculations Division on the ground floor and lodgings for Institute personnel on the second floor. In 1893 the pathology-anatomy building was completed, a new structure for autopsies and the burning of animal corpses was begun (equipped with modern technologies to burn the diseased bodies of even large animals without requiring them to be first cut apart), and work began on a new, two-story structure for the Physiology Division.[85] When a new division of General Pathology was organized in 1894, it too acquired its own building. Ol'denburgskii himself initially financed this new construction, perhaps with the help of "special sums" at his disposal; in April 1893, however, the state allocated 150,000 rubles for expansion and modernization. These funds, Ol'denburgskii wrote gratefully in his report of 1893, had made possible the fundamental transformation of the Institute.[86]

The Tsar and his ministers, of course, were unlikely to expend such sums simply to satisfy the vanity of a chemistry professor, let alone the desire of his lesser colleagues to have their own buildings as well. Nor would these buildings themselves have changed the nature of the Institute. An essential element in both developments was the unanticipated "influx of scientific forces wishing to use the Institute's facilities."[87] As Ol'denburgskii wrote to the Minister of Finance in March 1893, "From the first days of the Institute of Experimental Medicine it turned out that the number of those wishing to work in it exceeded all our initial expectations and assumptions."[88] This was good public relations, but it was also the truth.

By 1893 the Chemistry Division, which led the way, boasted forty-one outside investigators; another twenty-five worked in the other divisions.[89] These *praktikanty* (the singular form is *praktikant*) justified not only the expanded laboratory facilities but also the building of lodgings, libraries, a cafeteria, and even a club on Institute grounds.[90] By 1892 the demand of this rapidly growing workforce for experimental animals had outstripped the ability of the market to provide them. The following year, Ol'denburgskii proudly reported that the Institute's independent animal "factory" had produced 591 of the 1,001 rabbits, 267 of the 475 guinea pigs, and all 271 white mice required for laboratory work.[91]

Who were these unexpected praktikanty and why did they come to the Institute? Typical was the contingent in Nencki's laboratory in 1893: thirty-two of forty-one were physicians.[92] They were not attracted solely by an interest in cholera, diphtheria, and cattle plague, by the Institute's well-equipped laboratories, or even, as Ol'denburgskii suggested in a self-serving letter to the

Minister of Finance, by the Prince's fine choice of division chiefs.[93] This new labor force was, rather, the product of an important victory won by Russian proponents of scientific medicine some years before. Unbeknownst to Prince Ol'denburgskii, policymakers had sent Russian physicians on a "detour through the lab" long before he had unwittingly occupied choice ground along that route.[94]

Praktikanty: The Unexpected Workforce

The praktikanty who beat a path to the Institute's door in such unexpected numbers, and in so doing transformed the very nature of the Prince's creation, were largely motivated by their quest for a quick doctoral degree.[95] The demand for this degree was rooted in developments that, for almost four decades, had been quietly transforming the institutional portrait of the well-educated physician.

The years between Russia's stunning defeat in the Crimean War (1855–56) and the founding of the Imperial Institute of Experimental Medicine had witnessed a qualitative improvement in the status of Russia's medical community and that of science within it. The defeat in Crimea was widely perceived within ruling circles as evidence of the military importance of medicine, and the reforms of the 1860s and the accelerating growth of Russian cities put qualitatively greater demands on medical institutions. Recognition of the desperate need for improved health services led to a significant increase in state funding for medical institutions, to the training of increased numbers of physicians, and, as Nancy Frieden has demonstrated, to increased leverage for the medical profession, enabling it to secure rights of assembly, association, and publicity denied to other professional groups.[96] A series of laws from the 1860s through the 1890s steadily raised the salaries and perquisites of medical personnel (the vast majority of whom were state employees).[97] By 1891 Russian physicians hardly constituted a uniformly well-paid profession, but a medical career had become sufficiently attractive for applications to outstrip places in medical schools, and there was considerable competition for choice posts in medical institutions and the state's medical bureaucracy.[98]

The leaders of Russia's academic medical establishment stressed that the improvement of Russian medicine required not just more but better-trained physicians. In his history of St. Petersburg's Military-Medical Academy, its director, the pathologist V. I. Pashutin, recalled the widespread perception in the post–Crimean War years that the "great revolution in the teaching of medical sciences in Europe" was inextricably connected with the triumph of "scientific positivism." "The application of exact physical and chemical methods to the study of biological phenomena" had replaced "the purely speculative orientation" in European medical schools, and Russia could not afford to lag behind. So, Russia's most prestigious academic medical institution revised its curriculum to provide students with a basic knowledge of scientific approaches, and its faculty was expanded to include young Russian scholars schooled in the spirit of scientific positivism.[99] The universities, too, received additional funding to train students in experimental science. "The study of nature develops so quickly," wrote Minister of Education I. D. Delianov in a directive of 1869, "and the significance in it of experiment moves forward with such enormous strides, that to remain in place is to fall behind." For Delianov, the intellectual discipline inherent to experimental science carried the added advantage that, by training young minds to think soberly, it discouraged political radicalism.[100]

A series of statutes from the 1860s through the 1890s provided impressive incentives for physicians to gain experience in the experimental sciences and to certify their scientific credentials with a doctorate in medicine.[101] By the 1890s an ambitious physician had many good reasons to seek this higher degree. The Ministry of Internal Affairs annually distributed grants to physicians seeking to "improve themselves scientifically" and planning to put that expertise to state service. These physicians were granted a service leave (*komandirovka*) of six months to two years for studies at the Military-Medical Academy, a university medical school, a university clinic, or a hospital in proximity to a university. While on leave, these physicians were paid their usual salary plus a small bonus.[102] Time spent on these studies counted as service time toward advancement on the Table of Ranks; should the physician successfully complete a doctoral degree and remain in state service, each year spent studying would count as two years toward advancement of rank. Physicians who distinguished themselves "by special abilities and success" during their studies were included on a list compiled by the Medical Department of those entitled to special consideration when scholarly opportunities or "higher vacancies" appeared.[103] Physicians availing themselves of study leaves were required to report annually on their activities, and those with a mere medical degree (*lekar'*) were *obligated* to demonstrate their assiduousness by taking an exam for a doctorate in medicine or another higher degree.

The financial and professional advantages of using this study leave to obtain a doctorate in medicine were also substantial. This degree conferred rank 8 on its holder, compared with rank 9 for a simple physician. Furthermore, the privileges of this higher rank were not dependent on the job a physician subsequently held or even on continued state service. If the physician died, his family's benefits were set by this higher rank—an important advantage, since a deceased physician often left his family impoverished. Amid the increasing competition for choice jobs in medical institutions, the law prescribed that "in the higher positions, other qualities of the candidates being equal, those with higher scholarly degrees should be preferred."[104] A doctorate in medicine also offered Jewish physicians refuge from several discriminatory laws. A Jewish doctor of medicine entering state service was entitled to the same rank as non-Jews in that position and was permitted to serve in any ministry in any location (a Jewish physician without a doctorate could serve only in the Ministry of Enlightenment or the Ministry of Internal Affairs and, in the latter case, was forbidden to live in St. Petersburg or Moscow).[105]

The great influx of student-physicians into St. Petersburg led, in 1898, to the establishment of a Society for Mutual Aid among Physicians Arriving in St. Petersburg for Scientific Improvement. This society helped the largely provincial arrivals make their way professionally, socially, and materially in the capital. It facilitated contact with scientific institutions, published their scientific works, sponsored general meetings, convened an Honor Court, offered loans and material aid, and rented out inexpensive rooms in its dormitory.[106] Among the more attractive institutional alternatives for these physicians, as one journal reported in 1898, was the Imperial Institute of Experimental Medicine, with its expansive grounds, rich laboratory facilities, and extensive (6,500-volume) library. Although not a pedagogical institution, the Institute accepted "those desiring to study scientific questions" and "for the comfort of these persons, called praktikanty," had built "special small apartments for rent" and even a club for its more sociable physician-investigators.[107] Many aspiring doctors of medicine chose the Institute for their studies, justifying its expansion in the early 1890s and populating its laboratories until the outbreak of the First World War.

The numbers of praktikanty in the Institute's scientific divisions for each year between 1891 and 1904 were as follows:[108]

1891	6	1898	110
1892	39	1899	77
1893	66	1900	101
1894	95	1901	91
1895	119	1902	126
1896	92	1903	121
1897	115	1904	121

More of these praktikanty worked in the Chemistry Division than in any other, but by the turn of the century the Physiology Division had joined Chemistry as the Institute's primary attraction, with sizable cadres coming to all divisions other than General Microbiology. I have complete data by division for the years 1898–1904 only:[109]

	1898	1899	1900	1901	1902	1903	1904	TOTAL
Chemistry	18	21	22	16	16	24	22	139
Physiology	12	13	9	16	23	16	19	108
Pathology-Anatomy	17	18	9	10	10	17	16	97
Epizootology	13	12	9	11	14	10	7	76
General Pathology	11	11	7	6	9	9	6	59
Microbiology	2	2	2	2	2	5	7	22

The working conditions for praktikanty, and the chiefs' different styles as scientists and managers, are evident in the *Rules for Outside Personnel Wishing to Work in the Facilities of the Imperial Institute of Experimental Medicine*, a document drafted by Pavlov in consultation with other members of the Institute Council in 1894, when the Council decided that the influx of praktikanty necessitated a standardized policy.[110] According to these guidelines, the aspiring praktikant first obtained permission from the division chief, then from the director and governing trustee. After paying 25 rubles into the Institute coffers, the praktikant gained the right to work in the laboratory and to receive free supplies for experiments at the chief's discretion. The laboratories operated from 8 A.M. to 7 P.M. during the fall and spring and opened an hour later in the winter. The praktikant could work during the summer months or holidays only with permission of the division chief and director, and the use of electricity after 7 P.M. was strictly prohibited (budgets were tight, and electricity, gas, and firewood were expensive). The division chief had great authority over the praktikant's research: "research on each question is conducted by mutual agreement with the Division Chief, who is informed of every deviation in the conduct or course of an experiment and who is generally kept informed of the fate of the work." Results could be published only with the chief's consent.[111]

To this list of general rules, each division chief added his own. Those for Uskov's division of Pathology-Anatomy chiefly concerned the care of equipment and safety (e.g., dead animals were to be autopsied only in the presence of the division chief or his assistant; injections and other operations with dangerous bacteria were to be conducted only by a physician, who could not delegate the job to an attendant). Those for Nencki's Chemistry Division concerned order, economy, and safety (e.g., "At least once a day the praktikant must inspect his animals. Animals remain in the laboratory only during an experiment. Animal corpses are investigated within 12 hours after death. After 12 hours the animal corpses are exterminated"). S. M. Luk'ianov's Division of General Pathology forbade "rebukes and crude comments, . . . loud conversation and generally any noise in the laboratory" and prescribed two hours in the day when praktikanty could make any "personal explanations" to the chief (only in "extreme cases" should the chief be approached at other times). The single rule for Physiology captured its cooperative ethos: "Every praktikant is required to participate in the work of his comrades, specifically when there is conducted a complex experiment or operation that requires a large number of assistants, greater than the constant paid contingent in the laboratory."[112]

The stream of praktikanty seeking doctoral degrees, and the splendid laboratory facilities built to accommodate them, provided unique resources for division chiefs at the Institute—resources that they used according to their own styles as scientists and managers.

Making Ends Meet

Founded on spacious grounds and possessing Russia's most lavish facilities for laboratory medicine, the Imperial Institute of Experimental Medicine nonetheless struggled constantly with financial difficulties. The radical expansion of facilities in 1891–94 qualitatively increased fixed costs (especially repairs and heat)[113] while the Institute's annual state allocation remained flat. The Institute thus became ever more dependent on its "special sums," that is, on income from one-time state grants, personal contributions, and the sale of various products and services. This, in turn, imparted an increasingly practical orientation, forcing the Institute increasingly, as Ol'denburgskii later put it, to relegate its "basic task" of scientific investigation to the "back-burner." This pressure, already evident in 1892, had by 1916 reached the point where Ol'denburgskii warned that his Institute was "in danger of losing the physiognomy of a scientific institution . . . to a certain degree becoming almost a factory institution working for the sale of its products."[114]

Council meetings in the years 1892–1904 featured constant discussions and warnings about the Institute's finances. In 1894, Shperk's successor as director, S. M. Luk'ianov, reported a deficit of about 38,000 rubles, which Princess Ol'-

denburgskaia (filling in for her husband as governing trustee) agreed to cover with her personal funds.[115] In 1902 the financial crisis was so severe that most divisions sustained a deep cut—especially Chemistry, where the recently deceased Nencki had been succeeded by a new and less powerful chief.[116] In 1905 Vinogradskii resigned after a brief tenure as director, ostensibly because budgetary restrictions prevented him from implementing plans for "fundamental changes" in the Institute's organization and facilities.[117] This situation was reflected in the operating budgets of the scientific divisions, each of which, with the exception of Physiology's, declined over the Institute's first fifteen years.[118]

The relationship of the divisions' budgets to their scientific achievements was, of course, complex, and Ol'denburgskii could in 1916 justly point to the "great scientific authority" that the Institute enjoyed internationally as a result of the investigations conducted there.[119] The point remains, however, that the constant budgetary shortfall—typical was 1898, when the Institute's state allotment of 131,000 rubles covered about half its expenses of 255,000 rubles—made the Institute and its divisions constantly dependent on their ability to cater to patrons and a market with decidedly practical interests.[120]

In the years 1891–1904 the state provided one-time grants on four occasions: in 1893, as we have seen, to finance the Institute's expansion; in that same year, an additional 3,500 rubles to broaden the activities of the Epizootology Division (i.e., to combat diseases among livestock); in 1899, 10,000 rubles to organize an epizootological station in the Zabaikal region; and in 1900, a two-year grant of 140,000 rubles to refurbish and expand technical facilities connected to the Institute's practical functions (this was used largely to build a modern two-story building for the production of vaccines, a new disinfection room, and a bath for the Institute's servants).[121]

The same practical orientation characterized the largest private contributors, who donated considerably more money to study disease among livestock than among people. The most generous donor was Graf Orlov-Davydov, a large landowner and high official of the Imperial Court, who donated over 100,000 rubles in several gifts from 1893 to 1900, earmarked almost entirely for combating cattle plague.[122] There were exceptions, but hardly on a scale to influence the Institute's budgetary profile. The mother of Tsar Nicholas II, Empress Maria Fedorovna, donated 3,000 rubles in 1899 to encourage scientific work on the "proximate cause of scurvy, which has acquired epidemic proportions in places suffering from poor harvests."[123] Vinogradskii donated over 9,000 rubles in 1901 and 1902 to equip his own bacteriological laboratory; 2,200 rubles were collected after Nencki's death to found a science prize in his

Table 1

"Special Sums" from Institute Sales and Services (in rubles)

	1895	1900	1905
Serums	57,182	30,060	28,195
Malein, tuberculin	3,906	6,371	7,479
Patients	3,023	5,959	4,665
Praktikanty fees	1,000	1,225	1,340
Anti-pest bacteria	545	612	207
Sales of *Arkhiv*	466	515	499
Dog tending	289	424	523
Analyses	153	92	33

Source: Data from Tsentral'nyi Gosudarstvennyi Istoricheskii Arkhiv Sankt-Peterburga (TsGIA SPb) 2282.1.396: 164.

Note: Malein and tuberculin were used for the diagnosis of illnesses in livestock. The great majority of the human patients were treated for rabies. The most important of the anti-pest bacteria was the Loefler bacteria of mouse typhus, used for killing field mice. Not included in this table is the income—about 2,000 rubles annually—from renting apartments to praktikanty (TsGIA SPb 2282.1.163: 112).

honor; and in 1905 the Grand Duke Konstantin Konstantinovich honored Pavlov's Nobel achievement by donating the capital for an annual 100-ruble prize for physiological research.[124]

On an annual basis, then, the Institute budget was balanced precariously on the sale of various products, particularly serums (Table 1). The Institute produced and sold anti-streptococcal, anti-staphylococcal, and anti-plague serums, but its most important medical product was diphtheria antitoxin, which was a highly profitable and well-publicized contribution to public health.

In early 1894 Ol'denburgskii reported that French and German producers could not satisfy what would certainly be an immense Russian demand for diphtheria antitoxin, that the serum would be relatively easy to produce, and that Nencki and Dzerzhgovskii in the Chemistry Division would supervise production according to the techniques developed by Emile Roux at the Pasteur Institute.[125] In October 1894 Director Luk'ianov informed the conservative newspaper *Novoe Vremia* (New Times) of the Institute's intention to supply all St. Petersburg's children's hospitals with the serum, a project that attracted contributions from numerous well-wishers.[126] One week later, *Vrach* reported that production was proceeding at "full speed" in twelve horses stabled at the Ol'denburgskii *dacha* (summer home) on Kamenoostrovskii Island, where the operation was supervised by one medical paraprofessional (*fel'dsher*) and several stable boys.[127]

By the following year more than twenty horses were being used for serum production, and output had achieved "significant dimensions." Physicians arrived at the Institute from throughout the country to be taught (without fee) how to produce the serum. Unable to meet the Russian demand, the Institute contracted with Hoechst Company, the leading commercial producer of the Behring-Ehrlich serum.[128] Although the Institute was earning much-needed income from the sale of its own antitoxin, its spokesmen were at pains to explain that it turned no profit from resale of the German product. Such "speculation" ran against the grain of common Russian and professional values, and the Institute made clear to *Vrach*'s correspondent that "far from pursuing the goal of trading for a profit, the Institute releases the serum at the same price at which it itself acquires it."[129]

I cannot pursue here the Institute's adventures with diphtheria antitoxin,[130] but it is important to note the precipitous decline in serum revenues in the years after 1895 (see Table 1), a plunge that caused "extreme financial difficulties."[131] This resulted from increasingly stiff competition from the Moscow Bacteriological Institute and especially from foreign firms. The Institute's director reported in 1907 that the major pharmacies in St. Petersburg sold over twenty thousand flagons of foreign serums annually, and many had ceased to stock those produced at the Institute. Many hospitals and regional medical administrations throughout the country had also discontinued purchases from the Institute, buying the less expensive serums of competitors. The Institute responded with price cuts in at least two years (1903 and 1907) but was unable to reverse its shrinking market position.[132] By this time, as the director conceded, "the budget of the Institute, unfortunately," depended "to a large degree on the sale of serums."[133]

The scientific divisions, then, began the 1890s with facilities comparable to their best European counterparts, but these did not necessarily remain comparable as time wore on. The improvement and expansion of facilities depended on securing outside funding, most commonly by providing a serum or service to the medical market.

Public Relations

Prince Ol'denburgskii's annual reports to his state patrons—the Tsar and the Ministry of Internal Affairs—demonstrated a keen sense of the criteria by which his creation was evaluated. Aware of his patrons' limited comprehension of and interest in the details of scientific research, the Prince described the scholarly accomplishments of the Institute in quantitative terms meaningful to a layman, along with telling qualitative details illustrating the foreign prestige, practical benefits, and political loyalty of the institution.

Every January Ol'denburgskii collected from each division head a prescribed form that left little room for embellishment beyond the bare facts needed for the patrons. The chiefs listed the personnel working in their laboratories, noted any change in physical facilities, and provided an inventory of supplies; they supplied a sentence or two on "the general orientation of scientific activity" and listed the publications, public courses and lectures, practical contributions, and research trips produced by their division. The Prince then submitted a short, four- to five-page report to the Tsar and a longer, approximately twenty-five–page report to the Minister of Internal Affairs.

The Prince's report to the Tsar (Alexander III until 1894, then Nicholas II until 1917) was standardized by 1893. After listing the division chiefs, Ol'denburgskii summarized the Institute's scientific achievements through the number of praktikanty and publications in each division. He then explained the "essence" of each division's research in a single sentence, adding another if the researcher had received foreign recognition (ranging from a publication by Vinogradskii in the protocols of the French Academy of Sciences to Pavlov's Nobel Prize). He also described the Institute's practical activities quantitatively: the number of courses offered and students instructed, rabies patients treated, microbes collected, flagons of serum produced, library volumes purchased, and issues of *Arkhiv* published.

Ol'denburgskii sometimes added texture to his reports by highlighting the financial contributions of the Institute's well-wishers (this, of course, demonstrated its efficacy and reputation) and mentioning specific services rendered to the imperial family and other important gentry. In 1893, for example, he noted that N. K. Shul'ts, the "woman-physician" who headed the Institute's subdivision on bacteriological pathology, had checked the water in the Winter Palace for cholera and that the Epizootology Division had attended to an outbreak of anthrax at the Dubrovsk stud farm of Grand Duke Dmitrii Konstantinovich.[134]

The Prince always closed with a phrase tying the Institute's activities to the prosperity of the Tsar and his Empire. In his report of 1898 he pledged the Institute's "service to the Throne and the Fatherland on the field of science, facilitating, in accordance with the philanthropic destiny of Your Imperial Majesty, the peaceful prosperity of the peoples entrusted to You by providence." The following year he emphasized the enduring nature of the Institute's scientific mission: "Working with all its forces in the broad academic direction indicated by its Statutes, and unrelentingly concerned with satisfying vital questions in the most complete manner possible, the Institute, it would seem, is justly confident that its efforts will not vanish without a trace, since that which is achieved for science, and through science for life, is achieved doubly, achieved enduringly." In 1903, he affirmed the Institute's "firm faith in its scientific calling and its equally firm hope that it will continue to serve Russian science."[135]

Ol'denburgskii's longer reports to the Ministry of Internal Affairs differed essentially only in their detail (e.g., all publications were listed) and scope (the detailed administrative and financial accounting even included the number of documents processed by the Institute bureaucracy). Addressing himself to presumably clear-eyed bureaucrats, the Prince omitted the declarations of fealty and patriotism with which he concluded his reports to the Tsar.

The Institute spared few opportunities to publicize its achievements within the broader medical community. *Vrach* frequently contained items about the Institute's scientific achievements, participation in antiepidemic campaigns, and sponsorship of prize competitions and courses; and the approximately 650 publications that appeared between 1892 and 1904 bearing the words "from the laboratory of the Imperial Institute of Experimental Medicine" spoke eloquently about the Institute's scientific contribution.[136] All of the Institute's divisions participated in the First All-Russian Hygiene Exhibit, held in St. Petersburg in 1893. The Exhibit itself was apparently uneven at best; for financial reasons, its sponsors rented booths even to those advertising such "undoubtedly unhygienic" products as patented corsets, plasters, and soaps.[137] The Institute's exhibits, however, were clearly designed to impress the public with the modern scientific approach to the causes and cures of disease. Pathology-Anatomy displayed a piece of a kidney taken from a cholera sufferer some hours before death; General Microbiology exhibited photographs of the organisms responsible for nitrification in various lands and explained the process by which these organisms were cultured; Epizootology displayed photographs of the microorganisms responsible for glanders (a disease common in horses) and tuberculosis; and Physiology exhibited dogs prepared surgically for the collection of pure samples of their gastric and pancreatic secretions.[138]

The Institute's reputation as the center of Russian medical research was expressed in the convening on its spacious grounds of the Twelfth International Medical Congress (1897), during which foreign medical luminaries could be photographed alongside Russian royalty (Figure 4).[139] When the Seventh Conference of the International Red Cross met in St. Petersburg in 1902, del-

FIGURE 4. *At the International Medical Congress of 1897, Prince Ol'denburgskii (far left) stands with a group of luminaries, including Rudolf Virchow (third from left). From* Niva, *1897, no. 37: 880*

egates visited not just the famous art collection at the Winter Palace and the Tsar's glittering residence at Tsarskoe Selo but also Russia's palace of experimental medicine.[140]

Public lectures also proved a most successful means to enhance the Institute's prestige. In September 1894 Institute Director Luk'ianov pressed division heads to deliver lectures to physicians on some aspect of their research. Five months later, he and Pavlov delivered a series of lectures on the Institute's grounds (at no charge) to large audiences comprised mostly of physicians and students at the Military-Medical Academy. *Vrach* applauded this "new step" by the Institute, pronouncing it "completely successful" and the lectures themselves as "satisfying an urgent need." "As one would expect, the lectures of both professors were completely successful and elicited enthusiastic applause."[141] Luk'ianov's lectures on experimental pathology and Pavlov's on the physiology of the digestive system became the basis for well-received books in 1897. Among the Institute's other courses for physicians were those in 1902 on the bacteriology and diagnosis of plague and a "Short Course on Bacteriology" in 1904.[142]

By the turn of the century, Prince Ol'denburgskii's annual report to the Tsar had become the occasion for a ceremonial gathering of influentials. The guest list included leading ministers and bureaucrats (including everybody on the

Medical Council of the Ministry of Internal Affairs), all faculty members of the Military-Medical Academy, the physico-mathematical faculty of St. Petersburg University, and, generally, those members of the state and academic elite with a possible interest in or influence on the Institute's fate.[143] This august gathering heard the Prince's brief account of the Institute's accomplishments for that year and another report on an achievement of particular moment. In 1900, for example, the Grand Duke Konstantin Konstantinovich and Minister of Internal Affairs D. S. Sipiagin were among those to hear a report on the etiology of scurvy—that is, an account of the good use to which the Institute had put the funds donated that year by the Tsar's mother. The following year, an audience that included President of the Council of Ministers I. N. Durnovo and Minister of Popular Enlightenment I. V. Meshchaninov heard a description of the Institute's new facility in Zabaikal to combat cattle plague.[144] The reputation of the Institute's special occasions as a gathering place for the state elite had its disadvantages as well: in the wake of the Revolution of 1905, the celebration of the Institute's new clinic for skin diseases ended most unceremoniously with the assassination of St. Petersburg's arch-conservative military governor, V. F. Von der Launits.[145]

Conclusion

Prince Ol'denburgskii's plan for a scientific institute in which a number of workshop scientists would investigate bacteriological subjects was, as we have seen, much transformed in the process of its implementation. The Prince's tactics amid unforeseen structural forces and unforeseeable contingencies produced instead an Imperial Institute of Experimental Medicine that offered investigators in a range of medically relevant fields Russia's first institutional opportunity to practice large-scale scientific research. The Institute presented a set of resources, challenges, and incentives to its division heads. Each began with relatively lavish facilities and a potential pool of motivated yet unskilled coworkers; with a budget that could be increased only by finding outside patrons or markets for their products; with institutional pressures to generate quantifiable achievements, win allies in Russia's medical world, and gain foreign recognition; and with ready access to the elite of St. Petersburg's medical community and state medical bureaucracy. The division chiefs responded to these circumstances in various ways, depending on their own abilities, aspirations, and scientific-managerial styles.

A combination of structural circumstances and contingencies had also placed Ivan Pavlov among them, putting him at the helm of Russia's best-equipped and best-financed physiological laboratory. Pavlov's scientific and managerial response to his institutional circumstances—and the laboratory system he thus created—would transform him in little more than a decade from one of the least accomplished members of the Prince's organizing committee to one of the most celebrated physiologists in the world.

Chapter 2

THE VISIONARY OF LOPUKHINSKAYA STREET

A passage from a period of fantasy to a period of logical work is unavoidable if a person does not wish to be torn away from intellectual life, because this is a natural passage, because this is intellectual growth . . . The intellectual organism has the same history as the physical organism. Milk is in its time a satisfactory food, but with the growth of the body it becomes insufficient, even harmful, if taken exclusively. A more complex diet is necessary if the organism is to become stronger and develop.

— IVAN PAVLOV, *Wondrous Are Thy Works, Lord* (1879)

Pavlov could never understand how it was possible to conduct research in physiology without making use of an appropriate theoretical conception, even though only a temporary one. To him every new fact that was revealed in the laboratory was like a link in a chain that was being pulled out of a dark hold. One link was connected with another and inevitably led to the appearance of the next. Research must follow some guiding idea and not be a blind investigation of different possibilities.

— BORIS BABKIN, *Pavlov* (1949)

At the time of his appointment to the organizing committee of the Institute of Experimental Medicine, Ivan Pavlov was a professional nonentity, yet he was also a man of ideas, a visionary. Just as his accomplishments of 1891–1904 would have been impossible without the resources that the Institute put at his disposal, so would they have been (literally) unthinkable without the scientific-managerial vision that animated his use of those resources.[1]

Pavlov's scientific practices involved all the elements identified in historical studies of experimental science, and in subsequent chapters we shall see the importance of, for example, the choice of organism, the commitment to specific techniques and technologies, unexpected empirical results, the institutional setting, and the social dynamics of the laboratory group.

Yet Pavlov was a resolutely and consistently vision-driven scientist. The striking continuities of more than two decades of research on digestive physiology (and the three decades of work on conditional reflexes that followed) testify to the overriding importance of Pavlov's basic notions about the organism, good science, and good physiology (his scientific vision) and about how best to produce good physiology with the resources at his disposal (his managerial vision). There were, to be sure, important junctures at which other elements in the "mangle of practice" played a decisive role, but Pavlov's vision is consistently evident in the design and choice of techniques and technologies, the interpretation of sometimes perplexing experimental data, the planning of subsequent trials, and his synthesis of results.[2]

This scientific vision was strikingly stable. The same identifiable notions of the organism, good physiology, and the good physiologist are recognizable parts of Pavlov's identity from the late 1870s (the earliest date for which relevant material is available) until his death in 1936. He did, of course, change his mind about various questions and even previously ascertained facts, and he expressed himself on general issues more robustly and confidently as he became increasingly successful and secure, but the aesthetic, philosophical, and professional preferences embodied in his scientific vision remained essentially unchanged. These were not subject to refutation by experimental data, and none of the important events and challenges in Pavlov's life from the 1880s onward shook his confidence in his fundamental ideas about nature and science. Quite to the contrary: from the time he became chief of the Institute's Physiology Division in 1891, Pavlov enjoyed a series of impressive successes that deepened his confidence in the general beliefs to which he had become accustomed and the techniques with which he acquired increasing facility. One telling reflection of this stability is Pavlov's repeated use of the same anecdotes, phrases, and metaphors throughout his many decades as a public scientific figure. He often refined and elaborated these over time, but their message remained unchanged.[3] This is not, of course, to say that Pavlov's scientific vision was monolithic, seamlessly consistent, or always expressed in the same terms.

It consisted, rather, of various elements acquired at different formative stages in his life, and these included, if not contradictory views, then contradictory moments.

Pavlov's scientific vision featured the interplay of two related contradictory moments: one concerning the nature of the organism, the other the physiologist's approach to understanding it. For Pavlov, the *organism* was a purposive, nervous machine, but a machine that was distinguished from all others by its complexity and by the subtle relationships among its parts; the *physiologist*, like the physicist and chemist, sought precise, determinist laws, but the nature of the organism necessarily distinguished *physiological* thinking from that in other sciences. The goal of physiological thinking was both to encompass the complex animal whole (in which processes beyond the control of the experimenter abounded) *and* to attain the precise, repeatable, determined results that were the hallmark of any true science.

In this chapter, I explore Pavlov's scientific vision both chronologically and analytically (touching only briefly on his managerial vision, which is best discussed in the context of his laboratory operation). After examining the origins of Pavlov's scientific vision in his experiences before becoming chief of the Institute's Physiology Division, I analyze this vision as a stable, if contradictory, set of views, impulses, and preferences in the years 1891–1904. We will then be prepared to examine (in Chapter 3) the laboratory enterprise that took shape when Pavlov brought this vision to the institutional setting on Lopukhinskaya Street.

Attaining Intellectual Adulthood: Passion, Professionalism, and a "Wild Episode"

Pavlov's scientific vision was shaped by Russia's radical essayists of the 1860s, who impressed upon him the transforming mission and power of scientific knowledge and a materialist worldview, and by his mentor in physiology, Ilya Tsion, who taught him the perspectives and procedures of professional experimental physiology in the early 1870s. Pavlov grappled explicitly with the sometimes contradictory values that he imbibed from these two sources, modifying and adapting his youthful passions as he developed his professional persona. The result was a Bernardian vision of physiology that incorporated also the emphasis on precision and quantification associated with contemporary German science and particular intellectual mannerisms rooted in Pavlov's own experiences and temperament.

Pavlov had been bound for the priesthood in the 1860s, when, as a student in the Ryazan Theological Seminary, he was entranced by the radical scientism propagated by Dmitrii Pisarev and other popular essayists of the day. Russia was then undergoing fundamental social changes under Tsar Alexander II, whose modernizing reforms shook the very foundations of traditional Russian society and initiated a period of unprecedented intellectual and social ferment. For the young Pavlov, as for many of his generation, science became the symbol and instrument of modernization—the path away from the obfuscatory metaphysics, injustice, and social backwardness of Church and Tsar toward a materialist worldview, positive knowledge, social and technological progress, and the rational control of human destiny. For the "men of the sixties" (*shestidesiatniki*), the biological sciences occupied pride of place in this project, both in the form of evolutionism, with its materialist, relativizing implications, and in the form of modern physiology, with its promise to explain scientifically the mysteries of the human organism, including its ideologically sensitive spiritual dimension.[4]

Seminarians were expressly forbidden to "read works of their own choosing, especially books that include ideas contrary to morality and Church doctrine," but the young Pavlov regularly procured precisely such works from Ryazan's new public library.[5] He would read virtually around the clock, and he engaged in heated intellectual discussions both with the members of his student circle (*kruzhok*) and with his family. Ivan was especially attached to Pisarev's slogans, such as "nature is not a cathedral but a workshop" and "a bootmaker is of more use than Shakespeare." These captured the contemporary radical ethos that Russia needed less philosophizing and more practical works and that a utilitarian, materialist science would play a central role in the industrial, social, and spiritual transformation of the country.[6]

The young Pavlov's other favorite nonscientific author of the time was a leading British popularizer of bourgeois Victorian values. This is only apparently paradoxical. Pisarev's ideas were, by mid-decade, already under attack by the populist left as too individualistic and ultimately apolitical. Pisarev and his followers, after all, stressed the development of an intellectual vanguard that, having discarded tsarist metaphysics in favor of positive knowledge, busied itself with the physiologist's frog and narrowly practical tasks. Pisarev's "realists" (captured or caricatured—depending on one's point of view—by Turgenev's Bazarov in *Fathers and Children* [also known as *Fathers and Sons*]) were "enlighteners" and modernizers rather than political activists with a well-articulated vision of Russia's future and tactics for its transformation.

It is not surprising, then, that Pavlov delighted both in Pisarev's essays and in the aphorisms of Samuel Smiles, whose *Self-Help* and *Lives of the Engineers* "caught the spirit of an achieving society."[7] Smiles emphasized the importance of individual character and self-discipline and preached "the gospel of work." As Thomas Hughes has observed, Smiles's heroic engineers imposed their character on a hostile and disorderly nature, thus promoting reliable, evolutionary social progress. Smiles's books were popular not only in Great Britain but also among modernizers in such developing countries as Italy, Serbia, and Russia.[8] Pavlov could recite from memory lengthy passages from Smiles's *Self-Help* (translated into Russian as *Samodeiatel'nost'*), a collection of essays and aphorisms on the centrality of character and the virtues of industriousness, perseverance, self-discipline, regularity, punctuality, and honesty.[9] "The crown and glory of life is character," Smiles wrote, which "is moral order embodied in the individual." "An economical use of time is the true mode of securing leisure"; "the common highway of steady industry and application is the only safe road to travel"; and "the art of seizing opportunity and turning even accidents to account, bending them to some purpose, is a great secret of success."[10] Just as Pisarev's essays pointed to science's central role in the modernization and westernization of postreform Russia, so perhaps did Smiles's aphorisms appeal to the young Pavlov as morally resonant guidelines for professional success in these newly emerging circumstances.

Many years later, Pavlov would recall with special emotion the impression created by three other works—each concerning physiology—that gained much attention amid the ideological struggles of the 1860s: Russian physiologist I. M. Sechenov's "Reflexes of the Brain," British popularizer G. H. Lewes's *The Physiology of Common Life,* and the lectures of French physiologist Claude Bernard.

Sechenov's essay, originally entitled "An Attempt to Establish the Physiological Foundations of Psychical Processes," drew upon his investigations of central inhibition to provide a determinist physiological explanation of apparently spontaneous, volitional human actions. Originally intended for publication in the radical journal *Sovremennik* (The Contemporary) and later published as a book, "Reflexes of the Brain" (1863) mobilized physiology against a key element of tsarist ideology—the existence of an immaterial, psychic aspect in man—and contributed mightily to the common association of physiology with political radicalism. Sechenov denied that his determinism negated morality. On the contrary, he insisted, people would always prefer good machines to bad ones, and an understanding of reflex mechanisms might be useful for the production of good "human machines"—that is, enlightened individuals inclined toward noble acts.[11]

Like Sechenov's essay, Lewes's *The Physiology of Common Life* delighted

Russian radicals, who found in it support for their position that science was seizing the study of human sensation from "the hands of metaphysicians and psychologists."[12] (Lewes's positivism also endeared him to liberal and conservative thinkers who were attempting to separate the prestige of science from materialist metaphysics.) The young Pavlov considered this work a "striking epigraph" to the arguments of Russian radicals. It also captured his imagination with its portrayal of the physiologist's approach to the animal machine. Many years later, Pavlov plucked Lewes's book off the library shelf of a friend, turned directly to an illustration (reproduced in Figure 5), and recalled that, as a youth, "I was greatly intrigued by this picture. I asked myself: How does such a complicated system work?"[13] Lewes had reproduced this sketch from a work by Claude Bernard, whom Pavlov also later remembered as an important early influence.

Bernard was a most prestigious figure within the Russian intelligentsia at this time. Among Russian physiologists, he was universally acclaimed as a prophet of their professionalizing discipline. Three successive professors of physiology at St. Petersburg University assumed that post after working in Bernard's laboratory, and the French physiologist's works were quickly translated into Russian.[14] Like Lewes, Bernard was venerated by political thinkers of various orientations who claimed him for their own view of science and the organism. Radicals emphasized Bernard's insistence on emancipating modern physiology from idealist metaphysics and so, for example, republished an essay of his that the censor found "pernicious" for its undermining of "esteemed . . . teleological truths."[15] Yet it was conservative intellectual N. N. Strakhov who first translated into Russian Bernard's *An Introduction to the Study of Experimental Medicine* (1865).[16] In Strakhov's view, this work ably distinguished true science, sober experimentalism, and positive knowledge from the materialist metaphysics of Russian radicals and the fashionable vulgarisms of Carl Vogt, Jacob Moleschott, and Ludwig Buchner. Many years later, Pavlov declared that Bernard was "the original inspiration of my physiological activity." In any case, the French physiologist's reputation in Russia lends credence to Pavlov's comment that Bernard's "remarkable lectures with such lively descriptions of biological experiments, the force and compelling clarity of his thought, and the charm of his investigative intellect attracted me in my youth."[17]

Abandoning his family's long-standing clerical tradition, Pavlov set off in 1870 for St. Petersburg University, where, after an initial stratagem designed to avoid an entrance examination in mathematics (a subject in which he always confessed his weakness), he enrolled in the physico-mathematical faculty. He

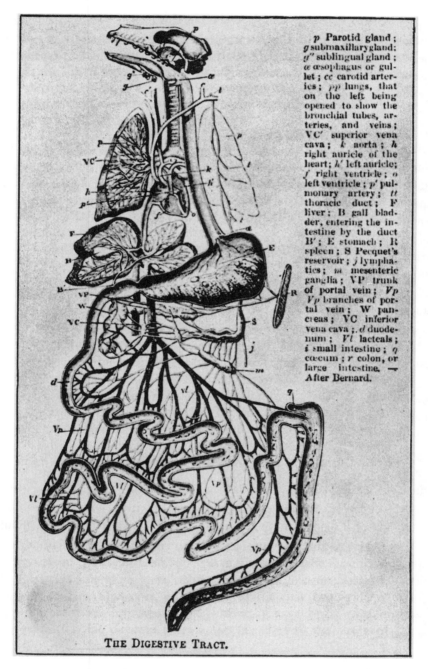

FIGURE 5. *G. H. Lewes's sketch of a mammal's internal organs. From G. H. Lewes,* The Physiology of Common Life, *vol. 1 (Edinburgh: Blackburn and Sons, 1859), 230*

participated in one of the University's many student circles, distinguishing himself (according to his future wife) as "the best-read and also the most heated and inexhaustible disputant." "Possessing a brilliant memory, he could recite by heart entire pages from Pisarev's articles and from his favorite book, Lewes's *Physiology of Common Life.*"[18]

As an undergraduate (1870–75), medical student (1875–79), and doctoral student (1880–83), Pavlov assumed the attitudes and aspirations of the professional scientist and struggled with the relationship between these values and those that had inspired him earlier. He retained the materialist sentiments and the ideal of a transformative science that he had imbibed in the 1860s, but quickly surrendered any radical political ideas earlier associated with these views. His growing appreciation of the distance between the radical ideals of his youth and his daily preoccupations as a professionalizing scientist contributed to an "identity crisis" in the late 1870s and early 1880s. He emerged from it with his scientism intact and still profoundly joined to his personal identity, but now through a moderate liberal vision of the practical benefits and civilizing mission of science.[19] He came to see his earlier passion for Pisarev and other radical theoreticians as a normal stage of youthful enthusiasms, a necessary and fruitful stage that one needed to traverse and transcend on the path to intellectual maturity.

As would be his practice throughout life, Pavlov generalized easily from his own experience. In a series of letters in 1879–80 to his future wife, Serafima Karchevskaia, he explained the differences between "the youthful mind" and the mature thinker in terms that clearly reflected the tensions attending his own professionalization. The youthful mind was excitable, passionate, indefatigable, and wide-ranging; "receptive, free, unprejudiced, bold" and constantly in search of novelty and a "general view." "We considered people without such a general view pathetic," he recalled of the circles in his seminary and early university years. "We could not understand how one could live without a general view of the world and life." Disdaining narrow specialization, the youthful mind addressed "issues from all possible sciences, philosophical questions about God, the soul, and so forth; about every fact of life." These mental characteristics, which reflected the physiological excitability of the young organism, constituted a good defense against the "tyranny of life's trivia" and were the preconditions of true knowledge and a "rational, worthy life."[20]

The thirty-year-old Pavlov did not, however, romanticize youth as a halcyon age of the intellect. Every positive attribute of the youthful mind reflected a corresponding deficiency. Its uninhibited scope and ambition—this "lighthearted stroll from one end of the universe to the other, as if it were a garden"—betrayed an ignorance of the "infinite difficulty" of attaining a real truth and of the time, energy, and means this required. Similarly, the young mind's receptivity and lack of prejudice reflected the absence of "deeply rooted opinions," and such opinions had their virtues.[21]

As the organism aged, every thinking individual was plunged into a "critical period" during which further intellectual development depended on finding a means to preserve the strengths and transcend the weaknesses of the youthful mind. For Pavlov, the best path to intellectual maturity was that upon which he had already embarked: the systematized, "specialized scientific studies" of the professional scientist.[22] By engaging in disciplined scientific work, one replaced the fading overseer of youthful excitation—the "authority of direct sensations"—with a new source of self-control: "conscious, systematized behavior."[23] The unsustainable general passions of youth yielded to the more mature satisfactions of specialized, disciplined investigation, which developed one's logical powers and inaugurated a new, more mature approach to knowledge.

> Once this is done, once the logical powers have acquired force, become strong, look how naturally the passage is made from the period of youthful intellectual excitation to that of sober thought. We allowed that in the transitional period excitability diminishes. But during the natural rational passage this does not end in a complete, irreplaceable loss. If the adult mind has fewer occasions to celebrate, be disillusioned, and so forth, then, of course, its rarer sensations are probably more profound. Novelty does not disappear; on the contrary, an infinite realm of the new opens up. Where is the end of human science? Does one not encounter at every step new understandings once one has trained oneself to think? Do not fear your final leave-taking from philosophy. Your active thought is stimulated by everything that falls upon your eyes; this poses questions, and these questions quickly reach the urgent need for all-encompassing principles, but do so seriously, with a firm tread, taking partial questions as a point of departure. To think in this way, to create in the intellectual sphere—this is the final goal of intellectual power; to train oneself to think, to put logical power into action in appropriate conditions—this is the final task of the epoch of human development. And true human happiness is guaranteed only to those who understand this task in timely fashion and devote to it all their time and effort. It is as if nature teases the young, excites their taste for the joys of intellectual life, opens the door and reveals the interesting, alluring kingdom of thought. But [into this kingdom] enters only the person who, entranced by its appearance, undertakes serious and difficult work in order to make himself worthy of it.[24]

Pavlov's first guide to this kingdom of mature thought—and the likely source of much of the wisdom he was sharing here with Serafima—was St. Petersburg University's new professor of physiology, Ilya Fadeevich Tsion (Elie de Cyon) (1843–1912).[25] Pavlov would recall in later years Tsion's "enormous impression upon all of us [aspiring] physiologists." "We were simply astounded by his masterful, simple presentation of the most complex physiological questions and his truly artistic ability to perform experiments."[26] In a letter to Tsion almost thirty years after their work together, Pavlov reminisced warmly that "for me, your lectures in the special course at the University and work in your laboratory are among the best memories of my youth."[27] Publicly, he commented simply that "I have the honor of being [Tsion's] student."[28]

Tsion's career, scientific practice, and outlook on physiology exemplify the larger historical point made by Frederic L. Holmes that, as physiology matured in the last decades of the nineteenth century, the work and perspectives of Claude Bernard and Carl Ludwig became "woven together into the texture of that discipline."[29] Only six years older than Pavlov, Tsion already possessed a "European reputation" for his scientific contributions, having collaborated in the 1860s with both Bernard and Ludwig on subjects related to the self-regulation of physiological processes. His first article, coauthored with Ludwig, demonstrated the reflex action by which the depressor branch of the vagus nerve ("Cyon's nerve") lowered the blood pressure by dilating the vessels. Working with Bernard, Tsion also discovered nerves that accelerated cardiac activity. He also taught a course on the circulatory and respiratory systems to a small group of doctors and interns that gathered in Bernard's laboratory at the Collège de France. The French physiologist was so impressed by the young Russian that he sponsored Tsion's successful candidacy for the French Academy of Science's first Montyon Prize for original research. Trained and highly praised by Western Europe's leading physiologists, Tsion was perfectly positioned for appointment to St. Petersburg University's expanding Department of Anatomy and Physiology—first, in November 1868, as assistant in its small physiological laboratory; then, in June 1870, as assistant professor. When Sechenov resigned his professorship in physiology at St. Petersburg's Medical-Surgical Academy, Tsion, armed with strong recommendations from Bernard, Ludwig, and Sechenov himself, also assumed that position in August 1872.

Tsion, then, was the very model of the modern experimental physiologist, a figure straight from the pages of the radical journals Pavlov had devoured just a few years before—except for his reactionary political views. A member of the conservative circles around the leading official ideologists M. N. Katkov and K. P. Pobedonostsev, Tsion and his associate N. I. Bakst—*privatdozent* in

physiology at St. Petersburg University—crusaded against the popular association of physiology with materialism and radicalism.[30] Thus Pavlov's mentor
sought explicitly to overthrow the image of physiology associated with Pisarev's essays and Sechenov's "Reflexes of the Brain"—works that had appealed
greatly to Pavlov's "youthful mind." (Sechenov was also popular among students for his gentle, kindly manner and ascetic lifestyle; in this regard, too,
Tsion presented a sharp contrast, acquiring the reputation of a brusque and
aggressive social climber with a taste for fine living.)

Tsion's lectures provided Pavlov's first systematic view of physiology. Here
he propounded a "physico-vivisectional" approach that incorporated what
he considered the best elements of two Western European traditions: the
"anatomical-vivisectional" orientation that had reached its apogee with Bernard
and the "purely physical" orientation associated with Ludwig and German
physiology. According to Tsion, practitioners of the anatomical-vivisectional
orientation had investigated the operations of animal organs through anatomical studies and vivisection—both of which retained a largely observational
rather than truly experimental character. These investigators had produced a
wealth of factual material, but their explanations of physiological phenomena
often rested upon the empty *deus ex machina* of "life force." "They ascribed to
this force the ability to govern and regulate all animal processes, to correct all
anomalies that developed, to separate substances that were useful to the organism from those which were harmful, and so forth."[31]

Proponents of the "purely physical" orientation, on the other hand, had attempted to explain physiological phenomena in terms of the same physical
and chemical processes that governed the inorganic realm. These scientists
had successfully banished from science the unscientific notion of "vital forces,"
but this orientation, too, proved unable "to satisfy all the demands of physiology as it penetrated deeper and deeper into the very depths of life phenomena." According to Tsion, "This orientation discovered its limits, on the one
hand, in its inapplicability to the investigation of several functions of the organism; and, on the other, in the increasingly visible fact that, while the laws
of the inorganic world were obligatory for organic phenomena, nevertheless
these chemical and physical laws, as a consequence of the particularities of organic phenomena, are usually manifested in forms that make their simple application far from sufficient for an explanation of these phenomena. All organic processes are in large part so complex, they are conditioned by the joint
action of so many varied forces, that it is often completely impossible to re-
duce these processes to the comparatively simple laws that lie at their foundation."[32]

The failure of purely physical models and the progress of physiological methodology had given rise to the "physico-vivisectionist" orientation with which Tsion associated himself. Practitioners of this orientation rejected the doctrine of vital forces and so preserved their science's "strictly scientific spirit." They continued to employ chemistry and physics in their approach to biological phenomena, but (in a Bernardian spirit) these were now clearly relegated to the status of "helpmate sciences." The modern physiologist rejected the earlier, "purely physical" view that physiology was merely the application of physics and chemistry to the organic world, and regarded "more attentively the particularities that characterize biological phenomena."[33]

Tsion saw the recent turn in physiology as owing much to methodological developments. Vivisection was being transformed from "a means of observation to a means of experimentation." "Vivisectionist techniques" were improving rapidly, allowing physiologists to separate and study in isolation the functions of individual organs. Furthermore (and here Tsion drew primarily upon the achievements of Etienne-Jules Marey and German physiology), by borrowing techniques developed in physics, physiologists could observe the resultant experimental phenomena more accurately, protecting themselves from "ruinous errors."[34] A host of modern apparatuses enhanced "the abilities of our sensory organs" by producing graphic representations of physiological processes and their relationships. Curves inscribed on graph paper constituted a "universal language comprehensible to all peoples"; not only did they provide "true representations, of the forms or the arrangement of objects" of investigation, but they also served "to relate the very act of the change of forms, of motion itself."[35]

These developments underlined an important truth about physiology. Because this science dealt with "the most complex and intricate phenomena occurring in nature,"[36] its progress depended primarily on the development of scientific methodologies. "In every physiological investigation, the precision of the method employed is of primary importance, and each newly discovered fact in physiology receives its citizenship rights only when it has been attained on the basis of a method, the strictness and precision of which permits no objections whatsoever."[37] The great progress of nineteenth-century physiology rested precisely upon its emancipation from "scientific systems," and its periodic changes in orientation resulted primarily not from metaphysical discussions but rather from "changes in the *methods* of investigation."[38]

Tsion accompanied his lectures with experimental demonstrations prepared in the Medical-Surgical Academy's physiological laboratory, which he expanded qualitatively for pedagogical and research purposes. The laboratory acquired two rooms equipped for vivisection, a third with apparatus for the analysis of blood, a fourth for chemical analysis, and special facilities for research on electrophysiology and the sensory organs. Tsion's updated laboratory included such modern technologies as Ludwig's kymograph and apparatus for analysis of the blood; Marey's sphygmograph, cardiograph, myograph, and polygraph; and other equipment developed by Hermann von Helmholtz, Eduard Pflüger, and Rudolf Heidenhain.[39] Tsion's pedagogical attention to the use of these modern technologies is clear in his two-volume guide to laboratory physiology—which his protégé Ivan Pavlov would, even forty years later, pronounce the best primer of its kind, together with Bernard's *Leçons de Physiologie Opératoire.*[40]

Pavlov joined a number of other young aspiring physiologists in Tsion's laboratory and developed there a passion for intricate experimental work, an appreciation of the role of methodology and technique in science, and the surgical skills necessary to Tsion's brand of experimental physiology. His evenings at the laboratory soon bore fruit, further acquainting him with the satisfactions of the "mature mind." In late 1874 and early 1875, Pavlov presented his first two scientific communications, each a collaborative work with another of Tsion's students that built upon their mentor's earlier scientific contributions. Pavlov presented the first work, on the nervous regulation of the heart, to the St. Petersburg Society of Naturalists; the second report, on the nervous regulation of the pancreatic gland, earned Pavlov and his collaborator a gold medal in a University competition.[41] By this time, Pavlov was nearing graduation from the University, and Tsion invited him to serve as his laboratory assistant at the Medical-Surgical Academy while pursuing there a medical degree, commonly seen as a prerequisite for a professorship in physiology.[42]

This, however, was not to be. "There occurred a wild episode," as Pavlov put it some years later, and Tsion, "this most talented physiologist," was "chased out of the Academy."[43] The conflagration that enveloped Tsion—and traumatized his protégé—resulted from the confluence of a number of factors: the changing relationship between the Russian state and academic institutions, faculty factionalism, widespread student discontent, anti-Semitism, Tsion's combatively conservative political views, his unusually high standards of academic rigor, and his personal truculence and lack of political finesse. For our purposes, the first public spark in this conflagration merits greatest attention: Tsion's address on "The Heart and the Brain" at the inaugural ceremony of the Medical-Surgical Academy in January 1873.[44]

The very title of this speech invoked Tsion's personal relationship with Bernard and identified his approach to physiology with that of the great

French physiologist. Tsion noted that Bernard had addressed this same subject in a well-known lecture of 1864 at the Sorbonne in which the French physiologist "attempted to provide a physiological explanation of some of the poetic forms expressing the dependence of the heart upon emotional states."[45] In his own speech, Tsion updated Bernard's analysis in the light of modern investigative techniques and recent scientific discoveries and joined Bernard's authority to his own crusade against the image of physiology as a radical, materialist science.

Referring to the work of Sechenov and the "men of the sixties," Tsion lamented the fact that "due to the unscientific spirit of several of the popularly disseminated essays on physiology, about no other science do there exist so many perverse views, so many false conceptions as about this science."[46] Science was not, and by its true nature never could be, an agent "destroying little by little all the ideals that mankind has created in the course of so many centuries, and which it has so long revered" (p. 2). The widespread belief that physiology would ultimately explain psychological and spiritual phenomena was but a naive response to the recent successes of "mechanical views of organic phenomena." Despite recent achievements in psychophysiology, the distance between the physiological processes accompanying psychological phenomena and the subjective content of thought and emotion would always remain unbridgeable. Those who thought otherwise resembled "a child who, seeing on the horizon the apparent contact of the sky with the earth, imagines that he need only reach that point in order to climb up into the heavens" (p. 21). Science did provide material relevant to a general worldview, but it served humanity not by resolving philosophical questions but rather by increasing productivity and "by regulating consumption in a more rational manner" (p. 5). For scientists, the search for truth itself, and the discovery of nature's harmonies, was a constant source of aesthetic pleasure (p. 3).[47]

Tsion developed these themes by discussing scientific developments in studies of the relationship between heart and brain. As Bernard had indicated, these two organs were united by nervous connections into an interactive unit. Physiology, then, had confirmed the poets' view that the heart was "an organ in which all human emotional (*dushevnye*) states are clearly reflected" (p. 7). The characteristics of the heartbeat varied with human emotions themselves. Indeed, because of "the involuntary nature of all changes in the heart and its vessels under the influence of emotional states, these [changes] are essentially the only true proof of the sincerity of our feelings" (p. 19). Explaining the use by contemporary physiologists of the sphygmograph and cardiograph to portray graphically the strength and frequency of the pulse and heartbeat, Tsion noted that these enabled a scientific analysis of "the sincerity of our emotions." This, in turn, might serve a series of practical purposes—for example, assessing which potential inheritors at the deathbed of a rich man were genuinely grief-stricken (and so facilitating a proper dispensation of the deceased's property) and determining whether a young lady's suitors were motivated by love or lust (and so helping her preserve her virtue) (p. 29). Like his attack on fashionable views of physiology as a materialist science, these examples were hardly calculated to endear Tsion to a student body accustomed to talk about the democratic essence of modern science and the equality of women.

For us, with an eye on Pavlov's developing scientific vision, another aspect of Tsion's speech bears emphasis: his attention to the nervous connections among organs, the relationship between emotional states and involuntary physiological processes, and the related limitations of mechanistic views. On the one hand, Tsion consistently described the heart as a pump and characterized circulatory processes in terms of mechanical laws.[48] This mechanistic language served an important professional and heuristic function: guarding physiology against a doctrine of "vital forces" that, as Bernard and others had argued, threatened to deprive physiology of the determinism necessary to any science.

Tsion emphasized, however, that this formulation did not mean the heart was *merely* a pump, as the "purely physical" school would have it. The nervous connections between the heart and the brain confirmed, rather, a more complex view long dismissed by mechanistic anatomists and physiologists as "poetic fantasy": "This small muscular sac is not only an ingeniously constructed pump" but also "an organ in which, as in a mirror, all the moods of our soul are quantitatively and accurately reflected."[49] The physiology of the heart, then, was inseparable from that of other organs—and from the emotional state of the organism.

In his speech, as in his lectures, Tsion propounded an updated Bernardianism enriched by developments in German physiology. He upheld Bernard's concentration on organ physiology and the relations between organs and his combination of determinism with attention to the "particularities that characterize biological phenomena." Tsion's Bernardianism, however, dispensed with all references to vital forces and incorporated the precision-oriented techniques adapted from the physical sciences by Marey and many German physiologists. The "purely physical" world served also as a source of heuristically useful models—for example, the model of heart as pump—but these were useful only if the physiologist bore in mind their inevitable limitations.[50]

Tsion emphatically rejected what he considered the metaphysical materialism of the so-called 1847 Group, adopting Bernard's agnosticism on ontological issues as the only stance compatible with positive science.

Indignant student reaction to the political tenor of Tsion's speech was compounded by his departure from a tradition at the Academy that they held dear: Tsion refused to guarantee a blanket "gentleman's C" to the medical students in his course on physiology in the 1873–74 academic year. He was apparently motivated by his conviction that students paid too little attention to the "theoretical" courses in the medical curriculum and by a widespread sentiment that the Academy's standards of academic rigor needed to be raised. (Two other professors joined Tsion that semester in refusing to grant the "gentleman's C.") One hundred and thirty second-year medical students failed Tsion's final exam and were informed that they would have to repeat his course the following year. Many students protested, and the factionalized faculty could not agree on an appropriate response. But the end of the academic year quieted the campus.

Over the summer, however, the populist journal *Otechestvennye Zapiski* (Fatherland Notes) published a polemical attack on Tsion as a mediocrity, a plagiarist, and a political reactionary whose very appointment to the Academy had been illegitimate.[51] Student protesters appeared en masse at his first lecture of fall 1874, heckling him and pelting him with eggs and cucumbers. One student leaflet put the issue this way: "Will we, 1,200 people, be defeated by Tsion, allow him to laugh at us? We will conduct a struggle to the end, until we chase him from the Academy."[52] A general boycott of Academy classes culminated in mass student demonstrations in late October 1874. The protests were suppressed by St. Petersburg's security forces, and many students were expelled and sent home. Student activists responded by spreading the boycott throughout the city's academic institutions. Never popular among the faculty, Tsion alienated administrators with several impolitic moves, and his support among even his conservative political allies proved thin. Strakhov, for example, confided that students had found an object "entirely worthy of their antipathy." Besides being a "yid," Tsion was "boastful, insolent [and] heartless" and "tried the patience of both students and professors with his infernal fussiness over his science."[53] In November 1874, the authorities at both St. Petersburg University and the Medical-Surgical Academy asked Tsion to take an extended leave. Never invited to return, he resigned formally from both positions one year later.

For Pavlov, of course, this was a most traumatic turn of events. His adored mentor had been humiliated and destroyed, and his own plans to assist Tsion in the Academy's physiological laboratory had evaporated. He was left without a patron and embittered at the liberal intelligentsia. Two days after the student demonstrations against Tsion were suppressed—on October 29, 1874—Pavlov delivered his first scientific report to the St. Petersburg Society of Naturalists. Three months later, in a clear protest against Tsion's forced departure, Pavlov and his collaborator did not attend the ceremony at which they were to receive a gold medal for research conducted in Tsion's laboratory. Tsion's successor as professor of physiology at the Academy offered Pavlov an assistantship in the laboratory there, but Pavlov indignantly refused.

Pavlov left no enduring record of his experiences during this "wild episode," but scattered comments reveal its searing effect. Even forty years later, in a public address entitled "The Russian Mind," he commented bitterly on Russians' attitude toward freedom of speech: "Do we have this freedom? One must say no. I remember my student years. To say anything against the general mood was impossible. You were dragged down and all but labeled a spy."[54]

"In Somebody Else's Laboratory"

From 1874 to 1890 Pavlov continued to work on the two topics he had taken up under Tsion's tutelage: the nervous control of the digestive glands and the circulatory system. In the 1870s he addressed the former subject in "On the Reflexive Inhibition of Salivation" (1877), "Material on the Physiology of the Pancreatic Gland" (1878), "Further Material on the Physiology of the Pancreatic Gland" (1878), and "New Methods of Implantation of the Pancreatic Fistula" (1879). He explored the nervous control of the circulatory system in "On the Vascular Centers in the Spine" (1877), "Material on the Innervation of the Circulatory System" (1882), his doctoral dissertation on "The Afferent Nerves of the Heart" (1883), and other articles on the nerves of the heart in 1883–88. In the late 1880s Pavlov resumed his research on the digestive glands in collaboration with the chemist Ekaterina Shumova-Simanovskaia, with whom he wrote two important articles: "The Secretory Nerve of the Gastric Glands of the Dog" (1889) and "The Innervation of the Gastric Glands of the Dog" (1890).[55]

As we can glean from scattered indications in Pavlov's correspondence, he had not forgotten the broader vision that inspired him in the 1860s. For example, in one letter of 1880, in which he grappled with the uncomfortable similarity between Dostoevsky's portrayal of Ivan Karamazov and his own inner life, he observed, "Where is the science of human life? Not even a trace of it exists. It will [exist], of course, but not soon, not soon."[56] The distant prospect

of such a science, however, held little allure for the "mature mind" of this professionalizing scientist and Tsion protégé in the 1880s and 1890s.

For our purposes here, two other aspects of Pavlov's life in these years are
especially important: his two stays in Western Europe and his laboratory experiences.

In 1884, Pavlov became one of three recently minted doctors of medicine
chosen by the Military-Medical Academy for a coveted two-year study trip to
Western Europe.[57] He decided to divide his two years between Heidenhain's
physiological laboratory in Breslau (now Wrocław, Poland), where he had
spent the summer months of 1877, and Carl Ludwig's Physiological Institute
in Leipzig. Ludwig's laboratory had become a mecca for scientifically oriented
physicians and aspiring experimentalists from around the world, and the more
than two hundred who spent time in Leipzig from the 1840s to the 1890s are
often referred to as Ludwig's "students." Among the laboratory's other visitors
in 1884–86 were such future luminaries as Johns Hopkins University pharmacologist John J. Abel and anatomist Franklin Mall, Columbia University
physiologist Franklin Lee, and University of Helsingfors (Helsinki) physiologist Robert Tigerstedt.[58]

The record of Pavlov's stay in Heidenhain's and Ludwig's laboratories is
scanty, but several things are clear. For one thing, these visits did not substantially change the investigatory paths on which Pavlov had embarked under
Tsion's direction. Pavlov had first visited Breslau in 1877 to demonstrate to Heidenhain some experiments related to his own argument for nervous control
of the pancreas.[59] Heidenhain had remained unconvinced and raised several
objections, which Pavlov sought to address in subsequent investigations. His
return visit was apparently intended to continue these discussions. Pavlov
journeyed to Leipzig in part, no doubt, to garner the career benefits of an association with the world's leading physiologist, but also to familiarize himself
with the physiological techniques for which Ludwig was famous, to use a specific apparatus necessary for his ongoing investigations, and to read the latest
scientific literature in the Physiological Institute's renowned library.[60] Although Ludwig assigned his Russian visitor a research topic, that assignment
reflected mutual interests: Pavlov pursued questions arising from his doctoral
dissertation by using an apparatus that measured the quantity of blood
pumped by the heart. (Interestingly, this apparatus was originally conceived
by a previous Russian visitor.)[61]

Pavlov's experiences in Western Europe influenced him substantially. First,
his stay in Ludwig's laboratory seems to have reinforced Tsion's earlier emphasis (originating, perhaps, in Tsion's own work with Ludwig) on the use-

fulness for physiology of graphic representation and precise measurement. Second, Pavlov was clearly impressed by particular animal technologies developed by the two European physiologists: by Ludwig's "isolated heart" and Heidenhain's "isolated stomach," both designed to facilitate physiological experiments on an intact and functioning organ. Third, after his trip to Europe, Pavlov used mechanistic imagery more frequently and self-confidently, identifying such imagery firmly with physiology's status as a determinist science. In this respect, however, his European experience was contradictory. Pavlov had earlier imbibed mechanistic imagery from many sources—including from Bernard and Tsion, both of whom were also sensitive to the limitations of simple mechanistic interpretations of physiological phenomena. Pavlov recognized a substantial difference between Ludwig and Heidenhain in this regard and always identified thereafter with what he perceived as Heidenhain's more specifically *physiological* style of experimentation and interpretation. I return to this point below.[62]

Pavlov himself later emphasized another consequence of his stay in Western Europe. "My trip abroad was precious to me," he recalled, "mainly because it acquainted me with a type of scientific worker, such as Heidenhain and Ludwig, who placed their entire life, all its joy and grief, in science."[63] In his eulogy to Heidenhain in 1897, Pavlov drew upon his personal experiences in Breslau and Leipzig to describe this ideal "scientific type."

> As a teacher, Heidenhain was a charming personality—completely simple, attentive, always extremely interested in everything, and rejoicing in the successes of his pupils. His expansiveness, his activeness, united the entire laboratory. Experiencing deeply every work conducted in his laboratory, he interested everybody else in it, so we all lived not only for our own interests, but also for the successes and failure of the entire laboratory. Above all this, he has still another invaluable quality: he preserved into his old age his naive childlike spirit, his heartfelt goodness . . . I also noticed this rare quality in another teacher, Ludwig. How were they able to preserve it? Very simple, gentlemen! They spent their entire life within the walls of a laboratory— among books, apparatus, and experiments—where there is one virtue, one joy, one attachment and passion: the achievement of truth.[64]

Pavlov wholeheartedly adopted the persona he identified in these two scientists—as inspiring laboratory managers, wholly dedicated scientists, and childlike spirits. His use of the word *childlike* testifies to the resonance of Ludwig's and Heidenhain's examples with Pavlov's own preoccupation with the passage from youthful passions to the mature mind. Here were two great scientists who had successfully negotiated that transition, preserving their youthful passion as professional scientists by devoting their lives to systematized, disciplined, laboratory work.

In Breslau and especially Leipzig, Pavlov encountered a new type of physiological laboratory: larger, comparatively well-equipped enterprises that drew upon the increasing cadres of aspiring scientists and scientifically oriented physicians made available by the growing prestige of scientific medicine. Ludwig's laboratory combined a pedagogical mission with a system of production that made efficient use of mostly young and inexperienced investigators from around the world. As W. Bruce Fye has indicated, Ludwig's coworkers were "usually recent medical graduates who came to Ludwig's institute to receive training in the techniques of modern experimental physiology and to participate in research under his direction."[65] Laboratory coworkers attended Ludwig's lectures and enjoyed free access to the Physiological Institute's rich library. Often about ten coworkers worked in the laboratory simultaneously in a setting that, according to one coworker, combined "a form of military discipline" and a "skillful division of labor" with an inspiring "esprit de corps."[66] Ludwig usually assigned coworkers a research topic from among the great variety of subjects and approaches that interested him. As one former coworker put it, "Each man had his own clearly defined problem, and the problems were as distinct as the men . . . It was remarkable how many different forms of research he could supervise at the same time and keep them all clearly in mind." In 1885, for example, Stolnikov worked on the rate of blood flow, Tigerstedt on the latent period of muscle, von Frey and Gruber on the metabolism of isolated muscle, Bohr on gases' entry into and exit from blood in the lungs, Lombard on reflex processes in the spine, Smith on histology, and Abel and others on issues in physiological chemistry.[67] Aided by a mechanic and two assistants, Ludwig supervised this work closely, checking experimental protocols and editing (often rewriting completely) the published reports. These reports bore the master's "characteristic means of expression," but authorship was credited to the coworker alone.[68]

Russian institutions offered no possibility for an even remotely comparable operation. In the 1870s and 1880s, Pavlov was associated with Russia's leading medical investigative institution, the Military-Medical Academy, but the laboratory resources there were comparatively paltry. The Academy's professor of physiology—Tsion in 1872–74 and Ivan Tarkhanov in 1876–94—directed a small, sparsely equipped workshop where three or four students worked on topics assigned by the professor. The Academy also housed several small laboratories that had been created by Sergei Botkin, professor of therapeutic clinical medicine, as part of his effort to bring scientific medicine to

Russia. Two of these laboratories were devoted to medically related investigations in chemistry and bacteriology; the third, for animal experiments in pharmacology, was managed by Pavlov from 1878 to 1890.

Here Pavlov discovered a talent and enthusiasm for running a laboratory and an opportunity to emulate the Heidenhain-Ludwig model, albeit with some important and frustrating limitations. For one thing, the laboratory was "small and poorly equipped"; if investigators needed any complex apparatus aside from an old kymograph, they had to build it themselves.[69] Furthermore, the hygienic conditions were "nasty," to which Pavlov attributed the death of his dogs and rabbits after any complex operation. "So, it's not my fault," he wrote to Serafima in 1882 of the slow progress on his dissertation. "I'll take up other experiments that do not require the animals on which I operate to remain alive."[70]

Another important limitation was that the laboratory was emphatically not Pavlov's. Botkin rarely appeared there, but he did assign dissertation topics to the one or two physician-investigators who arrived each year. These assignments concerned a variety of topics that Botkin considered important to clinical medicine. Most of them concerned the action of various pharmacological agents on bodily organs, although a number did not (e.g., he assigned one investigator to explain the coating of the tongue). As laboratory manager, Pavlov's job was to guide research on these diverse topics to a successful conclusion. This required him to expend an enormous amount of time and effort on work that he often found dull and pointless. As he confided to Serafima in 1881, "It is a shame that I will have to conduct the experiments of others . . . This would be all the same to me if only they suited me. But, you know, these themes come from Botkin, sometimes entirely, and they are entirely incoherent; my participation in such experiments is made entirely mechanical, very boring."[71]

Despite his private sentiments, Pavlov supervised the completion of about fifteen doctoral dissertations—mostly on topics far from his main interests—and apparently succeeded, at least in the eyes of some of his charges, in living up to the image of the ideal scientist that he had acquired in Breslau and Leipzig. One coworker in the Botkin laboratory later wrote, "Recalling this time, I think that each one of us shares the feeling of most lively gratitude to our teacher, not only for his talented guidance, but, mainly, for the extraordinary example that we saw in him personally as a man devoted entirely to science and living only for science, despite the very difficult material conditions—conditions of literal poverty—that he and his heroic [wife] were required to endure."[72] Both Pavlov's image of the ideal scientific manager and a lingering

sense of frustration at the time spent on this job are evident in his later am-
bivalent comment about his tenure in the Botkin laboratory: "I worked there
without distinguishing between what was mine and what belonged to others.
For months and years all my laboratory work went to participation in the work
of others."[73]

Pavlov expressed this frustration clearly in his job inquiry to the dean of the
newly created Tomsk University in 1888. Almost forty years old, financially
strapped, and without prospects, he reviewed his achievements and asked for
a professorship in physiology or, failing that, in pharmacology or general pathology (as "other purely experimental sciences"). Noting that he had worked
with Tsion, Heidenhain, and Botkin and that he could provide letters of reference from such leading lights of Russian medical science as Botkin, Sechenov,
and V. I. Pashutin, Pavlov gave voice to his sense that, in the absence of proper
facilities, precious time was slipping away: "And meanwhile my time and
strength is not spent as productively as it should, because it is not at all the
same to work alone in somebody else's laboratory as to work with students in
one's own laboratory."[74]

His situation soon grew worse, and then improved radically. In 1889 Pavlov
placed second in competitions for rare vacancies in physiology at both Tomsk
and St. Petersburg universities.[75] His most powerful supporter, Botkin, died
shortly thereafter. The following year, however, Pavlov's time of troubles came
to an end: in mid-1890 he was offered a position at both Tomsk University and
the Military-Medical Academy, in each case as an assistant professor of pharmacology.[76] Several months after accepting the position at the Academy, he
acquired a second laboratory as chief of the Physiology Division in the Imperial Institute of Experimental Medicine.

Two developments in 1889–90 signaled the use Pavlov would make of the
Physiology Division's rich facilities. First, his research in collaboration with
Shumova-Simanovskaia opened up new investigative horizons. Second, with
Botkin seriously ill, Pavlov apparently had the opportunity to assign dissertation topics to two new physician-investigators in the Botkin laboratory. He directed both N. Ia. Ketcher and B. V. Verkhovskii to undertake theses designed
to follow up his own work with Shumova-Simanovskaia on the nervous mechanisms of digestive secretion.

According to his good friend D. A. Kamenskii, Pavlov was initially ambiva-
lent about accepting a position at the Institute. For one thing, "the atmosphere
around Prince Ol'denburgskii was not especially comfortable for sensitive
people, and even many who sought work reconciled themselves to this setting
with difficulty."[77] As an assistant professor of pharmacology at the Military-

Medical Academy—and a forceful, explosive personality himself—Pavlov already had a good income and a small laboratory and so could have spared himself the aggravation of an overbearing, unpredictable boss in a new institution with uncertain prospects. Yet Pavlov and his family had endured years of material privation, often living separately with friends and relatives in the 1880s when they were unable to afford an apartment. They were still in debt and no doubt savored the material comforts that the income from two positions (especially the high salary from the Institute) could provide. So, Pavlov decided to accept the new position at the Institute and to maintain his assistant professorship at the Military-Medical Academy. He prevailed upon Ol'denburgskii to exempt him from the Institute's prohibition against division chiefs holding outside positions, and, "not without pressure from his [Pavlov's] friends," he assigned one of his first assistants to work on Ol'denburgskii's beloved tuberculin.[78] Still, Pavlov confided to Kamenskii that he would "not work long at the Institute, that once his material situation . . . improved somewhat he would remain only at the Military-Medical Academy."[79]

His sentiments changed when Nencki's plans for the Chemistry Division and the unexpected influx of praktikanty led to a radical expansion of Institute facilities—and to a fundamental change in its very nature. As Kamenskii recalled, "Pavlov became convinced that one could acquire here all necessary means for scientific work, that the physicians working with him would not need to spend their own money on experiments, that everything would be provided to them—both dogs and feed—and, mainly, that he would have here many coworkers. All this bound him to the Institute."[80]

The new chief of the Institute's Physiology Division, then, was hardly an eminent scientist. But he was a veteran investigator with developed ideas about what constituted good physiology and a keen awareness of the advantages of a large laboratory group for producing it.

The Mature Mind: Physiological Thinking and the Animal Machine

Like Tsion, Pavlov propounded an updated Bernardianism that incorporated the demand for quantitative precision associated with German physiology.[81] For both mentor and protégé, this general framework defined the tasks and approaches of the professional physiologist. That common framework, however, accommodated differences in intellectual style and tendency: for Tsion, a critique of materialist metaphysics; for Pavlov, a continued attachment to the materialism and scientism he had absorbed in the 1860s.

Throughout his career, Pavlov's vision and rhetoric bore a striking similarity to those expressed by Bernard in *An Introduction to the Study of Experimental Medicine* (1865).[82] For both physiologists, the organism was a purposive, complex, specifically biological machine governed by determinist relations. The physiologist's task was to uncover these unvarying relations, to control experimentally or otherwise account for the "numberless factors" that concealed them behind a veil of apparent spontaneity. Physiology would thus attain mastery over the organism and give birth to an experimental pathology and therapeutics that would revolutionize medicine.[83]

In this spirit, Pavlov always insisted on results that were *pravil'nye*, a word that means both "regular" and "correct," capturing his view that in physiological experiments, the two meanings were one and the same. Pavlov also shared Bernard's commitment to organ physiology, which addressed a level "high" enough to encompass the vital, purposive activity of complex organic machines but "low" enough to discover the unvarying laws without which physiology would not qualify as a science. Also like Bernard, Pavlov was acutely conscious that the physiological phenomena he studied often resulted from the constant and complex interactions among organ systems.[84]

Pavlov also shared Bernard's position that the physiologist, in his laboratory research, disavowed ontological questions, seeking only proximate causes and leaving the question of essences to philosophers. This did not mean, of course, that Pavlov had no ontological commitments or inclinations, but in the laboratory, as the province of the "mature mind," these emerged primarily as methodological issues.

Like Bernard, Pavlov wrestled with the relationship between physiology and the sciences just "below" it—physics and chemistry. For both physiologists, the central question here was not so much philosophical as methodological. Although they sometimes expressed their views (usually cautiously and tentatively) about the ontological relationship of physicochemical and physiological processes, both were more concerned with defining the physiologist's specific goals, approaches, and interpretive procedures. Pavlov's response to this set of issues constituted what he referred to as *physiological thinking* (*fiziologicheskoe myshlenie*).[85]

The terms of Pavlov's *physiological thinking* were set by a basic tension rooted in his conception of the organism and of physiology as a science: he sought both to understand the normally functioning organism in all its complexity *and* to express that understanding in *pravil'nye*, fully determined, and, when appropriate, quantitative terms. For Pavlov, as for Bernard and Tsion, the organism was a machine governed by determinist relations, but it was immeasurably more complex than any known machine and its parts were connected in subtle and extraordinarily sensitive relationships. A physiological understanding of the organic machine therefore required an integration of what Pavlov termed *analytical* and *synthetic* approaches. Organic phenomena could not violate physical and chemical laws, but because of their complexity they were rarely (at least, for the present) understandable through the same approaches and models that guided physics and chemistry. In Pavlov's view, the truthfulness of a physiological discovery was demonstrated by control (or prediction) of the phenomenon in question and, ideally, was expressed by a precise quantitative formula. He recognized, however, that the results of any experiment on an intact animal were always influenced by uncontrolled factors in the complex organic whole. Interpretations about the influence of these factors, then, always played some role in his assessment of experimental results. Pavlov's scientific vision is best sketched, I think, by exploring these different dimensions and the tensions they created. We will then be prepared to watch them play out in his actual laboratory practices.

Pavlov rarely expressed himself explicitly about philosophical issues, but he considered one bedrock position essential to physiology's status as a science: "The living organism is an indivisible part of nonliving nature and is subject to the very same laws." The basic laws of physics—the "constancy of matter, force, and elements"—were "entirely applicable to the living organism."[86] "Life," wrote Pavlov, "is the distinctive (*svoeobraznoe*) transformation of external matter into vital forces." Physiology's task was to track that process from the moment external matter was taken into the body through the processes of digestion, muscular movement, and respiration. The "ideal" of physiology, then, was the "reduction (*svedenie*) [of physiological phenomena] to physicochemical forces."[87] When Pavlov was particularly enthusiastic about a scientific achievement, he sometimes noted (cautiously) that it might "serve as proof of the correctness of the physicochemical view of life."[88]

Pavlov often expressed this perspective through his use of mechanistic imagery: the organism was a machine, a factory, or a laboratory; the physician (or physiologist) was a mechanic; the heart was a pump, the nerves were telephone cables, and so forth.[89] As Pavlov explained in his lectures to medical students at the Military-Medical Academy, "as a physiologist, I will look at the human organism as simply a machine. Only such a view can be termed scientific . . . You must look at the human organism just as a mechanic looks at a machine that he must know and fix."[90]

Such mechanistic imagery expressed not only Pavlov's faith in the ultimate explicability of organic phenomena but also another of his basic beliefs: that scientific knowledge is ultimately expressed in the prediction and control of a phenomenon. He expressed this sentiment publicly as early as 1877, but expressed it most robustly in his speech to the International Medical Congress in Paris in 1900. Here Pavlov recounted the history of physiologists' attempts to keep alive dogs that had undergone a double vagotomy:

> At every stage of the cultivation of our question has not the animal organism revealed itself to be simply a machine, an extraordinarily complex one, of course, but all the same as submissive and obedient as any other machine?
>
> [By cutting both vagus nerves] we inflicted colossal damage on this machine, many of its small and extremely important parts were broken, the relations between them were profoundly changed, the machine had become useless, its complete destruction seemed imminent.
>
> What did the physiologist do when confronted with this destruction? Skillfully, patiently, over the course of an entire century, he continued to analyze, one after another, all the disturbances which he himself had produced by his vulgar contact with this delicate machine; and these disturbances were enormous in number; nevertheless, he attempted to discover the meaning and significance of each of them to the general work of the machine.
>
> As a result, a gradation in the importance of damages was uncovered: some immediately elicited the complete collapse of the entire machine; others brought it to death more or less gradually; still others, finally, elicited only an irregularity in its work.
>
> And so, guided by these various categories of facts, it became possible to save the machine . . . And physiology in the final analysis triumphed over all these difficulties; the repaired animal machine . . . continued its work with complete success . . . Our power over this animal organism is continually increasing.[91]

Here, in characteristic fashion, Pavlov joined the notion of control over the animal machine to the medical mission of experimental physiology. "The practical physician is called upon to fix a machine that nobody understands as is necessary," he explained. "If one fixes a watch then he knows how it is built and, of course, his activity is entirely purposive. The same is demanded of the physician—to fix that which is broken, but in a machine about which there is not complete information."[92]

For Pavlov, however, this reductionist/analytical moment was always in tension with a holistic/synthetic one. The organism was "an extraordinarily complex" and "delicate machine"; its parts were constantly interacting in far more subtle ways than did a cog and wheel. "The animal organism," he once remarked, "is a closed machine, and there is no real beginning and end to that machine. Every beginning would be artificial." The physiologist, therefore, always confronted a "chain of undefined moments," a series of distant and often indiscernible causes and effects.[93] Knowledge of any single part or physiological process—and the correct interpretation of any experimental data—therefore required a grasp of the organism as a whole: "Only having in view the whole, the normal course of work in one or another section of the organism, can we without difficulty distinguish the accidental from the essential, the artificial from the normal, easily find new facts and often quickly notice errors. The idea of the general, joint work of the parts throws a clear light on the entire investigated area."[94]

This holistic moment rested, then, not on an ontological commitment to emergent properties but rather on Pavlov's sense of how a complex organic machine operated and how the physiologist must proceed in order to understand it.[95] Like Bernard, Pavlov regarded determinism as the sine qua non of true science and perceived the notion of immaterial, spontaneous "vital forces" as a threat to physiology's scientific status. So, whenever he noted the limitations of mechanistic explanations, he always closed the door firmly to neovitalist interpretations. To Pavlov, "life force" was always an "empty phrase."[96]

There was also an aesthetic dimension to Pavlov's methodological holism. As his longtime coworker A. F. Samoilov discussed in a perceptive commentary on Pavlov's investigative profile, the chief was simply most comfortable when dealing with a recognizably alive and vital organism.

> The sphere of phenomena in which he felt light and free encompasses the entire animal as a whole in its connection to the environment that surrounds and acts upon it, and it is this inclination that expresses the strong biological inclination of Pavlov's gift. He valued most the experiment on an intact, unnarcotized animal, on an animal with a normal reaction to an irritation, on an animal that was energetic and cheerful. I remember how he once looked with satisfaction at a dog with an esophagotomy and a gastric fistula as it ran happily into the room in anticipation of the pleasure of sham-feeding. He petted and stroked the dog and said more than once, "And where did people get it into their heads that there is a qualitative difference between us and animals? Do this dog's eyes not sparkle with joy? Why not investigate the phenomenon of joy in the dog; here it is much more elementary and therefore accessible."[97]

This passage in Samoilov's reminiscences follows his account of Pavlov's negative, even irate reaction to the growing literature on the physiology of cells and tissues. On discovering Samoilov immersed in journals of this orientation, Pavlov paged through them impatiently, threw one indignantly onto the table, and remarked, "Yes, if you work upon such questions and upon such ob-

jects, you will not go far. I wish that I had never seen all this." Although un-
persuaded by Pavlov's arguments, Samoilov noted sympathetically that "all
these investigations that concerned isolated parts of the body seemed to him
too isolated from the animal mechanism as a whole, from the whole organism; they seemed too . . . abstract [and] untimely."[98]

In his public comments, Pavlov was more balanced in his appraisal of this
increasingly influential tendency in physiology. He noted, for example, that
the great Heidenhain had been "a cellular physiologist, a representative of that
physiology which must necessarily replace our contemporary organ physiology and which one can consider a precursor of the last step in the science of
life—the physiology of the living molecule." The organism arose from a cell
and so "everything that exists in the organism existed in the cell"; an organ was
an "association of cells" and so "its qualities and activity depend on the qualities and activity of the cells that comprise it."[99] Thus, just as he did not recognize any ontological principle separating physiological laws from those of
physics and chemistry, so he recognized none that rendered the operation of
organs essentially distinct from that of cells and tissues.

Yet for Pavlov (as Samoilov observed), attempts to understand the animal
machine from the perspective of cells and tissues were "untimely," incapable,
at least for the foreseeable future, of providing synthetic knowledge about the
functioning of the normal, intact organism. The cell constituted "the beginning, the bottom of life," but it was the "middle of life"—the organs—that still
offered the best opportunity for physiologists to discover determinist truths
about the organism.

In one passage of his *Lectures on the Work of the Main Digestive Glands,*
Pavlov characterized organ physiology as follows:

> This is not a question of the essence of life, of the mechanism and chemism
> of cellular activity, the final resolution of which will remain for an innu-
> merable series of scientific generations an engrossing but never completely
> satisfied desire. In our, so to speak, sphere of life, in organ physiology (as op-
> posed to cellular physiology) . . . one can truly and soberly hope for a com-
> plete elucidation of the normal connection of all the separate parts of the
> apparatus (in our case, of the digestive canal) with one another and with the
> objects of external nature standing in special relation to them . . . At the level
> of organ physiology we are disengaged from such questions as: What is the
> peripheral ending of the reflexive nerves and how does it receive one or an-
> other irritant? What is the nervous process? How, as a result of which reac-
> tions and what molecular structure does one or another ferment arise in the
> secretory cell? And how is one or another digestive reagent prepared? We

[organ physiologists] take these qualities and these elementary activities as givens and, seeking their rules, the laws of their activity in the complete apparatus, can within certain limits govern that apparatus, rule over it.[100]

In this programmatic statement, Pavlov positioned himself among past and present researchers on digestion. Ever since Friedrich Tiedemann and Leopold Gmelin's treatise of 1826 on the chemical changes undergone by various foods as they passed along the digestive tract, a series of scientists—from Justus von Liebig to Carl Voit—had investigated the intermediary chemical reactions responsible for the processing of food in the organism. In their quest for an understanding of digestion, these scientists tackled such questions as: What is the elementary composition of albumin, fibrin, and casein? What is the chemical nature of gastric juice? How are proteins converted to peptones (soluble, partially digested proteins)? Even Claude Bernard, that champion of a specifically physiological approach to life processes, concentrated his research on digestion on the chemical properties of pancreatic and gastric juice; and, as F. L. Holmes has demonstrated, his theoretical ruminations reflected his belief that a theory of digestion must rest upon a chemically based explanation of the successive changes that food substances underwent in the body.[101]

It was this scientific orientation—and an analogous approach to neurophysiology—that Pavlov had in mind when he explained in the above-quoted passage from *Lectures* that "at the level of organ physiology we are disengaged" from issues concerning cellular and subcellular processes. Knowledge about the chemical composition of "reagents or pure ferments" and their effect on the component parts of food was certainly useful, but by itself it provided only a very "abstract" understanding of the physiological act of digestion itself. If one assumed the perspective of a medical practitioner, one was struck by the "large gap between such knowledge, on the one hand, and the physiological reality and empirical rules of dietetics, on the other."[102]

The organ physiologist sought to close this gap by comprehending "the entire, real course of the digestive act." He regarded lower-level processes as "givens" and directed his attention to the laws of "the complete apparatus." Chemistry provided a valuable tool, but it did not set the physiologist's agenda, which revolved around such questions as: Why are the reagents poured on the raw material in one order rather than another? Are all reagents always poured into the digestive canal for every food? Does the flow of each reagent fluctuate, and how, why, and when? This specifically physiological program would generate the basic framework within which knowledge generated by physiological chemists would acquire meaning.

Pavlov saw this as but one instance of a general principle: a scientific, medically relevant grasp of animal physiology required a combination of synthetic and analytical approaches. The task of analysis is "to become acquainted as much as possible with every isolated part" of the living machine; the task of synthesis is "to evaluate the significance of every organ in its true and vital dimension," "to study the activity of the organism as a whole and of its parts in strictly normal circumstances."[103]

Analytical physiology rested primarily on the *acute experiment*—that is, on experiments conducted on a freshly operated-on animal. Physiologists removed organs or severed nerves and observed the effects, or they opened up the animal's body to permit observation of and experimentation on individual structures. The "highest expression" of this analytical tendency was the isolated heart developed in Ludwig's laboratory. By completely removing the heart from the animal while maintaining the conditions necessary for the heart to function, this procedure allowed experimenters to investigate the heart's operations in complete isolation from other organs.[104]

Such procedures had yielded valuable information, but their usefulness was limited in two basic ways. First, the value of the analytical knowledge itself was compromised by the difficulty of separating the results of the operation from the functions of the organ under investigation. Acute operations were accompanied by bleeding, pain or narcosis, and infection and so by "huge physiological changes in the body." So sensitive and interconnected are the parts of the animal machine that these changes "in the great majority of cases are not linked with the direct traumatization of a given organ; so, the influence of the operation constitutes an action at a distance."[105] Such operations probably excited the inhibitory nerves throughout the body, distorting physiological phenomena in both gross and subtle ways often unrecognized by the experimenter. "Many physiological phenomena can completely disappear . . . or be manifested in an entirely distorted form."[106] The second limitation was even more fundamental. By their very nature, acute experiments on isolated organs could not yield synthetic knowledge—that is, an understanding of how the activities of "separate parts are interlinked in the normal operation of the living machine."[107]

This synthetic task, which comprised the end goal of physiology (and the fundamental link between physiological findings and medical practice), could be addressed only by what Pavlov eventually termed *chronic experiments* on normally functioning animals. In several chapters of this book, I closely examine Pavlov's chronic experiments on the digestive glands and the particular interpretive challenges they presented.

I should note here, however, that Pavlov's commitment to this synthetic task was firmly rooted in his scientific vision and preoccupied him from at least the late 1870s—that is, *long before* he acquired the laboratory facilities he considered necessary for its resolution.[108] This commitment had, by the late 1870s, generated specific interpretive problems and procedures that would emerge full-blown in his later work on digestion and conditional reflexes. In 1877, while investigating the nervous control of the circulatory system, Pavlov became convinced that curarizing a dog prior to experiments rendered the results "abnormal." He therefore decided to use "unpoisoned, unharmed" animals, which he trained to remain "completely peaceful" during experimental procedures.[109]

Two years later, Pavlov wrote about one difficulty encountered during experiments on such conscious and unpoisoned dogs: the results were influenced by "the possible psychic moods of the animal." One dog, especially, manifested a much higher blood pressure than expected, which Pavlov attributed to a "psychic or physiological state" caused by the animal's particular response to the experimental setting: the dog had failed to urinate. "A well-trained animal is accustomed to refraining from urination in the room," Pavlov observed. After the dog had urinated in the courtyard, however, "further measurements gave figures corresponding exactly" to those in other experiments. Other problems encountered in experiments with intact animals included the "extraordinary excitability" of some dogs and a common "fear of the unknown" (i.e., of the experimental setting). One dog, "despite long training, nevertheless would begin to cry out desperately each time it was fastened or unfastened" on the experimental table, and this psychic state was reflected in the dog's blood pressure.[110]

This early work, then, manifested the same distinctive tension that characterized Pavlov's later investigations: that between his determination to encompass the complex, animal whole—in which the psyche and other uncontrollable processes abounded—and his scientific ideal of absolutely precise and repeatable results. In his investigations of blood pressure, Pavlov sought both to minimize the effect of the animal's psyche on his experiments and to use his understanding of that effect as a source of interpretive flexibility. He clearly believed that, in principle, the results of various experimental trials should be absolutely predictable and identical. In practice, however, "far from all the experiments gave identical results," since the psyche of an intact animal caused "disturbances in the course of the curve."[111]

The importance of experimenting on an intact, normal animal was a central theme in the landmark article on the role of appetite in gastric secretion

coauthored by Pavlov and Shumova-Simanovskaia in 1890, on the eve of
Pavlov's appointment to the Imperial Institute of Experimental Medicine. The
authors portrayed the differences between their results and those of previous
investigators as a result of the distortions inherent to acute experiments: "It
becomes obvious," they wrote "that the usual, traditional setting of physiological experimentation on an animal that has been poisoned . . . and recently
subjected to a complex operation is a serious danger that—and this is especially important—is inadequately recognized by physiologists"; during such a
procedure "many physiological phenomena can completely disappear from
the eyes of the observer or are manifested in an extremely distorted form . . .
The time has come for physiology to begin the search for means of experimentation in which deviations from the norm in the experimental animal . . .
will be as negligible as possible."[112]

For Pavlov, the demands of synthetic physiology required a fundamentally
different approach to experimentation and interpretation than that employed
in physics, a point he made by contrasting the scientific styles of Ludwig and
Heidenhain. Pavlov admired both physiologists and acknowledged their specific contributions frequently in his own work. But he used the limitations of
Ludwig's approach to illustrate the perils of a one-sided reliance on approaches and models drawn from physics, while Heidenhain emerged as "the
physiologist's physiologist" and as a model with whom Pavlov clearly identified.

In an article on vivisection in 1893, Pavlov outlined three rules that were
characteristic of physiological investigation and distinguished it substantially
from investigations in physics. In order to ascertain the true relations between
complex, interconnected physiological phenomena, the physiologist must be
alert to "the slightest details of the experimental setting," must conduct numerous trials of the same experiment, and must vary the form of the experiment.[113] A one-sided reliance on the "physics method frequently ends in failure—as is well-known from the history of physiology." Here Pavlov drew
upon his own experience working with Ludwig, "who to the very end of his
very fruitful activity" retained "the distinctly physical tendency in physiology,
as the successor to Volkmann, Weber, and others." "One must recall," continued Pavlov, "some of the works of the Leipzig laboratory—or, even better,
work there a bit—in order to easily notice this quality of Ludwig's activity.
Here experiments are generally conducted stingily, the small details of experiments are not especially taken into account, but the result of each experiment
is given a quantitative expression with the aid of clever and more or less exact

instruments—and then this quantitative material is subjected to careful abstract processing in a study."[114]

Heidenhain, on the other hand, working in Breslau, had an entirely different approach. "He brings a mass of animals before him, attentively studying the circumstances of the experiment, constantly varying the form of the experiment, and not concerning himself especially with writing constant protocols with the quantitative dimension of the data. The task is considered completed when, finally, the basic result becomes entirely sharp and constant." The history of physiology testified to the superiority of Heidenhain's approach: "Many results of the Ludwig laboratory were systematically revised by the one in Breslau. The author of these words [Pavlov himself] was a witness to a most touching scene in which the 70-year-old Ludwig complained through his tears about his alleged persecution by the Breslau laboratory."[115]

Eulogizing Heidenhain at a meeting of the Society of Russian Physicians in 1897, Pavlov told this same basic story, embellishing it with details about Heidenhain's refutation of Ludwig's theories of urine formation, lymph formation, and digestive absorption in the organism. The basic idea of Heidenhain's experimental criticism was this: that "simple physicochemical notions about the essence of these processes do not at all correspond to reality." For example, Ludwig had interpreted urine formation as a simple physicochemical process of filtration and diffusion, but Heidenhain had demonstrated experimentally that this process owed much, instead, to "the active participation in urine formation of the epithelial cells."[116]

Characteristically, Pavlov added that the failure of simplistic mechanistic models did not justify unscientific vitalist conclusions. Heidenhain's refutation of Ludwig's findings had encouraged "people with metaphysical tendencies to assert the inapplicability of the physicochemical perspective to the analysis of living phenomena, and the necessity to turn, during the investigation of life, to a special vital, spiritual principle." Yet Heidenhain was himself a "struggler for the physicochemical theory of life" and had offered an elegant analogy, described by Pavlov thus: "Imagine that there stands on the bank of a river a man who is unfamiliar with the action of steam, and he sees a canoe and a steamboat. Initially the two might seem identical to him, but then his observations begin to reveal various differences between them: the canoe moves with the rapidity of the water, but the steamboat sometimes moves more quickly than the water, sometimes less quickly, and, finally, even against the current. It has an independent force."[117]

Just as Heidenhain had demonstrated the inapplicability of simplistic physics-like explanations, so his very investigative style (in contrast to Ludwig's) exemplified a specifically physiological style of investigation. In Pavlov's words:

> [Heidenhain] had a special method of work. Beginning work, he conducted experiment after experiment each day, even two experiments a day. Initially, he did not keep protocols of the experiments, but merely, being constantly himself present, observed [the experiment] for any small detail, mastering the smallest condition, and, in this manner, finally made himself the master of the fundamental condition. Only then did he write the phenomena into the protocols, without the least slackening of his attention to everything that was transpiring. Such a method is especially important for the physiologist. We are not physicists, who can extract the numbers from an experiment and then leave in order to calculate the results in an office. The physiological experiment must always depend on a mass of the smallest circumstances and surprises, which must be noticed at the time of the experiment, otherwise our material loses its real sense.[118]

Pavlov apparently made regular use of this account when explaining the nature of physiological thinking in his lectures on physiology at the Military-Medical Academy from 1895 to 1924. Here too, Pavlov portrayed Ludwig, "creator of the physicochemical orientation in physiology," as repeatedly suffering the refutation of his mechanistic theories by the more specifically physiological thinking of his Breslau colleague, and reduced to tears "because his truths turned out to be false truths."[119]

The Purposive Nervous Machine

In Pavlov's thinking, the determined regularity of physiological processes expressed their "purposiveness" or, as he sometimes put it, their "adaptiveness."[120] The identity of these two notions was rooted in Pavlov's adaptationist views. Like Bernard, Pavlov used the term *purposiveness* to denote the coordinated activity of the organism as a whole in the interests of survival. Purposiveness was reflected in "the precise links of the elements of a complex system both among themselves and, as an entire complex, with the surrounding circumstances." For Pavlov, contemporary evolutionary theory sanctioned the view that all physiological processes in the organism—down to their most minute details—had a definite purpose. Over time, the parts of the organism had become perfectly adapted one to the other, and the organism as a whole had become perfectly adapted to its environment. "The grandiose complexity of higher organisms, like lower ones, continues to exist as a whole only as long as all its constituent parts are subtly and precisely linked, balanced both with each other and with surrounding circumstances. The analysis of this balancing of the system comprises the first task and goal of physiological investigation."[121] In Pavlov's view, Darwin had demonstrated that evolution produced complex organisms in which all parts and processes had been forged into a perfectly functioning, purposive machine. Yet, as Pavlov once complained in response to a critic, proponents of the "physicomechanist doctrine" continued stubbornly to find in the word *purposiveness* a retreat from objectivity toward teleological thinking.[122] Adaptationism flourished within every post-Darwin evolutionary community, but Pavlov's interpretation was solidly grounded in a Russian evolutionary tradition that incorporated Darwin's theory within a set of well-developed pre-Darwinian evolutionary views that emphasized the "harmony of nature."[123]

As will become evident in Part II, this notion of purposiveness/adaptiveness played a central role in Pavlov's approach to digestive physiology. Here let me merely offer one illustration of its place in his reasoning about specific physiological issues. In his youth, Pavlov had been impressed by the dire consequences of his uncles' excessive alcohol consumption, and in the 1890s he strongly supported campaigns to banish alcohol entirely from Russian life. Nevertheless, when one coworker discovered that alcohol excited gastric secretion, Pavlov concluded that this substance must serve some purpose in the organism. "Mankind's instinct has taken him too far in the use of alcohol, but, on the other hand, the initial use of alcohol was elicited by the demands of the organism."[124] This kind of reasoning created a compelling rationale for interpreting the details of experiments in terms of the purposiveness of the animal machine.

For Pavlov, the lawful functioning and subtle interconnection of organ systems and the precise adaptation of organisms to their environment could result only from nervous mechanisms. In his doctoral dissertation, he explicitly embraced *nervism,* which he defined methodologically as "the physiological theory that attempts to extend the influence of the nervous system to the greatest possible number of the organism's activities."[125] Indeed, throughout his scientific career, Pavlov worked almost exclusively on the nervous control of organ systems: the circulatory and digestive systems from the early 1870s to the late 1880s, the digestive system alone in the 1890s, and the higher nervous system itself from 1903 to 1936.

Pavlov's nervism, however, was much more than a methodological principle. The coordination and precision of organic processes in the living machine were, for him, only comprehensible as the product of nervous control. Before

William Bayliss and Ernest Starling's disorienting discovery of secretin in 1902, he found it difficult to even conceive of humoral mechanisms as a truly physiological process.[126] As we shall see, he confidently explained the most varied phenomena as the result of both established and hypothesized qualities of the nervous system. For Pavlov, the specific excitability of the nervous system, the far-reaching effects of even localized excitation and inhibition on distant organs, and the existence of still undiscovered nervous structures (such as parallel secretory and trophic nerves) always provided an initial explanation of puzzling phenomena.

The influence of Pavlov's nervism on his choice and conceptualization of scientific problems is evident from the very beginning of his scientific career. For example, in 1880 he confided to Serafima that "for my future work I have a very, very daring and important idea." "I am making the proposal: are there no nerves governing the very production, the very formation of the blood. An extremely important thing, of course. Blood is such an important liquid in the organism, and to know the mechanisms upon which its formation depends means to have something big. How our understanding would change of many disease processes, how much rational treatment would gain. But there is no real basis for success except for several analogies. One can try anything; but in this case it is good to remember that a half year's work is at risk. But God favors the bold—and I will probably begin."[127]

This inspiration proved short-lived, and Pavlov wrote his dissertation instead on the nervous control of the heart. In the mid and late 1880s, his research involved establishing the existence of nervous control over various, mostly digestive organs. His correspondence from March 1888 testifies to the animating power of nervism. When he was able to establish that the vagus nerve controlled the pancreatic gland, he celebrated. When his initial experiments failed to reveal vagal control of the gastric glands, he concluded, "It turns out that the vagus nerve does not act; it will be necessary to move on to other experiments, to other nerves." When stimulation of the vagus nerve yielded no consistent results, he hypothesized that "the lack of success results from this: aside from the nerves eliciting secretion there are others acting in an opposite manner—these halt the activity of the glands and usually take primacy over the former [nerves]." After separating the excitatory and inhibitory branches of the vagus, he reported happily that "the vagus [now] drives pancreatic juice superbly. I also thought about the gastric glands, but it turns out—no, I now need to try another nerve that goes to the stomach."[128]

Throughout the 1890s, Pavlov's nervism continually shaped the choice and development of his laboratory's lines of investigation. As his longtime coworker V. V. Savich observed,

> I. P. [Pavlov] had little interest in chemism; all his ardor was directed toward "nervism," to the nervous connections of the organism. So, one of his students (Mironov) investigated the causes of the secretion of the mammary glands. After the severing of all nervous connections, secretions continued: it was clear that here one was dealing with the humoral connections of the organism. And this alone was enough to abandon this question forever! But all nervous mechanisms were investigated to the end.
>
> This tendency of I.P.'s explains very well why he halted in midstream in the analysis of the action of acid upon the secretion of pancreatic juice. Nervous connections were hypothesized a priori, and it was sufficient for one control experiment to turn out well in order to consider the thing entirely proven.[129]

We shall see in Chapter 7 that, just as Pavlov's attachment to nervism framed his experiments and conclusions about the digestive system, so did it contribute—after Bayliss and Starling's discovery of a humoral mechanism for pancreatic secretion—to his abandonment of digestive physiology for the investigation of conditional reflexes.

Theory, Fact, and Interpretation

In his public comments, Pavlov consistently denigrated the role of theory, preferring to speak "the most elegant language of facts." "Physiology," he explained to his medical students, "is not a theoretical science. It is built entirely on facts." "Facts are one thousand times more important than words," he continued. "If you understand facts, you understand everything." Rarely did he refer to "my theory" of anything, and he used the word *theoretician* only pejoratively. In Pavlov's rhetoric, *theories* and *theoreticians* were abstract, disconnected from reality, and incompatible with the concerns of the mature mind.[130]

He consistently referred to the interpretive dimension of physiology in more modest terms that emphasized its close relationship to experimental results and its distance from philosophical systems. For example, he promised readers of his *Lectures on the Work of the Main Digestive Glands* "an idea increasingly embodied in the form of tenable and harmoniously linked experiments." Elsewhere, he referred positively to "the idea linking these facts," "the view of the laboratory," "basic conclusions," "an entirely clear generalization," and "the verbal meaning of facts."[131]

On one occasion, when asked for an explanation of a reported laboratory result, Pavlov responded, "Explanations are cheap; an explanation is not science. Science is distinguished by absolute prediction and mastery, and our calculations are clear and proven. One can propose as many explanations as one likes." When pressed by a second interlocutor, he retreated a bit: "I do not, of course, have anything against explanations in science. Of course, it is impossible to do without an understanding, an explanation of the facts. I wanted only to emphasize that explanation is not the goal of science, but rather its means."[132]

We should not overinterpret this episode, but it does highlight two important points. First, Pavlov adopted here his usual public posture of the hard-headed experimentalist, the "man of facts" disdainful of speculation. Yet his notion of physiological thinking left ample room for the role of interpretation, and he was well aware of the complex relationship between fact and interpretation in his own experimental practices. As Pavlov explained to his medical students, "The physiologist cannot explain all acquired facts in a $2 + 2 = 4$ manner."[133] In one of his favorite formulations, the physiologist *ulovlivaet fakty*— that is, "detects facts" or "catches facts," as one might catch someone's eye or catch someone's drift. Second, Pavlov defended here a very high standard for convincing scientific results: "absolute prediction and mastery." This sentiment echoed Bernard's comment in *An Introduction to the Study of Experimental Medicine* that "true science exists only when man succeeds in accurately foreseeing the phenomena of nature and mastering them."[134] For both physiologists, the experimentalist had mastered a phenomenon only when he could reproduce it at will. For Pavlov, the verbal analogue to this was *not* an elaborate "theory" but rather an "exact scientific formula."[135]

This brings us to an important tension between Pavlov's Bernardian vision and his experimental practices. As is well known, Bernard denied the usefulness of mathematical averages or means to characterize physiological phenomena. These, he argued, were inappropriate to the search for unvarying, determinist laws. In *An Introduction to the Study of Experimental Medicine,* he made this point by describing his own relentless search to reconcile (rather than average) conflicting results. (See, for example, the description of his path to the discovery that an animal could be made diabetic by puncturing the fourth ventricle of its brain.) In Bernard's view, the moral of these narratives was clear: a "negative fact" did not negate a "positive fact" but rather challenged the determinist physiologist to perform further experiments that reconciled apparently variant results and, finally, gave him "absolute prediction and mastery" of the phenomenon.[136]

Bernard's examples, however, all concerned questions that could finally be answered with a simple yes or no—for example, does puncturing the fourth ventricle of the brain result in diabetes? Beginning in 1894, however, Pavlov's synthetic physiology involved the analysis of secretory *patterns* in experimental trials that never culminated in a simple yes or no answer. In such trials, "absolute prediction and mastery" was impossible to attain and very difficult even to define. As we shall see, Pavlov's attempt to uncover the determined, precise glandular responses elicited by different foods required him to make constant interpretive judgments about the "essential" similarities or differences between rows of data. Furthermore, he needed to do so without either an articulated statistical method or the possibility of controlling the "numberless factors" that, he had good reason to believe, were constantly skewing the data in his experiments on intact animals. No guide, not even the redoubtable Claude Bernard, had prepared him for the interpretive challenges presented by this task.

Chapter 3

THE LABORATORY SYSTEM

The wealth of resources, the opportunity to work in it around the clock for an entire year, the enormous quantity of entirely healthy animals with complex operations, the ability to acquire every day a portion of all digestive juices in fresh form . . . and, finally, a marvelously trained service staff—such is the external side of the activity of the [Pavlov] laboratory.

As for its inner life, here you see a very interesting phenomenon: it is as if its numerous workers, having been earlier occupied with various specialties having nothing in common with physiology—surgeons, therapists, pediatricians, oculists, and so forth . . . —are transformed upon entering the laboratory into physiologists and come to work amiably together on one general task that excites them all, under the untiring guidance of the professor, who is at once the most experienced among them, an indispensable teacher, a devoted assistant, and one's best comrade.

—VLADIMIR BOLDYREV, doctoral dissertation,
St. Petersburg Military-Medical Academy (1904)

Immediately after assuming control of the Physiology Division at the Imperial Institute, Ivan Pavlov transformed himself from a workshop physiologist into a factory physiologist. In the 1870s and 1880s he had worked essentially "alone in somebody else's laboratory," conducting his research in a small room with sparse equipment, sometimes with an assistant or an occasional collaborator. In these years Pavlov pursued one line of investigation at a time and, most important, he alone conceived, conducted, and interpreted experimental trials. Pavlov's "small world" at the Institute, however, was a site of large-scale social production with a specific social-cognitive division of labor. Pavlov himself, of course, occupied the central role in this system of production—as its creator, manager, and governing intellect—but his achievements cannot be understood as if he were working alone. These were inseparable from "the unique manner in which experimental work in Pavlov's laboratory was organized."[1]

In this chapter I explore the synthesis of institutional circumstances and Pavlov's scientific-managerial vision in the social-cognitive structure and dynamics of his laboratory. We will then be prepared, in Part II, to explore these dynamics in greater detail with respect to specific junctures in the laboratory's research. As is so often the case with Pavlov, Claude Bernard's ruminations provide a good point of departure.

As the product of Bernard's reflections on his own practice, his *Introduction to the Study of Experimental Medicine* addressed the epistemological, craft, and, one might say, psychological challenges that confronted the individual investigator. For Bernard, the same individual devised, conducted, and interpreted an experiment and so must adopt the fundamentally different "qualities of mind" required at each juncture. When devising an experiment, the experimenter must have a preconceived idea; when observing its results, he must become a passive "photographer of phenomena"; when ascribing meaning to these results, "reasoning intervenes, and the experimenter steps forward to interpret the phenomenon." The fluid dynamics of experimental trials made these stages "impossible to disassociate" in practice, but they remained conceptually distinct. "In the experimenter we might also differentiate and separate the man who preconceives and devises an experiment from the man who carries it out or notes its results. In the former, it is the scientific investigator's mind that acts; in the latter, it is the senses that observe and note."[2] Thus, even the blind naturalist François Huber had "left us admirable experiments which he conceived and afterward had carried out by his serving man, who, for his part, had not a single scientific idea. So Huber was the directing mind that devised the experiment; but he was forced to borrow another's senses. The serving man stood for the passive senses, obedient to the mind in carrying out an experiment devised in the light of a preconceived idea."[3]

The managerial vision animating Pavlov's laboratory operation involved the transformation of Bernard's "qualities of mind" into a highly rationalized division of labor. Pavlov himself (in principle) assumed control over the qualities that Bernard had credited to Huber's "directing mind," while using his coworkers—like Huber's servant—as extensions of his own senses, as largely "passive photographer[s] of phenomena."[4]

The obvious challenges inherent to this transformation of an epistemological issue into a managerial policy underscore the close relationship between scientific and managerial visions. Our story, in fact, involves not one factory but two. Pavlov referred to the digestive system alternately as a "chemical factory" and a "laboratory." He ascribed to it the very same qualities that he did to successful human endeavors: both were "precise, regular, and purposive." In laboratory investigations, the factory metaphor expressed and guided the search for *pravil'nye* results—for precise, repeated (or "stereotypical"), and purposive patterns in the glandular responses of dogs to varying quantities of different foods. Pavlov never referred to his laboratory as a factory—to do so would have demystified the ethos that helped make it hum; but his management style and rhetoric clearly expressed his belief that "the marvelous mechanism" of the digestive system would reveal its secrets only to a laboratory endeavor that matched its most essential qualities.[5]

Forces of Production

Everything is in the method, in the chances of attaining a steadfast, lasting truth . . .

—IVAN PAVLOV, "Experimental Psychology and Psycho-pathology in Animals" (1903)

THE FACTORY SITE

Even before the Institute formally opened in 1891, Pavlov and several of his associates from the Botkin laboratory set to work in Russia's best-equipped physiological laboratory. His Physiology Division occupied five rooms in the single wooden building that housed all the Institute's scientific divisions. Pavlov used the smallest room for surgical operations and the four larger ones to house animals and conduct experiments. In addition to a laboratory budget more than five times greater than that of any other Russian physiologist, he had the use of two attendants, one paid assistant, and a growing number of praktikanty.[6]

However lavish by Russian standards, these facilities quickly came to seem cramped and inadequate. Designed as a workshop for a handful of men, they were soon swarming with praktikanty—twelve in 1892 and seventeen in 1893—and the animals for their experiments.[7] Furthermore, Pavlov's experiments with the Eck fistula (which linked the portal vein with the inferior vena cava)[8] soon convinced him that it was difficult or impossible to maintain the aseptic standards necessary for successful surgical operations in the single room available for them. As in Pavlov's years in the cramped Botkin laboratory, animals perished during the operations designed to prepare them for experiments.

The influx of praktikanty that caused these problems also cemented Pavlov's loyalty to the Institute. As manager of Botkin's clinical laboratory in the 1880s, Pavlov had supervised the doctoral research of about fifteen physicians but, as we have seen, had been frustrated by Botkin's scientific-managerial style. Botkin assigned to the physician-investigators in his laboratory a wide variety of topics—ranging from the pharmacological action of various substances to the mechanism of the coating of the tongue—complicating immeasurably Pavlov's task of seeing their research to a successful conclusion. Now, however, Pavlov was master in his own house. With a clear research program that extended far beyond the resources of any single investigator, he immediately began to assign praktikanty to topics of his own design.

The space problem was resolved in 1893–94 when an unexpected contribution from Alfred Nobel enabled the Physiology Division to become the Institute's third scientific section to acquire a separate building.[9] Perhaps motivated by the Institute's highly visible efforts against the cholera epidemic that swept through his Baku oilfields in 1893, Nobel in that same year asked his nephew Emmanuel—one of several Nobels who built an oil-based industrial empire in Russia—to relay his intention to donate 10,000 rubles to the Institute.[10] This was an unconditional gift, but the ailing sixty-year-old philanthropist did express the hope that the beneficiary would address two subjects that he found particularly pressing. Would transfusions of blood from a young, healthy animal (Nobel suggested a giraffe) revivify an ailing animal of the same, or another, species? Could the stomach of a healthy animal be transplanted to an ailing one with salutary effect? Emmanuel added a short cover note (mentioning Alfred's "interest in physiology") and forwarded his uncle's letter to Prince Ol'denburgskii, who, after receiving the Tsar's permission, accepted the gift in August 1893.[11] Some months later, Pavlov, who was temporarily filling in for the absent director, Eduard Shperk, formally thanked Nobel and informed him that, since the gift carried no conditions, it would be applied to the general needs of the Institute. The money was, in fact, already being used to finance a two-story stone addition to Pavlov's laboratory (Figure 6).[12]

Perhaps to justify his division's use of Nobel's money—and perhaps genuinely excited by Nobel's ideas—Pavlov assigned one praktikant, V. N. Geinats,

FIGURE 6. *The two-story stone laboratory building financed by Alfred Nobel's gift. On the* far right *is Pavlov's wing of the building in which all scientific divisions were originally housed. Courtesy of Academy of Sciences Archive, St. Petersburg branch*

to develop a surgical procedure for uniting the circulatory systems of two different dogs. Pavlov developed great hopes for this project and, in one enthusiastic moment, announced that he would soon turn the entire laboratory to such "sewing." The operation, however, failed repeatedly and was reluctantly abandoned.[13]

A more enduring consequence of Nobel's gift was the new quarters, constructed under Pavlov's close supervision and completed in 1894, more than doubling the space at his disposal and allowing him more fully to implement his vision of the physiological enterprise. The basement became a full-service kennel with individual cells for experimental animals; the first floor provided three more rooms for experiments; and the second floor housed a surgical and recovery complex that embodied Pavlov's commitment to investigating the normal functioning of organs through what he termed *physiological surgery* and the *chronic experiment* (Figure 7). This expression of Pavlov's holism was central to the forces and relations in his laboratory, so I discuss it briefly here.

For Pavlov, the *chronic experiment* allowed the physiologist to investigate normal physiological processes, which, Pavlov claimed, were too often distorted during an *acute experiment* (a term he used synonymously with *vivisection*). In contrast to acute experiments, which were conducted on animals immediately after an operation from which they would eventually die, chronic experiments began only after the animal had recovered from an operation and regained its "normal" physiological state. Acute experiments had their uses— and Pavlov employed them himself—but they yielded only "analytical" knowledge, not a "synthetic" understanding of the organism at work. Shortly after the completion of his new building, Pavlov explained to the Society of Russian Physicians that acute experiments conducted on a freshly operated-on and bleeding animal that was either writhing in pain or heavily narcotized so distorted normal physiological processes that they led inevitably to "crude errors." It was impossible to untangle the results of the operation itself from normal physiological processes. In chronic experiments, on the other hand, "the physiologist counts on the animal living after the removal of parts of organs, after the disturbance of connections between them, the establishment of a new connection, and so forth"—in other words, after a surgical procedure that afforded permanent access to the physiological processes of an animal that had been purposefully altered but remained essentially normal.[14] The surgical and recovery complex, then, embodied a long-standing element of Pavlov's scientific vision: his view that the exploration of normal physiological processes— specifically, the responses of the digestive glands to various stimuli (e.g., teas-

FIGURE 7. *Laboratory room adjoining the surgical and recovery complex. Courtesy of Academy of Sciences Archive, St. Petersburg branch*

ing with food, the act of eating, or the passage of various foods through the digestive system)—was both necessary and possible.

The "normalcy" (*normal'nost'*) of the experimental animals undergoing chronic experiments was, then, central to laboratory work and a source of authority for Pavlov's arguments vis-à-vis both physicians and other scientists. Physicians who drew on clinical experience to dispute the laboratory's results could be reminded, in the sympathetic tones of a fellow medical man, that they encountered an impossibly complex mass of interconnected phenomena in their daily practice and that these could not be disentangled outside the laboratory.[15] Similarly, when the experimental results of other scientists conflicted with Pavlov's own, these could be explained (and either reconciled or dismissed) by reference to the physiological abnormalities that resulted from their crude, acute experiments.

This notion of "normalcy" inevitably entailed a series of "interpretive moments." Pavlov acknowledged, as we have just seen, that physiological surgery and chronic experiments involved some departure from normal physiological relations ("removal of parts of organs, . . . disturbance of connections between them, the establishment of a new connection, and so forth"). Given that the laboratory setting itself, to say nothing of the surgical operations performed there, always had *some* effect on the dog's behavior and reactions, how was one to determine whether the dog remained acceptably "normal"? For example, were a dog's digestive processes functioning normally if, after an operation, its appetite diminished, it accepted only one kind of food, or it lost weight? It therefore fell to the experimenter, the praktikant—within, as we shall see, a matrix of social relations in the laboratory—to answer such questions and to affirm the normalcy of an experimental dog. Or, to affirm its *lack* of normalcy. Pavlov and his coworkers were, after all, dealing with a large, complex organism, and *pravil'nye* results were inevitably difficult to obtain. Feeding two different dogs the identical quantity of the same food *always* produced somewhat different secretory results, and sometimes radically different ones. Even the results of identical experiments on a single dog varied. Pavlov (following Bernard) saw these variations as reflecting the "numberless factors" that concealed determined regularities behind a veil of apparent spontaneity. So, when two dogs yielded strikingly different results, one animal was pronounced relatively "normal" and the other relatively "abnormal." Divergent results with a single dog were handled similarly.

I discuss the social-cognitive dynamics of such interpretive moments in Part II; for now I need only note that the notion of normalcy was simultaneously a laboratory goal, a reservoir of interpretive flexibility, and a source of authority for the laboratory's knowledge claims. To the outside world, Pavlov's laboratory consistently represented its experimental dogs as normal—happy, energetic, and long-lived. Within the laboratory, however, Pavlov and his coworkers struggled constantly to create and define normalcy while also exploiting fully the interpretive flexibility afforded by such judgments.

Because chronic experiments depended on the animal's surviving surgery, Pavlov conceded no essential difference between physiological surgery and clinical surgery on humans. In his speech to the Society of Physicians entitled "The Surgical Method of Investigation of the Secretory Phenomena of the Stomach" (1894), and more extensively in *Lectures on the Work of the Main Digestive Glands* (1897), he proudly presented the plan of his surgical ward—"the first case of a special operative division in a physiological laboratory."[16] Dogs were washed and dried in one room, narcotized and prepared for surgery in a second, and operated on in a third. A separate room was devoted to the sterilization of instruments, the surgeon, and his assistants. Separated from the surgical ward by a partition were individual recovery rooms for dogs. These were well lit and ventilated, heated with hot air, and washed by means of a water pipe with minute apertures—enabling rooms to be "copiously syringed from the corridor without [the assistant] entering the room."[17] Figure 8, a

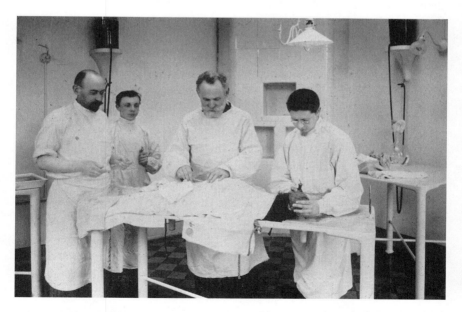

FIGURE 8. *Surgical creation of a dog-technology, 1902. From* left to right: *assistant A. P. Sokolov; attendant I. V. Shuvalov; I. P. Pavlov; praktikant Ia. A. Bukhshtab. Courtesy of Academy of Sciences Archive, St. Petersburg branch*

posed photograph, communicates this essential identity between physiological and clinical surgery; looking at the photo, one realizes with a slight start
that the patient is a dog.

For the physiologist to master nature's most complex phenomena, Bernard
had argued, his workshop must be "the most complicated of all laboratories."[18] In this spirit, Pavlov explained to the Society of Russian Physicians in
1894 that the demands of chronic experimentation—of this qualitative extension of the physiologist's grasp on the organic whole—required a radical expansion of the laboratory's physical plant. "In the final analysis the very type
and character of physiological institutes should be changed; they should definitely include a surgical section answering the demands of surgical rooms in
general."[19] For physicians in his audience, this was yet another of Pavlov's constant injunctions that they use their social connections to secure greater financial support for physiology; for Russian physiologists, it was a reminder
that only Pavlov possessed the resources to practice what he preached.[20]

THE WORKFORCE

In the years 1891–1904 about one hundred people—his coworkers (*sotrudniki*)—worked in Pavlov's laboratory. A small minority—about ten—were
permanent or semipermanent staff: the chief (*zaveduiushchii*), the assistants
(*pomoshchniky*), and the attendants (*sluzhashchie*). The great majority were
temporary investigators, praktikanty (Figure 9).

As chief, Pavlov provided the laboratory's scientific-managerial vision and
ruled in firm patriarchal fashion. He hired coworkers, assigned research topics, performed complex operations on dogs, participated in the praktikanty's
experiments as he saw fit, edited and approved completed work, and rewarded
success and punished failure. His were the governing ideas in the laboratory,
and he tolerated no alternatives. He was also the spokesman for the laboratory's achievements, defending his coworkers and explaining the broader significance of their work when they delivered papers or defended dissertations
to outside audiences. Pavlov himself wrote articles on a wide range of specialized subjects, including the nature of pepsin, the effect of hunger on the
stomach, and the effects of a double vagotomy. But most important were his
periodic publications synthesizing laboratory results and explaining their significance for physiology and medicine. In the years 1891–1904 these published
works included "Vivisection" (1893), "On the Surgical Method of Investigation
of the Secretory Phenomena of the Stomach" (1894), "On the Mutual Relations
of Physiology and Medicine in Questions of Digestion" (1894–95), *Lectures on*

FIGURE 9. *The chief and his workforce in 1904. Unless otherwise noted, the coworkers are praktikanty.* Bottom row, from left to right: *E. A. Ganike (assistant), G. A. Smirnov (member-coworker), I. P. Pavlov (chief), S. V. Parashchuk, L. F. Piontkovskii, V. N. Boldyrev, B. P. Babkin, A. P. Sokolov (assistant), Ia. A. Bukhshtab, N. M. Geiman, I. S. Kadygrobov, V. P. Neelov, M. A. Arbekov.* Top row, from left to right: *P. V. Troitskii, B. G. Berlatskii, L. A. Orbeli, I. S. Tsitovich, V. V. Savich. Courtesy of Academy of Sciences Archive, St. Petersburg branch*

the Work of the Main Digestive Glands (1897), "The Contemporary Unification in Experiment of the Main Aspects of Medicine, as Exemplified by Digestion" (1899), "Physiological Surgery of the Digestive Canal" (1902), and, during the laboratory's transition to research on conditional reflexes, "The Psychical Secretion of the Salivary Glands (Complex Nervous Phenomena in the Work of the Salivary Glands)" (1904).[21]

The great majority of the workforce consisted of temporary investigators, the *praktikanty*, drawn to the Physiology Division by the changes in Russian medicine described in Chapter 1. Most came to Pavlov's laboratory between the ages of twenty-five and thirty-five, during their first decade of work as practical physicians, and they lacked training in physiology beyond that provided in a single medical school course. Many were military physicians, and all but one were male.[22] They entered the laboratory from a wide range of medical settings: of the 75 percent for whom information is available, twenty-eight were physicians in St. Petersburg's hospitals and clinics (twelve of these in the clinics of the Military-Medical Academy), thirteen served in hospitals and clinics outside the capital, ten were rural physicians, and nine worked for the Medical Department of the Ministry of Internal Affairs. The praktikanty were drawn almost entirely from the diverse middling social stratum known in Russia as the *raznochintsy*. Their nationality is often difficult to determine, but clearly the great majority were Russian and a disproportionate number were Jewish. Praktikanty usually spent two to three years in the laboratory, during which time about 75 percent wrote dissertations, defended them at the Military-Medical Academy, and received their doctorates in medicine.[23]

The nature of this workforce—young and transient, largely untrained in physiology, and intent on gaining a quick doctoral degree—facilitated Pavlov's use of its members as extensions of his own eyes and hands. Consider long-time coworker Boris Babkin's perspicacious description of the most numerous contingent among the laboratory's praktikanty, military physicians pursuing their doctorate at the Military-Medical Academy.

> About sixty or seventy of them were enrolled [in the Academy] yearly, remaining for two years. During the first year they had to pass their examinations for the degree of doctor of medicine—a repetition of the state examinations—and during the second year they had to work in one of the academic clinics or laboratories, presenting the results of their clinical or experimental investigations in their M.D. [doctoral] thesis. The majority of the doctors attached to the Academy were regimental doctors who had had no opportunity to work in the hospitals and to refresh their knowledge and perfect their medical skill. The greater part of the Russian army was stationed on different strategic borders, far from any cultural center, even of the most modest kind. Because of this, many of the military doctors, especially those who had been stationed for a long time in some dreary little town, were very backward in medicine and even more in science.[24]

"Very backward in medicine and even more in science," these physicians provided the basic human "material," as another observer put it, for the production process.[25]

The praktikanty were not, of course, an undifferentiated mass, and at special junctures in laboratory production—when the chief was engaged in "retooling"—he sometimes employed coworkers for their special expertise. For example, Pavel Khizhin's surgical skills and training played a critical role in the creation of a key dog-technology, and Pavlov's later transition to research on conditional reflexes owed much to perspectives he imported by recruiting the praktikanty Anton Snarskii and Ivan Tolochinov. These exceptions, however important, also illustrate the rule: when the physiology factory was operating normally, the praktikanty served as skilled hands.[26]

The praktikanty conducted thousands of experiments in Pavlov's laboratory, painstakingly collecting, recording, measuring, and analyzing the dogs' secretory reactions to various exciters during experimental trials that often continued for eight or ten hours at a time. The strains of this work are clearly and poignantly evident in an obituary for Iulian Iablonskii, Pavlov's praktikant and assistant in 1891–94, who died in 1898 after a protracted mental illness: "Increasingly fascinated by physiology, he soon decisively abandoned the clinic for the laboratory. For entire days he sat, collecting digestive juices, making calculations, and later, as an assistant to the professor, making necessary preparations for experiments and complex operations. In his third year . . . there appeared the first signs of over-exhaustion, and then a sinister mental illness. Undoubtedly already ill, the deceased defended his dissertation and was sent to the provinces."[27] Iablonskii's fate was unique, but the rigorous work process he endured was not.

Pavlov also had at his disposal each year two paid assistants and one unpaid "member-coworker," who provided a relatively stable supervisory stratum amid the transitory praktikanty. Although they conducted scientific research, their principal task was to incorporate praktikanty into the laboratory's productive process—to inculcate in them the laboratory's procedures and culture, facilitate the smooth progress of their work, and keep the chief informed of their abilities, progress, and problems.[28] All but one of these assistants were physicians with a developing specialty of some use to the laboratory. V. N. Massen, a gynecologist, established the laboratory's initial aseptic and antiseptic procedures; N. I. Damaskin and E. A. Ganike were biochemists, and A. P. Sokolov brought a background in histology. Damaskin and G. A. Smirnov came to the laboratory with doctorates already in hand, whereas Massen, Iablonskii, and Sokolov acquired their doctorates for theses researched there. None possessed a broad physiological education beyond that acquired at Pavlov's side. As long-standing members of the laboratory, Ganike, Sokolov, and Smirnov became bearers of its institutional memory.

This was especially true of Ganike. Arriving at the Physiology Division in 1894 from the collapsed Syphilology Division, he remained Pavlov's close collaborator until the chief's death in 1936. Ganike's background in chemistry and his "unusual technical inventiveness" made him the laboratory's resident technician and problem-solver.[29] He was also Pavlov's all-purpose right-hand man and chief supervisor. Ganike handled the laboratory's budget, supervised its chief money-making enterprise, and drafted its annual reports for the chief's approval. He also enjoyed a close personal relationship with Prince Ol'denburgskii. When Pavlov was absent or busy, it was Ganike, whom the Prince addressed with the familiar *ty*, who represented the Physiology Division at meetings of the Institute's governing Council. Self-effacing, intensely private, and devoted to Pavlov, Ganike left only the skimpiest of memoirs, but Babkin, who worked with him closely, has left the following portrait.

> Ganike was an exceptional person. He was extremely original and at the same time one of the most modest, cultured, well-bred and honorable of men. He was a bachelor and lived in the Institute of Experimental Medicine. He worked at night and slept most of the day, arriving at the laboratory about 3 or even 5 in the afternoon. He was very musical and played the violoncello in the laboratory at night to an accompaniment provided by a mechanical device. For his accompaniment he cut out notes in paper tape and inserted this in a special mechanical piano, which was set in motion by an electric motor, while he himself played the solo part on his violoncello.[30]

That Pavlov, who demanded punctuality and regularity from his coworkers, accepted Ganike's nocturnal ways speaks volumes about the taciturn assistant's value to the laboratory.

The other long-term workers in the laboratory were the attendants charged with caring for the dogs and preparing them for experiments, assisting during surgical procedures, troubleshooting at the bench, keeping the laboratory in order, and other miscellaneous tasks. Several attendants worked in the laboratory for many years, accumulating important craft knowledge. One praktikant recalled that two attendants, Nikolai Kharitonov and a certain Timofei, thus became "indispensable participants in each experiment, and such active participants that they were not so much helpers as, rather, almost the directors." Another praktikant wrote of Kharitonov and a younger attendant, Ivan Shuvalov, that their accumulated experience with the sometimes puzzling behavior of dogs and fistulas enabled them to "provide in many cases absolutely invaluable assistance." They also became the chief's valued assistants during surgical operations. Pavlov's wife, Serafima, later recalled that when Kharitonov was absent "it was as if Ivan Petrovich [Pavlov] had lost his hands."[31] When Kharitonov grew too old, Shuvalov assumed this task, which required some knowledge of the irascible chief as well.

> It was not easy to assist Pavlov when he was operating. He did not like to call out the name of the instrument he wanted at a given moment or to say what he would do next, and at the same time he was extremely impatient. The instruments were handed to him by the very able young laboratory attendant, Vania [Ivan] Shuvalov, who knew the operational procedures perfectly and handed Pavlov the required instrument at the right moment. But [in Shu-

valov's absence] the assistants, especially the newcomers, often failed to give Pavlov the help he wanted or did so at the wrong time. Then he would push the assistant's hand away and say: "I speak with my hands—you must get used to that," or he would begin to mutter irritably: "Well, hold this, hold this!" or some such words. He had no patience with new assistants . . . and they would feel altogether at a loss during an operation and would give him even less help than they were capable of.[32]

In the group photograph in Figure 8, Shuvalov stands ready to assist the chief.

The laboratory workforce, then, consisted of the chief, the assistants, the attendants, and the praktikanty—all with their prescribed roles. Before exploring their interaction in the laboratory's productive process, however, I must introduce its last, and by no means least, participant.

THE LABORATORY DOG AS TECHNOLOGY AND ORGANISM

At the center of the productive process were laboratory dogs modified by ingenious surgical procedures to Pavlov's investigative ends. In the physiology factory, these dogs were simultaneously technologies, physiological objects of study, and products.[33] I will defer discussion of dogs-as-products and explore here their dual character in the production process itself.

Laboratory dogs were technologies (or "intermediate products") created in the laboratory to produce something else—as in a factory that makes machines for the manufacture of another product. As Bruno Latour puts it, "you cannot make the facts if you do not have the machines, any more than you can make iron without the big furnaces and the big hammers."[34] Laboratory dogs were particular kinds of "machines" designed and produced in the laboratory to generate particular kinds of facts. As with any technology, their existence and design influenced the organization and nature of the work process. As intermediate products, these dogs also created "local knowledge" and rendered problematic the replication of laboratory results by others.[35] Physiologists incapable of creating, say, a dog with an isolated stomach could reproduce the laboratory's experiments only by acquiring a dog from Pavlov or journeying to St. Petersburg. These dogs were also pedagogical technologies, serving as "wonderful material in all regards for teaching" and so were "no less indispensable for university laboratories than the most important physiological apparatuses" (Figure 10).[36]

I distinguish between the laboratory dog as "technology" and as "physiological object of study" to emphasize that it remained a living, functioning, and infinitely complex organism. Designed to perform "normally" in labora-

FIGURE 10. *Dog-technologies on display.* From left to right: *Mysh', with intestinal and gastric fistulas; Zhuk, with intestinal and gastric fistulas and with artificial connection between stomach and intestines; Mal'chik, equipped in the same manner as Zhuk; Kurchavka, equipped with esophagotomy and fistula for service in the "small gastric juice factory," where he reportedly produced a liter per day for nine years (see Chapter 8); an unnamed dog, equipped and employed in the same manner as Kurchavka, which after two years of factory service escaped from the laboratory. Courtesy of Academy of Sciences Archive, St. Petersburg branch*

tory experiments, the laboratory dog possessed biological attributes that often complicated its use as a technology for the production of *pravil'nye* facts. This tension between dog-as-technology and dog-as-organism was rooted both in the laboratory dog's "lifestyle" and in the confrontation between its biological complexity and Pavlov's scientific vision. We will be prepared to explore this tension after a closer look at the principles and practices of physiological surgery.

The varied operations performed in Pavlov's surgical ward to produce a laboratory dog for chronic experiments were developed to satisfy three basic criteria: (1) the animal must recover to full health and its digestive system must return to normal functioning; (2) the product of the digestive gland must be rendered accessible to the experimenter at any time for measurement and analysis; and (3) the reagent in that glandular product must be obtainable in pure form, undiluted by food or the secretions of other glands. Pavlov, a convinced "nervist," believed that the digestive system could function normally only if surgical operations left intact the basic nervous relations that controlled physiological processes.[37]

The simplest and most common operation was implantation of a fistula to draw a portion of salivary, gastric, or pancreatic secretions to the surface of the dog's body, where it could be collected and analyzed. Fistulas were not original to Pavlov's laboratory; for each digestive gland, however, he and his coworkers refined the operation to meet the three criteria enumerated above.[38] This proved relatively simple with the gastric and salivary glands. Gastric and salivary fistulas diverted only a small portion of glandular secretions to the surface, so any disturbance to normal digestive processes was presumably minimal; both could be opened or closed at the experimenter's discretion, and neither resulted in any substantial pathological symptoms.[39]

The creation of a "normal" dog with a pancreatic fistula, however, posed great difficulties. Pavlov himself had devised one procedure in 1880 and assigned several praktikanty to improve it in the 1890s, but he conceded even in 1902 that, despite "much labor and attention," the pancreatic fistula left much to be desired. The problem lay in the complex "physiological connections of this gland" and in the constant leakage of pancreatic juice from the fistulized dog. Escaping pancreatic ferments macerated the abdominal wall, causing ulceration and bleeding, and the chronic loss of pancreatic fluid undermined the dog's health in dramatic and mysterious ways. Animals often suddenly fell ill a few weeks or even months after the operation, losing their appetite and developing various nervous disturbances; sometimes "acute general weakness" was followed by fibrillations and death. Conceding that the pancreatic fistula was "not ideal," Pavlov insisted that its usefulness was nevertheless clear in "the numerous, clear, indubitable, and decisive results of investigations." The "normalcy" of these dog-technologies, however, always remained problematic.[40]

A second standard operation was the esophagotomy, which Pavlov and his collaborator E. O. Shumova-Simanovskaia had used in combination with the gastric fistula in 1889–90 to demonstrate the ability of appetite to excite gastric secretion and to obtain pure gastric juice from an intact and functioning dog. The esophagotomy involved dividing the esophagus in the neck and causing its divided ends to heal separately into an angle of the skin incision. This accomplished the "complete anatomical separation of the cavities of the mouth and stomach," allowing the experimenter to analyze the reaction of the gastric glands to the act of eating. Food swallowed by an esophagotomized dog fell out of the opening created between the esophagus and the neck, rather than proceeding down the digestive tract (see Figure 25 in Chapter 8). Because the dog chewed and swallowed but the food never reached its stomach, this procedure was termed *sham-feeding*. Sham-feeding an esophagotomized dog equipped with a gastric fistula gave the experimenter access to the gastric secretions produced during the act of eating. The experimenter then collected these secretions through the fistula at five-minute intervals, later measuring them and analyzing their contents. This dog-technology allowed the experimenter to collect virtually unlimited quantities of gastric juice and to analyze the secretory results of the act of eating. Because ingested food never reached the stomach, however, this dog-technology did not permit investigation of gastric secretion during the second phase of normal digestion, when food was present in the stomach.

This task was addressed by the complex dog-technology that soon became both a symbol of Pavlov's surgical virtuosity and the source of the laboratory's cardinal theoretical achievements. In 1894, after a series of frustrating failures, Khizhin and Pavlov created "the remarkable Druzhok" with an "isolated stomach" (or "Pavlov sac") (see Chapter 4). The isolated-stomach operation was difficult and complex, but the principle behind it was simple. The goal was to create an isolated pocket in part of a dog's stomach and to do so in such a way that, after the dog's recovery, the entire stomach continued to work normally while the "small stomach" could be studied separately (Figure 11). As Pavlov explained to the Society of Russian Physicians:

> The stomach is divided into two parts; a large part, which remains in place and serves as the normal continuation of the digestive canal; and another, smaller part, completely fenced off from the rest of the stomach and having an opening to the surface, through the abdominal wall. The essential thing in this operation is that in one part of this small stomach the fence [separating it from the large stomach] is formed only of mucous membrane while the muscle and serous layers are preserved, because there [through these layers] passes . . . the vagus nerve, which is the main secretory nerve of the gastric glands. In this manner we acquire in an isolated part of the stomach a completely normal innervation, which gives us the right to take the secretory activity of this part as a true representation of the work of the entire stomach.[41]

Food, then, came into direct contact only with the large stomach, but it excited presumably normal gastric secretion in both the large stomach and the isolated sac. Because the isolated stomach remained uncontaminated by food and the products of other glands, the experimenter could extract pure glandular secretions through a glass tube and analyze the secretory responses to various foods during the "normal" digestive process.

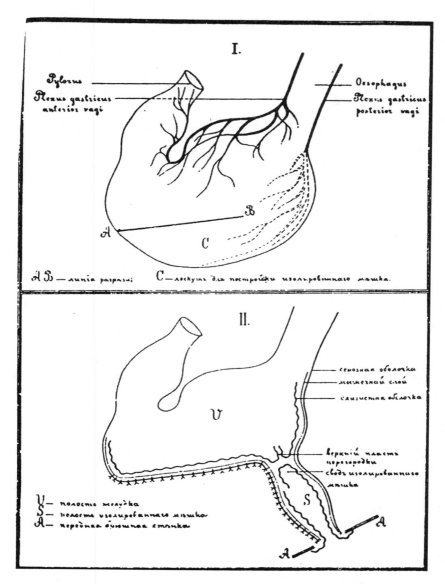

FIGURE 11. *The isolated stomach, as sketched in the appendix of Khizhin's doctoral dissertation. The top sketch describes the creation of the isolated stomach by an incision along the line AB, which creates a pouch composed of the membrane in region C. The bottom sketch portrays the result: the large stomach (region U) is separated from the isolated stomach (region S) by a barrier composed of mucous membrane (the wavy line in the bottom of region U). Vagal innervation is preserved, but food cannot cross from the large to the small stomach. A special fistula can then be inserted in region S for collection of secretions in the isolated stomach. From P. P. Khizhin,* Otdelitel'naia rabota zheludka sobaki, *Military-Medical Academy Doctoral Dissertation Series (St. Petersburg, 1894)*

Pavlov was not the first to create an isolated stomach, but his substantial variation on those developed by Rudolf Klemensiewicz (1875) and Rudolf Heidenhain (1879) reflected his nervism. Heidenhain doubted that central nervous mechanisms played an important role in gastric secretion, and so the "Heidenhain stomach," or "Heidenhain sac," involved transection of the vagus nerves. Pavlov thought this transection rendered the Heidenhain sac abnormal, and he therefore modified Heidenhain's operation, making the procedure "more difficult" but preserving vagal innervation. Two related assumptions were built into the "Pavlov sac": that exciters of the gastric glands did not act locally (in just one part of the stomach) but rather generally (distributing any excitation to the small sac as well), and that mechanical stimulation of the stomach wall played no role in gastric secretion (since such stimulation was exerted by food on the large stomach but not on the isolated sac). These assumptions contradicted a loose consensus among physiologists and a firm one among physicians, yet they were central to Pavlov's claim that what took place in the isolated sac mirrored normal digestive processes.[42]

Pavlov cultivated the image of laboratory dogs that, after recovering from these surgical operations, led normal, "happy" lives. The reality was somewhat different. For one thing, many dogs died as Pavlov and his coworkers developed new surgical procedures. About twenty perished before Khizhin and Pavlov successfully created an isolated stomach, and an untold number were sacrificed during attempts over more than a decade to perfect the pancreatic fistula. Survivors usually developed fatal conditions long before their natural lifespan had expired: the isolated stomach would slip or become infected; the pancreatic fistula would lead to various illnesses, spasms, softening of the bones, and a terrible death. In 1897 Pavlov referred proudly in his *Lectures on the Work of the Main Digestive Glands* to the fact that one dog (Druzhok) had lived for two and one-half years with an isolated sac. He assured his readers that the operation did not bring in its wake "any sensory unpleasantness, to say nothing of danger to the life of the animal operated upon."[43] By the time this work appeared, however, it was obvious to Andrei Volkovich, the praktikant then working with Druzhok, that this assertion was—at least in spirit—untrue. The severe deterioration of Druzhok's isolated stomach had rendered the dog useless for experimentation. Furthermore, the erratic functioning of Druzhok's gastric glands led Volkovich to speculate that the abnormal manner in which the dog had been fed for years (through a fistula) had caused the glands to "gradually atrophy." Pavlov himself observed in 1898 that, after acquiring an isolated sac, dogs tended "to lie on their backs with their legs up," apparently because they experienced "unpleasant or painful sensations when

FIGURE 12. *Out for a walk on the Institute grounds. Courtesy of Academy of Sciences Archive, St. Petersburg branch*

in a normal position."[44] Thus the relative normalcy of these dogs, like that of the dogs with a pancreatic fistula, remained a matter of interpretation.

Whether or not they were "happy and normal," the dog-technologies used in chronic experiments lived much longer than those consumed in acute experiments, and this facilitated a relationship with experimenters that sometimes resembled that between pet and master (Figure 12).[45] Each dog received a name and manifested an identifiable personality (*lichnost'*). This simultaneously rendered the dogs both more and less normal. On the one hand, what better testimony to a dog's normalcy than a recognizable personality? On the other, personalities varied and that of any single dog inevitably influenced the results of experimental trials, making the results, if not "abnormal," at least somewhat idiosyncratic and subject to interpretation.

Had laboratory dogs been simple, ideal mechanical technologies, the praktikant's task would have been relatively straightforward: turn them on under conditions prescribed by the chief, then measure and analyze the secretory results. The contradictory nature of these laboratory animals, however, and the drive to gain from them *pravil'nye* results, inevitably entailed a series of interpretive moments. The "numberless factors" presumed to conceal *pravil'nye* results were frequently identified with a dog's psyche and individual personality.[46]

Pavlov characterized the psyche as a "dangerous source of error" in experiments on digestive secretion. The dog's "thoughts about food" threatened
constantly to introduce the "arbitrary rule of chance" to experiments and so
to produce "completely distorted results." Only through the "complete exclu-
sion of psychic influence" could experimenters uncover the otherwise factory-
like regularity of the digestive machine. So, praktikanty conducting chronic
experiments came to work in separate, isolated rooms and were enjoined to
"carefully avoid everything that could elicit in the dog thoughts about food."
Such procedures, however, could not, even in principle, exclude the psyche
from chronic experiments, since a dog's personality and food tastes shaped the
"psychic secretion" that constituted the first phase of its response to a meal.
In chronic experiments, then, the idiosyncratic psyche acquired "flesh and
blood," and the results were incorporated into descriptions of the *pravil'nye*
processes of the digestive machine.[47]

So familiar were the secretory consequences of a dog's psyche and personality that, as praktikant I. O. Lobasov wrote in 1896, "it is taken as a rule in the
laboratory to study the tastes of the dogs under investigation." Some dogs had
pronounced food preferences; others refused the horse meat offered in the laboratory or ate it without enthusiasm. "In such picky dogs sham-feeding with
an unpleasant or even undistinguished food produces an extremely weak [se-
cretory] effect." Inattention to the individual "character" (*kharakter*) of dogs,
Lobasov continued, explained the inability of some Western European scientists to elicit gastric secretion by teasing a dog with food.

> Dogs exhibit a great variety of characters, which it is well to observe in their
> relation to food and manner of eating. There are passionate dogs, especially
> young ones, who are easily excited by the sight of food and are easily subject
> to teasing; others, to the contrary, have great self-possession and respond
> with great restraint to teasing with food. Finally, with certain dogs it is as if
> they understand the deceit being perpetrated upon them and turn their back
> on the proffered food, apparently from a sense of insult. These dogs only re-
> act to food when it falls into their mouth . . . Certain dogs are distinguished
> by a very suspicious or fearful character and only gradually adapt to the lab-
> oratory setting and the procedures performed upon them; it stands to rea-
> son that the depressed state of these dogs does not facilitate the success of
> experiments. The age of dogs is also important in determining their char-
> acter: the older the dog the more restrained and peaceful it is, and vice versa.[48]

The acknowledged importance of the dog's psyche and individual character
made these factors not only the "main enemy" of the experimenter seeking
pravil'nye results but also his "best friend" when attempting to reconcile con-

flicting data with laboratory doctrine. As we shall see in Chapters 4 and 5, judgments about the individual particularities of one's experimental animal invariably played a part in the interpretation of data. By their very nature, then, laboratory dogs constantly generated not "clean" quantitative data but rather complexities engendering interpretive flexibility. This was fully recognized in the laboratory, raising a question about the work process there. How could the praktikant—a mass of "borrowed senses" fit only for simple observation— render the interpretive judgments appropriate only for the chief's "directing mind"?

Relations of Production

With a good method, even a rather untalented person can accomplish much.

— IVAN PAVLOV, *Lectures on Physiology*

WE, THE LABORATORY

The workforce in Pavlov's physiology factory was bound together by an authoritarian structure and cooperative ethos. The chief's administrative authority was absolute: he hired and fired, assigned research tasks, decided when a task had been satisfactorily completed, and determined whether a praktikant would receive his doctorate. His intellectual authority was also, of course, considerable—by virtue of his knowledge, experience, and administrative power. The atmosphere of free, cooperative inquiry in the laboratory permitted coworkers to disagree openly with Pavlov on scientific questions, although the chief's legendary temper could make this extremely unpleasant. Laboratory *glasnost* both suited the spirit of scientific inquiry and socialized the laboratory's cognitive process, serving as one of several means by which the chief's "directing mind" presided over the interpretive moments inherent to experimental trials. Institutional realities and the career trajectory of the praktikanty—who lacked physiological education and were chiefly interested in quick doctoral degrees—shaped the results of this mixture of authority and cooperation. Praktikanty came and went, but "we, the laboratory" remained.

The laboratory's cooperative ethos was embodied in Pavlov's addition to the Institute's statutes of one rule specific to the Physiology Division: "Every praktikant is required to participate in the work of his comrades, specifically when there is being conducted a complex experiment or operation demanding a large number of assistants, greater than the constant paid contingent in the laboratory."[49] Praktikanty frequently paid tribute to this ethos in memoirs and acknowledgments, of which the following, written in 1894, is typical: "My

fervent thanks to the profoundly esteemed professor Ivan Petrovich Pavlov, according to whose thought and guidance this work was conducted; and whose
active participation and precious help greeted its every step . . . [My thanks
also] to all the laboratory comrades, who always came to my aid enthusiastically as a result of both their personal goodwill and the principle of broad mutual aid that reigns in Professor Pavlov's laboratory."[50]

Laboratory *glasnost* meant that scientific issues were openly discussed
among praktikanty in general meetings and in one-to-one sessions with
Pavlov. As one praktikant recalled, "everybody felt himself a member of one
common family and learned much, studied much, knowing the course of the
work of his comrades. No secrets were permitted."[51] The interaction of these
two dimensions of laboratory life—*glasnost* and Pavlov's immense authority—was central to the productive process, allowing Pavlov to direct, monitor,
and process the research of the fifteen or so praktikanty who worked for him
at any one time and to incorporate the observations of their "borrowed senses"
into ongoing laboratory traditions.

Pavlov always openly and proudly acknowledged that the data for his own
general works were obtained almost entirely by his praktikanty, whom he
credited by name for specific results and technical innovations. He himself,
however, took credit for the laboratory's methodologies, thus implicitly assuming much credit for his praktikanty's achievements. "With a good method,"
he once remarked in a lecture, "even a rather untalented person can accomplish much."[52] Furthermore, the concepts that gave these results meaning belonged to "the laboratory." As Pavlov put it in the preface to *Lectures:*

> In the text of the lectures . . . I use the word "we," that is, I speak in the per-
> son of the entire laboratory. Citing constantly the authors of specific exper-
> iments, I discuss jointly the experiment's purpose, sense, and place among
> other experiments, without citing the authors of opinions and views. I think
> it is useful for the reader to have before him the unfolding of a single idea
> increasingly embodied in tenable and harmoniously linked experiments.
> This basic view that permeates everything is, of course, the view of the lab-
> oratory, encompassing its every fact, constantly tested, frequently corrected,
> and, consequently, the most correct. This view is also, of course, the deed of
> my coworkers, but it is a general deed, the deed of the entire laboratory
> atmosphere in which everybody gives something of himself and breathes it
> all in.
>
> Looking upon everything the laboratory has accomplished in our field, I
> value especially the participation of each separate worker and therefore feel
> the need on this occasion to send to all my dear coworkers, scattered

throughout the broad expanses of our motherland, heartiest greetings from the laboratory which they, I hope, remember as it does them.[53]

This goes to the heart of the division of labor and intellectual property. For Pavlov, "we, the laboratory" involved the collective work of all its personnel over the years, but he himself provided its stable personal and interpretive identity (the others were soon "scattered throughout the broad expanses of our motherland"). The experiments belonged to the praktikant, but the "basic view" or "single idea" that united them and gave them meaning belonged to "the laboratory"—that is, to Pavlov himself. At the same time, his constant references to "the laboratory's view" and to the experiments of various praktikanty gave Pavlov's conclusions greater authority, portraying them as the results of collective thinking and independent experimentation by numerous individuals on countless dogs.

These values were embodied in the highly standardized structure and language of laboratory dissertations. These invariably began with a review of previous literature that developed into a rationale for "Professor Pavlov's proposal" that the praktikant investigate a particular issue in a particular manner. The literature review almost invariably obeyed the following sequence: first, a statement about the fundamental importance of methodology; second, summaries of earlier research conducted in various laboratories; third, a statement about the cardinal methodological achievements of the Pavlov laboratory; and fourth, summaries of recent research, almost exclusively that produced in the Pavlov laboratory. The impression created is captured nicely by the words with which one of Pavlov's favorite praktikanty concluded this section of his thesis: "To the author of the present work fell the happiness of participating in the elaboration of a small part of this great task: Professor Ivan Petrovich Pavlov proposed that . . . "[54]

In the body of the dissertations, the word *I* appears almost exclusively with reference to specific observations or to the actual process of conducting experiments; either the passive voice, or *we*, or the name of the chief himself is attached to conclusions and ideas. So, for example, A. S. Sanotskii (1892) writes that "I tested the influence of teasing with meat" and refers to "my observations," "my experiments," and so forth, but "we have a right to conclude," "we come to the conclusion," and so on.[55] Pavlov's central role in the interpretive moments arising during the praktikant's work was acknowledged in standard phrases: "suggestion of the theme," "constant guidance and aid in word and deed," "constant participation and warm attention." The chief no doubt expected such phrases, yet this was not the empty rhetoric of obeisance. It reflected, rather, Pavlov's extraordinary energy and engagement and a production system that made him an active participant at critical junctures in the praktikant's work.

LINES OF INVESTIGATION

Pavlov's scientific-managerial vision meshed with the workforce at his disposal in a system of production that gave both the chief and most praktikanty what they most wanted. For Pavlov, the praktikanty were set to work on his own scientific vision, multiplying his sensory reach manyfold while enabling him constantly to monitor the work process and its results, to incorporate these into his developing ideas, and to convert them efficiently into marketable products. For the praktikant, this system provided a sometimes exciting investigatory experience and justified the confident expectation "that after one year in Pavlov's laboratory the thesis would be written and the degree of doctor of medicine would be received."[56] This doctoral thesis originated in Pavlov's scientific-managerial vision, by which he generated an endless series of topics that, within his laboratory system, could be quickly and successfully completed by a praktikant with no prior physiological training.

A fundamental, unalterable principle was that Pavlov assigned all research topics. One praktikant later recalled that Pavlov appreciated initiative among his coworkers, but "he could not give it a wide berth, since this would interfere with the development of his scientific idea, which proceeded according to a set plan." A coworker could express a desire or intention, and this might be sanctioned temporarily if it corresponded with Pavlov's plans. Otherwise, should the praktikant contest the point, "there arose an argument that rarely ended with the victory of the coworker." Another coworker recalled, "When a young scientist had matured and was able to formulate his own ideas and plans for research, work with Pavlov became difficult. Subjects that had no direct relation to the work of the laboratory did not interest him, and often he would even refuse to discuss them." The laboratory was oriented almost entirely to production, and perhaps because of the nature of his workforce, the chief devoted little effort to pedagogy that did not have a direct productive function.[57]

By what rationale did Pavlov assign topics? Wandering into his laboratory in any year, one would find praktikanty engaged in a wide variety of subjects. If we look at the chronological development of research topics, however, Pavlov's "set plan" and the reason for his insistence on assigning research topics are readily apparent. In the years 1891–1904 the topics assigned to praktikanty reflected a standardized approach to the main digestive organs (the gastric glands, the pancreatic gland, and, somewhat less and somewhat later, the salivary and intestinal glands). Research on each organ followed a general sequence: establish the existence of nervous control over the gland, develop an appropriate dog-technology, identify the specific exciters of glandular secretion, establish quantitatively the *pravil'nye* patterns of glandular activity, and verify the "stereotypicity" (*stereotipnost'*) of these secretory responses. Research on the different glands proceeded in parallel, each providing models for research on the others. Alongside these principal lines of investigation, praktikanty were often assigned topics designed to fortify the Physiology Division's institutional position, explore possible new research paths, respond to critics of laboratory doctrine, or examine puzzling results that lay off the main investigative paths.

When these lines of investigation developed normally, Pavlov never assigned two praktikanty to the same topic simultaneously. This made good sense, since one praktikant's results were a necessary prelude to the research of the next along the standardized route of investigation. This practice also required the chief to interpret only a single set of experimental results at a time. Pavlov departed from it only three times: in assignments for work on the pathology of the digestive system in 1898–1900; on the psychic secretion of the salivary gland, beginning in 1903; and on the influence of nerves and "humors" on pancreatic secretion in 1902. In the first two cases, he was considering a major shift in the focus of laboratory work and quickly generalized initial results in a public speech hailing the dawn of a new era, not just for his laboratory but for physiology itself. In the third case, Pavlov was responding to the discovery of secretin—a major blow to the nervist views underlying his laboratory work.[58]

This scientific-managerial strategy can be illustrated by a brief look at assignments for work involving the pancreas. Before acquiring his laboratory at the Institute, Pavlov had traversed the first part of his standard investigatory path, demonstrating to his own satisfaction that the vagus and sympathetic nerves controlled pancreatic secretion. Animals with a pancreatic fistula, however, died unexpectedly and were still considered insufficiently "normal" for chronic experiments. The main task, then, was to improve this dog-technology. In the Physiology Division's first year (1891), Pavlov assigned two praktikanty to this objective—one to develop a better fistula, the other to explore various dietary means to keep animals with pancreatic fistulas alive. In 1894 and 1895, armed with the results of this research, Pavlov assigned new praktikanty to test likely exciters of pancreatic secretion. By this time, experiments with Druzhok had convinced Pavlov that the gastric glands responded to specific foods with specific secretory patterns; in 1896, he assigned an especially promising praktikant, A. Val'ter, to find similar patterns in the pancreatic gland. When Val'ter succeeded in doing so, Pavlov assigned A. R. Krever to confirm his results. Two other praktikanty elucidated mechanisms of nervous control.[59]

Two interesting observations about Pavlov's scientific-managerial style emerge here. First, Pavlov assigned Krever to verify Val'ter's results in 1898, a year *after* Pavlov had showcased Val'ter's results in his own *Lectures*. Indeed, Pavlov declared Val'ter's results "stereotypical" even before Val'ter had managed to complete his thesis, let alone before Krever's (as it turned out, tortured) confirmation of them. This raises an obvious question about the process and meaning of such verification.[60] Second, since Pavlov was satisfied by 1897 that research on the pancreas had confirmed that, like the gastric glands, it produced precise, purposive secretory reactions to various foods, laboratory research on this gland was slowing by the end of the century. New praktikanty were assigned instead to other topics (e.g., the study of intestinal secretions and the interaction of the glands). This changed suddenly in 1902 with Bayliss and Starling's announcement of a humoral mechanism for pancreatic secretion. Pavlov immediately assigned several praktikanty (Val'ter, P. Borisov, V. V. Savich, and Ia. A. Bukhshtab in 1902) to investigate this challenge to his nervism and to repair Val'ter's earlier findings in the light of this and other new developments.[61]

This great productive capacity and flexibility was an important advantage of factory production. Pavlov was able to develop concurrently his standardized line of investigation for each gland while also using incoming praktikanty to respond quickly to critics, new developments, and simply curious phenomena. No workshop physiologist could do so.[62] Furthermore, the chief's position in the factory afforded him a "panoramic view." Moving at will from one praktikant's work to another's, he could concentrate his own efforts on the key task of the moment while keeping his eye on synthetic possibilities. He confided to his son some years later that "I have turned this into a system. If I did not move simultaneously from one work to another I would never have been able to conduct one work as successfully as I now conduct tens of them."[63] I now look more closely at the managerial system by which Pavlov "conducted" the research of his praktikanty.

WORKING FOR PAVLOV

On entering the physiology factory, the praktikant was incorporated into a highly structured production system that harnessed his "borrowed senses" to

Pavlov's "directing mind." Little was left to chance. An attendant cared for the praktikant's dog and provided the craft skills necessary at the bench; an assistant socialized the praktikant into laboratory culture, familiarized him with necessary procedures and interpretive models, and supervised his work; and when experimental results proved baffling, "all physiological difficulties were solved by Pavlov or his assistant."[64]

Typically, a physician wishing to work in the laboratory made his application directly to Pavlov, who interviewed and quickly decided whether to accept him. Sometimes the laboratory was filled to capacity and a strong letter of recommendation was necessary for an applicant to gain admission.[65] Pavlov was chiefly concerned in the interview to begin sizing up the applicant's ability and to establish that the praktikant would be completely at his disposal.

Once accepted, the praktikant was assigned to an assistant, under whose watchful eye he spent several weeks or even months familiarizing himself with laboratory procedures. This lengthy period both facilitated his socialization into laboratory life and gave Pavlov and his assistant an opportunity to determine an appropriate work assignment. As Babkin observed, "This lengthy ordeal to which the worker had to submit was partly due to the fact that, according to Pavlov, one of the most difficult tasks which devolved on him as laboratory chief was the choice of problems for his coworkers. He gave most careful thought to each question that he was planning to investigate with a new collaborator and worked out a preliminary plan in his mind, but all this required time."[66] Babkin's choice of words here—his reference to problems that Pavlov "was planning to investigate with a new collaborator"—is most appropriate.

The praktikant's socialization involved all aspects of laboratory culture. During his first few weeks in the laboratory, he observed the experiments of other coworkers and imbibed general laboratory values. For example, on arriving thirty minutes late to the laboratory one day, the new praktikant I. S. Tsitovich found his assistant, A. P. Sokolov, waiting for him. "With his very first words Sokolov criticized my half-hour tardiness. I was a little insulted by such captiousness, which I ascribed to hostility on his part. Later I became convinced that his criticism was fully deserved, since Ivan Petrovich and the entire laboratory worked like the mechanism of a watch. With the laboratory's strict discipline my lateness really could not be justified."[67] Tsitovich also learned, to his surprise, that a mere praktikant had the right to disagree with the chief, and he cheerfully engaged in his first exchanges with Pavlov on scientific developments.

After a few weeks the praktikant received his own dog, either that of a departing investigator or, if a new animal was required, one prepared surgically by Pavlov or an assistant. The choice of dog reflected Pavlov's decision about which line of investigation the praktikant would pursue. Under the assistant's eye, the praktikant now familiarized himself thoroughly with the appropriate techniques. He also read the "relevant literature," which consisted almost exclusively of reports of previous work in Pavlov's laboratory, thus further familiarizing himself with the chief's expectations. Pavlov sometimes stopped by his bench for a moment or two and conferred with the assistant about the praktikant's progress.

When both assistant and chief judged the praktikant ready for work and had sized up his abilities, Pavlov assigned him a specific task. Work began under careful supervision. Tsitovich's recollection is typical: "The assistant related to me in great detail how and what I must observe, how to take notes on the experiment, how to avoid extraneous influences [on the dog]."[68] Chronic experiments demanded a great deal of patience and self-discipline, often compelling the praktikant to sit virtually motionless for hours. (Pavlov later liked to recount an anecdote about walking into an experimental room only to find both dog and praktikant asleep on the job.)[69] The ability to endure these lengthy periods of observation and collection was the chief obstacle between the praktikant and his doctoral degree. Possessing a surgically prepared dog and an expertly defined topic and guided by attendants, assistants, and the chief himself, "all that was necessary for a doctor's success was that he should perform his work carefully, bringing to it all his concentration and understanding."[70]

The relationship between observation and interpretation, however, is rarely that simple, especially within a context that locates the two in different persons. In Pavlov's physiology factory, this relationship was shaped by two interactions: that between Pavlov and the praktikant and that between the praktikant and his laboratory dog(s).

PAVLOV AND THE PRAKTIKANT

Pavlov's presence permeated the laboratory daily. Unless he was lecturing or attending a faculty meeting at the Military-Medical Academy, he arrived at the laboratory between 9:30 and 10:00 A.M., immediately checking the coatrack in the entrance hallway to ascertain who was present and who was not. "He never missed a day at the laboratory and did not like anybody to be absent or late."[71] As praktikant A. F Samoilov recollected, "When in the mornings he entered, or, more correctly, ran into the laboratory, there streamed in with him force and energy; the laboratory literally enlivened, and this heightened businesslike tone and work tempo was maintained until his . . . departure; but even then, at the door, he would sometimes rapidly deliver instructions regarding what remained to do immediately and how to begin the following day. He brought to the laboratory his entire personality, both his ideas and his moods. He discussed with all his coworkers everything that came into his mind. He loved arguments, he loved arguers and would egg them on."[72]

Pavlov spent his mornings and afternoons attending to the work of one or more praktikanty—observing, commenting, and participating in experiments if moved to do so. About fifteen praktikanty were working in the laboratory simultaneously and Pavlov managed to make himself a presence in the work of each, although he singled out one or two whose work interested him especially at any given time.[73] At the very least, the chief dropped by occasionally to check the protocols; if the experiments proved exceptionally interesting he often worked alongside. "From the moment that a problem was assigned to a worker, Pavlov took a most active interest in it and inquired about its progress almost daily. Often he would sit for an hour or more in the worker's room observing an experiment. He would examine the protocols and often remembered the figures previously obtained better than did the worker himself. Finally, if he was especially interested in the work, he would participate in the experiments himself."[74]

The memoir literature makes clear that Pavlov used his sessions with praktikanty to exercise a steady influence on both the course of the experiments and the interpretation of their results. As L. A. Orbeli recalled:

> In regard to the correctness of the [experimental] protocols, Ivan Petrovich was very demanding. He did not limit himself to asking how things were going. He would take the notebook with the protocols and begin to look through it. He might ask one of the workers how much juice he had acquired over a quarter of an hour. He would then take the notebook and check. If the verbal answer conflicted with the notes in the protocols, even by several tenths [of a cubic centimeter], the session would end with a dressing down. He knew how to retain in his memory for several days or weeks the most minor details of a work, and sometimes would recall that "at such and such a time an experiment yielded such and such figures." This extraordinary demandingness, perspicacity, and attention to the protocols; this extraordinary memory for all the details of the work conducted in his laboratory, was Ivan Petrovich's exclusive quality.[75]

Aside from these one-to-one sessions, there were frequent laboratory-wide discussions, which Pavlov would initiate sometimes in the Division's common

room and sometimes by drawing others into his discussion with a single praktikant in the laboratory. V. P. Kashkadamov, who worked in the laboratory
from 1895 to 1897, recorded the following recollection:

> Not less than once a week he would confer with each of us and attempt to
> draw all the workers into these discussions. Thanks to this we were always
> aware of all the work being conducted in the laboratory. All facts were sub-
> jected to an all-sided discussion and to the strictest criticism. If the slight-
> est carelessness, inattentive relationship to work, or hurried conclusion was
> revealed Ivan Petrovich would hurl himself upon the guilty party and crit-
> icize him sharply. Such sharpness, especially at first, offended me, and I re-
> acted to it very painfully. Then, when I became convinced that Ivan Petro-
> vich's rage cooled in fifteen minutes and he forgot about it entirely, relating
> to the guilty party as he had previously, I came to regard it much more
> calmly.[76]

Orbeli recalled similarly that, excited about a new fact or observation gathered
at a praktikant's side, Pavlov would wander from room to room, informing all
the coworkers about the event and its significance. "Having established an important proposition or having noticed a new fact, he would call everybody together and begin a public discussion on the spot. This habit (thinking publicly) facilitated the precision of his ideas and thoughts, and also attracted the
coworkers to the work."[77]

These discussions also helped the chief direct the work of his subordinates
and unite the laboratory behind a single perspective: "each scientific fact,
achievement or error was heatedly discussed at our daily general meetings . . .
Everybody knew what others were working on, what interpretation to ascribe
to new facts, how one could interpret them otherwise, what perspectives were
revealing what results."[78] In the great majority of cases, Pavlov's guidance was
exercised smoothly, as his greater authority, knowledge, and commitment allowed him to dominate free-ranging discussions and shape the interpretation
of data.

Sometimes, however, the praktikant proved less pliable, eliciting the chief's
intolerant, even belligerent, reaction to results and interpretations that contradicted his own views. For example, in 1901 a self-confident praktikant, V. N.
Boldyrev, showed Pavlov the protocols of some experiments that apparently
contradicted the laboratory's doctrine of purposiveness. Boldyrev had not fed
his dog for an entire day but observed that, nevertheless, the pancreatic gland
secreted periodically. This seemed to contradict Pavlov's view of the factory-
like response of the digestive glands to specific exciters. The result was an "extraordinarily stormy scene." Pavlov hollered that Boldyrev was obviously a

sloppy observer, that he must have had food in his pockets or smelled of food or made some inadvertent movement that excited the dog. The scene ended with Pavlov literally chasing Boldyrev out of the laboratory. Yet the stubborn praktikant returned and repeated the experiment with another dog. The result was identical, as was Pavlov's response. Boldyrev then sat with the dog for twenty-four straight hours, with the same result. Finally, Pavlov joined Boldyrev and confirmed his observation—which was soon incorporated into laboratory doctrine.[79]

The memoir literature contains several such examples, always with Pavlov exploding and then finally surrendering to the force of scientific facts. In any case, as this literature also makes clear, it was a rare praktikant who stood up to Pavlov's authority and legendary temper and who was as committed as the chief to a particular interpretation of laboratory results. Furthermore, it was the chief who decided which data and perspectives revealed by a praktikant's research would be pursued—and which would not.[80]

Pavlov's most direct intervention in the work of the praktikant was his editing of all reports, articles, and dissertations. This allowed him to shape the interpretation of data, to incorporate the praktikant's work into the laboratory's institutional memory, and to project a unified laboratory voice into the broader scientific and medical communities. On drafting one of these "literary products,"[81] the praktikant was invited to Pavlov's office in the laboratory, where he was treated to sweet tea, black bread, and Ukrainian bacon while he read his draft aloud to the chief. (In the case of a dissertation, this continued for two hours a day over about two weeks.) Pavlov sat with his head back and his eyes closed, frequently interrupting with questions or corrections and "sometimes revising all through, most attentively, before publication. He even wrote some of them himself."[82]

Each literary product was edited to a particular style. Reports read to the Society of Russian Physicians, for example, were no more than ten minutes long, with a simple presentation of data and conclusions. When one praktikant submitted a draft in which he polemicized with other scientific traditions and elaborated future research perspectives, Pavlov reacted negatively: "'What is this? What have you scribbled about here? Let me see this!' With a highly skeptical look he took my notebook and leafed through it. 'Well, what have we here!' and tore out about one-half of it. 'Words, little brother, are just words—empty sounds. Just give the facts, *this* will be valuable material.'"[83]

Pavlov's editing lent a highly standardized structure and content to laboratory publications. By the mid-1890s, discussions of previous research and issues in digestive physiology—even the language itself—were almost identical from one literary product to the next. (The exceptions were written by the few people who came to Pavlov's laboratory with well-developed scientific interests and inclinations.)

This editing reached deeply into the content of the praktikant's product. Babkin later recalled one revealing detail about Pavlov's editorial preferences.

> One of [Pavlov's] favorite expressions was "quite definite." An experiment had to show "quite definite" results, and if the results were indefinite then the worker had to ascertain the reasons for this. Pavlov was never satisfied with half measures. Either some wrong technique had been employed or the phenomenon was more complex than the experimenter had imagined. In the latter case it was necessary to change the plan of attack, taking the new factors into consideration. In both his own and his students' publications Pavlov tried as far as possible to avoid such expressions as "it would seem" and "probably." In other words, he avoided "suggestive results." He was a determinist by conviction and believed that every phenomenon had its cause.[84]

As editor, then, Pavlov "processed" results, pressing the praktikant to offer "quite definite" conclusions and offering helpful interpretations to this end. The praktikant himself, with little physiological training, needed to explain quite complex phenomena in a short period of time and knew that he would not receive his doctorate until he had done so to Pavlov's satisfaction. The chief's suggestions, then, seldom fell upon deaf ears.

So, all scientific works were "filtered" through Pavlov, who took this final opportunity to relate the praktikant's data to laboratory doctrine. A common recollection about this filtering process is worth pondering: "He loved not to read, but to hear the work, immediately elucidating inexactnesses, demanding explanation and confirmation of the material through experiments. There frequently arose heated discussions, during which Ivan Petrovich, using his brilliant memory, would refute the figures and propositions offered by the writer of the dissertation."[85] This curious point—that Pavlov remembered the data better than did the praktikant himself—arises repeatedly in the memoir literature.[86] It appears suspicious, even absurd, on the face of it—however prodigious Pavlov's memory—when we consider that he was usually supervising the work of some fifteen praktikanty conducting hundreds of experiments, each generating columns of data.

I am inclined, however, to accept these recollections as essentially accurate and as an important reflection of Pavlov's scientific style. He could not, of course, remember *all* the experimental data, but neither was he equally interested in them all. Just as he considered the research of some praktikanty more important than that of others, so he considered some experiments more telling than others. Contrary to his carefully cultivated image, Pavlov was a deeply intuitive thinker. Like Bernard, his notion of experimental reason left ample room for the "preconceived idea"; and like Gerald Geison's Pasteur, he confidently identified the "signal" amid the "noise."[87] Pavlov carried with him an ideal "template" of what good experimental results along his main lines of investigation should look like. When he observed results that fit this template, he remembered them well and so was quite capable of citing such data to refute or amend interpretations of other experiments that fit his preconception less snugly.

This highlights a critical point for reading the praktikanty's literary products: Pavlov was the coauthor of each. Throughout the praktikant's tenure in the laboratory—during his initial socialization, the meetings with assistant and chief at the bench, the give-and-take of general laboratory discussions, and his editorial sessions with Pavlov—his "borrowed senses" constantly confronted the chief's "directing mind." In the dissertations, this confrontation was often reflected in detailed physiological explanations downplaying results that threatened long-standing laboratory doctrines and emphasizing those that affirmed them. Reading these dissertations, one sometimes notices that their argumentation "changes direction"—that data and prose running counter to, say, the notion of a purposive pattern in pancreatic secretions suddenly shift and take the opposite direction; or, more commonly, that tentative suggestions become "quite definite" conclusions. This, I think, testifies to Pavlov's hand and to the deeper significance of Babkin's observation that "all physiological difficulties were solved by Pavlov or his assistant."[88]

Appreciation of Pavlov's role brings us back to the interpretive moments inherent to the chronic experiment. I now turn to the second critical interaction in the laboratory.

THE MEN AND THEIR DOGS AT WORK

We have seen that the tension between laboratory dogs as technologies and as intact organisms created a series of interpretive moments in chronic experiments. As technologies, the dogs were expected to yield *pravil'nye* results. For example, one dog's gastric glands were expected to produce the same pattern of secretions in response to 200 grams of meat from one meal to the next, and this secretory curve was expected to be "essentially" the same as that produced by another dog. Pavlov and his praktikanty also recognized, however, that, as an intact organism, each dog possessed a psyche and a distinctive personality and that these influenced experimental results. The praktikant's task, then, went far beyond collecting, measuring, and analyzing digestive fluids; he had also to assess the normality and personality of his dog(s) and interpret his results accordingly—with Pavlov's help and until gaining Pavlov's approval. Reviewing the doctoral dissertations produced in the laboratory reveals several features of this interpretive process.

In keeping with Pavlov's scientific vision, a praktikant necessarily assessed the "normalcy" of his dog. This assessment rested in part on such objective indicators as the animal's maintenance of a stable weight and temperature, but it was not limited to these. The word *happy* (*veselyi*) occurs regularly in attestations of normalcy. For example, Sanotskii (1892) assured his readers that, having recovered from their operations, "the dogs were happy and energetic, possessed a marvelous appetite, and gave at a glance the general impression of completely normal animals."[89] Attesting to the full recovery of his dogs from the implantation of the troublesome pancreatic fistula, Val'ter (1897) noted that they "create the impression of entirely normal, well-fed, happy animals." The dog on whom most of his conclusions were based, Zhuchka, "ate its food enthusiastically," ran a normal temperature, and "produced the impression of a healthy animal enjoying her life."[90] Sometimes, as in a dog with a pancreatic fistula, the praktikant knew the operation had fundamentally disrupted the dog's digestive system and would eventually lead to its death. He then needed to attest that the dog was "sufficiently normal" to generate trustworthy data. To this end, Bukhshtab (1904) described the medical ups and downs of his dog Lada, who suffered from both a pancreatic fistula and transection of the nerves between its stomach and intestines. Bukhshtab related that Lada actually gained weight and "felt good" but had lost some of its "former stamina": "It would become exhausted from standing in the stand, and ate unenthusiastically after the end of the experiment; therefore, the next day its weight declined. Therefore, we began to conduct experiments, not every day, but with breaks of a day or two, to allow the dog to recover and preserve its health and weight longer." Despite these efforts, Lada's "ability to withstand various external influences was lessened." The animal developed mouth ulcers, refused food, and lost weight, finally dying three months after its nerves were transected. Bukhshtab concluded that his data were valid, however, since experiments with Lada were conducted only when the dog was "in complete health."[91]

The praktikant also needed to identify the dog's personality (*lichnost'*), character (*kharakter*), or individuality (*individual'nost'*) and to interpret experimental results accordingly. "Professor Pavlov has many times told those working in his laboratory that knowledge of the individual qualities of the experimental dog has important significance for a correct understanding of many phenomena elicited by the experiment," wrote one coworker in 1901. "During the conduct of our experiments we always kept this in view."[92] Here the praktikant drew on observations concerning the dog's ease in adapting to the laboratory settings, its reaction to teasing with food, its preference for certain foods, the relative quantity of its secretory reactions, the consistency of these reactions from day to day, and so forth.

This assessment of the dog's personality was often invoked in interpreting experimental data. For example, Vasil'ev (1893) noted that his two dogs produced markedly different secretory reactions, perhaps owing to their differing ways of life before entering the laboratory: one was a "simple street dog" and so ate any food readily; the other was "obviously a hunting dog, judging by the breed and by its neural temperament." Krever's dog Sokol (1899) was "distinguished by the great sensitivity of its digestive canal" and so easily disturbed that it had to be taken for calming walks between experiments. Even the possible effect of these walks played a role in the interpretation of experimental results. Zavriev's Volchok (1900) was "very cowardly, reacting to every manipulation with panicky terror." Kazanskii's Laska (1901) was "peaceful, happy, and affectionate" and "very greedy for food. It trembled at the sight of the food bowl and burst off the stand, almost tipping it over."[93]

Kazanskii's other dog, Pestryi, was entirely different.

> As for particularities in Pestryi's nature, we can note that he was not distinguished by greed for food. He never threw himself upon the food being brought to him; he always ate calmly, unhurriedly, but with visible appetite. During the initial experiments he did not eat raw meat enthusiastically, as a consequence of which the quantity of juice in the first hour sometimes was less than during the second (a little); but then having become accustomed to meat he began to eat it enthusiastically. He was happy and always obedient during the experiments; but was also distinctively nervous and easily offended. It was enough to raise one's hand at him for him to begin to squeal, bark and grumble . . . Pestryi initially leaned toward the pieces of meat and sausage offered him [in teasing experiments]; but then, as if he had been offended or had understood the deception, he would turn away from the food offered him in that way.[94]

Here Kazanskii invoked Pestryi's personality and relative apathy toward food in order to reconcile experimental data with laboratory doctrine. According to the "stereotypical secretory curves" (constructed earlier through interpretation of experiments with Druzhok), the rapidity of gastric secretion elicited by a meal of raw meat should peak in the *first* hour, not the second (as was sometimes the case with Pestryi). This rapid secretion during the first hour, however, owed much to "psychic secretion," which, according to Kazanskii's argument, was muted by Pestryi's particular character. Similarly, Pestryi's changing disposition explained the different results in presumably identical experimental trials (sometimes secretion peaked in the first hour, sometimes in the second). Finally, laboratory doctrine held that appetite itself rather than the mechanical effects of food on the nerves of the mouth, generated the initial "psychic" phase of gastric secretion. This could usually be demonstrated by teasing animals with food and observing the secretory results. Pestryi, however, often failed to produce this secretory response, instead turning away from the food "as if he had been offended or had understood the deception." Kazanskii's voracious Laska would of course respond both to feeding and to teasing with a more copious "psychic secretion" than would the restrained Pestryi, and their differing "psychological profiles" were necessarily borne in mind when constructing a single, stereotypical curve from the differing data produced by the two animals.

Such interpretive moments constituted an "industrial secret" well-known to those who worked on the factory floor but largely unappreciated by consumers familiar only with its finished products.

Conclusion

The fundamental social-cognitive dynamics of Pavlov's laboratory remained essentially stable, not only in the years 1891–1904 but also, apparently, in subsequent decades. This does not mean, of course, that these dynamics were always the same during every juncture in laboratory work. That could never be true of any complex social situation, nor of such a complex process as experiments on living animals. Pavlov's scientific vision exercised a controlling influence on the planning, performance, and interpretation of experiments, but unforeseen empirical results, the emergence of equally plausible alternative interpretations, or the emergence of new technologies sometimes briefly acquired "the upper hand."

The social-cognitive structure of "we, the laboratory" gave the chief constant control over the interpretive moments in experimental work and only rarely presented him with authoritative alternatives to his own judgment. This was true to such an extent that when I deal in detail with evolving laboratory interpretations (in Part II), the praktikant who actually conducted the experiments often falls into the distant background. Even when he does not, phrases such as "Khizhin concluded" must almost always be read as "Khizhin and

Pavlov concluded." Yet this regularity, too, was upset at rare junctures—and with important results. As we have seen, social and cognitive processes in the Pavlov laboratory were tightly interwoven. In Part II we shall see how Pavlov's "single idea" underwent a steady development as he controlled the interpretive moments in experiments on digestive physiology, and how a rupture in standard social relations—the emergence of two praktikanty as more than skilled hands—played a key role in changing an important feature of laboratory doctrine.

The stomach of an enlightened person has the best qualities of a good heart: sensitivity and gratitude.

—ALEXANDER PUSHKIN

Without a dog, you'll catch no rabbit.

—Russian folk saying

In *Lectures on the Work of the Main Digestive Glands* (1897), Ivan Pavlov offered his readers "the unfolding of a single idea increasingly embodied in tenable and harmoniously linked experiments." In this part of the book I explore the process by which this "embodiment" occurred—that is, the production of knowledge claims in the laboratory.

Given the complex, dynamic nature of experiment and interpretation in the laboratory, a process in which key elements and their relationship were constantly changing over time, a clear chronological narrative would necessarily sacrifice the detail that makes the story convincing and meaningful. My narrative, therefore, sometimes passes over the same terrain two or even three times, focusing on different objects and relationships during each pass. It seems useful here, then, to offer the reader a general road map of developments.

In the years 1894–97, Pavlov developed his mature conception of the digestive system as a "complex chemical factory." For Pavlov, the digestive system incorporated a capricious psyche and *pravil'nye* nervous-chemical mechanisms into a single, *pravil'nyi* mechanism. (Recall that *pravil'nyi* [plural, *pravil'nye*] means both "regular" and "correct," capturing Pavlov's view that, in physiological experiments, the two meanings were one and the same.) This unifying conception drew upon two main lines of inquiry: (1) delineation of the role and relationship of these two main exciters of the glands (the psyche and nervous-chemical mechanisms), and (2) development of the "characteristic secretory curves," which described quantitatively the precise, regular, and purposive responses of the glands to various foods.

The laboratory view of these two subjects, and the factory metaphor itself, emerged largely on the basis of experiments conducted by Pavlov and four praktikanty on two dogs—that is, on one "template dog" for each of the main digestive glands. Under the chief's close supervision, three praktikanty—Pavel Khizhin in 1894, Ivan Lobasov in 1895–96, and Andrei Volkovich in 1897–98—studied the gastric glands through experiments on Druzhok, the first dog to perform successfully with a Pavlov sac. Their counterparts for the pancreatic gland were Anton Val'ter and his dog Zhuchka, in experiments of 1896–97. Pavlov synthesized these results in his *Lectures on the Work of the Main Digestive Glands.* On the eve of that work's publication, Volkovich attempted to verify the results obtained with Druzhok by experimenting on a second dog, Sultan. Only after publication of Pavlov's *Lectures* were Val'ter's results for the pancreas tested through experiments on other dogs.

The first three chapters of Part II are devoted to these developments. In Chapter 4 I explore the ways in which praktikanty used Druzhok to analyze the role of the psyche and to generate data about secretory patterns, and describe some of the interpretive choices inherent to this work. In Chapter 5 I examine the emergence of the "factory" metaphor and its relationship to the evolving interpretation of experimental data from 1894 to 1897. I then turn in Chapter 6 to Pavlov's presentation of the experimental data in *Lectures* and to the laboratory's verification of some of its key knowledge claims shortly after publication of this synthetic work. Here, too, I examine Pavlov's rhetorical strategies in his master work and his use of its pages to communicate an appealing vision of the relationship between laboratory physiology and clinical medicine.

The capricious psyche that played a central role in studies of the gastric and pancreatic glands would soon disappear from Pavlovian physiology. In Chapter 7 I describe the transformation of the laboratory view of the psyche in the years 1897–1904 and analyze the process by which a relatively minor line of in-

vestigation—studies of the salivary glands—gave rise to a new era in laboratory research.

   This process is interesting not just because it gave birth to the now familiar
Pavlov of salivating dogs and conditional reflexes but also because it owed
much to a rupture in standard social relations in the laboratory. In research on
the gastric and pancreatic glands, the laboratory functioned as a largely closed
world, with the chief confidently and authoritatively presiding over the interpretation of experiments conducted by praktikanty who lacked specialized expertise. In research on the salivary glands, however, the laboratory was exposed
to outside perspectives by two key praktikanty who brought with them, from
psychology and psychiatry, specialized expertise about "the mind of the
glands" and whose collaboration with Pavlov therefore acquired an unusual
form. The fundamental shift in the laboratory view of the psyche, then, owed
much to a rupture in the laboratory's long-standing social relations.

   Each of these developments reveals the central role of Pavlov's scientific vision and the basic tension within it: his attempt both to understand the normally functioning organism in all its complexity *and* to express that understanding in *pravil'nye,* fully determined, and, ideally, quantitative terms. This
tension lent an inherent interpretive dimension to even the most apparently
esoteric experiments, to say nothing of efforts to "embody" in them the chief's
"single idea." Pavlov grappled not only with the influence of his dogs' personalities and moods on experimental results but also with the relationship between his Bernardian notion of determinist physiology and the results of experiments that, however ingeniously executed, invariably produced somewhat
varying results. These issues were built into the most dramatic empirical expression of his single idea—the characteristic secretory curves—and forced
Pavlov to make rhetorical decisions in the presentation of his data to outside
readers. Yet the basic tension within Pavlov's scientific vision would also prove
a main source of his appeal as a scientist who bridged the gap between laboratory physiology and clinical medicine.

   The ideal source with which to begin a study of data processing in the physiology factory would be the laboratory notebooks kept by praktikanty for each
dog. Were these notebooks available, we could follow the data and their interpretation through progressively more processed forms—that is, from experimental protocols to doctoral dissertations to published articles to the chief's
synthetic works. These notebooks, however, have disappeared (most likely the
bulk of them were burned for fuel during the siege of Leningrad). I begin, then,
with the next "highest"—but still relatively unprocessed—source: the doc-

toral dissertations. This source is not ideal, but by comparing data and interpretations as they moved "upward" from the dissertations toward Pavlov's synthetic statements, we can understand the general contours of interpretive practices.

Finally I should note again, as in Chapter 3, that Pavlov was deeply involved in the interpretation of all experiments, and the conclusions advanced in the dissertations and publications of his praktikanty can almost always be taken as the chief's own. So, except in special circumstances that are noted, phrases such as "Sanotskii concluded" should be read as "Sanotskii and Pavlov concluded."

Chapter 4

THE REMARKABLE DRUZHOK

> The happy choice of an animal, an instrument constructed in some special way, one reagent used instead of another, may often suffice to solve the most abstract and lofty questions.
>
> —CLAUDE BERNARD, *An Introduction to the Study of Experimental Medicine* (1865)

> We must painfully acknowledge that, precisely because of its great intellectual development, the best of man's domesticated animals—the dog—most often becomes the victim of physiological experiments. Only dire necessity can lead one to experiment on cats—on such impatient, loud, malicious animals. During chronic experiments, when the animal, having recovered from its operation, is under lengthy observation, the dog is irreplaceable; moreover, it is extremely touching. It is almost a participant in the experiments conducted upon it, greatly facilitating the success of the research by its understanding and compliance.
>
> —IVAN PAVLOV, "Vivisection" (1893)

Pavlov had good reason for gratitude toward the dog, particularly toward one mongrel of setter and collie that the laboratory named Druzhok (Little Friend).

The chief and his coworkers worked with many scores of dogs, and Pavlov frequently invoked their sheer quantity to lend authority to his conclusions; but these dogs were not created equal. For one thing, some possessed physiological and temperamental attributes that made them, in Claude

Bernard's phrase, a "happy choice" of experimental animal. They varied greatly in their "understanding and compliance," their excitability and food tastes, and their ability to generate *pravil'nye* results. For another, dog-technologies were endowed by their laboratory creators with different surgical modifications and occupied different places in the chronological development of the laboratory's lines of investigation.

Temperament, technology, and timing combined to make Druzhok, as E. A. Ganike put it, "the basic object for the study of digestive processes in the stomach."[1] As the laboratory's first successful dog-technology with an isolated Pavlov sac, Druzhok provided the first opportunity for sustained chronic experimentation on the "normal" functioning of the digestive system. By virtue of the chief's satisfaction with the results of these investigations, Druzhok became the "template dog" for the gastric glands. That is, this dog's secretory responses were enshrined in characteristic secretory curves that described the reactions of the gastric glands to various foods, and these curves became the standard against which the gastric responses of other dogs were subsequently assessed. Druzhok also served as the model for Zhuchka, who became the template dog for the pancreatic gland.

Druzhok's years of laboratory service (1894–97) coincided with the emergence and development of the "single idea" that became "increasingly embodied" in laboratory experiments: that of the digestive system as a complex chemical factory. This single idea emerged from the two main lines of inquiry pursued through experiments on Druzhok: the attempt (1) to delineate the roles and relationship of psychic and nervous-chemical exciters of the gastric glands and (2) to describe the precise, purposive work of those glands in characteristic secretory curves.

In this chapter I examine the design and creation of this dog-technology and introduce the experimental processes and interpretive choices that emerged when Pavlov and three praktikanty used Druzhok to address these two issues. This provides the background for exploring, in Chapter 5, the process through which results with "the remarkable Druzhok" (and with his fellow template dog, Zhuchka) were fashioned into the image of a digestive factory that greeted readers of Pavlov's *Lectures on the Work of the Main Digestive Glands*.

Before Druzhok: Nervism and the Problem of "Psychic Secretion"

Druzhok's laboratory career was prepared by the laboratory's previous attempts to extend the doctrine of nervism and grapple with the mysteries of "psychic secretion."

In his last two years in the Botkin laboratory, Pavlov had attempted to demonstrate nervous control over the pancreatic and gastric glands. Success came more easily with the former, and in an article of 1888 Pavlov presented his experimental evidence for recognizing the vagus as the "secretory nerve of the pancreas."[2] Persevering in the face of initially negative results, Pavlov and his collaborator, the chemist Ekaterina Shumova-Simanovskaia, finally established that the sway of the vagus nerve extended to the gastric glands, and they announced their results in articles of 1889 and 1890.[3]

The point of departure for these investigations was Pavlov's disagreement with the consensus among physiologists that central nervous mechanisms played no role in gastric secretion. The leading authority on this subject, Rudolf Heidenhain, recognized two exciters of gastric secretion: the mechanical effect of food in the stomach on the gastric membrane and the subsequent absorption of the products formed by the contact between food and gastric secretions (which, in turn, acted in an undetermined manner on the glands). In his authoritative contribution to L. Hermann's multi-volume *Textbook of Physiology* (1883), Heidenhain had concluded that "the result of numerous experiments makes indubitably clear that nerves coming to the stomach from without possess no noticeable, direct influence on secretion."[4] In Heidenhain's view, the negative results both of his own experiments (in which he stimulated the vagus nerve while observing the gastric membrane through a special mirror inserted through a fistula) and of those conducted by other investigators (usually involving the severing of the vagus nerves) were especially compelling, since similar techniques had convincingly demonstrated the presence of nervous mechanisms in the salivary and pancreatic glands.[5] Other physiologists had followed Heidenhain's lead, attributing little or no role in gastric secretion to central nervous mechanisms.[6]

Heidenhain did, however, suggest that this conclusion would have to be revised if contradictory reports about one phenomenon were confirmed: a "true proof" of central nervous control would be confirmation of "the frequent indications that the mere sight of food is sufficient to elicit secretion in hungry animals." In a well-known work of 1852, Friedrich Bidder and Carl Schmidt reported that teasing a hungry dog with food elicited gastric secretion; and in 1878 Charles Richet observed that, in one human subject suffering from complete blockage of the digestive tract, the chewing of food "with an intensive taste" (such as sugar or lemon) elicited a flow of gastric juice. The reliability and meaning of these results were, however, subject to doubt. For example, in 1867 Moritz Schiff repeated Bidder and Schmidt's experiments and concluded that the secretion elicited by teasing did not act on albuminous substances and so hardly qualified as normal gastric juice. In 1876 Braun reported that stimulating a dog's mouth with vinegar and ether failed to elicit gastric secretion. For Heidenhain, especially given the negative results of experiments on the connection between the vagus and the gastric glands, the meaning of Richet's results—in which the patient did not merely see the food but actually chewed it—was ambiguous. Perhaps some vasomotor reflex was excited by chewing and this had the incidental result of eliciting a slight secretion.[7]

Heidenhain's experiments on a dog with an isolated sac confirmed his skepticism, for two reasons. First, although the vagus nerves to the Heidenhain sac were cut, the sac produced plentiful gastric secretion, seemingly disproving any important secretory role for the vagus nerves. Second, when Heidenhain's dog-technology chewed and swallowed food, gastric secretion ensued only after "a more or less long time"—that is, only after the food itself had reached the stomach. The simple act of eating, let alone of merely seeing food, did not seem to routinely elicit gastric secretion. "I think," Heidenhain concluded, "that the complete insufficiency of direct proofs should at least lead to caution regarding hypotheses of the existence of secretory nerves."[8]

Taking Heidenhain's discussion as their point of departure, Pavlov and Shumova-Simanovskaia attempted to provide a "true proof" of central nervous control over gastric secretion. They sought to establish "the fact of the excitation of gastric secretion without any direct action upon the mucous membrane of the stomach" and to reproduce "this fact in the sharpest and most constant form."[9] They did so in the following manner. Dogs were esophago-tomized and equipped with a gastric fistula. After recovering from the operation, they were sham-fed and teased with food.[10] The right branch of the vagus nerve was then severed and the dogs were again sham-fed and teased with food. Finally, the left branch of the vagus was also severed and the same experiments were performed—now, presumably, on dogs with complete vagal enervation.

The authors acknowledged some difficulties with this procedure. Most important, the dogs lost a great amount of saliva through the esophagotomy, which led to their "progressive exhaustion" and eventual death. Still, by feeding them solids through the gastric fistula and liquids through a cannula inserted in the anus, Pavlov and Shumova-Simanovskaia succeeded in maintaining "several animals" "for a rather significant time."[11] The "normalcy" of these animals, however, was open to question.

Pavlov and Shumova-Simanovskaia concluded that their experiments demonstrated "the sharp and constant influence of the central nervous system on the secretion of gastric juice." "The results of experiments could not have met our expectations more."[12] They reported that in every one of about

twenty experiments with seven different dogs, sham-feeding with meat elicited gastric secretion. (Milk and soup, however, elicited none.) Furthermore,
sham-feeding always failed to elicit secretion after the vagus had been completely severed. The authors conceded, however, that whether the vagus nerves
were intact or severed, "teasing the animals with the sight of meat not once
gave us a sharp result."[13]

This concession, although offered somewhat casually in the article, revealed
an important ambiguity in their results. As we have seen, Heidenhain had written that a "true proof" of central nervous control would be gastric secretion
elicited not by the act of eating but by the "mere sight of food." By Heidenhain's criterion, then, Pavlov and Shumova-Simanovskaia had failed to establish the existence of nervous control.

The authors (and I attribute the physiological argumentation in the article
mostly to Pavlov) dealt with this problem in an interesting rhetorical fashion.
They explained that they had adopted their methodology—that is, their reliance on sham-feeding esophagotomized dogs—for three main reasons.
First, the esophagotomy prevented saliva from reaching the stomach (and so
forestalled the possible objection that the saliva itself elicited gastric secretion). Second, the feeding of an esophagotomized dog reproduced experimentally the conditions of Richet's observations on a patient with a blocked
digestive tract. And third, "By feeding an esophagotomized dog through the
mouth we counted upon uniting and possibly strengthening both investigated
influences (the psychic and the reflected irritation from the surface of the
mouth)."[14] In practice, by "uniting" these two mechanisms, the authors
downplayed the negative results of the teasing experiments. Had they been
less committed to proving central nervous control, they might have concentrated on the sharply different results with sham-feeding and teasing, perhaps
even confirming Heidenhain's skepticism about "psychic secretion." Instead,
they concluded, "What do we have in these experiments: a reflex from the surface of the mouth or psychic excitation? Probably both, but probably with the
predominance of the former, since merely teasing the animals with the sight
of meat not once gave us a sharp result" (p. 180). (After conceding the negative results with teasing experiments, the authors thereafter referred to the secretory effect of sham-feeding as a "reflex from the surface of the mouth" [e.g.,
p. 189].)

Emphasizing their positive results with sham-feeding, Pavlov and Shumova-Simanovskaia attributed the negative results of previous investigators
to faulty methodology, which had disrupted the normal digestive processes of
their experimental animals (pp. 195–96). Even the great Heidenhain had been

led astray by the abnormalcy of the Heidenhain sac. Convinced that the vagus nerves played no important role in gastric secretion, the Breslau physiologist had severed them in his version of the isolated sac. Therefore, "normal [physiological] relations do not exist" (p. 190), and the Heidenhain sac produced distorted secretory phenomena that reinforced the false assumption built into it. For example, Heidenhain reported that gastric secretion began only twenty to thirty minutes after the dog had ingested food (and this reinforced his view that central nervous mechanisms played no important role), but in Pavlov and Shumova-Simanovskaia's esophagotomized dogs (which they took as more "normal" in this respect), gastric secretion began within five or six minutes of eating.

This time lag also created a problem for the authors' nervist interpretation, since it was "strange," they admitted, in view of the rapidity of nervous transmission. Pavlov would later term this time lag "the hidden period" (*skrytyi period*), capturing his assumption that some unseen nervous process was indeed taking place. The authors emphasized that this time lag was a "rather determined" (i.e., consistent) phenomenon, indicating that it had "a definite goal and a precisely operating mechanism" (p. 180).

Encouraged by the perspectives opened up by this work, Pavlov suggested that Nikolai Ketcher, a physician who arrived in the Botkin laboratory in 1890, write his doctoral thesis on the mechanism by which sham-feeding elicited gastric secretion: did this occur through a "reflex from the surface of the mouth" or through psychic excitation?[15] (In other words, did this phenomenon meet Heidenhain's criterion for establishing nervous control over gastric secretion?)

Ketcher sought to separate these two processes experimentally through several procedures that presumably excited one possible mechanism while bypassing the other. He probed the dog's mouth with a stick, forcing the animal to chew on it; this elicited no secretion. He force-fed the dog, "using tweezers to cram pieces of meat down into its very pharynx and forcing it to swallow them without chewing"; a plentiful secretion resulted. In Ketcher's view, these experiments indicated that "neither taste sensations nor the processes of chewing and swallowing are capable of eliciting the reflexive secretion of gastric juice." He also experimented with various foods, attempting to link their different physical and chemical properties with the results obtained during sham-feeding. Perhaps only albuminous foods that required pepsin—the proteolytic (protein-digesting) ferment in gastric juice—for their digestion elicited gastric secretion? In Ketcher's experiments, however, sugar, potato, and bread elicited "no less gastric secretion than did meat." Solid foods always elicited a greater secretion than liquids.[16]

Ketcher concluded that sham-feeding elicited gastric secretion through a combination of mechanical irritation of the nerves of the surface of the mouth and psychic excitation. Solid foods irritated the nerves of the roof of the mouth, generating reflexive "taste sensations," which in turn excited the psyche. Liquids did not irritate the nerves of the mouth mechanically but nevertheless elicited gastric secretion, so perhaps this secretion was "not reflexive, but rather psychic."[17] Thus Ketcher insisted on the role of the psyche, but he enjoyed no greater success than had Pavlov and Shumova-Simanovskaia in obtaining consistently positive results by granting a dog "the mere sight of food."

Having begun his research in the Botkin laboratory, Ketcher completed it in the gleaming and comparatively spacious new quarters of Pavlov's Physiology Division at the Imperial Institute of Experimental Medicine. Praktikanty were already streaming into the laboratory, and the chief assigned two of them to this line of investigation. One, N. P. Iurgens, repeated Pavlov and Shumova-Simanovskaia's acute experiments on the role of the vagus; the other, physician and surgeon Anton Sanotskii, repeated and elaborated Ketcher's chronic experiments on the "role of various factors in the secretion of gastric juice."[18]

In sharp contrast to previous experiences in the Botkin laboratory, Sanotskii (and Pavlov) found that, in the "suitable hygienic conditions" of the Institute's new facilities, the majority of esophagotomized dogs survived the operation, recovered weight rapidly, and returned to their "normal" state. "The dogs were happy and energetic, possessed a marvelous appetite, and produced at first glance the general impression of completely normal animals."[19] Only in rare cases did a dog "become rapidly exhausted and die." Depending on their weight, the dogs received one to two pounds each of meat and bread daily, plus a bottle of condensed milk.[20] They still lost a great deal of saliva, and sometimes lost weight and became "exhausted." Sanotskii assured the readers of his thesis, however, that at the first sign of a progressive loss in weight all experiments ceased.[21]

Sanotskii achieved an important result that had earlier eluded Pavlov, Shumova-Simanovskaia, and Ketcher: teasing six different dogs with meat, he consistently elicited a psychic secretion. The quantity, proteolytic power, and duration of psychic secretion differed from dog to dog and from day to day with the same animal.[22] Here, then, was Heidenhain's "true proof" of nervous control. "A more or less lively representation (*predstavlenie*) of food," without any physical contact whatsoever, was itself a powerful exciter of gastric secretion. This psychic excitation clearly acted on the gastric glands through the vagus nerves, demonstrating central nervous control of gastric secretion.

Why, then, had previous experimenters consistently failed to elicit psychic secretion in their animals? First, they had used dogs that were insufficiently "normal"—that is, dogs that had never fully recovered from their operations and were constantly sickly and losing weight. Second, they had used dogs that were insufficiently hungry. Sanotskii, on the other hand, had deprived his dogs of food for eighteen to twenty-four hours before an experiment.[23]

In the light of his success with teasing, Sanotskii revised Ketcher's conclusions, discarding entirely any direct role for the mechanical excitation of the nerves in the surface of the mouth. In so doing, he reinterpreted the results of one of his predecessor's key experiments. For Ketcher, the secretion of gastric juice as a result of force-feeding indicated that mere mechanical stimulation of the mouth, even without the animal's "complicity," caused gastric secretion. Sanotskii, however, thought it much more likely that force-feeding created some degree of "psychic excitation."[24] He followed this up with a series of experiments in which the dog was force-fed various inedible substances with physical resemblances to food. For example, force-feeding with a water-soaked sponge failed to elicit secretion. Even a sponge soaked in meat broth failed to produce a psychic secretion ("although the mechanical irritation was combined in this case with . . . the same type of taste that occurs when the animal is swallowing meat").[25] Conversely, a dog was sham-fed pieces of meat soaked in mustard and coated in salt, and the absence of psychic secretion demonstrated that mechanical irritation of the nerves of the mouth was powerless to produce gastric secretion unless accompanied by psychic excitation. As in all experiments of this type, there were exceptions—and an explanation for them.[26]

Using esophagotomized dogs and dogs with a Heidenhain sac, Sanotskii also confirmed that a second, weaker gastric secretion resulted from the presence of food in the stomach.[27] Only his general conclusions concern us here. First, he affirmed Heidenhain's view that this second secretion resulted from absorption of food products into the stomach wall. Second, whereas Heidenhain had suggested several possible mechanisms for this secretion, Sanotskii argued that, like psychic secretion, this process was "also nervous" and probably dependent on the action of food products on the sympathetic nervous system.[28]

Sanotskii concluded, then, that there were two "special mechanisms" of gastric secretion: a "distinctive [or idiosyncratic (*svoeobraznyi*)] psychic process" transmitted through the vagus nerves and generating a "very active" product (i.e., juice with a high pepsin content); and a second mechanism, "also nervous," that was excited by the process of absorption in the stomach and was probably dependent on "a sympathetic nerve." This latter mechanism produced a "relatively very weak" gastric secretion—that is, it had a low pepsin content.[29]

The important role of the dog's psyche, the experience of working with intact animals, Sanotskii's medical background, and Pavlov's drive to link laboratory investigations to medical practice all fostered making a connection between the dog's "more or less lively representation of food" and the human appetite. Sanotskii noted that the results of his experiments promised to give the very notion of "appetite" a more defined physiological character.

> The absence of appetite while ingesting food resembles, to a certain extent, the forcible insertion of food substances through the surface of the mouth and throat in our experiments; similarly, the behavior of animals during sham-feeding leaves no doubt that the influence of this feeding on the stomach can with complete justice be identified with the effect of a normal feeding with a big appetite. The concept of appetite would thus acquire a more defined character and acquire, so to speak, a material form. A greater or lesser appetite while ingesting food would mean . . . a more or less plentiful secretion of gastric juice and, consequently, a more or less rapid, successful, and complete digestion of food substances in the stomach.[30]

In the same spirit, Sanotskii suggested that clinical investigators and pharmacologists concerned with digestive complaints should explore two separate sets of explanations and palliatives, corresponding to the two different mechanisms of gastric secretion. Shortly thereafter, Pavlov assigned one praktikant to explore the possibility of using laboratory findings to treat a defective appetite and related problems.[31]

By late 1893, then, Pavlov and his coworkers had established the importance of psychic secretion and so had made a strong argument for nervism in the physiology of the gastric glands. They had distinguished two mechanisms of gastric secretion and had noted that these gave rise to products of varying proteolytic power. These developments, in turn, rendered available dog-technologies inadequate for generating synthetic knowledge about normal digestive processes in an intact animal. Food ingested by an esophagotomized dog did not reach the stomach (so the second phase of gastric secretion was missing entirely) and the Heidenhain isolated sac was now viewed as hopelessly ab-

normal because of its lack of vagus innervation. A new dog-technology was
needed.

Druzhok's Laboratory Birth

In the fall of 1893 Pavlov began working with Pavel Khizhin on the creation of
a dog with a vagally innervated isolated stomach. (As we saw in Chapter 3, this
sac was "isolated" from the stomach but presumably retained its normal nervous connections with it.) The physiology factory was "retooling," and the
chief chose a praktikant particularly well qualified for this task. For one thing,
Khizhin was a known quantity as Ol'denburgskii's trusted physician and as a
fellow loyalist throughout the tense tuberculin episode of 1891. For another, he
was an experienced surgeon, having studied surgical technique as a medical
student under V. A. Basov (the first Russian to successfully implant a gastric
fistula in a dog) and having performed numerous operations at the Ol'denburgskiis' clinic in Ramon (see Chapter 1).

The creation of a vagally innervated isolated sac involved a long, complex
operation demanding surgical skill, the manual manipulation of slippery tissues, and improvisation and teamwork.[32] Many of the craft skills necessary to
surgeon and assistants were gained only through experience with numerous
failures. "To describe" the operation, Khizhin later wrote, was "immeasurably
easier than to actually accomplish it."[33] In Khizhin's description of their initial attempts we see some of the more mundane but considerable difficulties
involved: "Personal experience convinced us of how difficult it is to hold in
one's hands such a gentle and slippery organ as the stomach wall, at the same
time using one's fingers to straighten the profusely bleeding parts of the
wounded area—which are stubbornly doubling over and collecting in the
folds of the mucous membrane—in order to place the bleeding vessel under
the attendant's tweezers" (pp. 16–17). The operation took about four hours to
complete and required at least two people—the surgeon and an "experienced
and skillful attendant." The basic conception of the "isolated stomach" was
Pavlov's, but both Khizhin and another praktikant, Aleksandr Samoilov, contributed important surgical suggestions (such as Samoilov's suggestion on
how to ease the difficulty described by Khizhin above [p. 15]).

The surgical formation of the flap that would become the isolated sac was
an especially difficult stage of the operation (see Figure 12 in Chapter 3). In describing it, Khizhin emphasized the importance of coordination between surgeon and assistant: "The great delicacy of this moment consists in the danger

of damaging the layers of the stomach wall that contain the [vagus] nerve branches while one is separating the mucous membrane [to create the flap for the isolated sac]. The success of this separation depends to a great degree, again, on the attention and skill of the assistant, who again grasps the separated part of the stomach membrane with two tweezers, raises it up, guides it, and in so doing gives the operator the ability to attentively follow the effect of each short stroke of the scalpel" (pp. 16–17).

Khizhin devoted seven dense pages to his description of the operation (frequently even advising readers which hand to use for what), but at several points conceded that he could not adequately convey all the important information in prose. For example, the actual formation of the sac from the surgically created flap of the stomach wall was the "most difficult" stage in the operation, "not so much in its technical implementation as in its description" (p. 17).

From fall 1893 through March 1894 the laboratory team endured repeated failures. At least sixteen dogs died under the knife or shortly thereafter. Autopsies revealed that seven bled to death, six died of peritonitis, one from heart failure, and two from "unknown causes" (p. 21). In October 1893 one dog survived the operation but died soon thereafter. An autopsy revealed that its isolated stomach had not held. In November, a second dog survived long enough to undergo several experiments and to receive a name, Tsygan (Gypsy). The secretion from its isolated sac, however, ran milky rather than clear, indicating contamination by food products from the large stomach. In late January 1894 another dog, Gordon, survived for two weeks with its isolated sac intact—long enough for Khizhin to conduct a series of experiments. This sufficed for Pavlov to laud the operation publicly as another indication of the dawning of a new era in physiology defined by surgical technique. In his short address to the Society of Russian Physicians in March, however, he also acknowledged his need for advice about how to overcome continuing surgical difficulties.[34]

At this point, the director of the Institute fell ill and Pavlov temporarily assumed his duties. Burdened with unfamiliar administrative tasks and assured by local pathologists that the isolated-sac operation was doomed to failure, he considered abandoning the procedure. He confided to his wife, Serafima, shortly thereafter that his "thoughts became temporarily divorced from the laboratory." Meanwhile, the panic-stricken Khizhin, his leave time drawing short without a completed dissertation in sight, despaired of ever receiving his doctorate.[35]

Finally, on April 2, 1894, nature and laboratory technique combined to produce "the remarkable Druzhok." The name "Little Friend" expressed the relief,
gratitude, and exhilaration that attended the dog's recovery from the operation with his isolated stomach intact. Four days after the operation Druzhok
was "energetic and happy." By April 9 his appetite was "excellent," and on April
13 he took his first postoperative stroll around the grounds.[36] The next day
there began five months of experiments, all conducted, according to Khizhin
and Pavlov, while the dog was in "excellent health."[37]

Druzhok's laboratory "birth" was later dramatized in reports and reminiscences by the major actors—Pavlov, Khizhin, Ganike, and Samoilov. Each
maintained a curious silence on one point: who actually performed the operation? Pavlov certainly conceived the basic idea for the isolated stomach, which
was immediately labeled "partial resection of the stomach by the Heidenhain
method, as altered by I. P. Pavlov," or the "Heidenhain-Pavlov sac."[38] Given the
great drama and importance of Druzhok's birth and the frequent stories in the
memoir literature about Pavlov's undeniable surgical wizardry, the silence
about this triumphant moment is especially puzzling. Pavlov himself stated
explicitly that he operated on Gordon, but he never made the same public
claim about Druzhok.[39] Both Samoilov and Ganike later wrote dramatic passages about the tensions, failures, and emotional ups and downs surrounding
Druzhok's creation, but when their narratives reach the actual surgical event
each writer adopts the passive voice.[40] In both his report to the Society of Russian Physicians and his thesis, Khizhin credits both Pavlov and Samoilov with
specific surgical innovations important to the ultimate success of the operation, but he too is silent about the actual operation, commenting only that
"having acquainted me with this task [creation of the isolated sac] and continuing to remain personally at the head of this affair, professor I. P. Pavlov did
me the special honor of assigning me to the immediate concerns regarding
implementation of the above-described program."[41] This ambiguity contrasts
sharply with the routine acknowledgments in later praktikanty's doctoral theses that Pavlov had created the isolated sac in their dogs.

It seems likely, then, that Khizhin operated on Druzhok while Pavlov was
mired in administrative chores, his thoughts "temporarily divorced from the
laboratory." If this is so, it provides an interesting example of the allocation of
intellectual credit in the laboratory: the basic conception of the operation was
allotted to Pavlov and credit for important aspects of its implementation was
awarded to Khizhin (and, secondarily, to Ganike and Samoilov).Unlike the
normal experimental duties of a praktikant, this breakthrough operation was

too important to be credited to anybody other than Pavlov or, since this was apparently impossible, to "we, the laboratory."

Druzhok's isolated sac, combined with the now doctrinal view of the importance of the psyche, made the dog not a mere site of glandular secretion but rather an active subject whose "character" (*kharakter*) demanded the attention of experimenters. Other dogs had earlier revealed an individualized character—for example, Sanotskii noticed that psychic secretion differed from dog to dog—but an esophagotomy had separated this character from other digestive processes in the stomach. That is, since food ingested by an esophagotomized dog never reached the stomach, all gastric secretions were psychic secretion; there was no need (and no possible way) to separate the influence of the dog's psyche from the presumably purely nervous-chemical effects elicited by food in the stomach. As the first dog with a fully functional isolated sac, however, Druzhok now confronted experimenters with the complex relationship between psyche and glandular response and with the difficult experimental problem of separating one from the other.

Work with Druzhok required considerable patience. Most experiments lasted about five hours, and some as many as ten, during which Khizhin strove to avoid exciting the animal by movements or sounds. After an attendant brought Druzhok into the room, Khizhin first ascertained that the dog's gastric glands were at rest. Experiments could commence only after a glass tube inserted in the aperture to Druzhok's isolated stomach revealed no secretion for at least thirty minutes.[42] Khizhin then teased or fed Druzhok, waited about five minutes for the first drops of gastric juice to appear in the special fistula that ran from the isolated sac, and collected the subsequent secretions at fifteen-minute intervals. Secretion continued for only about two hours when Druzhok was merely teased with food, but when Khizhin fed the dog milk, bread, meat, or mixed food, the experimenter had to remain as still as possible for five to ten hours at a stretch.[43]

Druzhok, too, needed to be "understanding and compliant." The dog's willingness to lie quietly on a table during the long trials greatly facilitated the experimenter's success. Otherwise, given the length of experiments, both investigator and dog would soon tire, undermining the precision of experimental results. For one thing, Khizhin noted, the dog's exhaustion "could hardly fail to be reflected in the course of secretory activity." For another, an "exhausted or simply bored animal" would inevitably make jerky movements that disrupted the collection of gastric secretions. Given the small quantity of secretions from the isolated sac (the surface area of which was estimated to be only

10 to 20 percent of that of the intact stomach), failure to catch even small quantities of secretion in the collecting cup could skew the results significantly. Furthermore, if the dog insisted on standing "the experimenter himself would become quite exhausted by having to continually hold the collecting cup or cylinder under the aperture of the tube that projects from the sac, which would not at all facilitate the precision of the data received." Druzhok adapted rapidly to this requirement, lying peacefully on the table and "taking no particular interest in anything" during the experiment. Better yet, he greatly facilitated the research by frequently sleeping for five to seven hours at a stretch.[44]

Although Khizhin attested to Druzhok's "excellent health," he admitted that one problem with the isolated sac raised a question about the dog's normalcy. "From the very first days of its proper feeding" after recovery from surgery, "the gastric juice that flowed from the aperture of the isolated sac began energetically to eat away (to digest) the tissue surrounding the aperture; this subjected the animal to pain, and required it to tend constantly to this region, to assiduously and almost continually lick it with its tongue, which, nevertheless, helped things little; and in a rather short time the area around the isolated sac can become one large sinusoid ulcer . . . The single palliative measure, so far, is the careful collection of the juice during each act of digestion." Khizhin quickly added, however, that "aside from this problem, the animal generally moves quickly and eats superbly, having quickly regained its former weight, energy, and cheer (bodrost')."[45] As we shall see below, however, this and other questions about the dog's normalcy would serve throughout Druzhok's laboratory life as a source of interpretive flexibility.

In April and May 1894, while Pavlov tended to administrative duties,[46] Khizhin conducted about sixty experiments, the great majority of which were designed to address the "first task" put to him by the chief: to compare the functioning of the Pavlov and Heidenhain sacs and to evaluate the former's reliability as a "true mirror" of normal secretory processes in the large stomach.[47] To this end, Khizhin conducted two types of trials. In the first he fed Druzhok either bread or mixed food (a combination of various quantities of milk, meat, and bread), measuring every fifteen minutes the quantity, proteolytic power, and acidity of the secretions elicited in the Pavlov sac. In the second type of trial he teased Druzhok with "the sight and smell of meat," again measuring the secretory response (p. 41). Khizhin compared his results to those obtained by Heidenhain, by previous coworkers in the Pavlov laboratory who had used a dog with a Heidenhain sac, and by another coworker who had studied the secretory reaction of other dogs to sham-feeding.

Khizhin concluded that a series of "characteristic particularities" demonstrated the superiority of the Pavlov sac to the Heidenhain sac as a mirror of normal gastric secretion (p. 46). Three of these particularities were attributed to the Pavlov sac's distinguishing characteristic, its vagal innervation: it produced a psychic secretion (absent in the Heidenhain sac), had a shorter "latency period" (due to the onset of psychic secretion), and elicited gastric secretion of higher proteolytic power (because psychic secretion was rich in ferments). The fourth particularity was even more intriguing: the *quantity* and *quality* of gastric secretion in the Pavlov sac, unlike that in the Heidenhain sac, varied independently, "each pursuing its own goals." That is, the amount of gastric juice and its proteolytic power did not rise and fall together. Within Pavlov's scientific vision, the regularity of this phenomenon attested to the existence of a determinist relationship with a particular purpose and to the existence of a specialized nervous mechanism.

Khizhin concluded, then, that the Pavlov sac was ideal for studying normal gastric secretion. The chief had succeeded in creating "an isolated sac that preserves completely the normal relations of secretory innervation and consequently possesses the ability to precisely and truly express everything that occurs in the stomach under any set of conditions; in a word, Prof. Pavlov's isolated sac is a pure and true mirror of the stomach in which one can with complete clarity observe the activity of this organ in all its smallest details" (p. 48).[48]

On June 2, 1894, just as Khizhin was completing his experiments on the qualities of Druzhok's new stomach, Pavlov completed his tenure as acting director of the Institute. Three days later, the chief first mentioned Khizhin's research in a letter to his wife, who was summering with their son in the countryside: "I am now delighting in total concentration on Khizhin's experiments. An enlivening of our projects is inevitable—success after success, not only new but downright beautiful." Two days later he postponed a scheduled visit to his family: "Khizhin's work is flowing so successfully, and is of such gripping interest, that I want to see it all myself."[49]

The encouraging performance of Druzhok's new stomach certainly promised "an enlivening of our projects," but what exactly did Pavlov find "not only new but downright beautiful"?

"Like Clockwork": Peptone, the Psyche, and the Interpretive Moment

Having established the virtues of his dog-technology, Khizhin conducted a series of experiments from June 1 through about June 8, 1894, to test Druzhok's response to various substances that, when present in the stomach, might reasonably be expected to excite gastric secretion. In an attempt to avoid arousing the dog's appetite—and so to avoid a psychic secretion that would obscure the nervous-chemical effects of these substances themselves—Khizhin used a cannula (a long, thin, hollow tube) to introduce material directly through the throat into the stomach. In this manner, he tested Druzhok's secretory response to water, acids, alkalis, salts, starch, egg white, and a commercial peptone prepared by the Parisian pharmaceutical factory Chapoteaut.

Khizhin (and Pavlov) had good reason to believe that peptone excited the gastric glands. As we have seen, Sanotskii's experiments had indicated that the first, psychic phase in gastric secretion produced a large quantity of pepsin-rich juice, which greeted the food mass as it arrived in the stomach. Khizhin (and Pavlov) reasoned that whatever substance or substances excited the peripheral nerves of the stomach membrane (initiating the second, nervous-chemical phase of secretion) must be either a common component of food or a common product of the contact between food and the pepsin contained in psychic secretion. They reasoned further that a dog's usual meal of mixed food contains albuminous substances and that these substances might be converted to peptone by the pepsin in psychic secretion. So, this peptone might well be the exciter of the second phase of secretion, in which case peptone placed directly in the stomach should itself excite the gastric glands.[50]

In his thesis (which he completed four months after these trials), Khizhin published the protocols of three experiments—conducted on June 1, 3, and 4—that seemed to dramatically confirm this hypothesis. In each trial, the introduction into Druzhok's stomach of a solution of Chapoteaut peptone (most likely mixed with gastric juice and water to simulate the products of the first, psychic phase of secretion) resulted in an abundant flow of gastric secretion. These protocols are reproduced here (Experiments LVII and LXI) with only minor simplifications.[51]

Experiment LVII
1 June [1894]

At 12:47 A.M., when there was no secretion from the isolated sac, poured into Druzhok's stomach by means of a gastric cannula a solution of 10 grams of Chapoteaut peptone + 100 cc of distilled water + 400 cc of gastric juice.

When the cannula was withdrawn several drops of the solution fell on the dog's tongue, and it began to lick its lips.

The first drop of juice appeared [in the tube running from the isolated sac] *ten minutes* after the pouring. The subsequent course of secretion was as follows.

TIME	QUANTITY (CC)	HOURLY QUANTITY (CC)	ACIDITY (HOURLY, %)	PROTEOLYTIC POWER (HOURLY, IN MM, BY METT METHOD)
12:47–1:02	1.3			
1:02–1:17	3.8			
1:17–1:32	5.2			
1:32–1:47	4.7	15.0	0.529	3.58
1:47–2:02	4.4			
2:02–2:17	4.4			
2:17–2:32	3.4			
2:32–2:47	2.0	14.2	0.547	3.75
2:47–3:02	1.5			
3:02–3:17	1.0			
3:17–3:32	0.1			
3:32–3:47	0.2	1.8	—	5.62
3:47–4:02	0.2			
4:02–4:17	0.0			
4:17–4:32	0.0	0.2	—	—
TOTAL		31.2	0.511	3.93

— = no data.

Source: P. P. Khizhin, *Otdelitel'naia rabota zheludka sobaki,* Military-Medical Academy Doctoral Dissertation Series (St. Petersburg, 1894), 130–31.

Note: Occasional comments on the appearance of the gastric secretion (e.g., "completely transparent," "transparent with mucus," "opaque with mucus," and "mucous") are omitted from the table.

Experiment LXI
3 June [1894]

At 2:04 P.M., when there was no secretion from the isolated sac, poured into Druzhok's stomach by means of a cannula a solution of 20 grams Chapoteaut peptone + 60 cc distilled water + 80 cc of gastric juice.

The first drop appeared *11 minutes* after the pouring. The subsequent course of secretion was the following.

TIME	QUANTITY (CC)	HOURLY QUANTITY (CC)	ACIDITY (HOURLY, %)	PROTEOLYTIC POWER (HOURLY, IN MM, BY METT METHOD)
2:04–2:19	1.2			
2:19–2:34	3.4			
2:34–2:49	5.8			
2:49–3:04	4.6	15.0	0.493	2.50
3:04–3:19	4.5			
3:19–3:34	2.9			
3:34–3:49	2.0			
3:49–4:04	2.6	12.0	0.511	3.29
4:04–4:19	0.9			
4:19–4:34	0.9			
4:34–4:49	0.6			
4:49–5:04	0.4	2.8	—	5.0
5:04–5:19	0.1			
5:19–5:34	0.0			
5:34–5:49	0.0	0.1	—	—
TOTAL		29.9	0.511	3.04

Source: P. P. Khizhin, *Otdelitel'naia rabota zheludka sobaki,* Military-Medical Academy Doctoral Dissertation Series (St. Petersburg, 1894), 130–31.

Note: Occasional comments on the appearance of the gastric secretion (e.g., "completely transparent," "transparent with mucus," "opaque with mucus," and "mucous") are omitted from the table. The values given as the "total" for acidity and proteolytic power do not seem to make sense (perhaps these were printing errors), but this is not relevant here.

In his discussion of these results, Khizhin emphasized both the dramatic excitatory effect of peptone and the striking similarities between Druzhok's secretory responses in the two experiments. Not only the totals, he emphasized, but also the data for hourly and even fifteen-minute periods were "remarkably similar and almost identical, like clockwork."[52]

It was this aspect of Khizhin's results, I suspect—this precise, repeated, mechanical regularity of secretory responses to a specific exciter—that moved Pavlov, in his letter to Serafima of June 5, to describe them as "downright beautiful." These superbly *pravil'nye* results promised to be only the beginning. On June 4, the day after the second of these trials, Khizhin began his first experiments on Druzhok's secretory response to feedings with raw meat—an important development in the direction of experiments ("an enlivening of our projects"). In his trials of April and May, Khizhin had fed Druzhok mixed food and bread to test the functioning of the isolated sac. But the meat experiments that began on June 4—and which, together with similar experiments with milk, consumed the entire month of July—were clearly designed to compare patterns of secretion for different foods. The search for distinct secretory patterns characterizing the digestion of different foods, and the attempt to describe these patterns in characteristic secretory curves, became a central theme of Khizhin's thesis and of Pavlovian digestive physiology.

An epilogue to these peptone experiments highlights the role of the interpretive moment. The results that delighted Khizhin and Pavlov in June 1894 were discarded one year later as meaningless artifacts. Based on a series of experiments in 1895, Pavlov and Ivan Lobasov, the next praktikant to work with Druzhok, concluded that peptone did not, in fact, excite gastric secretion. Lobasov attributed Khizhin's results to impurities in Chapoteaut peptone and to the inadvertent excitation of Druzhok's appetite.[53]

The differing conclusions reached by Khizhin and Lobasov each owed something to the interpretive choices created by the acknowledged role of the psyche in digestive secretion. Khizhin had acknowledged in his thesis that, because it was impossible to feed Druzhok without the dog noticing, he could not exclude the possibility that Druzhok's secretory response to peptone placed in the stomach resulted from psychic excitation.[54] "Despite every effort and contrivance to deny the animal an understanding of the qualities of the liquids introduced" through the cannula, a few drops almost always fell on the animal's tongue, which "in a hungry dog can immediately serve as a powerful source for the development of psychic excitation."[55] Indeed, in his experimental protocol of June 1 (Experiment LVII), Khizhin noted that a few drops of the peptone solution had fallen on Druzhok's tongue, causing him to lick his lips. This could have been interpreted as an indication that Druzhok's secretory response owed something to the arousal of his psyche. Khizhin, however, chose to attribute this secretion to the action of peptone on the mucous membrane of the stomach. Conversely, Lobasov acknowledged that placing peptone directly in Druzhok's stomach sometimes *did* elicit a small gastric secretion, but he attributed this secretion not to the presence of peptone but rather to the arousal of Druzhok's psyche by the procedure itself.[56] Thus, for both Khizhin (and Pavlov) in 1894 and Lobasov (and Pavlov) in 1895, the acknowledged power of the psyche to elicit gastric secretion was simultaneously an "enemy" of precise, conclusive results and a "friend" in the flexible interpretation of discordant data. As we shall see in Chapter 5, this was all the more true of the much more complex interpretive issues involved in the construction of characteristic secretory curves.

By the time Lobasov had refuted Khizhin's results, Pavlov, I think, had already glimpsed in them the outlines of a secretory apparatus that produced precise, repeatable, and distinctive responses to different exciters; the resultant investigations had already produced much confirming evidence. In his *Lectures* of 1897, Pavlov publicly acknowledged his earlier error about peptone; but, if my suspicion is correct, these "downright beautiful" results had already given rise to a view—one superbly compatible with Pavlov's scientific vision—that a careful, methodologically sound investigation of the gastric responses to various foods would discover distinctive, stable, and determined secretory patterns.[57]

The June experiments with various substances placed in the stomach also led Khizhin and Pavlov deeper into the mysteries of the connection between psyche and glandular response. Both later expressed "astonishment" at their discovery that placing egg white directly in Druzhok's stomach did not elicit gastric secretion. Pavlov later recalled that "it was to be expected, from a priori reasons, that if the gastric juice were specially adapted to act on the proteids, these substances would also prove to be chemical stimuli of the mucous membrane of the stomach."[58] This "wholly unexpected" negative result was followed by an equally intriguing discovery on June 8: when Khizhin placed egg white in Druzhok's stomach and then teased the dog with meat, the secretory result differed substantially, both in duration and proteolytic power, from that of either part of this process (sham-feeding or insertion of egg white directly into the stomach) considered separately. So, although egg white placed in the stomach itself failed to excite gastric secretion, in normal digestion this foodstuff was apparently "ignited" by psychic secretion, with distinctive secretory results. This cast in sharp relief the close relationship between the first, psychic phase of gastric secretion and the second, nervous-chemical phase.

Beginning in June 1894, these two related quests—the attempt to distinguish between psychic and nervous-chemical mechanisms and the search for *pravil'nye* secretory patterns—would frame Druzhok's experience in the laboratory and tax the experimental ingenuity and interpretive imagination of the men who worked with him.

Druzhok's Appetite

Khizhin inherited the laboratory's working definition of *psychic secretion* from Sanotskii's investigations of two years before. This first stage in gastric secretion was caused by an "idiosyncratic [or, distinctive (*svoeobraznyi*)] psychic process" elicited by "lively and clear representations about food."[59]

As investigations with Druzhok proceeded on both phases of gastric secretion, the word *pravil'nyi* ("lawful," "regular," or "correct") came to occupy the same central place in discussions of the second, nervous-chemical phase of secretion as did *svoeobraznyi* ("distinctive" or "idiosyncratic") in discussions of the first, psychic phase. This defined a set of experimental, interpretive, and doctrinal challenges. A central task was to distinguish between and analyze these two mechanisms in the process of gastric secretion and to use this information to understand gastric secretion as a fundamentally *pravil'nyi* process. According to the emerging "laboratory view," an idiosyncratic ghost inhabited the digestive machine, but the machine remained *pravil'nyi* nevertheless.

The acknowledged power of the idiosyncratic psyche created extreme difficulty in isolating its effects from those of the *pravil'nyi* nervous-chemical processes in the stomach. As Sanotskii put it with respect to his efforts to determine whether food excited gastric secretion by acting directly on the wall of the stomach, "One must repeat the experiment many times to be certain that the secretion actually results from this contact, and not from some unintended and unforeseen influence on the animal's psyche—from an influence under which there appears in the hungry animal a representation about food. Therefore, it is best initially to regard every positive result [i.e., every secretion] with suspicion, and in subsequent experiments to guard oneself with every precaution against the possibility of errors."[60] This difficulty was exacerbated by the fact that this psychic influence differed from dog to dog and day to day. For example, teasing one dog with meat elicited only 3.25 cubic centimeters of gastric juice in forty-five minutes, but teasing another in precisely the same manner elicited 15 cubic centimeters in just five minutes. The idiosyncratic, individualized nature of the psyche, then, created a series of interpretive moments, during which Sanotskii's analysis of data rested explicitly on subjective judgments about the character and mood of his experimental animals. For example, confronted with the experimental result that force-feeding elicited gastric secretion in one dog but not in another, Sanotskii explained that the former animal "was always distinguished by a rather strong impressionability toward everything connected in any way to food."[61]

Khizhin's experience with Druzhok underscored the key role of the dog's psyche and temperament. The secretory results of teasing experiments, he discovered, "depend to a great degree on the individuality of the animal, and also on some other causes which give to this very individuality rather broad fluctuations."[62] To conduct meaningful experiments, then, one needed to understand and adjust to Druzhok's particular character. For example, Khizhin wrote the following about one series of experiments in which he sought to elicit "appetite juice" by teasing Druzhok.

> In view of the fact that, as he proved more than once, Druzhok possesses an unusual impressionability and a broad self-esteem, we needed to approach this teasing with special delicacy; otherwise—as actually occurred—having noticed that we are only teasing him (by crudely and immediately snatching a piece [of food] away when he reaches for it)—he turns his snout away and does not even want to look at the things around him. Therefore, in order to attain our goal and interest Druzhok with the teasing, we would carry in a plate with meat, milk, and bread, and place it near his snout, avoiding even the appearance that we wish to tease the dog; we cook a piece of it on the gas flame, pour the milk and cut the bread; under such delicate conditions the dog immediately took an interest in our activities and began to get disturbed—inhaling energetically, stretching out its snout in order to draw closer to the food, began to intensely smack its lips, to swallow from the plentifulness of saliva; at the very same time juice appeared in the tube introduced into the aperture of the isolated stomach.[63]

As Pavlov put it in a speech of 1894 to the Society of Russian Physicians, if the dog "guesses" it is being deceived, teasing will not produce a psychic secretion. "The dog is an intelligent animal and is angered by this ruse no less quickly than a person would be."[64]

Attempting to distinguish between the psychic and nervous-chemical components of secretory responses, Khizhin conducted a series of experiments during which foods were introduced directly into Druzhok's stomach through a cannula. The results were sometimes puzzling. For example, Druzhok's secretory response to milk was *greater* when this substance was introduced directly into his stomach than when the dog ingested it normally. This seemed to contradict laboratory doctrine, which regarded total secretion as the arithmetical sum of psychic secretion and nervous-chemical secretion. If A (psychic secretion) + B (nervous-chemical secretion) = C (total secretion), how could B be greater than C? Khizhin's (and Pavlov's) interpretive response exemplifies the manner in which the psyche served as a highly flexible explanatory variable. One could not, Khizhin explained, exclude the possibility that

Druzhok responded with a psychic secretion even to feeding with a cannula: a bit of milk almost always fell on the dog's tongue during this procedure, and even when it did not, Druzhok's "extraordinarily subtle sense of smell" virtually guaranteed that the milk would exercise some influence on the dog's psyche.[65] Even so, why should Druzhok produce a *greater* secretion when fed with a cannula than when fed normally?

Khizhin reasoned as follows. Laboratory dogs in general, and Druzhok in particular, seemed to dislike milk, responding to it with little or no psychic secretion. Perhaps the procedure itself—the prolonged, awkward, and disturbing activities associated with shoving a long tube down Druzhok's throat—produced a stronger psychic reaction than did feeding the dog this relatively unappetizing food.[66] If this explanation were pursued, however, it might have raised basic problems about some accepted laboratory procedures and explanations; perhaps for this reason, Khizhin did not develop it.[67] He contented himself with the observation that the quantitative results for normal feeding and for feeding with a cannula were "rather similar" and showed "absolutely no difference" in proteolytic power.[68]

As this example illustrates, Khizhin's discussion of experimental results reflected a number of implicit choices regarding which data to discuss, which apparent discrepancies to explore, and which apparent discrepancies to ignore. With so many experiments and so many data at his disposal, Khizhin (and Pavlov) thereby gave selective evidential and interpretive weight to certain results and to certain perspectives that developed from discussion of those results. This is true to some extent for all intellectual endeavors, of course, but the sheer quantity of data produced in Pavlov's laboratory gave him and his coworkers great flexibility in this regard.

Khizhin candidly acknowledged the tentative nature of his conclusions, especially those pertaining to the distinct roles of psychic and nervous-chemical factors. The existence and power of psychic secretion was unquestionable, but given that Druzhok could not be fed without his noticing, all conclusions about the secretory effects of nervous-chemical factors remained in a "shadow" of doubt. Even the fundamental tenet that "the mucous membrane of the digestive canal possesses specific excitability"—that the nerves of the stomach responded only to specific exciters and thus differently to different foods—remained only hypothetical. "We must always remember that the psychic excitation of the animal has perhaps not been entirely excluded, and, consequently, we must be very cautious in our conclusions."[69]

This admission condemned Druzhok to further operations. In September 1894—while Khizhin hurriedly composed his doctoral thesis and his successor, Ivan Lobasov, waited in the wings—the large stomach of the laboratory's prize dog was fitted with a gastric fistula. Lobasov could now test the nervous-chemical response to the presence of various foods in the stomach by feeding Druzhok directly through the fistula, bypassing his mouth (and presumably his psyche) altogether. This relatively simple surgery, however, was not performed optimally: the aperture of the fistula emerged not in the very center of Druzhok's belly, where the dog would be least likely to notice activity around it, but a bit left of center. Toward the end of Lobasov's experiments, in April 1896, Druzhok underwent another surgical procedure, an esophagotomy, to permit experiments on the dog's secretory response to sham-feeding. The addition of a fistula and esophagotomy to Druzhok's isolated stomach reflected Pavlov's investigative drive to separate the different mechanisms of digestive secretion—that is, to study the psychic and nervous-chemical phases in isolation.

By the time Lobasov inherited Druzhok, Khizhin's research had given rise to the factory metaphor for gastric secretions and to the related view that specific foods generated distinctive, characteristic secretory curves. (These developments are discussed in Chapter 5.) Lobasov's task, then, was to distinguish between the psychic and nervous-chemical mechanisms of Druzhok's secretory responses to various foods and to explain the role of these mechanisms in generating "the entire complexity and characteristicity" of secretory responses to each.[70] As Lobasov put it in his thesis, he was assigned, first, "to detect (*ulovit'*) and distinguish in each case of the secretory work of the stomach during eating what relates to psychic secretion, and to elucidate to what degree the typicality of secretion with various sorts of food is determined by the participation of the psychic moment."[71] Second, he was to elucidate the specific exciters of "secretory work" in the second, nervous-chemical phase of digestion.

Lobasov worked with Druzhok from February 1895 through October 1896, during which time, he wrote, the dog was "with rare exceptions, in complete health."[72] The problem that Khizhin had earlier noticed with Druzhok's isolated sac had, however, grown more serious. If the dog was fed after experiments and permitted to roam freely, his gastric secretions "ate away terribly at the skin of the belly around the opening of the isolated sac; this disturbed the dog terribly, threatened his health (bleeding from the eaten-away arterial vessels and a deep eating away of tissues) and little by little consumed the mucous membrane of the isolated sac."[73] Like Khizhin, Lobasov sought to vitiate this problem by collecting Druzhok's gastric secretions until they had slowed to a trickle. This, however, was a long, difficult, and tedious procedure: gastric secretions could continue up to ten hours after a feeding, and one can imagine the difficulties involved in getting Druzhok to stand still all this time while a praktikant or attendant tried to wipe the leaking juice from his belly.

In any event, Lobasov conceded that Druzhok's postprandial secretions were not cleaned up diligently, to which he attributed a most distressing result: "the surface of the isolated sac diminished little by little, and sometimes, through our negligence, diminished sharply." Lobasov estimated that, by late 1896, the mucous membrane of Druzhok's isolated sac had shrunk by about one-quarter or one-third. When comparing his results with Khizhin's, he corrected for this shrinkage by multiplying his data by an appropriate coefficient. Because "the distinguishing characteristics" of secretory reactions to various foods remained unchanged and because "experiments were conducted only when the dog was absolutely healthy," he professed himself satisfied that Druzhok was nevertheless producing "normal" secretions.[74]

In his experiments on the nervous-chemical phase of digestion, Lobasov went to new lengths to deactivate the psyche, "to obstruct the appearance in the dog of thoughts and fantasies about food."[75] Experiments were conducted in a separate room, into which "only a little noise" penetrated from outside (p. 71). During the experiment, "nobody entered, insofar as possible, and the observer himself maintained complete quiet in order not to attract the dog's attention" (p. 27).

Yet the problem remained: how to feed Druzhok without his noticing? Lobasov attempted to do so stealthily through the gastric fistula. This, however, was a clumsy procedure requiring two to five minutes, during which time Druzhok frequently noticed the fumbling activity around his stomach and responded with a psychic secretion that ruined the experiment (pp. 45–46). Finally, Lobasov arrived at a solution: a long glass tube, designed to fit inside the gastric fistula, was filled with food and hidden inside the room. Druzhok was placed on the table and the experimenter waited for him to fall asleep. The tube was then quickly inserted into the fistula, and the food was pushed into Druzhok's stomach with a plunger. The procedure required only about twenty or thirty seconds, and the psychic moment seemed reliably absent.[76]

Lobasov used this technique to identify the psychic and nervous-chemical components of secretory responses. For example, placing milk directly in the dog's stomach yielded an "abundant chemical secretion." By comparing these results with those obtained when Druzhok ingested milk normally, Lobasov reached the conclusion that "secretion with milk is mainly chemical." That is, the quantity and quality of secretion elicited by these two modes of ingestion were essentially "identical," and each lacked the distinctive signs of a psychic secretion (i.e., rapid secretion and high proteolytic power).[77] Here Lobasov acknowledged some differences—when Druzhok ingested milk normally there was a "certain participation of the psychic moment," expressed in the "somewhat higher" proteolytic power of secretion—but appealed to the "identical" secretory curves for both modes.[78] Both normal feeding and ingestion through a cannula generated curves in which the amount of secretion in the first hour was low (indicating that psychic secretion was absent), followed by a much larger and steady secretion of "chemical juice" in the second, third, and fourth hours.

Here, too, Lobasov relied on his reading of the idiosyncratic psyche to explain puzzling results. He noted, for example, that in the experiments of other praktikanty with other dogs, sham-feeding with milk sometimes *did* elicit a strong psychic secretion. Seeking to reconcile these results with his own, Lobasov noted that these previous experiments were usually conducted on dogs that had just completed a period of relative fasting after an operation. Because these dogs were relatively hungry, he suggested, they responded with a robust psychic secretion even to such an unappetizing meal as milk. His dog had been better fed before experiments, and "the more the dog has been fed, the choosier it is about food." This reconciled earlier results with his own, which he pronounced "typical."[79]

As with the peptone experiments, a brief epilogue illustrates the difficulties of separating the presumably capricious psyche from the presumably *pravil'nye* nervous-chemical mechanisms and underlines the interpretive flexibility that this difficulty created. In 1897 a third praktikant, Andrei Volkovich, inherited Druzhok. Confronted with troubling experimental data, he suggested that even stealthily feeding an unknowing Druzhok through a cannula might elicit a psychic secretion. Seeking to explain the puzzlingly high proteolytic power of the secretory response to nonfat milk placed directly in the stomach of an apparently oblivious Druzhok, Volkovich suggested that even this procedure "is not of indifference to the dog, directing its psyche to the food; or the passing of the food along the lower portion of the digestive tract is capable of eliciting some sensations in the dog's brain, the consequence of which can be some quantity of strong [psychic] juice."[80] In other words, perhaps the psyche could be excited by processes in the stomach itself. This suggestion, which might have undermined basic elements of the laboratory view and laboratory methodology, apparently sufficed to put aside the anomalous data, but it was not pursued further.

I have here only sketched the general contours of the laboratory's experimental and interpretive confrontation with Druzhok's appetite. We have seen the difficulties involved in distinguishing between psychic and nervous-chemical mechanisms and have been introduced to the dual role of the psyche as "enemy" of precise results and "friend" in explanations of varied data. This sketch will, I hope, acquire texture and definition in Chapter 5 as I explore the relationship of these themes to the development of the laboratory's central synthetic knowledge claim.

Druzhok's Final Service

By 1897 "the remarkable Druzhok" had labored long and well, leaving a considerable legacy to the laboratory. As the first long-lived survivor of the isolated-sac operation, he had afforded Pavlov and his coworkers their closest look at the "normal" operation of the gastric glands. Sustaining the subsequent implantation of a gastric fistula in 1894 and an esophagotomy in 1896, Druzhok had permitted the laboratory to grapple with the puzzling interaction of psychic and nervous-chemical mechanisms. He had remained throughout "understanding and compliant": almost always swallowing the proffered portions of meat, bread, and milk (he had, in truth, drawn the line at nonfat milk, and after one experiment with this unpleasant substance refused milk altogether), sleeping peacefully through long experiments, and producing results that the chief was soon to immortalize in a grand synthetic work.

The dog's labors, however, were not yet over. Just as his survival of the isolated-sac operation had revolutionized one line of investigation, so did the illness that finally resulted from his operations—and from more than three years of experimental work—contribute to the launching of another.

By fall 1897 it was clear to Andrei Volkovich, the third and last praktikant to work with Druzhok, that the laboratory's prize animal was unwell. Druzhok's isolated stomach had deteriorated so severely as to render him useless for experimentation. Furthermore, his gastric glands had begun to function erratically, leading Volkovich to speculate that the abnormal manner in which the dog had been fed since his esophagotomy (through a fistula, without the benefit of appetite) had caused his glands to "gradually atrophy."[81]

The chief had assigned to Volkovich the task of comparing Druzhok's secretory responses with those of the laboratory's second dog with an isolated sac, a male setter named Sultan (I examine Volkovich's procedures and results in Chapters 5 and 6). After several months of experiments, however, Sultan, too, became sick. His illness was first manifested as a gradual increase in the volume of his secretory responses to food. Volkovich initially "paid little attention, since catarrh is frequent among laboratory dogs," but the problem worsened and eventually altered the dog's secretory response to various foods. Sultan also became "unusually greedy," devouring food indiscriminately, yet losing weight. When blood began to flow from the isolated sac, his illness was diagnosed as "an ulcer of the small stomach with hypersecretion," and this was confirmed by autopsy after he died of acute peritonitis. Volkovich speculated that Sultan's ulcer may have been caused by experiments with extremely cold and gassed milk, or perhaps from the constant irritation of the dog's mucous membrane by the rubber tube inserted into the isolated sac.[82]

In any event, the laboratory "acquired, completely unexpectedly, the opportunity to observe a clinical case in a laboratory setting—the development of an ulcerated stomach with hypersecretion." Experiments on the distorted glandular responses of the sick Sultan to various foods led to the conclusion that hypersecretion was "a neurosis affecting the peripheral endings of the reflexive-secretory nerves." This neurosis was characterized by "disturbance of the usual relationships between the psychic and reflexive phases of secretion, with the sharp predominance of the latter" and so by the disturbance of the "strict purposiveness" of glandular work.[83]

The various medical problems with Druzhok, Sultan, and other laboratory dogs inspired the chief to launch a new line of investigation: the experimental pathology and therapeutics of digestion. Pavlov saw this as a logical step, in keeping with Bernard's vision of the laboratory as a place not only for generating physiological knowledge but also for developing new clinical practices.[84] The chief was clearly excited about the possibilities of this new line of investigation, and he departed from his usual managerial practice by assigning several praktikanty to this topic simultaneously. In the years 1894–97 only two dogs, Druzhok and Sultan, had been equipped with an isolated sac for the study of normal secretory patterns; beginning in 1897 at least seven more were thus equipped for this new line of research.[85] Over the next few years, praktikanty experimented with various means of causing, diagnosing, and treating gastric ailments in dogs.[86] In a major address to the Society of Russian Physicians in 1899, Pavlov claimed that this research, however minor its practical results as yet, reflected the dawning of a new era in which the laboratory would become a central site for therapeutic developments. This task, however, proved more complex than Pavlov had envisioned, and he soon abandoned it completely.[87]

In the context of this new line of investigation, Pavlov spoke freely about the illnesses of his laboratory dogs, drawing upon these as a source of authority in discussions of pathology. For example, in the article "Laboratory Observations of Pathological Reflexes from the Abdominal Cavity" (1898) he confided that "almost all" the dogs with an isolated sac tended to lie on their back with their feet up, indicating they experienced "unpleasant or painful sensations" when in their normal posture. This comment is especially interesting because not only does it flatly contradict his assertion in *Lectures* that the operation did not result in "any sensory unpleasantness," but it also reveals information relevant to judgments about the dogs' "normalcy" that was consistently absent from prior laboratory publications. (It may also explain why the laboratory apparently did not, in these years, photograph dogs with an isolated sac.) Pavlov noted that among the "great number" of dogs with an isolated sac (about ten dogs by 1898), many presented "no serious deviations from the norm, but in several there were very sharp pathological symptoms." These animals could eat only small portions, he added, and one experienced paralysis of a paw when it consumed more. Another dog reacted to larger portions by refusing to eat, quickly reaching the brink of starvation before recovering unexpectedly. "Such phenomena appeared several times, but finally completely ceased."[88] In 1902 Pavlov intervened confidently in a clinical discussion of suppurating inflammation of the stomach, noting that "not one of the dogs upon which we have operated (and there have been very many of them) has failed to produce similar phenomena."[89] We cannot know, given available sources, which of these pathologies afflicted "the remarkable Druzhok."

Volkovich's and Pavlov's diagnoses of Druzhok's illness provide the last bit of available information for the dog's biography. The memoir literature tells us nothing about Druzhok's subsequent fate. Even if the atrophy of his gastric glands did not culminate in a fatal illness, this veteran experimental animal was incapable of living outside the laboratory without special care: he was unable to eat normally because of the esophagotomy, and leaked gastric juice from the isolated sac for hours after being fed through a fistula. Had special provisions been made for Druzhok after the end of his laboratory career, this would almost certainly have been mentioned in the memoir literature as an example of Pavlov's kindness to his experimental animals—especially in view of the antivivisectionists' attacks on him.[90]

Conclusion

Druzhok's laboratory career was the product of both Pavlov's long-standing scientific vision and the specific features of his physiology factory.

Long before 1894, as we have seen, Pavlov was committed to the investigation of the normal functioning of intact animals and to the creation of animal-technologies that would allow him to encompass normal physiological processes in the intact organism *and* achieve *pravil'nye* results. By the late 1870s he was already grappling with the relationship of the psyche to physiological processes (see Chapter 2). The nervism that so complicated Pavlov's isolated-sac operation—and distinguished Druzhok from Heidenhain's dog-technology—was also a feature of Pavlov's scientific vision from the very beginning of his career.

Yet Druzhok's career was inseparable from the resources and laboratory system of the Physiology Division. On the most basic level, some sixteen dogs perished before the isolated-sac operation was successfully performed; before 1891 Pavlov could only dream of such richesse. Druzhok's laboratory birth may also have been facilitated by the new surgical complex constructed in 1893–94 and by the new hygienic procedures developed by Pavlov's assistant V. N. Massen; and the dog's relatively long postoperative life certainly owed something to the Division's attendants and kennel facilities.[91] Most important, the line of investigation underlying Druzhok's creation and career involved the time, labor, and skills of numerous praktikanty and assistants—most directly, of Ketcher, Sanotskii, Khizhin, Ganike, Samoilov, Lobasov, and Volkovich.[92] We might imagine a scenario in which Pavlov, proceeding as a workshop physiologist with a single assistant, could have reproduced the necessary labor of all these men. Even the most generous scenario, however, would require significantly more time and would ignore the importance of specific skills (e.g., Khizhin's and Samoilov's surgical expertise) and the necessity at times for the simultaneous labor of a group of coworkers. Furthermore, as a factory physiologist, Pavlov cultivated several other lines of investigation simultaneously; these lines of inquiry and their influence on one another would also disappear in any workshop scenario.

The name *Druzhok* appears in several of the laboratory's doctoral theses but vanishes in the more processed articles of praktikanty and in the chief's synthetic works. The great value of this experimental animal, after all, resided in the generation of knowledge not just about his own psyche and glandular responses but about the digestive factory in all higher animals. We turn next to the process by which results with the laboratory's first template dog were shaped and presented to that end.

Chapter 5

FROM DOG TO DIGESTIVE FACTORY

The outer limit of physiological knowledge, its goal, is to express this infinitely complex interrelationship of the organism with the surrounding world in the form of an exact scientific formula.

— IVAN PAVLOV, "Lectures on Physiology"

A theory can be thought of as the fitting of a curve to a spray of data. One can always simply go from point to point, connecting the dots like those in a child's coloring book. But all that is left is a meandering line with little explanatory power; there is no way to predict how future points are likely to fall. Science is the search for neat, predictable curves, compact ways of summarizing the data. But there is always the danger that the curves we see are illusory, like pictures of animals in the clouds.

— GEORGE JOHNSON, *Fire in the Mind* (1995)

In December 1894, shortly after Pavel Khizhin had completed his dissertation, Pavlov addressed the Society of Russian Physicians about his laboratory's findings and their implications for medicine. This speech differed profoundly in tone and content from that which he had delivered nine months before, when he and Khizhin had been struggling unsuccessfully with the isolated-stomach operation. At that time, Pavlov had urged the physicians in his audience to appreciate the potential of the chronic experiment and to help him overcome repeated failures with the isolated sac. Now, emboldened by Khizhin's successes, he spoke as a confident factory physiologist. The chief invoked years of laboratory research, the names of sixteen coworkers, and, especially, results with "the remarkable Druzhok" in a systematic analysis of the digestive system and the relationship between experimental physiology and medical practice. He was fairly bursting with results, research perspectives, and general observations, and his speech proved too long for a single Society session (he completed it at the January 1895 meeting).[1]

In his December address, Pavlov introduced the "single idea" that now imparted a general direction to the interpretive moments in laboratory research: the digestive system was a "complex chemical factory" that responded precisely to the requirements for processing various foods. There could be no question, he assured his audience, that investigations of this factory would uncover "the very same subtlety and adaptiveness of work" that characterized other parts of the animal machine.

The most impressive empirical expression of this subtlety and adaptiveness was the "characteristic secretory curves" that described the distinctive responses of the main digestive glands to different foods. The factory metaphor and these curves had emerged together in the course of Khizhin's research and would develop together over the next three years, attaining their most refined expression in the chief's *Lectures on the Work of the Main Digestive Glands* (1897). Pavlov's central synthetic idea, then, can quite literally be discussed as "the fitting of a curve to a spray of data."[2] In this chapter, I explore this process of "fitting" and the curves and conceptions it involved.

The Digestive System and Other Factories

We cannot know precisely when Pavlov first formulated the factory metaphor, but the language pattern in Khizhin's dissertation points toward the final months of that praktikant's labors in mid-fall 1894. Throughout most of his text, Khizhin employs the laboratory's traditional lexicon, referring to the "phenomena" or "exciters" of gastric secretion. A new language emerges, however, in the general introduction, the conclusion to one critical section, the general conclusion, and the title of the dissertation—passages that, I suggest, were drafted or redrafted in consultation with Pavlov as Khizhin's labors drew to a close. Here Khizhin refers not to "phenomena" or "exciters," but to secretory "work" (*rabota*), as in the title of his thesis: *The Secretory Work of the Stomach of the Dog.*[3] From this time forward, the word *work* entered the standard lexicon of laboratory publications, most famously in Pavlov's *Lectures on the Work of the Main Digestive Glands*.[4]

This word represented the tip of the factory metaphor that Pavlov introduced in his speech to the Society of Russian Physicians in December 1894.

> The digestive canal is in its task a complex chemical factory (*zavod*). The raw material passes through a long series of institutions (*uchrezhdeniia*) in which it is subjected to certain mechanical and, mainly, chemical processing (*obrabotka*), and then, through innumerable side-streets, it is brought into the depot of the body. Aside from this basic series of institutions, along which the raw material moves, there is a series of lateral chemical manufactories (*fabriki*), which prepare certain reagents for the appropriate processing of the raw material.
>
> Anatomy and physiology have disassembled this factory into its component parts and have become acquainted with the significance of each. These lateral manufactories, which determine the function of the digestive canal, are essentially the glands and their ducts. Physiology has been preoccupied with the reagents prepared by separate chemical manufactories; it has become acquainted with their properties, has shown their relationship to the various constituent parts of food. Of course this knowledge is an enormous acquisition. But does it suffice for the physician, whose task is to fix this factory when it is damaged? Obviously not, for this is mere analytical data. What is the activity of this factory at full operation, how and by what is it brought into motion, in what manner does one part go into operation after another, in what manner does the work change in dependence upon the type of raw material, does the entire factory always operate with all its parts, or not? All these questions and many others arising upon inspection of our factory are, of course, not even close to resolution; many have barely been posed. Nevertheless, one cannot doubt that in the investigation of this subject we will find the very same subtlety and adaptiveness of work that strike us in other, better-studied areas of physiology.[5]

The factory metaphor represented the chief's considered interpretive response to Khizhin's results with Druzhok, as pondered together with the results of other laboratory research on the gastric and pancreatic glands. This metaphor had three general implications for laboratory investigations and interpretations.

First, the factory metaphor placed an emphasis on the digestive apparatus as a precisely coordinated system operating toward a single end. As is clear from the above extract from Pavlov's speech, this perspective defined a series of questions about the operation of the system. The subtle and adaptive work of the digestive factory required a subtle and adaptive coordinating mechanism. In Pavlov's thinking, the only conceivably satisfactory mechanism was the nervous system with its property of specific excitation. In its role as coordinator of the digestive factory, the nervous system, then, came to bear increasing explanatory weight as the often confusing, even paradoxical, data from chronic experiments piled up. As a result, during the interpretive moments in experimental work the nervous system acquired a number of hypothesized structures and properties.

Second, the factory metaphor expressed Pavlov's Bernardian determinism in a new, refined, exacting manner, channeling the search for *pravil'nye* ("regular" and "correct") results in specific directions. Long before 1894, as we have seen, Pavlov shared Claude Bernard's vision of the organism as a purposive, fully determined, specifically biological machine in which apparent spontaneity testified only to complexity and the "numberless factors" that veiled causal relations. What, precisely, did the factory metaphor add to this? Most important, it focused attention on the specific purposive, determined relation to be uncovered experimentally: that between the raw material and its processing—that is, between the ingested food and the responses of the glands to it.[6] This, in turn, put "pressure" on laboratory data, encouraging the quest for determined and precise secretory patterns.

This search for precise, "factory-like" patterns in the varying data generated by chronic experiments gave rise to a certain tension between Bernardian principles and Pavlov's experimental practices. As we saw in Chapter 2, Bernard had firmly rejected the use of average results or statistical techniques; this, he argued, was incompatible with the physiologist's determinist creed. In Bernard's narratives about scientific method, experimental trials always ended with a definitive, repeatable conclusion that reconciled earlier "positive" and "negative" results. These narratives, however, always involved questions that could be answered with a simple yes or no. Pavlov's factory metaphor pressed him to provide similarly definitive responses, but to a very different kind of question and using much different kinds of data. For all the laboratory's experimental ingenuity, chronic experiments on intact, complex dog-technologies always yielded data that varied to some extent from experiment to experiment. Especially as the factory metaphor developed, these data were all presumed to have significance, to require a "quite definite" explanation. The identification of *pravil'nye* results, then, always involved a comparison of varying data and a judgment about the relative similarities and differences in different experimental trials.

Pavlov did so without an articulated statistical method, which had not yet found a place in physiology and which, in any case, would have been difficult to incorporate within his scientific vision.[7] The tension between Pavlov's Bernardian ideal and the actual nature of his experimental results was, as we shall see, expressed in a homegrown mathematical logic/rhetoric through which the average data of varying trials were interpreted and presented as if they represented the results of a single, ideal experiment.

The pressure of the factory metaphor on the interpretation of experimental data is already evident in a subtle difference between Khizhin's characterization of glandular secretion in mid-fall 1894 and Pavlov's formulation (based on the very same data) a few months later in his speech to the Society. Khizhin, as we have seen, openly acknowledged the tentative nature of his conclusions, and in his formulation "the secretory work of the stomach possesses to a great degree the ability to adapt to various types of food." For Pavlov, however, "for each food there exists its particular work of the digestive factory."[8] The subsequent maturation of the factory metaphor is evident in Andrei Volkovich's then-standard formulation of 1898. "The work of the gastric glands," wrote Volkovich, "is distinguished by strict precision, lawfulness, and purposiveness, expressing the greatest capacity of adaptation to various types of food."[9] As we shall see, successive interpretations of experimental data reflected the corresponding tendency to find unvarying, stereotypical patterns in the varying results of chronic experiments.

A third, related consequence of the factory metaphor was that it highlighted the contradictory relationship between the idiosyncratic psyche and the *pravil'nyi* digestive machine. Previously, Pavlov could be true to his Bernardian determinism by simply recognizing the psyche's important role in shaping glandular responses to food and using this acknowledged importance to interpret experimental data. There was no need to confront the nature of the psyche, which was simply "black-boxed" as an important, poorly understood factor that might even lie outside the bounds of determinist physiology. The factory metaphor, however, incorporated the psyche fully within the digestive machine, transforming the long-appreciated relationship between the idiosyncratic psyche and the *pravil'nyi* digestive system into a contradiction. The acknowledged importance of the psyche remained a source of interpretive flexibility in laboratory trials, but an uneasy dualism developed—not between mind and body per se but between *pravil'nost'*—"regularity" or "correctness"—and capriciousness. How could a factory haunted by an eccentric ghost be purposive, precise, and regular? What did the power of this ghost leave for the nervous system, which supposedly imparted *pravil'nost'* to the digestive machine?

The image of factory production so permeated late nineteenth-century discourse that its appeal, especially during Russia's industrial revolution of the late 1880s and 1890s, requires little explanation. I should note, however, that this metaphor was more appealing to some Russian intellectuals than to others and had various possible meanings. Many Russian thinkers, on both the monarchist right and the populist left, saw the emergence of huge factories in St. Petersburg and other large cities as an ominous sign of western-style capitalism with its attendant evils (most commonly, the emergence of an urban proletariat, class struggle, and the destruction of cottage industry and Russia's communal social fabric).[10] From his days as a youthful admirer of Dmitrii Pisarev and Samuel Smiles, Pavlov, however, had a westernizing bent, and he instead associated factories with a set of positive attributes: with precise, powerful, efficient, effectively coordinated, modern production for a particular goal.

Pavlov apparently had no direct knowledge of factories, but his longtime acquaintance Dmitrii Mendeleev did.[11] Best-known in the West for his periodic table of the elements, Mendeleev also served, from 1893, as chief of Russia's Bureau of Weights and Measures and was a leading authority on economic, industrial, and technological developments. In this capacity, he wrote the article "Factories" in the twelfth volume of the authoritative Brokgauz and Efron encyclopedia, which appeared in 1894—on the eve of Pavlov's landmark speech to the Society of Russian Physicians.[12] Several nuances common to Mendeleev's article and Pavlov's speech suggest that Pavlov may have drawn upon the article (or, perhaps, on a personal conversation) in his conceptualization of the digestive factory.

In his article, Mendeleev emphasized the distinction between manufactories (*fabriki*), in which raw material was subjected to primarily mechanical alterations, and factories (*zavody*), which relied primarily on chemical processes that produced more profound "molecular transformations of substances." The latter, Mendeleev emphasized, were the more advanced, making chemical factories the cutting edge of industrial progress. This same distinction is evident in the passage from Pavlov's 1894 address, quoted above, in which he describes the glands themselves as "manufactories" and the digestive system as a whole as a "factory." (The glands simply produce digestive juice, whereas the actual chemical alteration of food occurs in the digestive canal.) Pavlov further noted that the digestive canal is a "complex chemical factory" in which raw material (food) "is subjected to certain mechanical and, mainly, chemical processing." According to Mendeleev's criterion, then, the digestive system matched the most advanced type of factory in existence.[13]

Mendeleev also observed that modern factory production differed from craft production, not only in the number of workers involved but also in "the presence of particular (specialized) knowledge demanding preliminary preparation, and also of machines and apparatuses that act, although under the guidance of people, mainly by means of the forces and phenomena of nature [as harnessed by science and technology], . . . for example, by means of furnaces, steam machines, and so forth."[14] For Pavlov, the controlling apparatus in the digestive factory, and the repository of its "specialized knowledge" (in the form of specific excitability), was the nervous system. As he explained in 1894, the precision of glandular processes was "of course possible only through the participation of the nervous system, of this regulator, coordinator, of the activity of various organs."[15]

It bears emphasis that Pavlov's notions about the factory were based not on any actual experiences but entirely on an idealized image. The factories of the time were rarely so regular and precise as Pavlov imagined; their workers and managers struggled with the same difficulty in obtaining *pravil'nye* results as did Pavlov and his laboratory coworkers.[16] Yet it was Pavlov's metaphorical, ideal image of the factory that guided him conceptually and that he glimpsed in the data generated in his laboratory. For Pavlov, only Bernard's "numberless factors" prevented experiments from yielding these same results with recognizably factory-like precision and regularity.

Pavlov did, one might argue, have experience with one factory—his own laboratory (though this is my own formulation and emphatically not his). The factory metaphor did, I think, gain commonsensical power for Pavlov from his own enterprise, which, as we have seen, he strove to make purposive, regular, and precise. Furthermore, by setting before Pavlov a "panoramic view" of results on different glands, his laboratory enterprise afforded him a view of digestive mechanisms as a system. So, for example, whereas Khizhin had demonstrated (at least temporarily) that peptone was the specific exciter of the gastric glands, another praktikant, Ivan Dolinskii, had discovered that the hydrochloric acid produced by the gastric glands served, in turn, as an exciter of the pancreatic gland. This made "the joint activity of various parts of the chemical factory" much more striking than it would have been had Pavlov been a lone investigator addressing one gland at a time.

We can appreciate another dimension of the factory metaphor by returning to the language of Khizhin's dissertation. The same passages that employ the word *work* also introduce another term related to the factory metaphor: the secretory response to different foods was not *shablonnyi* and digestion "is never accomplished according to any *shablon*."[17] A *shablon* was a template, the "carved form of a pattern or generally of curved shapes" used to produce the same shape time after time; the adjective *shablonnyi* (adverb, *shablonno*) was commonly used to describe inflexible, clichéd human responses to varied situations.[18] Khizhin's point was that the secretory curves produced by different foods were not produced by a single template; rather, they differed for different foodstuffs. In his thesis, Khizhin described the work of the digestive glands most frequently (seven times) as *pravil'nyi* and twice as proceeding *zakonno* (lawfully).[19]

Khizhin (and Pavlov) saw these qualities as expressing a deeper truth about glandular work: its *purposiveness* or, in more contemporary biological language, its *adaptation* (*prisposoblenie*) to the requirements for efficiently digesting any meal.[20] As we saw in Chapter 2, this interpretive framework reflected Pavlov's adaptationist views—views with deep roots both in Bernardian physiology and in Russia's evolutionist tradition.

The word *adaptation* was also usefully ambiguous. Beginning with Khizhin, it was employed in laboratory publications to imply that the different secretory curves expressed the purposive adjustment of glandular secretions to the differing chemical requirements for the optimal digestion of various foods. When this broad claim was challenged, the term *adaptation* was defended more narrowly as an expression of the narrower empirical claim that the glands simply responded differently to different foods. When Khizhin presented his findings to the Society of Russian Physicians in fall 1894 and emphasized the adaptiveness of the glands, one physician in the audience objected: "You used the word 'adaptation.' Why is this 'adaptation'; what advantage is there to the organism in the fact that with bread there is secreted a small amount of thick [i.e., pepsin-rich] juice, while with meat it is thinner and of a greater quantity; why is this advantageous—perhaps, to the contrary, it is not advantageous, perhaps there is here no adaptation?"[21] This concluded a long series of questions, directed mostly at experimental methodology and the reliability of the isolated sac, that this physician had posed.

Khizhin responded to the challenges about experimental issues, but the chief himself addressed the issue of adaptation in his comments a few moments later: "As for purposiveness—this is philosophizing. We are not talking about whether something is useful or not useful, but are, rather, showing an existing fact. And here the first lesson is that the stomach does not work *shablonno,* as one might think from textbooks. Its work is undoubtedly strictly adapted to each given type of food: for bread there is a special juice, for meat its special [juice]. We have before us the fact of the extremely subtle relationship of the stomach to the details of the digestive task; as for why this occurs—

this is another question that it remains to explain."[22] Here Pavlov fended off the physician's objection by defending the narrow definition (the "existing fact") of *adaptation*. His choice of words, however, reveals the deeper meaning of this fact for Pavlov: given his adaptationist views, the very existence of an identifiable pattern testified to an underlying purpose.[23]

It is an interesting reflection of the division of intellectual property in the laboratory that, although the factory metaphor clearly structured the interpretation of laboratory data and the language in which results were reported, this metaphor was never expressed directly in the praktikanty's reports and articles. Experiments belonged to the praktikanty, but this all-encompassing expression of the laboratory's basic view, which united experiments and gave them meaning, belonged to Pavlov himself. Introduced to the Society of Russian Physicians in his speech of December 1894, the explicit statement of the factory metaphor was reserved for Pavlov's public lectures to medical audiences and for his statement of the grand vision in his published *Lectures* of 1897.

The factory metaphor, then, emerged at the final stages of Khizhin's work and matured in subsequent years, problematizing the relationship of the capricious psyche to the *pravil'nyi* digestive machine, shaping and intensifying the search for factory-like experimental results and, inevitably, influencing the identity, construction, and meaning of the "characteristic secretory curves." We turn now to the dynamics of this last process.

Curve Construction

After acquiring "downright beautiful" results in his peptone experiments, Khizhin worked closely with Pavlov on a series of experimental trials in which he fed Druzhok various meals and measured the quantity and quality of the secretory responses in the dog's isolated sac (see Chapter 4). The main result of these trials was the construction of *characteristic secretory curves*.[24] In the original formulation, these curves dramatized the ability of the gastric glands "to adapt to various types of food" and became the conceptual and rhetorical cutting edge of Pavlovian digestive physiology.

Like the "normalcy" of experimental animals, the characteristic secretory curves were presented to the consumers of laboratory products as simple empirical facts, but they actually represented a series of interpretive and rhetorical decisions. The very process of the "fitting of a curve to a spray of data" incorporated the developing laboratory view of the operation of the digestive glands. Not surprisingly, then, as this view developed between Khizhin's initial work with Druzhok in 1894 and the chief's *Lectures* of 1897, the identity, main features, and meaning of these curves changed as well.

These curves originated in the data from a relatively small number of experiments on Druzhok. In his doctoral thesis, Khizhin reports that he experimented on Druzhok's response to normal feeding in the following trials: (1) five experiments each with 100, 200, and 400 grams of meat; (2) five experiments with 200 grams of bread; (3) three with 600 cubic centimeters of milk; and (4) forty-three with various quantities of mixed food.[25] ("Mixed food" was soon discarded as a meaningful category and these data dropped out of laboratory publications, never appearing in Pavlov's *Lectures*.) Khizhin conducted only three trials with milk because Druzhok disliked that substance and refused to ingest it altogether after his first encounter with the nonfat variety. As we shall see, the conduct and interpretation of trials with bread were also complicated by the food tastes of the laboratory's template dog.

Even identical trials with the same quantity of the same food yielded varying (sometimes widely varying) secretory results, and this, in turn, lent both an interpretive and rhetorical dimension to curve construction. As we have seen, Pavlov and his coworkers were fully aware that the results of any trial were influenced by Druzhok's personality and mood, yet they sought to find in the quantitative results of their trials a *pravil'nyi* pattern. This necessarily involved judgments about the relative importance or unimportance of similarities and differences between the data for the quantity and proteolytic power of secretions during various trials. Physiology at this time had no established traditions for making such judgments. Physiologists were largely unacquainted with the development of statistics and, in any case, it remained for them to negotiate the tension between statistical thinking and the particular determinist ideal enunciated by leading figures in their discipline. (Even in 1929, Halbert Dunn, having reviewed two hundred articles on physiology and medicine in U.S. journals, concluded that "in over 90 per cent [of the articles] statistical methods were necessary and not used.")[26] Unconstrained by a formal statistical methodology, Pavlov (no doubt like others in the same situation) developed his own, homegrown approach to making these judgments.

Curve construction entailed not only a particular logic but also an important rhetorical dimension.[27] As a rhetorical form, curves dramatized similarity and difference, significance and insignificance. As one praktikant put it, the curves "make it possible to distinguish the primary from the secondary, and to accent the general main features that primarily characterize the type of secretion with one or another food."[28] An understanding of the "main features" of the "meat curve," for example, revealed the essential identity between the results of two experimental trials with meat that might otherwise seem quite different. Similarly, what might otherwise seem only a slight difference between one trial with meat and another with bread could be considerably elevated in importance when judged against the background of the characteristic curves (Figures 13 and 14).

We can appreciate the general interpretive issues inherent to curve construction by asking ourselves the same question that Khizhin and Pavlov asked themselves in 1894: which of these curves are "essentially" the same, and which are different? For Khizhin and Pavlov in 1894, all the curves in Figure 13 were essentially the same, whereas those in Figure 14 fell into four subgroups: a, b, and c; d; e; and f and g. Especially in the absence of a single, conventional statistical methodology, one can choose a number of plausible groupings (or none at all), depending on which aspects of the curves one chooses to construe as important, how much "stretch" one is willing to grant within a category, how little stretch one allows between categories, and how much confidence one has in the data.

We can further appreciate the subjective dimension of such judgments by considering the seven curves in Figure 15. For Pavlov, in 1897, these fell clearly into three pairs (I have added one curve as a decoy). I suggest the reader attempt to group these curves, as this will highlight the interpretive moments involved. (Pavlov's groupings are provided in the Notes).[29] Such interpretive moments, always shaped by prevailing laboratory doctrine, were inherent to curve construction in the laboratory. Pavlov emphasized quantitative precision—and laboratory publications were permeated by numbers, curves, and various formulae—but the nature of and criteria for this precision were themselves worked out as experimental and interpretive tasks unfolded and were highly sensitive to interpretive decisions.

This is not to say that these groupings were plucked from thin air. Consider Khizhin's discussion of the curves in Figure 13 (amount of secretion). In his view, the total quantity of secretion varied according to the type of food ingested. A unit of weight of each type of food excited a different, "determined amount of juice secreted during its digestion."[30] Meat elicited the greatest amount of secretion per weight, followed by bread and then milk. The course of secretion, however—that is, the shape of each curve—was essentially the same for all food substances. In each case he saw "one and the same *pravil'nyi* course, which can be expressed in the form of a curve that attains its acme either during the first hour of the digestive act (the 'psychic type') or only in the 2nd or 3rd hour ('usual type'). Immediately after reaching its acme, this curve

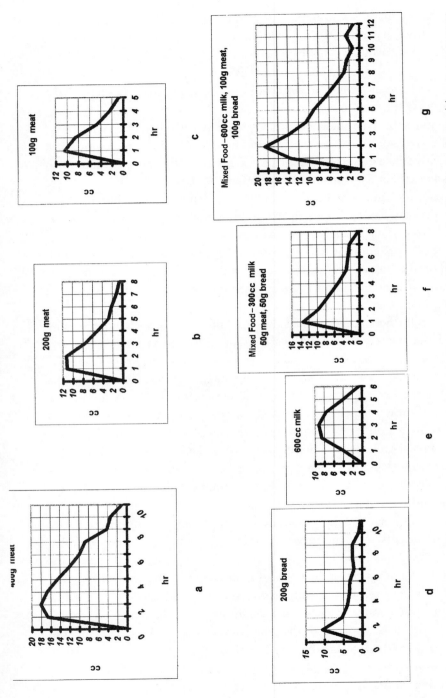

FIGURE 13. *Khizhin's characteristic curves (from average data) for amount of gastric juice secretion (in cubic centimeters) in response to meat, milk, bread, and mixed food, in the quantities indicated. From P. P. Khizhin, Otdelitel'naia rabota zheludka sobaki, Military-Medical Academy Doctoral Dissertation Series (St. Petersburg, 1894), appendix; based on average data on p. 65 (f & g), p. 71 (a–c), p. 88 (d), and p. 93 (e)*

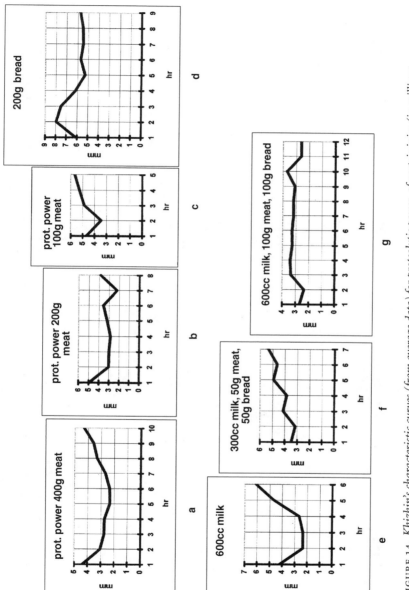

FIGURE 14. *Khizhin's characteristic curves (from average data) for proteolytic power of gastric juice (in millimeters, as measured by the Mett method) in response to meat, milk, bread, and mixed food, in the quantities indicated. From P. P. Khizhin, Otdelitel'naia rabota zheludka sobaki, Military-Medical Academy Doctoral Dissertation Series (St. Petersburg, 1894), appendix; based on average data on p. 65 (f & g), p. 78 (a–c), p. 88 (d), and p. 93 (e).*

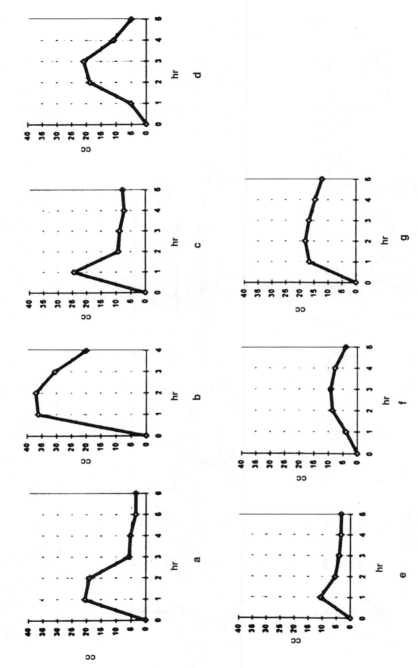

FIGURE 15. *Find the three pairs of stereotypical secretory curves. The three pairs (six curves) are from data in A. N. Volkovich, Fiziologiia i patologiia zheludochnykh zhelez, Military-Medical Academy Doctoral Dissertation Series (St. Petersburg, 1898), 25.*

begins to fall. This decline, accomplished gradually and progressively in the course of the remaining hours of the digestive act, finally attains 0" (pp. 116–17).

From this perspective, all the curves in Figure 13, as Khizhin finally concluded, "have much in common" (p. 117). The secondary differences among them—for example, the differing hours in which the curves reach their peak—do not distinguish the curves for different foods. For example, the curves for 100 grams of meat (c), 200 grams of bread (d), and 400 grams of mixed food (f) all reach their peak in the first hour; and those for 400 grams of meat (a), 800 grams of mixed food (g), and 600 cubic centimeters of milk (e) all reach their peak in the second or third hour. For Khizhin, this secondary difference among the curves resulted from a mere idiosyncrasy: the varying intensity of Druzhok's psychic response to different meals and so the varying amount of high-quantity, high-proteolytic power psychic secretion that these meals elicited. When Druzhok responded enthusiastically to a meal, the resultant psychic type of secretion reached its peak in the first hour; when he did not, a usual type of secretion peaked in the second or third hour. (I return to this point below.) Other differences between these curves—for example, their relative steepness and their variations in the fourth hour—are not considered significant here. Still others—for example, variations within any hour—are invisible, since the curves are plotted with hourly results (although measurements were taken every fifteen minutes).

The curves in Figure 14 describe the course of proteolytic power, and here Khizhin's analysis tended in precisely the opposite direction. That is, while he emphasized the similarities among the curves showing amount of gastric secretion produced by different foods (Figure 13)—constructing a common, *pravil'nyi* course of the quantity of secretion—he emphasized the *differences* among the curves showing the proteolytic power of secretions for different foods (Figure 14). This different interpretive direction reflected the state of the "laboratory view" at the time. For Khizhin, the curves in Figure 14 described the most important of the "sharply expressed distinctions" in the glandular response to different foods: the characteristic level and course of proteolytic power in the gastric juice elicited by each. These curves are constructed from the average hourly data from experiments measuring the proteolytic power of Druzhok's secretory response to various meals.[31] According to Khizhin, both the level of proteolytic power and its fluctuations during the digestive act—that is, both the height and the shape of the curves in Figure 14—were distinctive and characteristic for each food. So, although there are some secondary differences between curves a, b, and c, these "meat curves" are "essentially" the same and differ fundamentally from the "bread curve" (d), the "milk curve" (e), and the "mixed-food curves" (f and g).

"To each type of food," Khizhin explained, "there corresponds its completely special [level and] course of the proteolytic power of juice during the digestive act; this course is in each case so different from that for other foods, and is repeated with such consistency in experiments with one and the same type of food, that we are compelled to consider it characteristic for the given type of food" (p. 118). So, for example, 200 grams of each food produced gastric secretions of different proteolytic power, measured in millimeters by the Mett method (see Chapter 4): an average of 4.0 for meat, 6.16 for bread, and 3.05 for milk (p. 108). Furthermore, the hourly curves for proteolytic power obeyed different rules for different foods. For example (as is evident in curves a, b, and c in Figure 14), "with raw meat, in all the quantities we tested, the course of proteolytic power manifests a single general characteristic—the decline of the curve from the very beginning of the act of digestion; with small quantities of meat and short acts of digestion this decline is limited to the second hour—there follows a rise to a point that even surpasses the initial level; with a large quantity of meat and with longer acts of digestion, the duration of this initial fall of the curve is also increased and its rise is put off further and further, occurring usually in the last hours of the digestive act" (pp. 81–82). The curve of proteolytic power for bread (d) also had its "special character." It was very high in the first hour, rose even higher in the second, remained at that high level throughout the third and sometimes the fourth hours, declined only in the fourth and fifth hours (during which time it lost "about 15 percent" of its height), and retained that level from the sixth hour until the end of the digestive act (p. 91). The curve for milk (e) was "striking" because it was "completely opposite" to that for bread. Milk elicited juice of less proteolytic power than did any other food, and bread elicited the highest. Furthermore, the shape of the milk curve seemed precisely opposite to that of the bread curve: it began at a relatively high level, falling over the next hour to one-half its previous value and remaining at that lower level for the third and fourth hours; in the fifth hour it regained its previous height, maintaining that level even in the sixth hour. The sharp differences between the bread and milk curves raised the possibility that these represented "opposite poles of activity of the glandular apparatus of the stomach" (p. 95).[32]

Perhaps the reader did not initially group the curves in Figures 13 and 14 in the same manner as did Khizhin and Pavlov, but will now grant that their groupings are at least plausible. (This has been my experience and that of most people whom I have victimized with this exercise.) One may not be convinced, but at least one can "see" it. This illustrates the rhetorical dimension of the curves. They may not fall immediately into the Pavlovian groupings, but once the principles for grouping have been explained, these, combined with the curves' striking visual power, acquire some force. To that extent, the curves acquire "stretch value," enabling them to encompass results that differ in their particulars. As we shall see, the striking visual power of a curve, as opposed to a table of data, made the decision about which data to convert to curves (and which to leave as a mere table) a rhetorical judgment of some consequence.

Khizhin appended the characteristic curves to his thesis and invoked them constantly to help his readers properly "distinguish the primary from the secondary." In his report to the Society of Russian Physicians, he provided a visual display of these curves to convince his audience about his groupings. In his most highly processed literary product, the edited form of his doctoral thesis that appeared in the French edition of the Institute's journal, *Arkhiv Biologicheskikh Nauk* (Archive of the Biological Sciences), the curves were moved from the appendix to the text so as to better utilize (I think) their rhetorical power.

Let us now go a bit deeper and examine the interpretive decisions involved in the construction of these curves. These decisions are virtually invisible in the more processed literary products—in Khizhin's report to the Society and in Pavlov's *Lectures*—but Khizhin candidly acknowledged several of them in the body of his thesis, and still others are evident from the data he provided. Two important construction tools were a mathematical logic/rhetoric of "average and instantiation" and the interpretation of experimental results in the light of judgments about the dog's psyche and mood.

Khizhin's logic/rhetoric of average and instantiation took the following form. A series of experiments was conducted on the secretory responses to bread, meat, and milk. The results varied considerably, even among experiments with the same food. Khizhin then produced a summary chart of average results for each food, used these to construct a curve, and illustrated that curve with the results of a single "model" (*obrazets*) experiment. He explained and justified this procedure as follows: "This conclusion has been constructed (*postroen*) by us upon investigation of average arithmetical values, but it is also completely justified by data acquired through direct experiments."[33] This reporting procedure had a clear rhetorical content: it minimized the importance of varied results within a category (say, proteolytic power for meat) and focused attention on differences between them (say, between the proteolytic power for meat and bread). The difference was then dramatized by a model experiment that best illustrated the pattern. This reporting procedure also, of course, minimized the reader's ability to compare variation within a category with variation between categories. Khizhin's (and Pavlov's) justification of these curves on the basis of both average data and a single model experiment reflected the tension between Pavlov's determinist ideal and his resort to average data made necessary by the varied results of individual trials.

Khizhin frequently discarded or downplayed aberrant experiments and emphasized others "instantiating" and even exaggerating what he perceived as underlying patterns. For example, he dismissed the results of one trial on the secretory response to meat that showed a decline in proteolytic power during an hour when, according to the characteristic curve, it should rise. He explained that "one cannot ascribe to it decisive significance, since this decline occurred in only one experiment in five."[34] Conversely, however, he highlighted the results of another experiment on the secretory response to mixed food that showed an unusually pronounced distinction between two hourly periods that, according to the characteristic curve, were indeed distinct. Here he explained that "in one of our experiments . . . this distribution of the rapidity of the flow . . . attained the greatest degree of its expression" (p. 89).

We can appreciate the importance of such interpretive decisions by briefly exploring several examples of the construction of similarity and difference. Let us first consider Khizhin's comparison of the quantity of secretion in trials with Druzhok and with the short-lived Gordon. Khizhin found a "remarkable *pravil'nost'*" in the secretory responses of each dog; but the central problem was that this *pravil'nost'*—that is, the shape of the secretory curve—was significantly different for each animal (pp. 56, 59–60). For both dogs, the curve expressing the amount of secretion peaked in the first or second hour after feeding, then fell gradually to zero. However, for each dog, the curve reached its peak either in the first *or* in the second hour; and Gordon's curve usually peaked in the first hour, whereas Druzhok's usually did so in the second.[35]

Khizhin solved these problems by invoking the power of the psyche and defining two different types of secretory response: the "psychic type" (which reached its acme in the first hour) and the "usual type" (which did so in the second). The psychic type of secretion occurred "only in cases when the dog eats especially greedily, in a dog that has been especially starved or tempted by especially tasty food, when its representations of food are so lively and sharp that the mucous membrane of the stomach is already in an active state before arrival of the food."[36] Gordon had produced a psychic type of secretion in all eleven experiments because these were conducted before he had fully regained his "physiological balance" after his operation and the monotonous diet that followed it. Therefore, "he always ate with special greediness."[37] Druzhok, on the other hand, had manifested a psychic type of secretory pattern in only thirty-six of seventy-eight experiments. These thirty-six trials, Khizhin noted, were distributed unevenly across four months of experiments, constituting 61 percent of the trials in the first month, 52 percent in the second, 36 percent in the third, and 35 percent in the fourth. This *pravil'nost'* enabled Khizhin both to explain the results with Gordon and to reconcile them with those obtained on Druzhok: as Druzhok regained his physiological balance (and so became less excitable), he increasingly manifested the usual type of secretory pattern.[38]

Khizhin, then, used the interpretive flexibility afforded by judgments about normalcy and the idiosyncratic psyche to construct from data that varied widely, both between dogs and among trials with one and the same dog, a "remarkable regularity" attesting to the *pravil'nyi* work of the glands. A secretory pattern manifested by Druzhok in forty-two of seventy-eight experiments— and not once in twelve experiments with Gordon—was enshrined as the "usual type." As we shall see, Khizhin's identification of "psychic" and "usual" secretory patterns, like his "clocklike" results with peptone, soon reached a denouement that further illustrates the interpretive moments inherent to curve construction.

These are also evident in Khizhin's construction of the meat curve. In his thesis, Khizhin reported that he had conducted five experiments each with 100, 200, and 400 grams of meat. The results of these experiments, although recorded in the "Druzhok" laboratory notebook, do not appear in Khizhin's dissertation.[39] Here (as with other foodstuffs) Khizhin provides only the average results of all trials taken together and the "minimum" and "maximum" results for each series. That is, taking together the five experiments with each quantity of food, he tells the reader only (1) the average amount of secretion produced during each hour, (2) the largest amount of secretion produced in any single experiment during each hour, and (3) the smallest amount of secretion produced in any single experiment during each hour. This form of presentation, again, minimizes the reader's ability to compare the results of individual experiments and so to judge the relationship between the characteristic curve and the results of concrete trials. Even these incomplete data, however, reveal a considerable range in results.[40] For example, in the five experiments with 100 grams of meat, the amount of secretion during the first hour varied by about 250 percent (ranging between 4.9 and 13.1 cubic centimeters), in the second hour by about 45 percent (between 6.9 and 10.0 cubic centimeters), and in the third hour by 300 percent (between 2.2 and 6.6 cubic centimeters).[41] By using average data, however, Khizhin fashioned a single characteristic curve for the amount of secretion in response to meat.

For Khizhin, this same characteristic curve described the amount of secretion for other foodstuffs as well, and the psyche again provided a ready explanation for aberrant differences among them. For example, noting that the meat curve more consistently reached its peak during the first hour than did the curve for mixed food, Khizhin attributed this to the "decisive predominance" of the psychic type of secretion in experiments with meat. In other words, Druzhok preferred meat to mixed food, and the resultant psychic type of secretion in trials with meat obscured the "essential similarity" between the dog's secretory responses to these two meals. Once the influence of the psyche was taken into account, Khizhin argued, the quantitative aspect of the secretory response to meat preserved "entirely the same lawfulness in the course of its secretion as was demonstrated by us for mixed food, and the resemblance is here so great that there is almost a repetition of one and the same figures."[42]

Druzhok's particular response to bread also created interpretive challenges and flexibility. Experiments with bread were complicated by Druzhok's passion for this food: the dog "always ate bread with special satisfaction; pricking up its ears and attentively following each movement of the experimenter, it greedily seized each piece offered and excitedly awaited the next." This compelled a change in feeding procedure. "In order to give the dog the opportunity to thoroughly chew and swallow the relatively dry pieces of bread, we usually did not hasten to offer it piece after piece, and so the feeding lasted from three to four minutes." Perhaps because this procedure was a lengthy affair, Khizhin apparently conducted experiments with only one quantity of bread—200 grams.[43]

The results, again, ranged widely.[44] But more interesting is a subtle difference in Khizhin's interpretation of the data for quantity of secretion and the data for quality of secretion. To appreciate the significance of this interpretive move, the reader must keep in mind that, according to the laboratory view at the time, the curves expressing the quantity of secretion were essentially the same for all foods, whereas the curves expressing the quality (i.e., the proteolytic power) of secretion were fundamentally different for various foods. The quantitative results of the bread experiments, however, showed a sharp peak in the first hour, which threatened to distinguish the bread curve from the meat curve. Khizhin explained this difference as an artifact of his feeding method. Feeding Druzhok small pieces of bread over three or four minutes kept the dog in a constant state of excitation, "as if we were teasing him," and this produced a "quasi-psychic type of secretion" that gave the bread curve a sharp peak in the first hour and so seemed to distinguish it from the meat curve. Once this deceptive result of psychic excitation was taken into consideration, however, the quantitative curve for secretion during a meal of bread "generally presents identical characteristics to that with meat and mixed food."[45]

Significantly, Khizhin did not invoke this quasi-psychic type of secretion when interpreting the data on proteolytic power from these very same trials. According to laboratory doctrine, this quasi-psychic response would have raised not only the quantity of secretion (which Khizhin corrected for) but also the level of proteolytic power (which he did not correct for). Why this asymmetry in Khizhin's interpretation? I suspect the answer is this: invoking the psyche solved a problem in the former case but would have created a problem in the latter. With respect to the quantity of secretion, invoking the psyche explained away an aberrant difference between meat curves and bread curves; with respect to the quality of secretion, however, it would have undermined an expected, *pravil'nyi* difference between the curves of proteolytic power for bread and meat. Here again, we see that the acknowledged secretory role of the psyche served as a "friend" in the interpretation of discordant results.

Curve construction, then, involved a number of interpretive decisions that were themselves shaped by laboratory doctrine. This will become all the more evident when I examine the evolving identification, explanation, and presentation of these curves between 1894 and 1897.

The Haunted Factory: The Psyche, Curves, and Nerves

The development of the factory metaphor in the two years after Khizhin completed his thesis is evident in the language of I. O. Lobasov's thesis (1896), in its analysis of the psyche, and in its regrouping of the characteristic secretory curves. In Lobasov's dissertation, the word *work* appears routinely (at least seventeen times) in references to the secretory glands. The word *adaptive*, introduced and used twice by Khizhin, permeates Lobasov's thesis, in which "adaptiveness" is pronounced "the essential characteristic of [secretory] work." Lobasov refers six times to glandular work as *complex* (*slozhno*) and once notes its *subtlety* (*tonkost'*)—terms absent from Khizhin's thesis and first appearing in Pavlov's response to Khizhin's critics at the meeting of the Society of Russian Physicians. Lobasov refers once to the *end goal* of secretory work and, unlike Khizhin, uses unapologetically the terms *bread juice, milk juice,* and *meat juice* (terms that, again, first appeared in Pavlov's defense of Khizhin).[46] Lobasov considered the secretory curves, though sometimes obscured by innumerable complexities, fully regular and reliable. Khizhin's tentativeness is nowhere to be found: "the complexity and typical constancy of the secretory work of the stomach," writes Lobasov, "is expressed in the adaptiveness of the glands" (pp. 139, 158).

On inheriting Druzhok from Khizhin, Lobasov had received the following charge from the chief: "To demonstrate the consistency of this phenomenon [psychic secretion], to detect (*ulovit'*) and distinguish in each case of the secretory work of the stomach during feeding what relates to psychic secretion, and to elucidate to what degree the typicality of secretion with various sorts of food is determined by the participation of the psychic moment" (p. 33). The task, then, was to more fully incorporate the idiosyncratic psyche into the *pravil'nyi* digestive machine, a machine now believed to generate distinctive secretory responses to different foods. Clearly, the digestive factory could not work regularly, precisely, and purposively if the most important force in its operation, the psyche, was entirely capricious. Psychic secretion too must necessarily manifest, as Lobasov put it, some "consistency."

His efforts to elucidate that consistency and to separate psychic from nervous-chemical mechanisms were facilitated by the implantation in the laboratory's prize dog-technology of a gastric fistula (in September 1894) and an esophagotomy (in April 1895). As we saw in Chapter 4, the fistula was intended to enable Lobasov to feed Druzhok while bypassing the animal's eyes and mouth (and so, presumably, the psyche); the esophagotomy enabled him to test Druzhok's secretory response to the act of eating alone—in other words, to analyze psychic secretion in isolation from nervous-chemical mechanisms.

The effort to incorporate the psyche within the digestive factory is evident in Lobasov's reinterpretation of the characteristic secretory curves. His reasoning illustrates both the implications of the developing factory metaphor and the interpretive dimensions of curve construction.

As we have seen, for Khizhin (and Pavlov) in 1894, the course of the secretory responses to various foods differed with respect to proteolytic power but not with respect to quantity. For Khizhin (and Pavlov), the curves in Figure 13 are essentially the same, expressing "one and the same *pravil'nyi* course." They differ only—and unimportantly—in that some (the psychic type) reached their acme during the first hour, whereas others (the usual type) peaked in the second hour.[47] In the view of Lobasov (and Pavlov) in 1896, however, these same curves differed fundamentally for different types of food. Curves a, b, and c described the meat curve, d the bread curve, and e and f the milk curve. In sharp contrast to Khizhin, Lobasov concluded that "to each sort of food there is a special character of these changes in the rapidity of secretion of the juice and its qualities." These patterns—the very existence of which Khizhin had denied—were, for Lobasov, "repeated with such constancy and such *prav-il'nost'* that we have the right to distinguish between milk, meat, and bread secretion—and milk, bread, and meat juices."[48]

This regrouping of the curves reflected the development of the factory metaphor and the related attempt to incorporate the psyche within the digestive machine. The psyche remained capricious from experiment to experiment, but it had now also acquired a determinist moment: a standard, presumably *pravil'nyi* psychic response had been identified for each food. Previously, Khizhin had discerned both psychic and usual secretory responses to each of the different foods; Lobasov, on the other hand, incorporated a standard amount of psychic secretion (or "appetite juice") into the characteristic secretory curve for each food. So, for example, Khizhin had not perceived an important difference between the course of secretion in experiments with meat and with mixed food: both types of meal sometimes generated a psychic type of secretory response and sometimes a usual type. He saw the difference between the average hourly values for each as reflecting an idiosyncrasy in Druzhok's personality: the dog's usual enthusiasm for meat, expressed in a large psychic secretion that drove up the amount of secretion in the first hour.[49] Lobasov, on the other hand, saw this large psychic secretion as part of the characteristic meat curve, just as a small or nonexistent psychic secretion was part of the milk curve.[50]

Khizhin's experimental data, however, presented an obstacle here. His distinction between psychic and usual secretory responses to one and the same food rested upon what he (and Pavlov) considered an important variation in their data. When a dog was fed the same quantity of meat several times, this resulted in two different secretory curves: one of the psychic type (peaking in the first hour) and another of the usual type (peaking in the second hour). This same kind of variation was evident in experiments with other foods as well.

Lobasov (and Pavlov) destroyed this typology with a simple change in counting procedure. To understand this move, we must recall that a so-called hidden period, or latent period, of about five minutes elapsed between the initial excitation of the dog (say, by feeding) and the appearance of the first drops of gastric secretion in the fistula running from the isolated sac. The nature of this hidden period was a mystery (and an especially troubling one given the extreme rapidity attributed to nervous processes). Khizhin had begun measurements of the first hour of secretion immediately after excitation; the quantity of secretion in this first hour, therefore, was weighed down by the hidden period, during which nothing was secreted. Lobasov reanalyzed Khizhin's data and discovered that if one began the first hour only after the hidden period had ended—that is, only when the first drops of gastric juice appeared—this drove up the total for the first hour significantly at the expense of the second hour. In the great majority of experimental trials with meat, the curve for amount of secretion now reached a "characteristic" peak in this newly constituted first hour. The remaining differences among these trials could still be explained, as previously, by fluctuations in Druzhok's mood, hungriness, and so forth.[51]

Lobasov, then, attempted to incorporate psychic secretion more fully within the laboratory's emerging concept of the digestive factory, giving greater emphasis than had Khizhin to the *pravil'nye* features of this first exciter of gastric secretion. Meat elicited a strong psychic secretion, and so both the quantity and the proteolytic power of the meat curve rose quickly and almost always peaked in the first hour. The same was true of the bread curve. That these foods elicited a reliable psychic secretion was clear from experiments in which Druzhok was fed several small portions sequentially. After each feeding, the amount and proteolytic power of glandular secretion rose noticeably, testifying to psychic secretion.[52] Milk was the exception that illustrated the rule. Feeding Druzhok small sequential portions of milk did not generate a noticeable rise in either quantity or proteolytic power. "Eating milk does not facilitate the development of the psychic process," Lobasov concluded. Drawing upon laboratory experience, he observed that "dogs consume milk without noticeable satisfaction and often do not consume their entire portion."[53] As milk elicited little or no psychic secretion, the quantity and proteolytic power of milk juice was reliably low in the first hour.

In this analysis, the psyche is somewhat *pravil'nyi*. The nature of the psyche is not here an ontological but rather an operational question. It concerns not the nature of the mind and emotions but the tension between capriciousness and *pravil'nost'*—and the ability of the experimenters to acquire *pravil'nye* results. Insofar as the psyche has become a standard part of a physiological process (the digestive act), it is a determined process, dependent on the vagus nerves, presumed to be reliably present or absent in the secretory response to specific foods, and subject to physiological mechanisms. So, for example, comparing Druzhok's secretory responses to whole and nonfat milk, Lobasov identified a determinist mechanism in the glandular consequences of the psyche: fat, he concluded, inhibits the vagus nerves (through which the psyche excites the gastric glands) and so inhibits psychic secretion itself.[54] The psyche, however, has not been completely subsumed within determinist physiology. It remains sensitive to the character, mood, and food tastes of the dog, and it therefore manifests itself differently from trial to trial.[55]

As we have seen, Pavlov charged Lobasov with discerning the specific contribution of psychic secretion and nervous-chemical mechanisms to the generation of the characteristic secretory curves. To what extent were distinctive secretory responses to different foods the result of psychic secretion, and to what extent did they reflect the *pravil'nyi* response of specific nervous exciters in the mucous membrane of the stomach?

To this end, Lobasov repeated a number of Khizhin's earlier trials, taking advantage of Druzhok's gastric fistula and esophagotomy to distinguish experimentally between psychic and nervous-chemical reactions. For example, by feeding Druzhok through the gastric fistula, Lobasov presumably excluded psychic secretion and so could test the nervous-chemical response to meat in the stomach. After meat was placed directly in Druzhok's stomach, secretion began only after about twenty-five minutes and the rate of secretion and proteolytic power of the resultant gastric juice were substantially lower than when elicited by normal feeding. Meat, then, contained something—unknown "extractive substances"—capable of eliciting nervous-chemical secretion, although these substances were a less powerful exciter than appetite. Lobasov confirmed Khizhin's finding that egg white, bread, and starch elicited little if any secretion when placed directly in the stomach. These foods would apparently remain undigested, "unaltered until they rot," if the secretory process was not first "ignited" by psychic secretion (p. 62).

Lobasov's summary statement reveals the tension between these findings and the image of a *pravil'nyi* digestive factory governed by precise nervous mechanisms.

From all these data it follows that the psychic moment, as a condition determining the secretory work of the stomach glands, is significant to the highest degree. Analysis of the data . . . has led us to the conclusion that the first portions of the juice, which begins the processing (*pererabotka*) of the more or less crude food mass that enters the stomach, initially consists exclusively, and, subsequently, to a great degree, of psychic juice with great proteolytic power. But various sorts of food elicit psychic secretion to different degrees; this difference in the energy of psychic secretion is one of the moments conditioning the difference of secretory work with various sorts of food, especially in the first hours of the digestive period. The data from special experiments show the significance of the psychic moment to be even more important. With some substances (meat and so forth), the decline of the psychic moment has the result that secretion ceases to be characteristic, loses its strength (less juice is secreted and its proteolytic power is lower still) and digestion loses rapidity; other substances (like bread and so forth), with the decline of the psychic moment, elicit no secretion of juice whatsoever and are doomed to lie in the stomach unaltered until they rot. (p. 62)

The problem here is that the "characteristic" features of the secretory responses to specific foods rest largely on the dog's psychic response to them. These curves, then, might reflect not the work of precise nervous-chemical mechanisms but simply the ability of a food to arouse the dog's appetite. This, of course, would undermine the beauty of the secretory curves, reducing them to expressions of the dog's personality, emotions, and food tastes and leaving little of the precise, regular, and purposive digestive factory.

Lobasov immediately and vigorously disavowed such a conclusion. "But if we are convinced of the indubitable importance of the psychic moment as an exciter of initial secretion, we must at the same time be convinced that the psychic moment is not the only condition determining the secretory work of the glands of the stomach: it is only one link in a chain of such conditions—although it is the first link, both temporally and in importance" (p. 63).

Experimental trials indeed provided powerful arguments for the action of secretory mechanisms other than the psyche. For one thing, sham-feeding and teasing experiments indicated that psychic secretion lasted no more than four hours, but gastric secretion during the digestive act usually continued for about nine. Furthermore, sham-feeding experiments often generated a relatively simple curve for psychic secretion—a curve that rose quickly to a peak and then fell steadily to zero—and this contrasted with the more irregular curves generated by normal feeding. Finally, the proteolytic power of psychic secretion in sham-feeding experiments remained within a relatively narrow range, but the proteolytic power of gastric secretions during the actual digestion of various foods reached levels both above and below this range. This, again, testified to the action of additional mechanisms (i.e., mechanisms other than the psyche). This was further confirmed by the fact that, when placed directly in the stomach (and so presumably bypassing the psyche), both meat and milk excited a glandular response. Psychic secretion alone could not, therefore, explain the "distinctive, but entirely determined" fluctuations in the rapidity and proteolytic power of glandular secretion during the normal digestive act (pp. 63–64).

A series of experimental trials led Lobasov (and Pavlov) to the conclusion that three nonpsychic mechanisms were involved in the work of the gastric glands. The first, unknown "extractive substances" (chemical exciters), "develop from the component parts of food under the influence of psychic juice." These substances probably irritated the peripheral ends of the afferent nerves of the mucous membrane of the stomach, generating a reflex that was distributed to the gastric glands. Meat was apparently rich in these unknown substances and therefore elicited gastric secretion when placed directly in the stomach. Other foodstuffs, such as egg white and bread, perhaps produced these extractive substances only when "ignited" by appetite juice. Therefore, when placed directly in the stomach, egg white and bread remained "unaltered until they rot"; but when ignited by psychic secretion, each generated a curve that differed from that for psychic secretion alone. Water was also capable of acting as a weak extractive substance eliciting gastric secretion. Summarizing these findings, Lobasov concluded that "the mucous membrane turns out to be gifted with an extraordinarily subtle adaptability to various substances: not only is it capable of responding with a secretion only to specific irritations, but in each separate case it can alter the quality of the secretion, bringing into activity one or another reflexive apparatus" (p. 110). (In Lobasov's thesis, these ideas about extractive substances changed from possibility [p. 85] to fact [pp. 98, 150].)

The other two nonpsychic agents—starch and fat—were incapable in themselves of eliciting gastric secretion but presumably influenced the qualities of secretion initiated by other agents. So, although starch was not a "true exciter," when introduced into a stomach already brought into secretory activity by other agents it manifested "its active qualities, eliciting a heightened appearance in the secreted juice of the albuminous ferment." This explained the high proteolytic power of the secretory response to bread. Starch acted "upon the work of the glands reflexively, acting specifically upon the peripheral endings of the centripetal nerves in the mucous membrane of the stomach" (p. 109).[56] The final nonpsychic mechanism in gastric secretion was fat. Like starch, fat did not "in itself bring the glands of the stomach into an active state; one must propose that, like starch, it possesses only the ability to alter the activity of the glands." Fat diminished the amount of secretion and "significantly" lowered its proteolytic power. This explained the low proteolytic power of milk juice and the low quantity of secretion in meat juice compared with bread juice (pp. 111–12).

Before exploring Lobasov's use of these mechanisms to explain the characteristic secretory curves, I should note one consequence of the great explanatory weight he (and Pavlov) placed on the nervous system. Because the quantity and proteolytic power of gastric secretion varied independently—that is, the same substance could generate a large amount of gastric juice with low proteolytic power, and the quantity of juice could increase while proteolytic power decreased, or vice versa—a single exciter acting on a single nervous apparatus did not suffice. In other words, if the extractive substances of meat excited the nerves governing gastric secretion, this should lead to an increase in both quantity and proteolytic power; if a food was low in such substances, both quantity and proteolytic power should be low. The very nature of the secretory curves, however, seemed to refute this: for example, meat elicited a greater quantity of less powerful juice than did bread. In his earlier studies of the salivary glands, Heidenhain had dealt with this problem by hypothesizing that there were two sorts of nerve fibers: "secretory" fibers that controlled the amount of liquids (here, the amount of secretion) and "trophic" fibers that governed the amount of solids (here, pepsin). Pavlov and Ekaterina Shumova-Simanovskaia had used this hypothesis in their articles of 1889–90, and it performed a critical explanatory role in Khizhin's thesis, then in Lobasov's. Concluding his exposition of the three agents of "chemical secretion" (extractive substances, fat, and starch), Lobasov wrote:

> This permits one to propose that, like salivary secretion, the process of the secretion of gastric juice consists in two separate phenomena: the secretion of the liquid part of the juice (the mixture of hydrochloric acid) and the production of its solid constituent parts (mainly the ferment). From the fact that we can reflexively elicit and alter the dimensions of the latter phenomenon we must conclude that each phenomenon is governed by a separate reflexive nervous apparatus. In this matter, we must think that the innervation of the stomach glands is accomplished by two sorts of nerve fibers, which one may term the secretory and the trophic—to use Heidenhain's conditional terms in reference to the innervation of the salivary glands. (p. 109)

The operation of the digestive factory, then, rested on the purely hypothetical existence of these two sets of nerve fibers. (We shall see in Chapter 6 that additional nervous structures were posited to explain pancreatic secretion.) Unless such fibers existed, and unless both sets possessed specialized nerve endings that could respond differently to the various extractive substances in foods, the characteristic secretory curves either were meaningless artifacts or were inexplicable as the results of nervous processes.

In sharp contrast to such hypotheses about possible nervous mechanisms, experimental indications of possible humoral mechanisms were systematically discounted. For example, Lobasov discovered that an intravenous injection of Liebig's meat extract elicited gastric secretion in a dog with a Heidenhain sac. Rather than exploring the possibility that this secretion might result from a humoral mechanism, operating through the blood and independent of the nervous system, Lobasov (no doubt with Pavlov's aid) interpreted it as follows. Dogs respond to the intravenous injection of emetics not only by vomiting but also with salivation and gastric secretion. In the same manner, the gastric secretion elicited in a dog with a Heidenhain sac by an intravenous injection of meat extract was "not an expression of [the extract's] significance for gastric secretion in real circumstances" but, rather, reflected the animal's response to the "poisoning" of its blood by a foreign substance. Here, clearly, the interpretive moments inherent to judgments about normalcy combined with Pavlov's nervism to frame the interpretation of experimental results (pp. 95–97).[57]

Lobasov (and Pavlov) deployed the acknowledged exciters of the gastric glands in an ingenious explanation of the characteristic secretory curves. This explanation depended on a fundamental distinction between two basic types of gastric juice: (1) psychic juice, produced by psychic secretion—secreted rapidly and having high proteolytic power; and (2) a generic "chemical juice," produced by the action of unknown extractive substances on the secretory and trophic nerve fibers in the mucous membrane of the stomach—secreted slowly and having low proteolytic power. The curves for different foods reflected the "balance of forces" between these two juices at different hours in the digestive act, as modified by the special action of starch and fat.

Lobasov acknowledged that much remained to be explained about "the energy and quality of secretory work in each separate moment of digestion," but this schema did enable him to explain some of the main attributes of the characteristic secretory curves. Drawing upon sham-feeding experiments to establish the average and normal range of proteolytic power for psychic secretion (the average was 7.4 millimeters, the maximum about 8.0), and using experiments in which food was placed directly in the stomach to establish the range of proteolytic power for chemical juice (2.5 to 5.0 millimeters), Lobasov explained the results of experiments with particular foods as reflecting a particular mixture of these two juices, plus the action of starch and fat.[58] When the proteolytic power of gastric secretion during a digestive act lay between these two figures, it was presumably some combination of psychic and chemical juices (if closer to 7.4 millimeters, the mixture contained more psychic juice; if closer to 5.0, it contained more chemical juice). If the proteolytic power of a secretion was higher than 7.4 or lower than 2.5 millimeters, this presumably testified to the action of a special agent (starch in the former case, fat in the latter).

Using the available variables, then (the psyche, unknown extractive substances, water, starch, and fat), Lobasov explained the main features of the various secretory curves (Figures 14 and 15) as follows. Meat elicited a strong psychic secretion, generating the high rapidity and proteolytic power of "meat juice" in the first two hours. Meat also contained extractive substances that, beginning about ten minutes into the first hour, generated a large amount of "chemical juice." As the digestive act continued, this chemical juice predominated increasingly over the fading psychic secretion, leading to a relatively high quantity of juice (consisting of both psychic and chemical juices) with a gradually decreasing proteolytic power (as increasing amounts of chemical juice, with its lower ferment content, gradually decreased proteolytic power below the characteristic level for psychic secretion). The fat content of meat also served to lower the proteolytic power of meat juice. (The rising proteolytic power toward the end of the digestive act was, presumably, among the phenomena that remained to be explained.) Bread also elicited a strong psychic secretion and so produced a large amount of high-ferment juice in the first hour. Unlike meat, however, bread was relatively poor in chemical exciters. Therefore, as the digestive act progressed the amount of secretion fell sharply (as the fading psychic secretion was not compensated for by chemical juice) and the proteolytic power of that secretion remained high (since the appetite juice was not watered down by the less potent chemical juice). The starch in bread raised the proteolytic power of "bread juice" still higher—to a level that sometimes even exceeded that of psychic secretion. Milk elicited little psychic secretion, resulting in an initial secretion low in quantity and proteolytic power. Subsequently, however, the water (and, perhaps, some unknown substance) in milk elicited plentiful chemical juice, which raised the amount of secretion to its peak in the third hour. As a result of its fat content, the proteolytic power of milk juice was lower than that of chemical juice.[59]

I have already discussed the interpretive moments involved in laboratory explanations of this genre. Rather than multiply such examples, I want to emphasize another point: the tensions created within the factory metaphor by this account of the secretory curves. These tensions were not addressed directly by Lobasov (or any other praktikant); rather, they awaited their resolution in the chief's synthetic treatise of 1897.

First, there is a tension between different elements of the factory metaphor: between its purposiveness and the different mechanisms employed to explain its precision. What, for example, is the "purpose" of low-ferment digestive juice in the digestion of fatty foods or of high-ferment juice in the digestion of starchy foods? How is it purposive that the richer a food is in extractive substances—and thus the more chemical juice it elicits—the lower the proteolytic power in the early stages of digestion (since the resulting chemical juice is much lower in ferment than is the appetite juice with which it combines)? Similarly, what purpose is served by a large piece of meat eliciting gastric juice with lower proteolytic power than that elicited by a small piece of meat?[60] Other tensions are associated with the insistence on the "subtlety" and "precision" of the digestive factory. For one thing, as we have seen, these arguments were built on the average data of numerous trials that generated a wide range of results; yet, as the factory metaphor matured, increasingly strong claims were made about the subtlety and precision of glandular work. Furthermore, although this subtlety was linked rhetorically with the specific excitability of the nervous system—and all four mechanisms of glandular work were treated as nervous mechanisms—the most "characteristic" features of the secretory curves remained dependent on the psyche (without which, secretion ceased to be "characteristic" or was altogether absent).

Paradoxically, the comparison of results obtained with the Pavlov and Heidenhain sacs underlined the relatively great importance of the psyche, starch, and fat and the relatively small role played by the action of hypothesized extractive substances on the specialized nerves of the stomach. Lobasov noted that the Pavlov sac, with its intact vagus nerves, manifested the action of all four mechanisms, whereas the Heidenhain sac, with its vagus nerves severed, responded only to the action of extractive substances on the nerves of the stomach. The Heidenhain sac, however, produced "only small fluctuations within comparatively narrow limits" never attaining "such high numbers as was observed in juice from Druzhok." The difference between the relatively broad range of secretory responses in the Pavlov sac and the relatively narrow range in the Heidenhain sac could thus be interpreted as evidence of the relatively small part played by the action of specific extractive substances on the nervous mechanisms of the stomach in the "subtlety, precision, and purposiveness" of glandular work. Within Pavlov's scientific vision, however, such a conclusion was virtually inconceivable.[61]

A Template Dog for the Pancreatic Gland and a Last-Moment Verification of Results with Druzhok

In 1896 Pavlov assigned an especially promising praktikant, Anton Val'ter, to do for the pancreas what Khizhin had done for the gastric glands. Like Druzhok, Val'ter's dog Zhuchka was equipped with the latest laboratory technology and proved to be an ideal laboratory animal. Having recovered her health after implantation of the troublesome pancreatic fistula, Zhuchka quickly regained an excellent appetite and was soon "enjoying" her life. As Val'ter put it, "It is in the steadfastness of this dog, fully stabilized after the operation, that one must find the essential reason for the great *pravil'nost'* of the results obtained upon it."[62] Although Val'ter had not yet completed his dissertation when Pavlov's *Lectures* went to press, his results with Zhuchka already delighted the chief, who used them to display and explicate in his own text the characteristic secretory curves for pancreatic secretion. When Val'ter completed his thesis, it acquired a classical title, *The Secretory Work of the Pancreatic Gland,* reminiscent of Khizhin's earlier groundbreaking work. Val'ter and Zhuchka became counterparts to Khizhin and Druzhok.

The laboratory view of pancreatic secretion was essentially the same as that of gastric secretion, but research on the pancreas confronted additional complications. According to laboratory doctrine, pancreatic secretion, like gastric secretion, occurred in two stages. In the first stage, the hydrochloric acid in gastric secretion excited the pancreas; in the second, as food moved out of the stomach it excited the specialized nerve endings in the mucous membrane of the duodenum. The specific excitability of these nerves led the pancreas, like the gastric glands, to respond precisely and purposively to specific foods.

Unlike the gastric glands, the pancreas secreted three separate ferments, each of which acted specifically on either proteins, fat, or starch. In a much-praised dissertation of 1893, praktikant V. N. Vasil'ev (and Pavlov) argued that the ferment content of an animal's pancreatic responses changed over time as the pancreas adapted to a particular diet.[63] In other words, the pancreas not only adapted to the composition of a specific meal during a single feeding, it also underwent a "chronic adaptation" to long-term changes in the animal's diet. This added another variable to the experimental search for *pravil'nye* results, since it meant that different dogs, with their different digestive histories, would probably differ in their pancreatic responses to the same meal. The choice of a single template dog, then, gained added importance.

The influence of the psyche on pancreatic secretion was held to be indirect but omnipresent. There was no experimental evidence for psychic secretion in the pancreas (Val'ter speculated that it existed but was relatively insignificant). By virtue of its important role in gastric secretion, however, the psyche influenced the volume of hydrochloric acid produced in the stomach and therefore, because pancreatic secretion responded sensitively to even small amounts of hydrochloric acid, also played an important role in pancreatic secretion.

Val'ter's interpretation of experimental trials with Zhuchka led him to the same basic conclusions as had Khizhin's with Druzhok.

Under identical experimental conditions with the same food, the secretion of pancreatic juice after the dog has consumed milk, bread or meat is repeated with stereotypical precision. This identity—which is manifest in the course of secretion over time, in its quantity, and in the qualities of the pancreatic juice—testifies to the fact that the secretory apparatus of the pancreatic gland works with great precision and lawfulness. When the food is varied, the secretion of pancreatic juice acquires a different course, distinctive for each sort of food; by the sharpness of these distinctions and the consistency with which they are repeated, there is a typical work of the pancreatic gland for each food substance; this typicality is manifest not only in the course of secretion, but also in the fluctuations of the juice and its fermenting activity.

The fluctuations of the fermenting qualities of the juice manifest an evident adaptation to the type of food ingested by the animal: the juice is dominated by precisely that ferment which is necessary for digestion of the given foodstuff.[64]

Val'ter attributed the precise, stereotypical work of the pancreatic gland to "the play of a special mechanism, based on the specific excitability of the mucous membrane of the gastric-intestinal tract."[65]

A series of interpretive moments played an important role in Val'ter's construction of stereotypical curves. He conceded that his data showed "significant fluctuations" from trial to trial: for example, the amount of pancreatic juice elicited by experiments with the same quantity of milk varied from 37.25 to 60.5 cubic centimeters, with the same quantity of bread from 138.75 to 215.25 cubic centimeters, and with the same quantity of meat from 103.25 to 144.0 cubic centimeters.[66] These unstereotypical results, however, could be explained by the uncontrolled variables in chronic experiments on the pancreas and therefore, wrote Val'ter, "do not contradict our view." For one thing, Pavlov had recently demonstrated that the water content of an animal's body—and so its recent history of eating and drinking—influenced gastric (and therefore pancreatic) secretion.[67] Furthermore, three other uncontrolled variables influenced the amount and strength of the acid in gastric secretion that excited the work of the pancreas. The first was "conditions related to the movement of the acidic content of the stomach to the duodenum." Because hydrochloric acid was a powerful exciter of the pancreatic gland, even small influences on its accumulation in the stomach would have great consequences for the amount of pancreatic secretion.[68] Second, experimental animals with a gastric fistula sometimes manifested a tendency toward hypersecretion. This phenomenon, which often appeared erratically, also obscured the stereotypicity of pancreatic responses to specific foods.

The third important uncontrolled variable was the indirect influence of the psyche. Val'ter provided a lucid description of the challenges posed by the idiosyncratic psyche to the experimenter's craft and interpretive skills.

> In most cases of normal feeding the first and most powerful exciter of gastric juice is the animal's appetite: the passionate desire for and enjoyment of food. One must say that this psychic exciter is, by the very nature of things, difficult to subordinate to the control of the experimenter. Even in the purest form of application of this exciter, in "sham-feeding" experiments, the animal often eats with varied interest and therefore produces various quantities of gastric juice; the experimenter's craft in such cases consists in various tactics, for example, finding a suitable tempo of feeding, to hold the animal's interest at a specific level (usually, to arouse it ad maximum). The short-livedness of normal feeding greatly limits the use of such tactics; here it often occurs that, despite the precise observance of all conditions of the experiment, the animal eats the very same food on various days with varying degrees of enthusiasm. When the dog eats with abandon, there is secreted, at least in the first hours of digestion, much gastric juice; when the dog eats sluggishly, there is little. Since the work of the pancreas is tightly linked, by virtue of its acidic exciter, to the work of the stomach, therefore the quantity of pancreatic juice also fluctuates in the same manner. This is so aside from the possibility that the psychic moment has a direct effect upon the pancreatic gland.[69]

Just as Khizhin had used the psyche and other variables to fashion from highly varied data his stereotypical curves for the gastric glands, so Val'ter used them to construct similar curves for the pancreas. (I discuss Val'ter's data in detail in Chapter 6, when analyzing their presentation in Pavlov's *Lectures*.)

When *Lectures* went to press in fall 1897, Pavlov's data for the characteristic secretory curves came almost entirely from two template dogs: Druzhok for the gastric glands and Zhuchka for the pancreas. As the manuscript made its way through the typographer, however, the chief sought to remedy this situation somewhat.

Khizhin, Lobasov, Val'ter, and Pavlov had presented their work to the Society of Russian Physicians, and so the chief was well-acquainted with possible objections to their conclusions. It was most likely here that, as praktikant Andrei Volkovich reported, the laboratory's conclusions about the work of the gastric glands "elicited the objection that all these experiments were conducted almost entirely on one animal." So, Pavlov assigned Volkovich, then a new praktikant, to demonstrate that "the results obtained on Druzhok can be obtained on any dog."[70] On October 2, 1897, Pavlov successfully created an isolated stomach in a second dog, Sultan. The animal recovered from its operation "superbly" and was soon "very happy, and generally produced the impression of a healthy animal enjoying its life."[71] With Pavlov no doubt hovering impatiently over his shoulder, Volkovich conducted a series of trials on Sultan's secretory responses to meat, bread, and milk in mid-October.[72]

Pavlov's discussion in *Lectures* of the characteristic secretory curves for gastric secretion remained totally dependent on trials with Druzhok, but Volkovich's experiments on Sultan sufficed for the chief to add a single sentence to a late chapter of his own synthetic work. It is to that work—and, in good time, to that sentence—that we turn in Chapter 6.

On Curves and Claims

Pavlov was not the first physiologist to construct and deploy curves in an analysis of digestive secretions. He was not even the first to use curves to describe the secretory responses elicited from an isolated gastric sac. How did Pavlov's practices in this respect differ from those of his predecessors? The fundamental difference, I think, resides in the degree to which these curves "took over" as ideal representations of the "thick things" with which Pavlov was confronted.[73]

In the years before Pavlov's research, N. O. Bernstein, Michael Foster, and Rudolf Heidenhain, among others, used curves to describe various digestive secretions.[74] For these three physiologists the curves expressed only general (and still hypothetical) tendencies, and they made relatively weak claims based on them.

Heidenhain's curves, constructed from data of four experiments on a single dog with a Heidenhain sac, are presented in Figure 16. Heidenhain ascribed a very low level of specificity and precision to these curves. For him, as for Bernstein and Foster, the most important and definite conclusion from the curves was that the amount and fermenting power of secretion did not remain constant throughout the digestive act but rather changed over time. He was intrigued by the fact that the curves ascended twice (rather than just once) and that those for the amount of secretion seemed to move in an opposite direction from those for fermenting power. Although these curves represented four trials with the same animal (on different days), the differences between them did not trouble Heidenhain. He noted only that all four, in their broad outlines, manifested a common course of secretion. He also noted some differ-

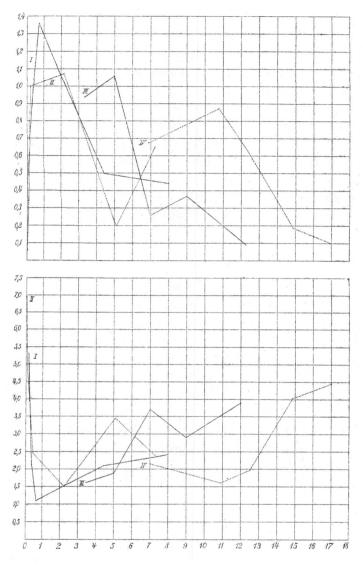

FIGURE 16. *Heidenhain's curves. Each curve is constructed from four experiments on the same dog.* Top: *Amount of gastric secretion. Vertical axis, cubic centimeters; horizontal axis, time in hours.* Bottom: *Proteolytic power of gastric secretion. Vertical axis, proteolytic power in millimeters; horizontal axis, time in hours. From Rudolf Heidenhain,* Handbuch der Physiologie der Absonderung und Aufsaugung, *in* Handbuch der Physiologie, *ed. L. Hermann (Leipzig: F. C. W. Vogel, 1883), 183*

ences between his data and Bernstein's, but he discussed these noncommittally.[75] He did not even mention the type of food that his experimental dog had ingested; this level of specificity, apparently, had no significance for him.

As we have seen, even in Khizhin's thesis of 1894 the Pavlov laboratory ascribed a much higher level of precision and specificity to its results and presented those results accordingly. Were we to approach the Heidenhain curves in the Pavlovian style, the first step would be to explain the differences among them (using, most probably, observations about the dog's differing mood on different days). We might decide to discard a trial or two on this ground. We could then calculate average data, construct a single ideal curve, and illustrate this ideal curve by the curve for the single concrete trial that most closely approximated it. The knowledge claims associated with this curve would then change over time as the factory metaphor matured, and the curve would finally be used to describe the purposive, precise, and regular work of the glands.

The comparison with Heidenhain highlights the importance of Pavlov's scientific vision and interpretive practices. These two great scientists worked with much the same dog-technology[76] and acquired basically the same type of data in their experimental trials. They differed fundamentally, however, in the confidence with which they fit a curve to their "spray of data" and identified that curve as an ideal representation of "thick reality." Pavlov certainly "trusted numbers," but, even more, he trusted his interpretation of numbers—his ability, through an understanding of the organism and a grasp of the experimental setting, to discern in his data the precise, regular, and purposive work of the digestive factory.

Chapter 6

THE PHYSIOLOGY
OF PURPOSIVENESS

The object of the experimental method is to transform this a priori conception, based on an intuition or a vague feeling about the nature of things, into an a posteriori interpretation founded on the experimental study of phenomena.

— CLAUDE BERNARD, *An Introduction to the Study of Experimental Medicine* (1865)

I think it useful for the reader that there unfolds before him, as it were, a single idea, increasingly embodied in tenable and harmoniously linked experiments.

— IVAN PAVLOV, *Lectures on the Work of the Main Digestive Glands* (1897)

Pavlov's *Lectures on the Work of the Main Digestive Glands* represented both a synthesis (or final processing) of six years of laboratory research and a rhetorically powerful literary product. Here Pavlov convincingly "embodied" his "single idea" in the great number and range of knowledge claims generated by the experiments of his praktikanty and offered a compelling picture of laboratory physiology and its relationship to medical practice.

The laboratory's most exquisitely refined literary product originated as a series of lectures that Pavlov first delivered to a medical audience at the Imperial Institute in 1895 and was subsequently refined through presentations at the Military-Medical Academy (Figure 17).[1] The form and con-

FIGURE 17. *Pavlov lecturing at the Military-Medical Academy. Courtesy of Academy of Sciences Archive, St. Petersburg branch*

tent of *Lectures* owed much to its origins as a lecture series for an audience of nonspecialists (primarily physicians) rather than as a monograph for physiologists.[2] Pavlov's voice is lively, informal, and authoritative as he draws upon his specialized knowledge to explain both the fundamentals of digestive physiology and, using his research as an exemplar, the proper relationship between laboratory and clinic.

Pavlov's lectures always featured experimental demonstrations, and he assures his readers that *Lectures* includes "all the experiments" performed before his live audiences.[3] During his lectures at the Military-Medical Academy, only a few students in the front rows could actually see the results of these experiments. Pavlov therefore chose several students to "witness" them for the rest of the class.[4] The reader of *Lectures* is also dependent on a witness—Pavlov himself, who selects data from his laboratory's vast "storehouse of information," asks the reader to believe that these are typical, and interprets them according to the laboratory's "single idea."

The six central lectures on the functioning of the digestive glands are framed by two others, Lectures 1 and 8, that address the nature of physiology as a science and its relationship to medical practice. In Lecture 1, "A General Overview of the Subject—Methodology," Pavlov introduces the factory metaphor, explains the cardinal importance of methodology for physiology, explicates the unique advantages of physiological surgery and the chronic experiment, describes his laboratory's dog-technologies, and presents his own Physiology Division as a model institution for physiological research. He synthesizes his laboratory's findings in the next six lectures.

In Lecture 2, "The Work of the Glands during Digestion," Pavlov presents the characteristic secretory curves for gastric and pancreatic secretion as the empirical embodiment of the "laboratory view." Lecture 3, "The Efferent Nerves of the Gastric and Pancreatic Glands," and Lecture 4, "A General Scheme of the Complete Innervative Apparatus . . ." explore the nervous apparatus that governs the purposive and precise work of the glands and introduce the central role of appetite. In Lecture 5, "The Place and Significance of Psychic (or Appetite) Juice in the Entire Work of the Stomach . . . ," Pavlov deepens his discussion of the psyche and debunks the widespread view—typifying "obsolete textbook doctrines"—that the crude action of mechanical irritants plays an important role in glandular secretion. In Lecture 6, "The Chemical Exciters of the Innervative Apparatus of the Gastric Glands . . . ," and Lecture 7, "Normal Exciters of the Innervative Apparatus of the Pancreatic Gland . . . ," he addresses a series of miscellaneous questions (ranging from the reliability of the isolated sac to the specific exciters of the pancreas) and reviews his general conclusions. The final lecture, "Physiological Data, Human Instinct, and Medical Empiricism," is devoted to the practical significance of the Pavlov laboratory's research for daily life and medical practice.

The text, then, moves first from the general to the specific—from a discussion of methodology and Pavlov's single idea to the characteristic secretory curves and the psychic and nervous mechanisms that generate them—and then outward to the relationship of physiology to medicine and daily life.

This chapter is structured in much the same manner. I first sketch the general picture of digestion that emerges from Pavlov's synthetic work, then examine the chief's final processing of experimental data to express his single idea. After a brief digression—to explore the manner in which the results obtained with Druzhok and Zhuchka were subsequently verified with other dogs—I explore the rhetoric in *Lectures* about the nature of experimental physiology and its relationship to medical practice.

The Digestive Factory

Pavlov's narrative is structured by the factory metaphor, which he introduces immediately as a replacement for "obsolete textbook doctrines." In *Lectures,* he has elaborated this metaphor slightly since its original formulation in 1894.

> By its basic task in the organism, the digestive canal is, obviously, a chemical factory (*zavod*) that subjects the raw material entering it—food—to processing (*obrabotka*), mainly chemical, in order to enable it to enter the juice of the organism and to serve there as material for the process of life. This factory consists of a series of compartments in which the food, according to its properties, is more or less sorted and either retained for some time or immediately moved to the next compartment. The factory, in its various compartments, is supplied with special reagents, obtained either from the nearest small manufactories (*fabrik*) built into the very walls of the factory—by industrial agreement, (*kustarnyi lad*), so to speak—or from more distant specialized organs, great chemical manufactories, which communicate with the factory through tubes that transmit reagents. These are the so-called glands and their ducts. Each manufactory supplies a special fluid, a special reagent, with definite chemical properties, as a consequence of which they [the fluids] act only upon certain component parts of food, which usually contains a complex mixture of ingredients. These properties of reagents are determined mainly by the presence in them of special substances, of so-called ferments.[5]

As in 1894, Pavlov has carefully preserved Mendeleev's distinction between a manufactory (*fabrika*)—for Pavlov, the glands—and a factory (*zavod*)—the digestive canal (and, with its outlying parts, the digestive system as a whole). Only in the factory is the food matter actually transformed by "mainly chemical" processes. Pavlov amplifies his earlier explication of this metaphor by reference to *kustarnyi lad*, referring to a common arrangement between Russian factories and cottage industries by which the former ordered precisely what they needed from the latter.[6] In Pavlov's view, this describes exactly the relationship of the digestive canal to the glands. Both the gastric glands ("built into the very walls of the factory") and the pancreatic gland (one of the "more distant specialized organs, great chemical manufactories, which communicate with the factory through tubes that transmit reagents") respond to specific, timely orders from the digestive tract, which thereby acquires precisely what it needs to process various foods.[7]

As in any efficient factory enterprise, the work of the digestive system is above all purposive (or adaptive) and therefore characteristic (or stereotypical) for each food, responding precisely and subtly to the requirements for optimal digestion. "The work of the glands," Pavlov writes, "is extremely elastic, and at the same time characteristic, precise, and completely purposive."[8] Or, in an alternative formulation, "The differentiation and variety of reagents authoritatively indicates the complexity, subtlety and adaptiveness of the work of the digestive canal to each partial digestive task" (pp. 21; 3). Specific phenomena that seem to lack purposiveness are puzzling. For example, it seems paradoxical that water excites the gastric glands, given that it lacks the proteins that are processed by the pepsin in gastric juice. On reflection, however, we see that this phenomenon, too, is revealed as purposive: "Water is very widely distributed in nature, and the instinct for it—thirst—is even more pressing and persistent than the desire for solid food. If a dry meal be eaten without appetite, thirst will compel one to drink water afterwards, and this fluid suffices to ensure the beginning and continuation of the secretory work of the glands" (pp. 129; 95).[9]

Pavlov draws constantly upon experiments conducted by his praktikanty to provide a detailed and unified picture of the coordinated work of this digestive factory (see Appendix A). Food first encounters the salivary glands, which supply reagents that prepare its descent through the digestive canal. As the food proceeds along the digestive canal, it is processed by reagents supplied by the gastric and pancreatic glands. This process is coordinated exclusively by the nervous system. First, appetite is aroused by the act of eating. The "passionate desire for food and the sensation of satisfaction, pleasure, derived from eating" is "the first and strongest stimulant of the secretory nerves of the gastric glands" (pp. 105–6; 75–76). This initial excitation is relayed through the vagus nerves to the gastric glands, where it generates a "psychic secretion" of indeterminate duration. This "appetite juice" is characterized by its rapid secretion and high proteolytic power. The second phase of gastric digestion begins five or ten minutes later, when the food excites the specialized nerves in the mucous membrane of the stomach, eliciting a "chemical secretion" of the optimal amount and proteolytic power for digestion of the ingested substance. The work of the glands is well coordinated: the moisture produced by the salivary glands excites the gastric glands, and the hydrochloric acid produced by the gastric glands excites the pancreas. As food proceeds from the stomach into the duodenum, it stimulates formation of the specialized nervous exciters in the mucous membrane, eliciting another precise secretory response. This system produces exactly the amount and quality of secretions necessary to process any given food. It is as if, as in any well-run factory operation, the glands "possess a mind."[10]

This purposive "mind" is rooted not in the psyche but rather in the specific excitability of the nervous system. Specific substances excite specific parts of the "nervous machinery" in specific ways, eliciting the precise secretion necessary for their digestion. The specific excitability of the nervous system, then, "lies at the foundation of the mechanism of the purposive activity of the organs" (pp. 92; 64).That the sensitivity and precision of the digestive factory could be the result of any other mechanism was inconceivable to Pavlov: "The basis of the adaptive capacity of the glands must of course first be sought in the nervous relations of these organs. One must turn to any other explanation only in the event of the complete failure of this one" (pp. 67; 45). Nowhere in *Lectures* does Pavlov find this necessary.

As presented in *Lectures*, the psyche is a complex actor—*pravil'nyi* ("regular" and "correct") in its general outlines but capricious in any individual experiment. Pavlov's laboratory had revealed it to be a constant, objectively measurable participant in the digestive process, but in any specific trial the activity of this "first and strongest" exciter of secretion depended on the personality, food tastes, and mood of the dog.

Pavlov introduces the psyche by relating the results of sham-feeding experiments. If such experiments are properly conducted, the act of eating elicits a strong response from the gastric glands. Furthermore, one can observe the relationship between the "greediness" (*zhadnost'*) with which a dog devours a meal and the quantity and quality of the secretory response. The "psychic moment," then, has "acquired a physiological character, that is, it has become compulsory, repeating itself without fail under defined conditions, like any fully investigated physiological phenomenon." If one regards this phenomenon from "a purely physiological point of view, one can say that it is a complex reflex" (pp. 104; 74–75).

Like all physiological phenomena, this complex reflex has a purpose.

> Its complexity is understandable, because the physiological goal in this case can be attained only by an entire series of actions by the organism. The object of digestion—food—is located beyond the body, in the external world; it can be acquired by the organism only by using, not just muscular strength, but also the higher functions of the organism—the animal's sense, will, and desire. Consequently, the simultaneous irritation by food of various sense organs—those of sight, hearing, smell, and taste— . . . provides the truest and most powerful impulse to the secretory nerves of the glands. Through the passionate instinct for food, persistent and untiring nature has tightly joined the search for and acquisition of food to its initial processing in the organism. (pp. 104; 75)

Having acquired the *pravil'nost'* of a physiological process, the psychic moment, commonly called appetite, "takes form as scientific flesh and blood, transformed from a subjective sensation into a precise laboratory fact" (pp. 104; 75).

The existence of the psychic moment as a "precise laboratory fact" depended, however, on proper experimental technique, and this, in turn, required the effective recognition and management of such imprecise qualities as the dog's character and preferences. For example, most dogs prefer meat to bread and some exhibit a marked preference for raw over boiled meat, but "sometimes one encounters dogs who incline more, with better appetite, to bread rather than meat, and in such dogs, contrary to the rule, sham-feeding with bread elicits more and stronger juice." A dog that has eaten within fifteen to twenty hours will perhaps react with a psychic secretion only to its favorite food, but once "prepared" by a two- or three-day fast, it will respond to any food with "a copious secretion of gastric juice" (pp. 103; 74). The results of teasing experiments (and, by extension, the action of the psyche in normal feeding) depended on "the degree of the desire to eat, and this depends on how much and how long ago the dog has eaten, and with what it is teased: is the food really interesting to it or does it regard it with indifference." Pavlov continues:

> It is known that dogs have no less varied tastes than do people . . . Among dogs, too, one encounters more positive and cold-blooded types, who are not in the habit of being teased by a dream, by that which is distant from their mouth, but rather wait patiently and peacefully until the food ends up in their mouth. Consequently, the experiment requires greedier dogs and dreamier (*mechtatel'nye*) animals . . . An extraordinarily important moment with which one must contend in these experiments is the cunning and touchiness of the animals. We rather frequently come upon dogs who quickly notice that they are being teased with food and become angry, stubbornly turning away from that which you are doing before them. It is always better, therefore, to conduct an experiment with teasing as if you are not even thinking of teasing the animal, but are simply preparing its food. (pp. 101–2; 72–73)

The intimate involvement of the psyche in the digestive process challenged the experimenter's skills and observational powers. The lack of vigilant management easily contaminated experiments on all aspects of glandular work. "When one is long occupied with the work of the gastric glands under various conditions one becomes convinced of the great danger posed to all experiments by the psychic secretion of juice. You must constantly, so to speak, conduct a struggle with this factor, constantly reckon with it, constantly guard oneself against it. If the dog has long not eaten, your every movement, your every exit from the room, every appearance of the attendant who feeds it, and so forth—all this can sometimes be an impulse to the work of the glands. The most unrelenting and careful attention is demanded to avoid this source of errors" (pp. 102; 73).

As presented in *Lectures*, then, the psyche is both an important factor in digestion and a source of authority for Pavlov's conclusions. In other laboratories, the capricious psyche led investigators astray; but in Pavlov's, it was expertly managed, acquiring "a physiological character" and "repeating itself without fail under defined conditions, like any fully investigated physiological phenomenon" (pp. 104; 74).

Purposiveness and Organ Physiology: The Logic/Rhetoric of Pavlov's Argument

In Lecture 1, the reader confronts Pavlov's concept of the digestive factory and its essential quality, which Pavlov introduces as a commonsensical truth. "Upon reflection, one must a priori recognize that each food, that is, each mixture of substances subject to processing, should encounter its own combination of reagents and their properties" (pp. 20; 2). In subsequent lectures Pavlov embodies this single idea in "tenable and harmoniously linked experiments." Or, in Bernard's formulation, he deploys his experimental data to transform this "a priori conception" into an "a posteriori interpretation founded on the experimental study of phenomena." The analytical and rhetorical structure of his argument is rooted in Pavlov's vision of organ physiology and the logical form appropriate to it.

As noted in Chapter 5, a number of tensions exist between the factory metaphor and experimental data. How could the variations in psychic secretion from dog to dog and from day to day be reconciled with the presumably purposive and precise work of the glands? How could the notion of purposiveness be reconciled with the fact that most foodstuffs failed to excite nervous-chemical secretion when placed directly in the stomach? What was the digestive purpose of, for example, the relatively small quantity of high-ferment juice elicited by a meal of bread and the relatively large quantity of low-ferment juice elicited by a meal of meat?

In *Lectures*, Pavlov speculates on these and other issues,[11] but he finally defers their resolution to the distant future. He acknowledges that he could not demonstrate biochemically the purposiveness of any of the characteristic curves—he could not even identify (nor could any other physiologist of his day) the "extractive substances" in foods that excited the nerves in the mucous membrane of the stomach and the duodenum. He did briefly adduce several "striking" examples of the purposiveness of particular points on the curves but concedes that this issue remained "an almost untouched area for investigation." So, "the conviction of the purposiveness of the fluctuations of glandular work can at the present time be based mainly on general considerations and only in part on separate, more or less clear and indisputable, cases of it" (pp. 57; 36).[12]

Pavlov's argument for purposiveness therefore rests on the regularities he perceived at the organ level—on the identification of laws "in the complete apparatus." These laws were embodied in the stereotypical secretory curves, which displayed a *pravil'nost'* and precision that testified to the purposiveness of digestive work. As he puts it, "Their lawfulness testifies to their importance" (pp. 56; 35). This relationship between stereotypicity and purposiveness is evident in the passage in which Pavlov introduces the first set of characteristic curves. "The work of the glands, that is, the secretion of juices, presents a certain definite course: juice is not poured out at the same rate from beginning to end, nor does it flow by a directly ascending line . . . It is poured according to a certain special curve—this one rising more or less quickly, that one holding to certain defined points, and another one descending gently or suddenly . . . In view of the exactness of this curve and its stereotypicity one must recognize that one or another course of secretion does not exist by chance, but is necessary, useful for the most successful processing of food and the greatest good for the entire organism" (pp. 43; 23).[13] The very existence of these curves, then—the stereotypical repetition of the same patterns in experiment after experiment—testifies to the glands' "astonishing exactitude" (pp. 49; 29), to the "great accuracy and precision" with which they provide "just enough" juice of the precise proteolytic power necessary for the processing of a specific food (pp. 41; 21). These curves justify Pavlov's assertion—just ten pages after his admission that the purposiveness of any single curve is unproven—that "the sum of facts adduced, I hope, sufficiently justifies the conclusion made above, and now repeated again, that the work of the glands being investigated is very complex and elastic, and at the same time surprisingly precise and, of course, purposive, although at the present time we see only isolated indisputable cases of this purposiveness" (pp. 66; 44).

The conceptual and rhetorical importance of these curves is clear from their appearance at critical junctures in Pavlov's text. They first appear early in Lecture 2, where Pavlov converts into curves the experimental data for two experiments on the gastric response and two on the pancreatic response to the same quantity of the same food (pp. 43; 23–24). These convincingly illustrate the stereotypicity, and hence purposiveness, of the course of digestive secretion. The next two pairs of curves (also in Lecture 2) demonstrate this same point with respect to the proteolytic power of gastric and pancreatic secretions (pp. 50; 29–30). Pavlov closes Lecture 2 with a discussion of the distinctive characteristic curves for gastric and pancreatic secretions in response to meat, bread, and milk (pp. 57, 64; 36, 40, 42). He begins Lecture 3 by reviewing what the reader has learned "in the boring form of curves": that the glands "poured their juice, with regard both to its quantity and quality, in correspondence to the mass and type of food, providing specifically that which was most advantageous for the processing of a given sort" (pp. 66; 45). This sets up his discussion, in Lectures 3 and 4, of the psychic and nervous-chemical mechanisms that generate these curves. In Lecture 5 he uses one set of curves to prove that the Pavlov isolated sac faithfully reproduces the secretions of the large stomach, and a final set to demonstrate that the ordinary course of gastric secretion is precisely equal to the sum of the psychic and chemical phases of digestion (pp. 109, 116; 81–82).

As we have seen, these curves were not the simple empirical products of experiments that produced precisely the same results every time. Pavlov is surely overstating his case when he writes that, as a result of his laboratory's precise and careful techniques, "the course of secretion in identical conditions has become truly stereotypical" (pp. 42; 22). How, then, does he present these varied results to demonstrate that glandular secretion is stereotypical and thus purposive?

"Two out of Five, or about That"

In physiology, we must never make average descriptions of experiments, because the true relations of phenomena disappear in the average; when dealing with complex and variable experiments, we must study their various circumstances, and then present our most perfect experiment as a type, which, however, still stands for true facts.

—CLAUDE BERNARD, *An Introduction to the Study of Experimental Medicine* (1865)

In presenting his purposive characteristic secretory curves, Pavlov confronted the tension between the developed factory metaphor, his Bernardian notion of determinism, and the nature of his data. On the one hand, the factory metaphor invoked the determined precision and regularity of glandular responses—qualities that Pavlov indeed discerned in his data and sought to portray convincingly for his readers. On the other hand, however impressive they might be, the results of chronic experiments with complex and intact dog-technologies were never precisely the same from one trial to the next (let alone from one dog to the next). The characteristic secretory curves ultimately rested on precisely the kind of average data that Bernard rejected,[14] and these averages concealed substantial differences from one trial to the next.

Let us begin with the first two curves in *Lectures*. Here Pavlov introduces the reader to the stereotypical precision of the glands by reproducing data from two of Khizhin's experiments on the quantity of gastric secretion elicited by the digestion of 100 grams of meat (Figure 18) and two of Val'ter's on the quantity of pancreatic secretion elicited by the digestion of 600 cubic centimeters of milk (Figure 19). For rhetorical effect, these data are converted to curves. The reader encounters two, virtually identical pairs that demonstrate dramatically the regular, precise work of the glands.

In his thesis, Khizhin reported a total of five experiments on the amount of gastric secretion elicited by the digestion of 100 grams of meat. He provided the complete results for only one trial, giving also the average, highest, and lowest amounts of secretion for the five experiments, taken together, every hour.[15] These results do not permit us to fully reproduce the data for all five experiments, but they do make clear that Pavlov chose the two experiments whose results best illustrated stereotypical precision. For example, the total amount of secretion in the five experiments was 18.5, 23.9, 25.4, 30.2, and 34.2

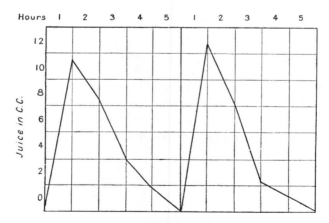

FIGURE 18. *Pavlov's two curves for the amount of gastric secretion after a meal of 100 grams of meat. From I. P. Pavlov,* Lektsii o rabote glavnykh pishchevaritel'nykh zhelez *(1897), in* Polnoe sobranie sochinenii *(Moscow: USSR Academy of Sciences, 1951), 2, pt. 2: 42–43; W. H. Thompson,* The Work of the Digestive Glands *(London: Charles Griffin and Co., 1902), 22–23*

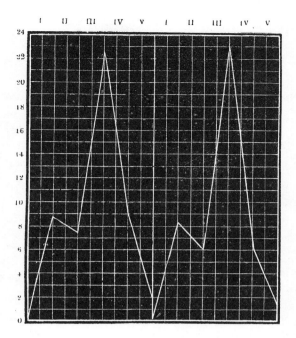

FIGURE 19. *Pavlov's two curves for the amount of pancreatic secretion (in cubic centimeters) after a meal of 600 cubic centimeters of milk. From I. P. Pavlov,* Lektsii o rabote glavnykh pishchevaritel'nykh zhelez *(1897), in* Polnoe sobranie sochinenii *(Moscow: USSR Academy of Sciences, 1951), 2, pt. 2: 43; W. H. Thompson,* The Work of the Digestive Glands *(London: Charles Griffin and Co., 1902), 24*

cubic centimeters.[16] In *Lectures,* Pavlov uses the second and third experiments, making much of their stereotypicity. Furthermore, Khizhin's thesis reveals that in two of the other three experiments, a greater quantity of juice was secreted in the second hour than in the first hour, yielding a different curve than the stereotypical one presented to the readers of *Lectures* (Figure 18).[17]

Val'ter's thesis provides more complete data, permitting a closer look at Pavlov's choices regarding pancreatic secretion (Figure 19). Val'ter conducted thirty-two experiments on Zhuchka's pancreatic response to the ingestion of 600 cubic centimeters of milk. As we have seen, he openly acknowledged the great variations in these results. For example, the total quantity of pancreatic secretion in these trials varied by almost 100 percent—from 37.25 to 72.25 cubic centimeters. Val'ter ascribed these variations to the uncontrolled variables in his chronic experiments (see Chapter 5). For reasons he did not explain, Val'ter discarded eight of his thirty-two trials. His data for the remaining twenty-four are presented in Table 2.

Table 2

Pancreatic Secretion Elicited by a Meal of 600cc of Milk

| | SECRETION (CC) | | | | | | | |
TRIAL	1ST HR.	2D	3D	4TH	5TH	6TH	7TH	TOTAL
1	8.75	7.5	22.5	9.0	2.0	—	—	49.75
2	8.25	7.0	13.0	10.25	0.75	—	—	39.25
3	11.0	6.75	14.0	12.75	5.75	0.25	—	50.5
4	8.25	5.25	6.5	17.75	5.25	9.5	0.75	53.25
5	8.25	6.0	23.0	6.25	1.5	—	—	45.0
6	8.0	8.25	8.5	11.0	2.5	—	—	38.25
8	7.25	9.5	12.5	11.0	2.75	—	—	43.0
9	5.25	12.0	20.25	8.25	2.75	—	—	48.5
10	8.75	5.0	15.25	18.5	0.25	—	—	47.75
11	10.75	11.25	17.5	10.5	1.5	—	—	51.5
12	8.75	7.25	14.5	16.0	6.75	—	—	53.25
13	7.0	16.25	25.5	11.5	2.25	—	—	62.5
14	7.75	8.75	24.0	14.25	17.25	—	—	72.0
15	9.75	15.0	31.0	16.5	—	—	—	72.25
18	11.0	21.25	12.75	13.0	—	—	—	58.0
19	8.25	7.25	31.0	6.75	—	—	—	53.25
20	8.0	26.75	23.75	2.0	—	—	—	60.5
21	5.75	6.25	14.25	11.0	—	—	—	37.25
22	6.25	5.0	30.5	6.0	—	—	—	47.75
23	6.5	6.0	25.0	8.5	—	—	—	46.0
24	9.0	6.5	12.25	11.75	13.25	0.75	—	53.5
27	6.75	7.0	10.0	13.0	11.25	1.25	—	49.25
31	7.5	3.75	18.75	8.5	—	—	—	38.5
32	9.5	6.75	25.5	5.5	—	—	—	47.25

Source: A. A. Val'ter, *Otdelitel'naia rabota podzheludochnoi zhelezy,* Military-Medical
   Academy Doctoral Dissertation Series (St. Petersburg, 1897), 180.
Note: I have reproduced minor computational errors in Val'ter's calculations of total
   secretion and have omitted information on the duration and ferment content of
   pancreatic secretion. The dashes which are reproduced from the original, apparently denote zero secretion.

In *Lectures,* Pavlov chooses trials 1 and 5, converts them into curves, and offers these to his readers as examples of the "truly stereotypical" work of the
glands. "The powerful impression of such an almost physical precision in a
complex organic process," he enthuses, "is one of the pleasant compensations
for sitting many hours in front of the glands at work."[18]

If we ponder Val'ter's data (Table 2), we can understand Pavlov's choice of precisely these two trials. Their results resembled one another in two important ways. First, they yielded much the same total amount of secretion (49.75 and 45.0 cubic centimeters). Second, they exhibited much the same course of secretion. That is, the amount of secretion in each began modestly in the first hour, fell slightly in the second hour, rose about threefold in the third hour, declined sharply in the fourth hour, and fell to almost zero in the fifth hour.

No other pair of trials fulfills these two criteria so well. Let us assume that Pavlov first chose trial 1 as his "template curve." Which other trials provide a good match? The total amount of secretion in trial 2 is too low. That in trial 3 is very close indeed, but here the amount of secretion rises much less sharply in the third hour and declines much less sharply in the fourth hour—hardly stereotypical. In trial 4 the total amount of secretion is, again, very close to that in trial 1, but the sharp increase in secretion comes only in the fourth hour. Trial 5 is a good fit, and Pavlov used it accordingly. The total amount of secretion in trials 6 and 8 is too low and the slope of these curves diverges markedly at several points from that in trial 1. Trial 9 fits trial 1 more snugly than does trial 5 in terms of total secretion, but the amount of secretion more than doubles in the second hour, contrasting sharply with the slight decline in trial 1. Trial 10 is again a good fit in terms of total amount of secretion, but the amount of secretion rises inappropriately in the fourth hour. Proceeding through Pavlov's "storehouse of information" in this manner, we can see that trial 5 provides the closest fit with trial 1, and of the others, only trials 19, 22, 23, and 32 offer plausible (if less convincing) alternatives.[19]

Pavlov was no doubt thinking about the selectivity of his choices when he interrupted his argument to concede, "Of course, not all experiments are so similar as those given, but if such a similarity is encountered in two experiments out of five, or about that, this cannot, in all justice, but be considered clear proof of the strict lawfulness of glandular work."[20] The phrase "two experiments out of five" refers, as we have seen, to his selection among Khizhin's experiments; the less exact "or about that" refers to his choice of two experiments (or, one might argue, of six experiments) from Val'ter's twenty-four.

It is not difficult to imagine Pavlov as he writes this passage—searching through Val'ter's data, finding just the right two experiments, and reflecting a bit about what he has done. He was, of course, engaging in a certain sleight of hand: assuring readers that he was presenting "typical" items from his "storehouse of information" while actually choosing those that made his case most convincingly. In my view, however, he signaled this to his readers, however elliptically, not only because it seemed the honest thing to do but also because he was essentially comfortable with his interpretation of experimental data.

He was, after all, following Bernard's dictum to present one's "most perfect experiment as a type"—that is, to choose the experiment that has been most effectively stripped of the "numberless factors" concealing the determinism of physiological processes. Pavlov was no doubt confident that, were he to show all his results to an open-minded and experienced physiologist and were he to have the opportunity to explain the complexity and difficulty of chronic experiments, the variations in mood and temperament from dog to dog and day to day, and the other "numberless factors" that obscured experimental results, such a physiologist would accept his choice of "model experiments." This was especially true because, although his raw data did not compel belief in a precise digestive factory, they did cluster in a manner that allowed one, if so inclined, to discern the contours of a factory concealed within.

In *Lectures* Pavlov moves immediately from acknowledging his selective choice of data to a reaffirmation of his central point. "There is every reason to think that the fluctuations encountered in various experiments are conditioned by often-overlooked differences in experimental conditions; that is, even in these fluctuations from experiment to experiment, the work of the glands is strictly lawful." This assertion suffices for him to reaffirm, two sentences later, that the "precision and stereotypicity of this curve" demonstrate the purposive work of the glands.[21]

Pavlov employs the same rhetorical tactic in his next two pairs of curves, which make the case for "the astonishing precision" of the gastric and pancreatic glands with respect to the fermenting power of secretion (pp. 49; 29). For the gastric glands, he chose two of Lobasov's experiments with 400 grams of meat. The results of these trials were not reported in the praktikant's dissertation, so I cannot analyze Pavlov's choices in detail here.[22]

The more complete data in Val'ter's thesis, however, permit a relatively close look at what would prove to be one of Pavlov's more controversial knowledge claims. In Pavlov's view, we should recall, the pancreas secreted three different ferments: a proteolytic ferment (which breaks down proteins) an amylolytic ferment (which breaks down starch), and a fat-splitting ferment. The precise and purposive response of the pancreas to meat, milk, and bread, then, was manifest in two ways. First, each specific food elicited a distinctive course and quality of pancreatic secretion, and second, the three ferments in the secretory response to any one food were not present in equal amounts (since the nature of the food demanded more of one ferment than another) and their secretory courses varied independently from one another.

In *Lectures,* Pavlov again draws upon Val'ter's data on the pancreatic response to a meal of 600 cubic centimeters of milk to make this point. He chose

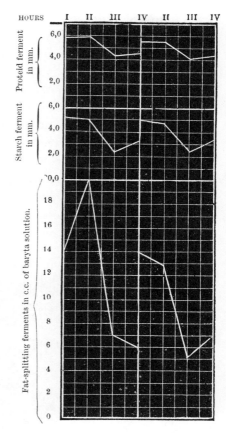

FIGURE 20. *Secretory curves showing ferment content in hourly portions of pancreatic juice after a meal of 600 cubic centimeters of milk. From From I. P. Pavlov,* Lektsii o rabote glavnykh pishchevaritel'nykh zhelez *(1897), in* Polnoe sobranie sochinenii *(Moscow: USSR Academy of Sciences, 1951), 2, pt. 2: 50; W. H. Thompson,* The Work of the Digestive Glands *(London: Charles Griffin and Co., 1902), 30*

the data from two trials of Val'ter's twenty-four, converted these into curves, and presented them to readers as an illustration of "the astonishing precision in [glandular] work: the glands provide precisely what is demanded, produce it consistently and to a hair's-breadth."[23] According to Pavlov, the curves for each ferment are distinctively different one from the other and vary independently (Figure 20). The secretion of each ferment, then, follows its own independent course, providing a distinctive mixture of ferments at each moment—a mixture suited to the optimal digestion of each different food.

Pavlov did not have a great many trials from which to choose here, and these curves proved the least rhetorically convincing in *Lectures.* Val'ter's thesis reveals that he had complete data for only three experiments of this type and that Pavlov chose the two that yielded results most closely resembling one another.[24] Interpreting these data through the lens of his factory metaphor, Pavlov saw the curves for the three ferments as fundamentally different and clearly independent one from the other. As we shall see, some other scientists would draw precisely the opposite conclusion from these same curves and would argue that the three ferments varied, rather, in parallel (see Chapters 9 and 10).

In *Lectures* Pavlov now moves from exemplary concrete trials to the characteristic secretory curves, the curves illustrating the distinctive responses of the gastric and pancreatic glands to various foods. I examined in Chapters 4 and 5 the construction of several of these curves; here I touch briefly on two revealing moments of curve construction. Reproduced in Figures 21 and 22 are two of the four sets of characteristic secretory curves from *Lectures*.

First, Pavlov "presents only one example" for each type of food, and he asks his reader to "believe that here, too, the precision is repeated no less than in the earlier examples."[25] By "one example," he means the results for feedings

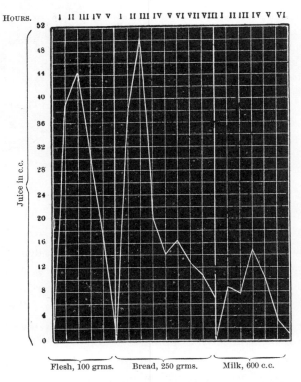

FIGURE 21. *Characteristic curves for the amount of pancreatic secretion with different diets. From I. P. Pavlov,* Lektsii o rabote glavnykh pishchevaritel'nykh zhelez *(1897), in* Polnoe sobranie sochinenii *(Moscow: USSR Academy of Sciences, 1951), 2, pt. 2: 64; W. H. Thompson,* The Work of the Digestive Glands *(London: Charles Griffin and Co., 1902), 40*

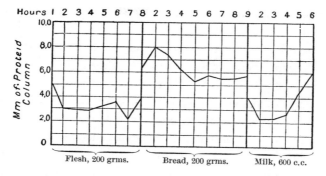

FIGURE 22. *Characteristic curves for the proteolytic power of gastric secretion after meals of meat, bread, and milk. From I. P. Pavlov,* Lektsii o rabote glavnykh pishchevaritel'nykh zhelez *(1897), in* Polnoe sobranie sochinenii *(Moscow: USSR Academy of Sciences, 1951), 2, pt. 2: 57; W. H. Thompson,* The Work of the Digestive Glands *(London: Charles Griffin and Co., 1902), 36*

with only one quantity of each food. These examples were carefully chosen (see Appendix B). Second, the tension between Pavlov's scientific vision and the nature of his data emerges in a curious manner: the stereotypical milk curve in Figure 21 *differs* from the two curves for concrete trials with milk already provided to demonstrate the "astonishing precision" of the glands (see Figure 19). This reflects the problem created by Pavlov's alternating reliance on average data and on data from concrete trials. The stereotypical secretory curve in Figure 21 was apparently constructed from Val'ter's average data; but when Pavlov sought two *concrete* trials to make his point about the precision of glandular work, he could not find two that both closely resembled one another and conformed to the average data. In effect, then, in *Lectures* Pavlov provides two *different* stereotypical curves for the amount of pancreatic secretion elicited by a meal of 600 cubic centimeters of milk. (This went unmentioned by reviewers and subsequent commentators, which perhaps tells us something about the way people tend to read curves and data charts.)

Verifying the Stereotypicity of the Template Dogs

Buried at the end of a long paragraph near the conclusion of Lecture 6, a paragraph devoted to demonstrating the reliability of the isolated sac, we find the following sentence: "Lately, another dog with an isolated sac constructed according to our method is stereotypically reproducing all the main facts collected on our first [such] dog" (pp. 148; 109). Pavlov is referring here to Volkovich's trials with Sultan, which had started after *Lectures* had gone to press.

Coming as it does several chapters after Pavlov's discussion of gastric secretion, this sentence does not disturb the general impression (not one reviewer raised the point) that the characteristic secretory curves—supported with data from "the experiments of Dr. Khizhin" and "the experiments of Dr. Lobasov"—are based on countless experiments on numerous dogs.

In sharp contrast to his rhetorical tactics at other junctures, Pavlov does not demonstrate the stereotypicity of results with Druzhok and Sultan by presenting secretory curves constructed from data generated with these two animals. Nor did Volkovich do so in his thesis. Although the praktikant appealed in his dissertation to a "comparison of the curves of secretion in both dogs," no such curves appeared there. Volkovich offered only a description of the general "characteristic features" of the dogs' secretory responses to various foods.[26]

Nor did Volkovich present the data on Druzhok and Sultan in a manner calculated to ease comparison. He offered separate charts for each dog; each chart contained selected data on the amount and proteolytic power of the dogs' secretory responses to meat, milk, and bread. Pavlov offers even less in *Lectures,* dispensing not only with the curves used so effectively in other cases to illustrate stereotypical similarities but also with all data from experiments with Sultan. Why did neither author provide separate curves for the two dogs and invite the reader to agree to their essential identity? Given Pavlov's rhetorical tactics in *Lectures,* it seems likely that they considered doing so. Had they used the average data provided in Volkovich's dissertation, these would have yielded the curves presented in Figure 23.

These curves would hardly have refuted the laboratory's basic view, and Pavlov no doubt saw them as "essentially stereotypical." (The reader has already encountered the top six curves in Figure 15 in Chapter 5, where I asked that the reader attempt to group them.) Clearly, however, these curves did not meet the laboratory's rhetorical standards for demonstrating factory-like precision and stereotypicity. Pavlov could indeed have argued that these two sets of curves manifested the same general features. The similarities in the two sets of bread curves are striking. Both curves for the amount of secretion elicited by meat reach their peak in the first or second hours, declining gradually thereafter; and both curves for the amount of secretion in response to milk rise slowly to a peak in the third hour, declining gradually thereafter. Other sets are more problematic. In any event, Pavlov and Volkovich apparently decided that the dramatic visual display provided by these curves would not effectively support their conclusion that the results with Sultan reproduced those obtained with Druzhok "with striking accuracy."[27] This did not mean, of course, that they could not interpret the differences in results in a manner that would up-

Amount of Secretion: Average values for Druzhok and Sultan

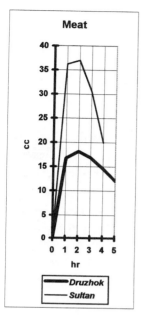

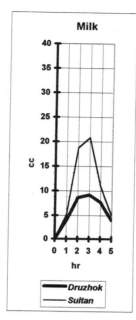

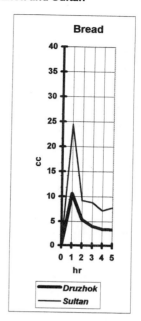

Proteolytic power of secretion: Average values for Druzhok and Sultan

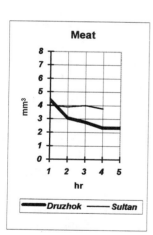

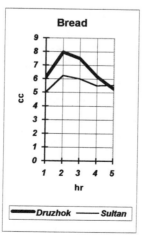

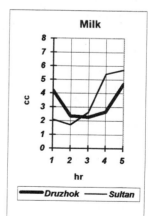

FIGURE 23. *Results with Druzhok and Sultan compared, based on data in Volkovich's thesis. These are average results for amount of secretion (in cubic centimeters) after a meal of meat (400 grams), bread (200 grams), and milk (600 cc for Druzhok; 550 cc for Sultan). Volkovich does not provide average results for other quantities elsewhere in his thesis. (Note that here proteolytic power is recorded as cubic millimeters.) From A. N. Volkovich,* Fiziologiia i patologiia zheludochnykh zhelez, *Military-Medical Academy Doctoral Dissertation Series (St. Petersburg, 1898), 25*

hold their essential stereotypicity and therefore the laboratory's general view.[28] Still, the differences in these curves dramatize the extent to which the characteristic secretory curves in *Lectures* reflect certain particularities in the personality of "the remarkable Druzhok."

Only after the publication of *Lectures* did Pavlov assign to Abraham Krever the unenviable task of verifying Val'ter's conclusions regarding the pancreas, conclusions that the chief had already publicly pronounced stereotypical and featured in his own synthetic work. Val'ter had openly conceded (in his thesis) the great variations within his data. Krever's results, obtained on five new dogs, were more varied still. He reconciled these data with Val'ter's conclusions through judgments about the reliability of various experiments and experimental animals. Of five experiments with one dog, Sokol, only one yielded results that could be "considered normal" according to Val'ter's schema. Of the remaining four, one had "a certain hypersecretory character" and the other three produced "a type of secretion that neither we nor Val'ter had ever observed." Krever (no doubt with Pavlov's counsel) interpreted these discordant data as the result of Sokol's nervous nature and the walks through the Institute courtyard by which the experimenter and attendants sought to calm the dog—factors that "made a new psychic excitation possible" and explained the "hypersecretory character of the experiment."[29] One experiment out of three on the second dog, Lyska, produced results that were "close to Val'ter's type, but with deviations." In this case, "Obviously, as a result of the recent operation, the physiological sensitivity of the stomach or the pancreatic gland was raised."[30] So Krever discarded the distorted experiments with Sokol and Lyska (but not those that produced results resembling Val'ter's). His other three dogs produced similarly mixed results: their secretory curves corresponded at several points with those that Val'ter had observed in Zhuchka, but diverged fundamentally at other points. These divergences, however, were ascribed to the vagaries of experiments with "several dogs with their individual personalities." When these factors were duly considered, Krever concluded, "our experiments confirm Val'ter's schema."[31]

Val'ter's findings underwent a second verification in 1902–4, when Iakov Bukhshtab examined pancreatic secretion in a context considerably complicated by Bayliss and Starling's discovery of a humoral mechanism of pancreatic secretion and by developments in the biochemistry of pancreatic ferments (see Chapter 9). These developments, Bukhshtab acknowledged, made it "difficult" to uphold Val'ter's scheme of purposive secretory curves. For much of the text, Bukhshtab's dissertation reads like a refutation of Val'ter's argument: the results with bread, meat, and milk did not show the same "constancy" as did Val'ter's, fluctuating by over 100 percent from experiment to experiment.[32] Here again, however, the idiosyncrasies of individual dogs provided an explanation that upheld the laboratory's central idea.

> Obviously, the stereotypical course of secretion that repeats itself in some experiments can be acquired when the stomach, pancreas, and other parts of the digestive system are consistently functioning in a precise and correct manner. We have already spoken above of the many circumstances that can disturb these [normal] relations. Not every dog presents a suitable object for the analysis of normal relations. It was Val'ter's especially good fortune to work on a completely healthy dog who lived in the laboratory in complete health for two years, who adapted completely to laboratory life and to the new conditions created by the implantation of a pancreatic fistula. Therefore, [Val'ter's] typology of secretions preserves its correctness even today. Dogs that do not possess the [fortunate] qualities of Val'ter's Zhuchka are not distinguished by that dog's same *pravil'nost'* in the work of the gland, but in these other dogs, too, we can find the very same characteristic features of secretion established by Val'ter.[33]

So, Zhuchka was accepted as "normal," as a "suitable object for the analysis of normal relations." The divergent results produced by Bukhshtab's dog were ascribed to its "unfortunate" nature and so were discarded. The isolated experiments with this dog that did produce results corresponding to those obtained earlier with the template dog were taken as confirmation of Val'ter's characteristic curves.

The Unification of Physiology and Medicine

In *Lectures,* Pavlov uses his own laboratory and research findings to offer his views not just of the digestive factory but of modern experimental physiology and its relationship to clinical medicine. Pavlov himself emerges as an exemplar of both the modern laboratory enterprise and a medically relevant, physician-friendly physiology.

The "Ladies and Gentlemen" whom Pavlov addresses at the beginning of each lecture were primarily the physicians, medical students, and medical bureaucrats of St. Petersburg, who comprised the most important constituency for laboratory physiology.[34] By 1897 Pavlov had accumulated more than twenty years of experience as an intermediary between the laboratory and medical practitioners—as a medical student, a member of the faculty of the Military-Medical Academy, the manager of three laboratories (Botkin's, and his own at the Academy and the Institute) that were largely populated by physicians, and an active participant in the Society of Russian Physicians.[35] The origin of *Lectures* as a series of addresses to a medical audience at the Institute, together with Pavlov's extensive experience with this audience, explains much about its form, tone, and rhetorical qualities.

Pavlov also, no doubt, had a second primary audience in mind: *Lectures* was his first synthetic statement to western physicians and physiologists. Having limited his previous western publications to relatively narrow subjects, and having for over a decade eschewed western scientific conferences entirely, Pavlov clearly intended *Lectures* to be translated into western languages and to secure for him a "European reputation."[36]

Pavlov describes in *Lectures* a close relationship between the nature of the laboratory and the quality of its products. Only a scientific enterprise that was itself purposive, regular, and precise could divine these same qualities in the digestive system. Previous investigators, bound by clumsy methodologies and vulgar generalizations, had inevitably failed to discern the subtlety and precision of the digestive system. Working crudely, they perceived glands that responded bluntly and indiscriminately to mechanical, chemical, or thermal agents. Pavlov's laboratory, on the other hand, had perfected surgical and experimental methods with a precision appropriate to their object of investigation. The data thus obtained enabled him to "banish—one hopes forever—[this] crude and fruitless idea" of the "bluntness" of the digestive glands by revealing "the contours of an artistic mechanism imbued, like everything in nature, with subtlety and internal purposiveness."[37] The distinctiveness of the laboratory's methodologies and technologies is underlined by the absence in *Lectures* of references to contemporary work in other laboratories. Pavlov praises Nicolas Blondlot and Rudolf Heidenhain for their contributions of earlier decades, but in the new era defined by physiological surgery and the chronic experiment, his own laboratory stands virtually alone.[38]

The very nature of Pavlov's laboratory enterprise enabled him to offer not merely his own interpretation but the "view of the laboratory," the result of thousands of experiments on hundreds of dogs by dozens of coworkers. "We, the laboratory," is on proud display in *Lectures,* where Pavlov praises his coworkers, refers to twenty-one of their publications, and constantly cites their experiments and specific discoveries. Like previous investigators, he and his coworkers had, of course, committed their share of methodological and interpretive errors. Their coordinated efforts, however, ensured the quality of the laboratory's products. The "laboratory view" was "constantly tested, frequently corrected, and, consequently, the most correct."[39]

Addressing himself to the medical practitioners in his audience, Pavlov emphasizes the necessarily symbiotic relationship of laboratory and clinic and offers a physician-friendly physiology that is respectful of traditional clinical wisdom. Effectively "triangulating" between mechanistic physiologists and physicians, Pavlov defends the empirical wisdom of the latter against the vulgarisms of the former while insisting on the indispensable role of the right kind of laboratory physiology. His perspective here differs sharply from the common view that "basic science should direct medical practice," a view that, as John Harley Warner has argued, often pitted laboratory knowledge against clinical experience.[40]

The physician, Pavlov explains—using a popular metaphor of the time—was the "mechanic of the human body." Only by gradually incorporating physiological facts would medicine eventually become that which it ideally should be: "the ability to fix the damaged mechanism of the human organism on the basis of a precise knowledge of it, the applied knowledge of physiology."[41] Yet the "strictly scientific resolution of therapeutic questions" still lay far in the future; the laboratory could not yet bring within its purview the complex issues that confronted the physician, whose "laboratory" was mankind itself: "Physiology, of course, cannot presume to forcefully guide the physician, because, lacking *complete* knowledge, it turns out constantly to be narrower than clinical reality. As compensation, however, physiological information can often elucidate greatly the mechanism of ailments and the inherent sense of useful empirical therapeutic practices. It is one thing to use a remedy without knowing its mode of action, and quite another—an immeasurably more advantageous situation—to know clearly what you are doing. The latter case will of course be more successful, with treatment that is better adapted to specific circumstances" (pp. 176; 133). Pavlov's specific examples of his physiology's contribution to medical practice, then, are almost always cases in which his laboratory had *confirmed* and *refined* accepted clinical (and even folk) practices. He adduces many cases in which, as adjudicated by his laboratory, the "instinct" of common people and their physicians enjoyed a "victory in the court of physiology" (pp. 176; 133).

These cases revolved largely around Pavlov's reaffirmation of the importance of appetite. Peoples and medical practitioners throughout the world had long understood that "food should be eaten with attention and satisfaction" (pp. 176; 133). The use of alcohol, spices, and condiments, the provision of a separate room for meals, and rules of etiquette that forbade serious discussions at the table all attested to a common understanding that it was important to facilitate "attention, interest, and pleasure in food" (pp. 178; 134). (Pavlov explains further that such habits were more extensively developed among the "more intellectual and well-to-do classes" for two reasons: "First, because among them there is greater intellectual activity and the various issues in life are more unsettling; and second, because food is usually provided in greater quantity than necessary [for the organism]. Among the simpler classes—where intellectual life is more elementary and there is a greater expenditure of muscular energy and a general insufficiency of nutrients—the interest in food is normally strong and lively without any special measures" [pp. 181; 137].)

Unfortunately, writes Pavlov, scientifically minded physicians had been led astray by physiologists in recent times. Blondlot had emphasized the importance of appetite in his *Traité analytique de la digestion* (1843), but in subsequent decades physiologists, unable to transform this psychic moment into a reliable laboratory phenomenon, had increasingly ignored it. They made "only minor mention . . . of psychic gastric juice, which was treated as a mere curiosity. Furthermore, great significance was ascribed to the mechanical irritation" of the digestive apparatus by food. Medical textbooks had followed suit, reducing appetite to "an insignificant subjective symptom"; and scientifically oriented practitioners, "attempting to find assistance in laboratory data and not finding there facts relating to appetite, naturally cooled toward it in their clinical practice" (pp. 179; 135).

The "progressive significance" of physicians' confidence in laboratory knowledge was "indisputable," but "here, as in all human affairs, there are errors and extremes. One must not forget that the absence of one or another phenomenon in given laboratory conditions does not mean that it is a fantasy." Scientists, Pavlov explains, frequently do not know all the conditions necessary to a particular phenomenon and so cannot reproduce it in the laboratory. Physicians should not, then, too readily relinquish their authority to the laboratory wisdom of the day, nor should physiologists ignore time-tested medical practices when framing and interpreting their experiments.[42]

In this account, Pavlov and the practicing physician emerge as partners and heroes. Pavlov had corrected the error committed by less skilled (and more mechanistic) experimentalists, restoring appetite to its rightful place in the digestive apparatus. In so doing, he had confirmed traditional clinical wisdom. Practicing physicians, while often avoiding any "direct" mention of appetite, had stubbornly continued to practice many therapeutic measures that were oriented toward its promotion. For example, although the laboratory had provided no evidence in favor of the use of bitters, which had consequently been "all but expunged from pharmaceutical lists," physicians continued to rely on them when treating patients. Now that Pavlov had affirmed the importance of appetite, this stubborn attachment to bitters was vindicated. The physician, with his worldly experience, and Pavlov, whose vision and experimental skills allowed him to affirm the physician's wisdom, could together enjoy a moment of triumph at the expense of mechanistic physiologists (pp. 182–83; 138–39).

Pavlov's laboratory findings also validated other common medical practices and supported a broad conception of the physician's authority. Physicians had long fed milk to grievously ill patients—and Pavlov now demonstrated that, in the absence of appetite, milk, with its rich chemical exciters, was indeed the most easily digested of foods. Physicians had long claimed that part of their art was to adapt their treatment to a patient's experiences and inclinations—and Pavlov now had proven scientifically the importance of individuality to the digestive process. And physicians had long emphasized the importance of proper regimen—and Pavlov now underlined the significance of managing the idiosyncratic psyche. Particularly in caring for people "who live in the great [urban] centers, where life is especially tense," the physician must not only have the authority to prescribe travel or a stay in a hydropathic institution but should also cultivate among the public "a correct view of the process of eating" (pp. 181; 137). This was particularly true for the Russian physician. "Especially in Russia's so-called intellectual classes, with its still utterly confused ideas about life in general, one often encounters a completely unphysiological, sometimes even contemptuously inattentive attitude toward eating" (pp. 181; 137).[43]

In his closing (and subsequently much-quoted) peroration, Pavlov invites physicians to engage in the necessary dialogue between laboratory and clinic by alerting him to the deficiencies in his own work.

> If the physiological data collected here help the physician to understand anything in the sphere of his activity and facilitate a more correct and successful approach to treatment, then the physician can secure for himself still greater advantage by informing the physiologist of any corrections . . . that, from his point of view, need to be made to the explanations offered here, and by indicating those new dimensions of the subject that have been revealed to the physician in the broad world of clinical observations but still remain outside the physiologist's field of sight. I believe profoundly that only by such a lively exchange of views between the physiologist and the physician will the goals of physiology as knowledge, and medicine as an applied science, be quickly and truly attained. (pp. 195; 186)

The style and content of Pavlov's research allowed him to assume this rhetorical posture with great authority. Working with intact animals, he was able to make specific recommendations about the treatment of intact patients and to appreciate the individual idiosyncrasies with which physicians contended in their daily practice. (For their part, one suspects, physicians would have been sympathetic to the inherent difficulties of obtaining precisely the same results in different experimental trials, as such was often the case in their own therapeutic experience.) As we shall see in Chapter 9, this fundamental similarity in their daily work had, for years, facilitated an exchange of experiences and advice between Pavlov and St. Petersburg's physicians.

Here, too, resided the great appeal of Pavlov's grand synthesis of the two basic moments at the heart of his scientific vision: his effort both to identify precise, quantitative, determined physiological laws *and* to encompass the intact, functioning animal. The joining of these two moments in *Lectures*, embodied thematically in the characteristic secretory curves and the powerful, idiosyncratic psyche, presented physiologist and physician alike with an impressive example of the union between the ideals of laboratory science and the daily complexities of medical practice.

The Dual Aftermath of Lectures

Lectures represented both a final processing of laboratory data and the laboratory's most important literary product. These two dimensions of the chief's synthetic work now take us in two different directions. As the laboratory's most refined and widely read literary product, *Lectures* played an important role in the historical fate of the laboratory's other products. This is the subject of Part III. As a stage in data processing, *Lectures* represented both the culmination of six years of intensive research and ground zero for a new beginning. Pavlov and his coworkers continued their research on digestive physiology for about six years, but in 1897–1904 investigations on the relatively minor salivary glands precipitated a revolution in Pavlov's approach to the psyche and in the very direction of his laboratory's research. It is to this development that we now turn.

Chapter 7

FROM THE MACHINE TO THE GHOST WITHIN

You know, this very same I. P. Pavlov used to talk all the time about "psychic juice" in the dog. And now I. P. Pavlov denies that the dog has a psyche.

> —I. R. TARKHANOV, remarks to the Society
> of Russian Physicians (1906)

The phenomena of the conditional reflex, if one gives them a psychological term, are precisely . . . associations.

> —IVAN PAVLOV, remarks to the Society
> of Russian Physicians (1904)

How did the scientist who established the importance of the idiosyncratic psyche in digestive physiology become the champion of an objective approach to mind? How, in other words, did "psychic secretion" become a "conditional reflex"—and the subject of an entirely new line of investigation that displaced digestive studies in the Pavlov laboratory?

This transformation is best understood as the product not only of specific experiments and Pavlov's scientific vision but also of the laboratory itself. The conceptual dynamics of this development owed much to a departure from Pavlov's standard managerial practice. Confronted with a problem that he recognized as psychological and therefore beyond his expertise, he recruited outside experts to help him resolve it, thus importing perspectives from contemporary psychology and psychiatry. The important role of insights from these two disciplines in the birth of research on conditional reflexes has been obscured by Pavlov's tale about this episode, a tale repeated uncritically by subsequent commentators.[1]

Vul'fson, Pavlov, and the Psyche's Choices

In the 1890s Pavlov's laboratory devoted relatively little attention to the salivary glands, but this line of research moved quickly along the standard investigatory sequence. In 1893 the chief assigned one praktikant, Sergei Ostrogorskii, to confirm nervous control over the glands.[2] A former praktikant, David Glinskii, working independently, accomplished the next step: development of an improved salivary fistula. Glinskii's fistula allowed the separate collection of fluids from the three salivary glands (the parotid gland and the two mucous glands—the submaxillary and sublingual). Using this fistula, Glinskii confirmed Claude Bernard's earlier observation that the various salivary glands responded differently to the same exciter and that these glands were "extraordinarily sensitive to the dryness of food." In 1895 Pavlov reported to the Society of Russian Physicians on Glinskii's results, finding in them evidence that, as for the pancreatic and gastric glands, "the work of the [salivary] glands is entirely purposive and individualized."[3]

In *Lectures on the Work of the Main Digestive Glands,* Pavlov devoted only a few pages to the relatively unimportant salivary glands, but affirmed that, like the main glands, they were controlled by nerves, responded only to specific exciters (and not to mechanical irritation), and secreted purposively. Saliva was the first fluid that greeted food on its entrance into the digestive factory. "It must, therefore, provide a good welcome to the entering substances." If an ingested food was dry, saliva moistened it; if it was hard and bulky, saliva covered it with mucin (mucus) in order to facilitate its passage down the narrow esophagus. Saliva also initiated the chemical processing of some foods (such as starch) and served as a "washing-out fluid" when injurious substances entered the mouth.[4]

Even more markedly than the gastric and pancreatic glands, the salivary glands responded to psychic excitation. This was long known to various investigators, and Pavlov presented it as a familiar observation: "Daily experience has shown us that the salivary glands become active even before food reaches the mouth. With an empty stomach, the sight of food or even the thought of it suffices for the salivary glands to immediately set to work; which is the basis of the well-known expression that 'one's mouth is watering.' So, a psychic event, the passionate longing for food, is undoubtedly an irritant of the nervous centers of the salivary glands."[5]

In sum, laboratory research on the salivary glands was very much on the standard Pavlovian track. By 1896 this line of investigation had reached the same place as had research on the gastric glands before Pavel Khizhin's experiments with Druzhok and research on the pancreatic gland before Anton Val'ter's experiments with Zhuchka. The next step was to use Glinskii's improved salivary fistula—just as Khizhin had used the isolated stomach and Val'ter the improved pancreatic fistula—to establish quantitatively the *prav-il'nost'* ("regularity" or "correctness") of salivary work.

Pavlov entrusted this task to Sigizmund Vul'fson, who arrived at the laboratory in 1896. Vul'fson was a typical praktikant: a young physician with no background in physiology, he required only a dissertation to complete his requirements for a doctorate in medicine. Pavlov assigned him "to establish precisely the exciters of salivation and to confirm their specificity and purposiveness."[6] Vul'fson accomplished this with skill and dispatch: he reported to the Society of Russian Physicians in October 1897 and again in March 1898, and shortly thereafter he completed his dissertation, *The Work of the Salivary Glands* (1898).[7] (The classical title of Vul'fson's thesis, like Khizhin's and Val'ter's before it, reflected the chief's judgment that it established the fundamentals of salivary secretion.)

Vul'fson proceeded much as had previous investigators working on the main glands. From March 1897 through February 1898 he experimented on four dogs with a varying combination of salivary fistulas (Milka, for instance, had a fistula of the two mucous glands; Ryzhaia a fistula of the parotid gland; and Vorona a fistula of both mucous and parotid glands). He analyzed the quantity and quality of glandular reactions to teasing with and feeding of various edible and inedible substances (including hydrochloric acid, urine, bisulfide of carbon, and a mustard oil solution). In the teasing experiments, Vul'fson followed the lead of previous laboratory researchers: "In order to heighten the effect of the psychic stimulation, the experiments were usually conducted in a manner creating the appearance that we wished to introduce these substances into the dog's mouth." Sometimes, to facilitate the dog's "better recognition" of a substance with which it was being teased, the substance was tinctured or placed in a distinctive container. Frequently these substances were placed in the dog's mouth immediately before the teasing experiment began "in order to leave in the animal a lively representation about them. Then we awaited the end of the secretion and began the experiment."[8]

Vul'fson identified a "strict purposiveness" in the work of the salivary glands.

The salivary response to edible substances varied in quantity according to the dryness of the food and was uniformly rich in mucin. Inedible substances elicited "about the same" quantity of saliva as did foods, but its mucin level was generally low and varied little from one inedible substance to another. This, Vul'fson observed, made good "sense." Mucin served to lubricate a food-stuff for its passage down the digestive canal, and inedible substances were not swallowed but, rather, ejected from the mouth. Mucin was unnecessary to this latter task, so inedible substances elicited instead a watery saliva that served to cleanse the mouth of any unexpelled remnants.[9]

The psyche's role in this "strict purposiveness" was especially evident in a number of "special facts." Most important, "psychic secretion is a complete reflection of the direct, purely physiological secretion, differing only in amount."

> The participation of the psyche in the process of salivation is indubitable. The very same results regarding quantity and quality of saliva that are acquired when substances come into direct contact with the roof of the mouth occur also when they are used for teasing alone. The very same dependence of the quantity of saliva upon the dryness of food that occurs during feeding is clearly manifest when the dog is merely teased with [dry food]. Just as introducing unpleasant substances into the mouth elicits salivation, so does a mere representation of the unpleasant substance elicit a rather strong secretion from both types of glands. Substances that do not excite the salivary glands by their direct contact with the mucous covering of the mouth also do not elicit secretion by their psychic influence. So, teasing with water, even when the dog is very thirsty, produces no effect. With respect to the composition of the secretion of the mucous glands, the reaction to teasing with food is no less accurate: a thick secretion rich in mucin; and in reaction to teasing with rejected substances the secretion is watery. But what is a simple physiological reflex in the case of the direct contact of substances with the roof of the mouth remains [in the case of psychic secretion] undefined and subject to reinvestigation.[10]

In contrast to gastric and pancreatic secretions, then, the psychic secretion from the salivary glands was essentially *identical* to secretion during the second, nervous-chemical phase of digestion. This psychic secretion manifested the same ability to differentiate among and respond purposively to substances that, in the gastric and pancreatic glands, occurred only in the nervous-chemical phase of digestion, when, according to laboratory doctrine, the specific excitability of the nerve endings in the mucous membrane of the stomach and duodenum generated distinctive responses to different foods.

Vul'fson observed in his report to the Society of Russian Physicians in October 1897 that the highly specific responses of the salivary glands involved not only "emotion, but also an element of thought—a representation about the nature of the external substances falling into the roof of the mouth."[11] The "adaptation" of salivary reactions to specific substances was "almost entirely of a psychic nature."[12] "The task of the psyche," he wrote, is "to sort out" objects, "to divide" them into two groups—accepted and rejected substances—in order to respond to each with the appropriate salivation.[13] The psyche exhibited great "scrupulousness," an unerring "judgment of particular circumstances," and the ability to "generalize."[14]

Two experiments were especially telling. The first concerned the salivary response to a combination of meat and mustard oil. When the dog ate meat alone, its saliva was rich in mucin, which facilitated the meat's passage down the digestive tract; when the dog was fed mustard oil (a most unpleasant substance), the glands produced a watery saliva to wash out the mouth. When mustard oil was added to meat, the result was not a simple summing of the responses to the two substances but rather a watery saliva appropriate to a rejected substance. The psyche, then, had transcended mere chemical responses to render a *judgment:* the combination of meat and mustard oil was categorized as a rejected substance.[15]

A second important experiment underlined the same point. "Rocks introduced into the dog's mouth do not elicit salivation, [but] sand elicits a great amount of saliva. What other than an evaluation of particular circumstances, and so the psychic moment, could determine this difference in salivation?"[16] Here again, the dog's salivary response was governed not by the simple chemistry of the substance but rather by an acute judgment rendered by the psyche. Rocks and sand were identical in chemical composition, but they presented the salivary glands with different tasks: the dog simply ejects rocks from its mouth (which requires no saliva) but must wash down its throat the small grains of sand that invariably remain after its attempt to expel the sand. This latter task required a mucin-rich saliva. Clearly, Vul'fson informed the Society of Russian Physicians in March 1898, further investigation of the salivary glands "must move to the psychology of salivation"[17]

A delighted Pavlov echoed these sentiments. He lauded Vul'fson's report (which, of course, he had closely edited) for developing the doctrine of specific excitability by demonstrating the "subtle and sharp adaptation of the salivary glands: everything edible elicits from the submaxillary gland a saliva rich in mucin, and everything rejected yields a watery saliva." It was especially interesting, he continued, that in the work of this first part of the digestive system "the participation of the psyche emerges clearly, so psychology almost entirely overshadows physiology . . . Substances rejected by the organism, which, it stands to reason, are united by a sense of the unpleasant, elicit watery saliva. Meat usually elicits saliva rich in mucin, but when mustard oil is added meat moves to the category of substances that elicit watery saliva—although it preserves its physicochemical qualities. Finally, the following fact also testifies to the dominance of psychology: the same types of saliva are secreted both when a tested substance is put into the mouth and when it is only used to tease the dog." Pavlov then added, "One can say directly that if in other cases we speak in jest, metaphorically, about 'the mind' of the glands, then in this case 'the mind of the glands' should be understood literally."[18]

The "mind" of the salivary glands buttressed Pavlov's general case for the importance of the psyche and his rejection of simple, mechanistic models of complex physiological processes, but the specific qualities of this mind also presented an obstacle to his standardized investigatory path. For years, he and his coworkers had recognized the importance of the psyche but had simply "black-boxed" it. In analyses of the gastric glands, for example, a dog's initial secretory reaction to foodstuffs was simply attributed to the influence of appetite (i.e., to the psyche). This psychic secretion gave way, in the second phase of the digestive process, to the specific nervous mechanisms that produced the characteristic secretory curves. Vul'fson's research, however, demonstrated the inapplicability of this schema to the salivary glands. Here, psychic secretion was essentially identical to the nervous-chemical secretion, differing only in amount. The purposive, precise, and specific reactions of the salivary glands to different foods resulted, then, not from the specific excitability of the nervous system (as presumably was the case with the gastric and pancreatic glands) but rather from the psyche's ability to, as Vul'fson put it, "sort out," "arrange," and "judge." The Pavlovian program for glandular physiology, then, seemed to lead directly into the psychology of the salivary glands.

Recognizing that here "psychology almost entirely overshadows physiology" and conceding his own lack of expertise in this area, Pavlov abandoned a long-standing managerial practice and turned to an outside expert.

Psychology to the Rescue: Snarskii Opens the Black Box

Anton Teofilovich Snarskii was a most atypical praktikant. The holder of both university and medical degrees, he entered Pavlov's laboratory to investigate a subject in which he possessed greater expertise than the chief.

Snarskii had joined the medical community's campaign against the cholera epidemic of 1892–93 and subsequently served as a military physician in the Far

East. In 1896 he was transferred to St. Petersburg, where he worked in V. M. Bekhterev's clinic for mental and nervous illnesses. Shortly thereafter he served as a clinician in the Alexander III Charity Home for the Mentally Ill. Given this background, Snarskii was quite familiar with modern theories and practices concerning the mind and mental illness when, in 1900, he became a praktikant in the Physiology Division. Here Pavlov assigned him to investigate the "mind" of the salivary glands.[19]

There is good reason to believe that Pavlov recruited Snarskii for this specific purpose. Snarskii's boss at the Charity Home was Pavlov's friend A. V. Timofeev. Timofeev had worked with Pavlov in the Botkin laboratory, receiving his doctorate for a thesis completed under Pavlov's guidance. After studying neurology and psychiatry with Jean-Martin Charcot in Paris, Timofeev returned to St. Petersburg in 1891. Declining Pavlov's offer of a post in the Physiology Division, he became instead director of the Charity Home, a position he occupied from 1891 to 1916. On Sundays, Pavlov regularly rode his bicycle (or, on snowy days, skied) to the Charity Home, which was located just outside St. Petersburg, in Udel'naia. After supper at Timofeev's home, the two would tour the institution together, during which time Pavlov was treated to sessions with especially interesting patients.[20] It seems likely that Pavlov met Snarskii through Timofeev, perhaps during one of these Sunday tours; that he recognized in Snarskii an appropriate partner for exploration of the psychology of salivation; and that the two conceived a mutually beneficial arrangement: Snarskii would lend his expertise to the laboratory's investigation of the "mind of the glands" and in the process would in a year or two earn his doctorate.[21] Snarskii became the second of three Charity Home employees (each of whom had previously worked in Bekhterev's clinic) to be recruited for work in Pavlov's laboratory in the years 1897–1902.[22]

This atypical praktikant produced a most atypical doctoral dissertation. Unlike those of the vast majority of praktikanty, Snarskii's thesis drew extensively on scientific authorities outside the laboratory. He cited physiologists who had addressed the biology of purposive behavior (including Jacques Loeb and Ivan Sechenov), the Russian zoopsychologist V. A. Vagner, and a number of Russian and Western psychologists (including G. I. Chelpanov, William James, T. A. Ribot, and Wilhelm Wundt).

Snarskii mobilized these authorities to criticize Vul'fson's (and, implicitly, Pavlov's) conclusion that the psyche actively "chooses," "sorts out," "arranges," and "judges." By the standards of contemporary psychology, he insisted, Pavlov's "mind of the glands" did not deserve the label *mind*. Like the contemporary psychologists whom he cited, Snarskii distinguished among a wide variety of mental qualities that involved a broad range of capacities. He concluded that psychic secretion reflected not high-level processes such as will, choice, and judgment but rather the relatively low-level process of "visual associations."

Snarskii developed this line of argument by constantly juxtaposing his views to Vul'fson's. He basically accepted Vul'fson's methodology and empirical results, but rejected his conclusion that "the psyche determines the corresponding work of the glands."[23] This, Snarskii insisted, was a psychological conclusion—and an amateurish one at that.

> The method is entirely exact and, insofar as the problem concerns physiology, it provides entirely clear and lawful conclusions. But these conclusions enter the sphere of zoopsychology, and here, it seems to me, the author's conclusions are too hurried and constructed on an inexact and contradictory terminology, which, in the sphere of psychology, too, should be completely exact and definite.
>
> Apparently, [Vul'fson] was fascinated, on the one hand, by his discovery of the purposive activity of the salivary glands, and on the other, by a certain participation of the psychic activity of the dog in the act of salivary secretion. This leads him to the conclusion that "the psyche determines" and that the dog, too, makes a "choice." (p. 4)

Echoing Sechenov's famous argument in *Reflexes of the Brain,* Snarskii pointed out that "entirely purposive and complex acts are characteristic of even decapitated animals," which obviously lack a psyche and are incapable of "choosing" anything. (Sechenov had used the example of a beheaded frog that purposively withdraws its leg from a drop of acid.)[24]

What, then, did Vul'fson mean by the word *psyche?* Was he referring to "purely reflexive or instinctive" actions or to "conscious activity"?[25] This was a specifically psychological question, and neither Vul'fson nor Pavlov had ever addressed it systematically. Snarskii's response to it led him to greatly diminish the active role attributed to the psyche during salivation.

> Judging by [Vul'fson's use of] the words "determines," and, especially, "choice," the author had in mind the latter [i.e., conscious activity]. It is proposed that in each case of indirect irritation the dog consciously makes an evaluation of the given substance, compares the sensation with previous ones, thinks it over (perhaps even vacillates) and finally, confirming only one of various possibilities, makes a choice—that is, a higher, conscious, willful, and voluntary act. In other words, it makes the decision to secrete one or another saliva. Such, after all, is the meaning of the words "choice" and "decision" in psychology. One can hardly acknowledge such a subtle and complex activity for each secretion of saliva, especially since such activity is possible only

in the presence of representations in the precise meaning of that word—that is, of *memories* of past sensations. In our experiments the psychic activity of the dog is expressed in a much more elementary form: each time the dog *sees* before it an irritating object it only *recognizes* it and, consequently, receives a so-called *linked* representation (as distinct from a *free* representation, which alone is an independent part of consciousness). (p. 4; emphasis in original)

The same psychic secretion that for Vul'fson and Pavlov represented the psyche's active "judgment" and "determination" was for Snarskii a mere "linked representation" independent of volition.[26]

Citing research by Ribot, Snarskii noted that various sensory impressions were easily linked with specific foods and that various types of representations thus became associated with one another in various patterns. Visual impressions were grouped into "complex aggregates" and auditory impressions into a "consequential series." Snarskii's analysis of the various animal senses led him to the conclusion that a dog's recognition of substances was based largely on "visual associations." For taste, these associations were formed with the "tactile sensations from the roof of the mouth." Edible and inedible substances were recognized through "visual representations, for example, of the vessel, of a certain shape, or of the tinting of the fluid in a certain color" (p. 8). So, the phenomenon of psychic secretion involved both the recognition of objects and "a corresponding salivary reaction as a consequence of the established visual association."

From a "psychological point of view," Snarskii continued, one must ask what "rung" this phenomenon occupied in the "general chain of psychic phenomena." Following such "authoritative teachers of psychology as Wundt and Harald Hoffding," Snarskii concluded that psychic secretion was the result of "the simplest processes that unite new impressions with preceding impressions—that is, elementary memory." This process of recognition by means of "newly established associations"—what Wundt termed *recognition anew*—was of course a more complex act than those resulting directly from elementary animal appetites. This did not, however, justify Vul'fson's sweeping attribution to psychic secretion of higher-rung psychological qualities. "To draw from this entirely elementary act the conclusion that the dog makes a 'choice' about which saliva to secrete in the given case is to make an unfounded logical leap. Direct recognition does not rise [to the level of] even a free representation, to say nothing of the long chain of psychic acts, such as the formation of concepts, judgments and conclusions, which must precede a conscious choice and decision" (p. 9). Snarskii offered a better explanation.

As long as there is direct irritation, the secretion of saliva is only a simple reflex. When the dog recognizes a previous irritant . . . it repeats a habitual reflex; but repeats it automatically, without any participation of conscious,
active will. Schematically, this would be expressed as follows: there is established a common reflexive arc between the direct irritant, on the one hand,
and the act of salivation on the other. We can imagine that the centripetal
end of this arc is, so to speak, split; and therefore the very same salivary reflex can be received by the representations associated directly with the irritation. But this very circumstance—that the secretion of saliva as the result
of association is 'a complete reflection' [these and subsequent quotes are
from Vulfson's thesis] of the very same act during direct irritation, speaks
clearly against the possibility of 'a choice': the act is accomplished entirely
stereotypically, automatically, through a well-trod path. The consciousness
of the dog plays no 'important' role; it 'chooses' nothing and in itself does
not 'determine' the activity of the salivary glands. (pp. 9–10)

Snarskii here used one of Pavlov's favorite words, *stereotypical,* to make his
point: psychic secretion resulted not from judgment or any other volitional
process but from an association between the irritant, the act of salivation, and
associated (perhaps visual) representations. Or, "schematically," the reflexive
response to the food irritation is split: part continues as the usual salivary reflex, while part is received by and becomes associated with other (say, visual)
irritations. This process was rooted not in the dog's emotions but in its appetites—that is, in a physiological form characterized by "stability, consistency, automaticity, and the absence of the novelty and indecisiveness that appear together with consciousness." "We are obviously dealing with purely vital,
physiological needs," wrote Snarskii, which represent the same type of "elementary form of affective life" as the need for sleep and sex (p. 10).

Snarskii used this schema throughout his thesis to reinterpret experimental results that Vul'fson and Pavlov had attributed to the psyche's choices. For
example, when a substance, such as acid, that elicited a salivary reaction was
tinted black and poured into a dog's mouth, "there is apparently established a
visual association between the caustic taste and the substance's color; correspondingly, a reflex is established." If black-tinted water was then poured into
the dog's mouth several times, however, the salivary response ceased, since the
roof of the mouth was not irritated by the water. If black-tinted acid was again
poured into the dog's mouth and the experimenter then feigned putting black-
tinted water into the dog's mouth, salivation again occurred: "It is clear that
the dog, on the basis of the just-established reflex, repeats the very same reflex,
defending the mucous [membrane] of the mouth from the corrosive action of

the acid. Two signs—acid and color—turn out to be linked, and the dog reacts identically to one and the other—without judging and without choosing" (pp. 46–47). As with a child who, after a painful experience with a physician, cries on seeing him again, "Everything is settled by a single sign accidentally linked with the subject" (p. 47).

The "psychic moment" in salivation, then, was most definitely not, as Vul'f-son (and Pavlov) had concluded, "liberated from the guardianship of physiology." Rather, wrote Snarskii, "I think that the psychic element is a later superstructure, established through experience, and is therefore not independent but subordinate" (p. 48). Psychic secretion was a "reflex" that was entirely accomplished in the subcortical region of the brain, outside the brain's conscious centers in the cortex (p. 50).[27]

For Snarskii, the salivary response, with its psychic "superstructure," was a means by which organisms adapted to their environment. The "adaptiveness of reflexes to the uses of the organism" arose "very early in the chain of living beings" and was even manifest, as Loeb had demonstrated, in the purposive reflexes of denervated ascidians.[28] The voluminous salivary response to such inedible substances as acid was "clearly linked to the need to preserve the mucous membrane of the mouth from harmful influences; one can see in this an instinctive adaptation of the reflex in the sense of self-preservation."[29]

"According to the principle of Professor Wundt," Snarskii concluded, one should adopt "the simplest explanation possible." This explanation, it seemed to him, was that the psychic activity of the dog during the work of the taste apparatus came down to two things: (1) "the elementary act of the formation of associations regarding one or another substance" and (2) "recognition anew" (Wundt's term) of these substances by means of "established, mainly visual associations." Salivation during direct feeding was a "purposive reflex," and salivation in response to teasing was but "repetition, the automatic reproduction of an established reflex; an act completely devoid of any element of conscious choice."[30]

Snarskii completed his dissertation by September 1901 and defended it before a three-person committee at the Military-Medical Academy in January 1902. According to the short published record of the defense, both Pavlov and Val'ter (now a *privatdozent* at the Academy) praised the thesis, though not extravagantly. Val'ter commented on its "mass of scientific facts" and their usefulness for further investigation, but he criticized Snarskii for his skimpy presentation of experimental data. Pavlov emphasized the "very important fact" that such a small organ as the salivary glands manifested "psychic activity" and noted Snarskii's identification of the precise nerves in which "the physiological reflexes begin." The third reader, Bekhterev, was most critical, noting that Snarskii had not explained several discrepancies among the results of his own experiments and between his data and those obtained by Vul'fson.[31] According to Pavlov's later recollection, Bekhterev also remarked to the candidate, "Your duty and mine is to teach physiologists psychology." Here, perhaps, Bekhterev was alluding both to Snarskii's psychologically informed revision of Pavlov's long-standing approach to the psyche and to the tensions between the Bekhterev and Pavlov laboratories resulting from the increasingly close proximity of their research.

Snarskii had drawn upon contemporary psychology to challenge the lay, black-boxed notion of the psyche that had governed laboratory interpretations throughout the 1890s. He did not, however, indicate a path forward that was compatible with the chief's experiences, laboratory setting, and scientific-managerial vision. Pavlov knew virtually nothing about the discipline of psychology and was uncomfortable with the philosophical problems, procedures, and terminology associated with it. Unless he retooled completely, he was utterly incapable of supervising praktikanty along this line of investigation. Pavlov's expertise, rather, lay in chronic experiments with dogs, and in his view the requirements of good science included precise, repeatable, quantifiable facts. So, even if he accepted Snarskii's basic conclusions, he lacked a methodology by which psychic secretion (or associations or complex reflexes) could be addressed in a manner consistent with his notion of good physiology.

The chief again looked outside his laboratory for "a person with whom one could go further."[32]

A Context for Change: Russian Biological Psychology at the Turn of the Century

Snarskii was the tip of an iceberg, a representative of fundamental changes occurring in the second half of the nineteenth century in Russian professional discourse on the biology of mind. Until Snarskii's arrival, Pavlov and his laboratory had been largely insulated from these changes: the chief had his own, lay ideas about the psyche, and his praktikanty had little incentive or authority (either institutional or intellectual) to dispute them. Moreover, Pavlov's approach to the psyche had proven quite effective for his purposes, contributing much to the successful investigations of the gastric and pancreatic glands. As we have seen, however, this situation changed after Vul'fson's research. Convinced that, in investigations of the salivary glands, "psychology overshadows physiology," Pavlov turned to an outside expert, and in so doing he imported into his laboratory developments in the world of professional psychology. As we shall see shortly, a second outside expert, Ivan Tolochinov, would soon pick up this line of investigation where Snarskii left off.

During the 1860s and 1870s, discourse on the relationship of biological and mental processes had been deeply embedded in larger political and ideological issues. Sechenov's "Reflexes of the Brain" (1863) had created a sensation because the author's schema for explaining higher mental functions on the basis of reflex physiology had resonated with the materialist philosophy of the radical "men of the sixties." Sechenov's essay, and others like it, were associated with the radical critiques of free will, the Orthodox Church, and the tsarist state and with radicals' call for the modernization of Russian society on the basis of positive knowledge, modern technology, and westernization. Arguments about the biology of mind filled the pages of Russia's "thick journals" from the 1860s through the mid-1870s, challenging the tsarist censor to devise a policy that would allow scientific discussions to flourish while suppressing the subversive ideas that seemed so often embedded in them.[33]

We have seen the echoes of this highly polarized discourse in Pavlov's biography—in his impassioned reading as a seminarian in the 1860s of works by Sechenov, G. H. Lewes, and radical essayists, which contributed to his decision to become a scientist, and in the negative reaction of leftist essayists and students to I. F. Tsion's speech "The Heart and the Brain" in 1873–74. Pavlov's letters to Serafima in the late 1870s and early 1880s about the weaknesses of the "youthful mind" and the necessary "leave-taking from philosophy" also reflected a more general development. As a young physiologist, he was grappling with the distance between the broader controversial issues that had first interested him in science and the constrained philosophical claims and more mundane daily practices of the professional scientist.

In the years between 1863 and 1900, Russian discourse on biology and mind had undergone a similar process of professionalization and depolarization. The "men of the sixties" had been replaced by the "men of the eighties."[34] Physiologists, physicians, and psychiatrists of varying ideological and political stripe—men such as V. F. Chizh, S. S. Korsakov, P. I. Kovalevskii, N. N. Lange, A. A. Tokarskii, and a now more philosophically restrained I. M. Sechenov—developed biological and experimental approaches to mind with an eye to medical practice and the assertion of medical authority. For the professional men of the eighties, the broader philosophical and political issues that had animated an earlier generation either faded into the background or lay outside the province of positive science altogether. For these professionals, materialism and idealism gave way, at least in their scientific activities, to methodological "realism." Liberals and conservatives, populists and monarchists, materialists and idealists were thus able to work together—in the laboratory or on the editorial board of a journal—toward common professional and scientific goals.[35]

By the turn of the century, then, it was neither radical nor uncommon to assert that what Pavlov termed *psychic secretion* might be explicable as a reflex. Indeed, Pavlov's longtime coworker Leon Orbeli recalled that when Pavlov discussed psychic secretion in his lectures on physiology at the Military-Medical Academy, his interpretation was challenged by his student audience. "I remember those lively discussions that would occur in our auditorium. The listeners often asked: but can't this event be explained as a reflex, just one from another sensory organ? Ivan Petrovich [Pavlov] then would present an entire line of facts that contradicted this explanation and compelled one to relegate 'psychic activity' to a special group of facts. This was in 1900–1901."[36] This same sentiment was expressed by some Russian medical reviewers of Pavlov's work. One commentator wrote in the *Russkii Arkhiv Patologii, Klinicheskoi Meditsiny i Bakteriologii* (Russian Archive of Pathology, Clinical Medicine, and Bacteriology) in 1900, for example, that what Pavlov termed *igniter* or *psychic* juice was probably "the effect not so much of psychic excitation as of a direct reflex."[37] Similarly, Pavlov's former praktikant Petr Borisov, who became professor of pharmacology and balneotherapy at Novorossiisk University in 1903, expressed his opinion in the same year that, notwithstanding the opinion of his "much-esteemed teacher," "the secretion of gastric juice during so-called sham-feeding is conditioned not by a psychic moment but rather by irritation of the taste nerves and a reflex from them to the glands of the stomach."[38]

Even closer to home, and of more direct impact, was the laboratory enterprise of Pavlov's colleague at the Military-Medical Academy, Vladimir Bekhterev (1857–1927). The lives of these two researchers and scientific entrepreneurs were intertwined for over fifty years—in mutually respectful relations from the mid-1870s until about 1906, when, their research interests having grown very close, a clash over experimental results precipitated an acrimonious competition that lasted until Bekhterev's death.[39] Pavlov and Bekhterev had been medical students together at the Military-Medical Academy (Bekhterev graduated in 1878, Pavlov the year after), after which both had enrolled in the Academy's doctoral program. Pavlov had acquired his doctorate in 1883 and Bekhterev in 1884; in the latter year both received a two-year scholarship to study in Western Europe. Each spent some time in Carl Ludwig's laboratory, at which point their paths diverged. Pavlov traveled to Rudolf Heidenhain's physiological laboratory, and Bekhterev pursued research and training in neurology and psychiatry with Paul Flechsig, Charcot, and Wundt.

Bekhterev was a quintessential man of the eighties. On his return to Russia he became chair of the Department of Psychiatry at Kazan University, where he studied the influence of the central nervous system on various internal organs (including the stomach), founded a psychophysiological laboratory, and became a leading member of the generation of Russian researchers in the 1880s and 1890s who developed biological approaches to the psychological and psychiatric phenomena related to medical practice.[40] In 1893, by which time Pavlov was an assistant professor of pharmacology at the Military-Medical Academy, Bekhterev was appointed professor of psychiatry and neuropathology there. In both his physiological laboratory and his clinical laboratory on nervous and mental diseases, Bekhterev pursued research on the cortical localization of various vegetative and sensory functions.

Pavlov and Bekhterev initially enjoyed friendly and mutually beneficial collegial relations. They worked together on a number of Academy committees—in 1901, for example, they jointly approved the statutes of the Russian Society of Normal and Pathological Psychology—and in the years 1895–1906 they served on the same dissertation defense committee thirty times. In these years Bekhterev routinely approved the doctoral work completed in Pavlov's laboratory, and Pavlov returned the compliment.[41]

In the 1890s and early 1900s, Bekhterev and his coworkers published many articles on the brain centers controlling various physiological processes, on various reflexive phenomena, and on experimental psychology. These publications, which often appeared in Bekhterev's journal, *Obozrenie Psikhiatrii, Nevrologii, i Eksperimental'noi Psikhologii* (Review of Psychiatry, Neurology, and Experimental Psychology), now touched directly on the psychic secretions that were so central to the Pavlov laboratory's research on digestive physiology.

The first public clash between Bekhterev and Pavlov occurred in 1899, when one of Bekhterev's coworkers, A. V. Gerver, delivered a report to the Society of Russian Physicians, "On the Influence of the Brain on the Secretion of Gastric Juice." Noting the Pavlov laboratory's claims about the importance of the "psychic sphere of the animal" in gastric secretion, Gerver argued that psychic secretion was "nothing other than a reflex transmitted to the gastric glands through the central nervous system."[42] He claimed to have found the psychomotor center in the brain that controlled this function. Pavlov's criticism of Gerver's report drew upon a psychological distinction that the chief had apparently conceived while pondering Vul'fson's results: "I want to note that the influence of the psyche on secretion is expressed not only in the form of desire, but also in the form of a thought. This is clear, at least, for the salivary glands. If one gives a dog dry bread, which the dog does not want, the saliva flows nevertheless; furthermore, not only an appetite for any food, but also a desire for a particular food is important. It is obvious, then, that the psychic processes of secretion of the digestive juices are expressed in two forms."[43] Because these two processes—desire and thought—were localized in different parts of the brain, Pavlov reasoned, Gerver could hardly have discovered a single psychomotor center that controlled psychic secretion. Bekhterev defended his coworker by insisting that, "according to contemporary science," the brain had psychic, sensory, and motor centers. Clinical experience, as well as scientific research, indicated that psychic centers involved "associational and thought centers," so it was entirely likely that there existed "special psychic centers for the secretory organs."[44] In 1900–1903 the Bekhterev laboratory produced several publications on these psychic centers and the reflexive processes that controlled them.[45]

While Pavlov argued with Gerver and Bekhterev and maintained his distance from contemporary psychological theory and practice, two outside experts recruited to his laboratory enterprise nevertheless brought with them the perspectives and experiences of the medical men of the eighties in general and of Bekhterev's laboratory in particular. Snarskii was the first bearer of this influence. We now turn to the second.

Tolochinov Brings an Analogy from Psychiatry

When Snarskii's experiments were completed, Pavlov again looked outside his laboratory for "a person with whom one could go further." He settled upon Ivan Filippovich Tolochinov (1859–1920). Tolochinov, like Snarskii, was a veteran of the Bekhterev laboratory who worked in Timofeev's Charity Home for the Mentally Ill. Unlike Snarskii, on entering Pavlov's laboratory Tolochinov had already received his doctoral degree, for a thesis on alterations in the nerve fibers of the brain during "paralytic imbecility." He had investigated this subject in both Bekhterev's and Timofeev's laboratories and had completed his thesis under Bekhterev in 1900.[46]

Joining the laboratory at Pavlov's request, possessing expertise that Pavlov valued, and with a doctorate in hand, Tolochinov enjoyed an unusual relationship of virtual equality with the chief. He did not require Pavlov's approval, and they worked together not as chief and praktikant but as colleagues. "He already was a doctor and worked with me purely from scientific interest," Pavlov later recalled. "Ivan Filippovich became very close to my heart, became a person very close to me."[47]

This relationship is reflected in the unusual first-person-singular language of Tolochinov's research report of 1902. In discussing the origins of his research, for example, he writes *not* the usual "Professor Pavlov suggested . . ." but rather "I had in mind to study in a more detailed manner the role of the psychic process in salivary secretion; to do so I conducted in the physiological laboratory of professor I. P. Pavlov . . . 90 experiments on 4 dogs with salivary fistulas."[48]

The details of the Pavlov and Tolochinov collaboration during this research are impossible to unravel from existing sources, but despite the men's subsequent dispute, the versions provided by each agree in their general contours. First, Tolochinov conducted most of the experiments, sometimes with Pavlov's participation. Second, Tolochinov was largely responsible for the initial experimental and conceptual breakthroughs; as Pavlov put it, "he was the first to put his hand to it; priority, one could say, belonged to him." And third, Pavlov fundamentally reinterpreted Tolochinov's findings—as reflected in his replacing Tolochinov's term for psychic secretion, *reflex at a distance,* with his own term, *conditional reflex*—and conceived an experimental line of inquiry to build upon Tolochinov's initial results.[49]

From November 1901 until June 1902, Tolochinov traveled from the Charity Home to Pavlov's laboratory several afternoons a week to conduct experiments for about three hours at a time. Boris Babkin, who entered the laboratory in 1902, later recalled that sometimes Tolochinov and Pavlov worked together and sometimes Tolochinov worked alone. Tolochinov later recalled that, to avoid "extrinsic influences," the experiments were conducted in a separate room "in the presence of only the laboratory attendants; from time to time, [the experiments] were checked by I. P. Pavlov."[50] Babkin provides the following sketch of Tolochinov as experimentalist: "He was an extremely quiet and withered man. It is impossible to remember without smiling the manner in which he carried out his experiments. He slowly waved a rusk in the dog's face a certain number of times. The dog licked its chops and saliva dripped from the fistula which had been made in one of its salivary glands for the purposes of these experiments. This was a conditional reflex. Then Tolochinov invariably tapped the dog gently on the head with the rusk, after which the animal was allowed to eat it—this was an unconditional reflex. This ritual never varied and was always conducted in the same methodical and melancholy fashion."[51]

Tolochinov initially intended "only to register more facts" about the psychic secretion of the salivary glands, and he proceeded within the post-Snarskii lexicon, using "purely psychological terms such as 'representation,' 'association,' and so forth."[52] His first experiments involved measuring the salivary response as he irritated the dog "at a distance" by waving various substances in its face— toast, fingers smeared with meat powder, open bottles of mustard oil and hydrochloric acid. He deliberately varied his trials in order to test the effect of irritating various sensory organs (e.g., with both the sight and scent of meat powder or mustard oil), irritating them at various distances and time intervals, and varying one quality of the irritating substance (e.g., by waving both dry and moist toast).

The initial trials, in October and November 1901, were largely aimed at establishing how to acquire a psychic secretion consistently. Tolochinov discovered that the dog must be very hungry and, as the laboratory had discovered long ago, that some dogs were better experimental animals than others. From October 1901 through March 1902, Tolochinov used three dogs: Ryzhaia (Redhead), Voron (Crow), and Pudel' (Poodle). Voron, a male prepared with fistulas of both the parotid and submaxillary glands, proved a poor experimental animal, producing a "generally poor secretory reaction" to every irritation "at a distance."[53] Ryzhaia, a female used previously by Vul'fson and prepared with only a parotid fistula, gave much better results. Experiments on Pudel' were conducted only after extirpation of the dog's frontal lobes on December 7, 1901—an indication that these experiments were conducted with one eye on the Bekhterev laboratory and its alleged psychic centers in the brain.

We have only the published protocols of these experiments and we must regard the commentaries to them with some skepticism, but these protocols, rearranged for our purposes, do suggest a radical change in experimental design in early February 1902, a date that fits Tolochinov's later description of the turning point in his investigations.[54] Before February 2, Tolochinov terminated his experiments without regard for the amount of secretion in the last trial. Even if he did notice earlier that the secretory response to the same irritant declined to zero with repetition, he had not made this phenomenon the subject of systematic experimental inquiry. Typical of earlier trials was this one with Ryzhaia on December 22, 1901:[55]

TIME	EXCITER	SECRETION (CC)
2:50	Wave dry toast, 2 min.	0.5
2:55	Wave moist toast, 2 min.	0.2
2:59	Wave moist toast, 2 min.	0.0
3:05	Wave dry toast, 2 min.	0.1
3:09	Scent of hidden dry meat powder, 2 min.	0.8
3:12	Scent of hidden moist meat powder, 2 min.	0.1
3:15	Scent of hidden dry meat powder, 2 min.	0.4
3:20	Scent of hidden moist meat powder, 2 min.	0.1
3:25	Scent of hidden dry meat powder, 2 min.	0.5

In this trial, Tolochinov was not interested in repeating each "irritation at a distance" until the salivary response declined to zero.

In about ten trials conducted on Ryzhaia from February 2 to February 20, however, Tolochinov did precisely this. Several times, as if to assure himself that he had discovered a real phenomenon, he repeated this twice—bringing the secretion to zero, renewing it by feeding the dog, and then bringing it to zero again. The first instance of such a procedure comes in the only experiment reported for February 2, 1902:[56]

TIME	EXCITER	SECRETION (CC)
2:47	Wave dry toast, 1 min.	0.2
2:50	Wave dry toast, 1 min.	0.1
2:53	Wave dry toast, 1 min.	0.0
[Tolochinov tests this again one hour later.]		
3:42	Wave dry toast, 1 min.	0.3
3:52	Wave dry toast, 1 min.	0.2
3:55	Wave dry toast, 1 min.	0.05
3:57	Wave dry toast, 1 min.	0.0

Over the next two weeks Tolochinov tested this phenomenon—soon to be termed *extinction*—with various "irritations at a distance." The following series, conducted on Ryzhaia on February 15, 1902, is typical of these more complex trials (p. 1277):

TIME	EXCITER	SECRETION (CC)
2:33	Wave dry toast, 1 min.	0.1
2:35	Wave dry toast, 1 min.	0.0
2.37	Wave dry toast, 1 min.	0.0
2:42	Sound of hidden toast scraped on plate, 1 min.	0.5
2:44	Sound of hidden toast scraped on plate, 1 min.	0.2
2:46	Sound of hidden toast scraped on plate, 1 min.	0.1
2:48	Sound of hidden toast scraped on plate, 1 min.	0.0
3:04	Wave dry toast, 1 min.	0.0
3:06	Sound of hidden toast scraped on plate, 1 min.	0.9
3:10	Sound of hidden toast scraped on plate, 1 min.	0.2
3:13	Sound of hidden toast scraped on plate, 1 min.	0.3
3:15	Sound of hidden toast scraped on plate, 1 min.	0.1
3:18	Sound of hidden toast scraped on plate, 1 min.	0.0

Tolochinov's later account indicates that this discovery of extinction proved to be a turning point—imparting "a new direction both to my work and to that of subsequent investigators in this sphere" (p. 1278)—precisely because of his background as a psychiatrist and former coworker in Bekhterev's laboratory. Explaining his decision "that it is irrational for a physiologist to resort to psychological terms for the explanation of salivation when the animal is shown foodstuffs," Tolochinov recalled:

> Resolution of this question presented no special difficulties because the influence at a distance of various factors upon the muscular work of certain organs of our body had been noticed long ago. In the experiments [on irritation at a distance] we are dealing with the very same influence on the work of the glandular organ.
>
> It was noticed long ago that in several patients knee reflexes sometimes result, not only from the blow of a hammer, but even when this instrument is merely waved with the intention to strike the lig. patel. propr.[patellar ligament], although this latter reflex is much weaker than the former . . . It is also remarkable that this phenomenon is to a certain degree involuntary; therefore it is most easily understood as a reflexive act from the brain cortex by means of waves of light, just as the reflexive response of the knee to a blow is the result of mechanical waves. This is the same type of phenomenon as the nictating reflex of the eyelid, which occurs, not only when the eyelid is touched, but also when any object, or the investigator's fingers, makes a more or less rapid approach to the eye.
>
> If one compares the irritation at a distance of the muscular organ and the glandular organ, the difference consists only in the special function of each, but the mechanism is one and the same.
>
> On these foundations, I proposed that the phenomena of salivation during irritation of the dogs at a distance by foodstuffs be considered *a reflex at a distance*, which was accepted by prof. *I. P. Pavlov*, who termed it a *conditional reflex* as distinct from the *unconditional* reflex received when the mucous membrane of the roof of the mouth is irritated directly by edible and inedible substances.
>
> My conviction of the truth of this new view . . . was further strengthened by the circumstance that the salivary reflex elicited at a distance obeyed the same basic physiological law as the nictating reflex of the eyelid or the knee reflex elicited at a distance. That is, it obeyed the law of extinction (*ugasaniia*) or decline (*otklonenie*) of the reflex, and, mainly, when certain conditions were observed, it was distinguished by an involuntary, fatal (*rokovoi*) character. If salivation sometimes did not occur in several dogs, I would explain this phenomenon by the declining irritation of a given center.
>
> In general, one gets the impression that the stronger the irritation at a dis-

tance, the more saliva is secreted; however, the amount of saliva during numerous types of irritation rarely was as great as with the direct irritation of the mucous membrane of the roof of the mouth by the very same foodstuffs, just as the knee reflex elicited at a distance does not acquire the same force as with the direct blow of the hammer. (pp. 1281–82; emphasis in original)

Like Snarskii, then, Tolochinov brought to Pavlov's laboratory professional experiences and expertise critical to the reevaluation of psychic secretion and foreign to the chief himself. For Tolochinov, the extinction of psychic secretion was reminiscent of his experiences in Bekhterev's clinic with the knee and eyelid reflexes. Bekhterev had devoted special attention to these reflexes and regularly demonstrated to physicians like Tolochinov the usefulness of the knee reflex as a diagnostic tool for nervous and mental diseases.[57] Just as Snarskii had drawn upon authorities in psychology to strip the "mind of the glands" of will and judgment, to portray psychic secretion as a simple "association" or "habitual reflex," so did Tolochinov draw upon clinical psychiatry to establish that psychic secretion behaved similarly to other "reflexes from a distance" that were "distinguished by an involuntary, fatal character."

The phenomenon of extinction also gave Tolochinov some firm ground on which to stand and observe other regularities governing his "complex reflex." For example, he soon discovered that an extinguished conditional reflex could be rekindled by exciting the unconditional reflex on which it depended—that is, for example, by feeding the dog in the following trials:[58]

TIME	EXCITER	SECRETION (CC)
4:16	Sound of scraping of toast hidden under experimental table, 1 min.	0.05
4:20	Toast fed to the dog (unconditional stimulus)	2.0
4:23	Sound of scraping of toast hidden under experimental table, 1 min.	0.2
4.25	Sound of scraping of toast hidden under experimental table, 1 min.	0.0
4:28	Sound of scraping of toast hidden under experimental table, 1 min.	0.05
4:30	Sound of scraping of toast hidden under experimental table, 1 min.	0.0

[The conditional reflex is extinguished.]

TIME	EXCITER	SECRETION (CC)
4:33	Toast fed to the dog (unconditional stimulus)	1.8
4:34	Sound of scraping of toast hidden under experimental table, 1 min.	0.4
	[The unconditional reflex has renewed the conditional reflex.]	
4:36	Sound of scraping of toast hidden under experimental table, 1 min.	0.2
4:40	Sound of scraping of toast hidden under experimental table, 1 min.	0.0
	[The conditional reflex is extinguished again.]	

From mid-February through early June 1902, Tolochinov used extinction as a background phenomenon for identifying increasingly complex regularities governing conditional reflexes.[59] He explored the secretory effects of alternating various irritants, of burning the dogs with hot wires on parts of their body (some within reach of the dog's mouth and so eliciting protective salivation, some out of reach), of feeding one dog in the presence of another (this inhibited secretion), and of denying the dog food for up to six days (the same regularities obtained, but in sharper form).

In July 1902 Tolochinov joined a delegation of Pavlov's coworkers that delivered reports to the Northern Congress of Physiologists in Helsingfors (Helsinki).[60] In his short report, Tolochinov unveiled the term coined by the chief—*conditional reflex;* explained how one obtained such a reflex in a laboratory dog; and briefly discussed the phenomenon of extinction. Reporting his discovery that a number of simple salivary reflexes originated with the trigeminal nerve that ran from the mucous membrane of the nose, Tolochinov explained: "This is why we oppose to the immediate and absolute reflexes—that is, to the reflexes of the mucous [membrane] of the nose and mouth cavities—all the other effects on the salivary glands that are ordinarily determined to be psychic effects and which we term conditional reflexes (*réflexes condition-nels*)."[61] To obtain a conditional reflex, Tolochinov continued, not the "avidity" but only the "attention" of the dog was required. The dog must be "more or less starving," and "to obtain the conditional reflex one must take advantage of an immediate reflex." Even a dog that was relatively indifferent to food could be made to yield a conditional reflex, "which weakens little by little and disappears if, having first allowed the dog to eat, one excites him again by showing the same food." Conditional reflexes could be obtained by exciting various sensory organs with food or inedible substances—by sight, by sound, or by smell; and the greatest secretory effect was obtained by exciting an ensemble of sensory organs simultaneously. Finally, Tolochinov noted that the conditional reflex responded sensitively to complex properties: for example, despite a "great desire" for meat powder, the dog did not salivate on being shown meat powder moistened with water (since ingestion of this moist substance did not require salivation).[62]

After the Helsingfors conference, Tolochinov returned to Pavlov's laboratory and conducted experiments there until early March 1903. These trials had two main objectives: first, to establish that the same regularities discovered principally with Ryzhaia could be reproduced on two other dogs, Milordka and Belka; and, second, to explore the effect on conditional reflexes of the ablation of the frontal cerebral lobes and the "salivary center" that Bekhterev and his coworkers claimed to have discovered. As we shall see, the first of these lines of investigation became one factor encouraging Pavlov to shift laboratory investigations toward the study of conditional reflexes. The results of the ablation experiments were buried until Tolochinov published them almost a decade later.[63]

Tolochinov himself took no further part in laboratory research. He returned to clinical practice in the Charity Home and, for years, failed to write up the bulk of his experimental findings for publication. Finally, in 1912–13, he infuriated Pavlov with a series of articles in which he included the protocols of his experiments, claimed a large share of the credit for the birth of conditional reflexes research, and criticized the direction that research had taken in subsequent years. Pavlov responded sharply, labeling Tolochinov's account a mixture of "fact and fantasy." In his own tale of the transition to conditional reflexes, discussed later in this chapter, Pavlov made much of Tolochinov's failure of will and personal betrayal.

Pavlov's Difficult Transition

Tolochinov's research, however significant, did not in itself determine Pavlov's decision to shift investigations from digestion to the psyche. The laboratory, after all, was constantly uncovering new phenomena and research possibilities, many of which were never pursued. We must also keep in mind that Pavlov was both an experimental physiologist and the manager of a large laboratory enterprise, so his evaluation of this new line of research necessarily involved both scientific and managerial decisions. Within Pavlov's scientific vision, the key question was: could investigations of the psyche, like research on digestion, generate precise, repeatable patterns that could be expressed quantitatively and interpreted according to their purposiveness? From an institutional and managerial perspective, the key question was: could this line of research consistently generate fresh dissertation topics that could be satisfactorily completed within two years by physiologically untrained physicians? Only when Pavlov decided that both questions could be answered affirmatively did he shift the focus of laboratory research.

His decision was also influenced by two scientific developments that made digestive physiology less appealing by introducing, as one praktikant later recollected, "a certain dissonance" into Pavlov's approach to that subject.[64] First, the discovery of secretin by William Bayliss and Ernest Starling in 1902 undermined Pavlov's nervist portrayal of digestive processes, reopening and complicating questions that he had considered closed. The secretion of pancreatic juice, the British physiologists announced, was elicited by this humoral agent and "does not depend on a nervous reflex."[65] The significance of this development for Pavlov's nervist schema and for the experimental program based on it was succinctly summarized by a defensive Russian reviewer in *Russkii Vrach* (The Russian Physician): "In any case, the conclusions of I. P. Pavlov's experiments now need review, in the opinion of Bayliss and Starling (most of all, in my view, it is their own experiments that demand a careful verification)."[66]

Pavlov had always found nervist explanations more aesthetically pleasing than the alternatives, and throughout the 1890s he had consistently avoided topics that might compel him to confront humoral explanations. V. V. Savich, a coworker at the Institute at the time of Bayliss and Starling's discovery, described Pavlov's reaction this way: "I. P. [Pavlov] immediately assigned me to verify Bayliss and Starling's data about secretin. We began in his presence the experiment that provided a complete verification of their views. I. P. stood for a bit, left for his study in silence and half an hour later emerged and said: 'Of course they are right. You know, we cannot pretend to have the sole right to the discovery of new facts!'"[67] There is no reason to question this much-cited account, but it does overstate Pavlov's acquiescence to this new development. In the laboratory's first public response to Bayliss and Starling's discovery, the praktikanty Borisov and Val'ter confirmed that secretin was capable of eliciting pancreatic secretion, but they also insisted that this secretion was relatively weak, that secretin also elicited secretion from other glands, and that its action depended on the participation of the nervous system. They therefore rejected

Bayliss and Starling's contention that secretin was a specific exciter of the pancreas.[68] (For more on these developments, see Chapter 9.)

In any event, Bayliss and Starling's "new fact" rendered digestive physiology much less appealing to the chief. The discovery of secretin—and soon thereafter of gastrin—complicated discussions of digestive secretion, which from that time necessarily involved analysis of the relative importance and interaction of nervous and humoral mechanisms. A number of Pavlov's praktikanty found this an attractive subject, but Pavlov himself did not.[69] He was also unenthusiastic about reviewing ten years of experimental results and interpretations in the light of now indisputable humoral mechanisms.[70]

New discoveries on the biochemistry and interaction of digestive ferments were also complicating Pavlov's portrayal of the digestive system. Perhaps most important, as we shall see in Chapter 9, these developments challenged the meaning of the regularities Pavlov and his coworkers had discovered at the organ level. Just as the discovery of secretin rendered thoroughgoing nervism obsolete in studies of digestion, so the new biochemistry of ferments signaled the limitations to Pavlov's beloved organ physiology. Discourse was moving away from the chief's strengths into areas in which he neither enjoyed technical supremacy nor spoke with special authority.

Pavlov's pronouncements and his pattern of praktikant assignments at the turn of the century suggest that he was wearying of the lines of investigation that had culminated in *Lectures* and was considering shifting laboratory research toward the experimental pathology and therapeutics of digestion. By 1900–1901, however, his enthusiasm for this field had cooled, leaving him open to alternatives, particularly those that would draw upon his developed expertise and laboratory resources.[71]

Pavlov may also have been attracted to the investigation of conditional reflexes by its potential to combine the passions of the "youthful mind" with the disciplined methodology of the professional scientist. We have seen that, as a seminarian in the 1860s, he had been excited by Dmitrii Pisarev's essays and Ivan Sechenov's "Reflexes of the Brain," but had discarded these "youthful" enthusiasms in his "leave-taking from philosophy." Pavlov later explained his subsequent decision to conduct research on conditional reflexes as in part the result of the "unconscious" influence of Sechenov, and he reflected that "the influence of an idea that is powerful by virtue of its novelty and truthfulness to reality, especially in one's younger years, is so profound, so enduring, and, one must add again, often concealed." There was no doubt some truth to this recollection (although, as we shall see below, it was also one-sided and self-serving). Pavlov never surrendered the interest in the mysteries of the human mind and human behavior that he had imbibed in the 1860s or his positivist faith that a scientific understanding of these subjects was the surest path to improving human society. Not until Tolochinov's experiments, however, had it seemed possible to address these subjects in a professional, truly scientific (mature) fashion.

Whatever the emerging difficulties with digestive physiology or the visceral attractiveness of this possible new line of investigation, the center of gravity in Pavlov's deliberations, in what he later described as his "difficult intellectual struggle," remained his scientific-managerial vision. The complexity of the scientific and managerial issues at stake made Pavlov's decision-making process slow and contradictory. As one longtime coworker later recalled, "Of course, I. P. [Pavlov] expected attacks on his new child and suffered through a great series of doubts and vacillations."[72]

So, judging from A. F. Samoilov's recollection about Pavlov's behavior at the Helsingfors conference in which Tolochinov delivered his report, one might date Pavlov's transition from July 1902 at the latest.

> In the corridor of the university building he [Pavlov] laid out for several people in a private meeting the facts he had established about conditional reflexes. He was in a state of great agitation. Relating the very first experiments, he already saw before him their subsequent fate and development. He saw before him a broad new field of investigations, again on an intact animal; investigations concerning the interaction between the animal and the external environment, that is, investigations on the very soil upon which his talent was greatest. He repeated: "Yes, we've got it, look what we've got!" And he added: "You know, there is enough work here for many decades—I will cease work on digestion, will leave it entirely for this new work."[73]

In this same vein, a praktikant who arrived at the laboratory in fall 1902 later recalled that "one could hear Ivan Petrovich frequently say 'Down with the physiology of digestion. And you all,' turning to his students, 'I will turn to the study of the nervous system.'"[74]

The response of Pavlov's colleagues in the Helsingfors corridor was, however, "restrained," and Pavlov's notes for his lectures on digestive physiology in September 1902 indicate that he analyzed psychic secretion precisely as he had before Tolochinov's (and even Snarskii's) work.

> Recent investigations reveal psychic phenomena: the dog, for example, can think, desire and express its feelings. It follows instructions, guesses, shows what is pleasant and unpleasant for it. Psychic phenomena in connection with the activity of the glands. We will build all our experiments upon the

attention of the dog, but its attention depends upon the impressions produced upon it by the object. We will tease the dog with meat powder. The salivary glands immediately begin to work. We will attract its attention with meat: although its desire to eat the meat is greater [than for the meat powder], it secretes less saliva since this is raw food. Adding dye to the acid in order to color it, and pretending that we want to pour it into the dog's mouth. The dog guesses that this is an unpleasant substance; it knows this from preceding experiments, and salivation [to wash the substance out] begins. It is enough to show any dark liquid for the saliva to flow, since the dog takes this for acid. But if we repeat this experiment several times, the dog will become accustomed to it and there will be less and less saliva, and finally not a drop, because the dog will understand the deception. The same with meat powder. This proves that this is a psychic phenomenon.[75]

This passage is especially interesting because Pavlov was here explaining psychologically (the dog "will understand the deception") the very same phenomenon, extinction, that had convinced Tolochinov that psychic secretion was a complex but determined reflex.

One looks in vain for a crucial experiment that convinced Pavlov that psychic secretion could be productively approached as a reflex and thus launched his new line of investigation. We are dealing, rather, with a complicated, nonlinear decision-making process involving a matrix of intellectual, managerial, and even personal considerations. Before exploring the main line in this process, we can orient ourselves with a few key dates.

In February 1903 praktikant N. M. Geiman delivered a paper to the Society of Russian Physicians that signaled a significant shift in the laboratory's longstanding rejection of mechanical exciters of the glands—a somewhat embarrassing shift that was important to the emerging new line of investigation. In April 1903 Pavlov delivered his first public address on conditional reflexes to the Fourteenth International Congress of Physiologists in Madrid. The pattern of praktikant assignments is also helpful. In 1902 only Tolochinov was pursuing the investigation of psychic secretion, but the new praktikant Geiman was assigned to clarify, using acute experiments, the mechanism of salivary responses to substances placed in the mouth. In 1903 only one of five new praktikanty was assigned to the new line of investigation (A. P. Zel'geim, who replaced the departing Tolochinov); but significantly, in October 1903, Pavlov pulled a favorite praktikant, Boris Babkin, off an important investigation of the pancreas and assigned him to work on conditional reflexes. In 1904, most of Pavlov's coworkers left for the front in the Russo-Japanese War. Still, he assigned one of his two new praktikanty, N. P. Tikhomirov, to the new line of in-

vestigation and devoted much of his high-profile Nobel Prize speech to research on conditional reflexes. In 1905 two of three, in 1906 three of four, and
in 1907 all new praktikanty were assigned to the new research. Pavlov's annual
reports to Prince Ol'denburgskii fit the same basic chronology. He first mentioned the new line of investigation in his report of December 1903. He listed
it last among the laboratory's research topics from 1903 to 1906, and as the only
topic in his report of 1907. It seems, then, that only in late 1903 did the new line
of investigation attain a status comparable to that achieved by experimental
pathology and therapeutics in 1898. That is, the research was sufficiently intriguing for the chief to tout initial results in a high-profile speech, and it was
beginning to change the pattern of praktikanty's assignments. Thereafter,
Pavlov seems to have become steadily more convinced of the efficacy of this
research, although he continued to hedge his bet in both his assignments of
praktikanty and his reports to his patron.[76]

Are Conditional Reflexes Conditioned? The Key Question
of Pravil'nost'

Pavlov was quite alert to the conceptual and rhetorical importance of language. In his second article on the new line of investigation (1904), he conceded that everything he had described physiologically as "complex nervous
activity" was "perfectly obvious from the subjective point of view." The purpose of his "physiological schema," he explained, was to provide "a basis for
the collection and exposition of additional new facts along a new path of investigation."[77] Proper language was so important to the socialization of praktikanty that, for at least a brief transitional period, Pavlov instituted a system
of fines to discourage lapses into the subjective lexicon of previous years.
Tolochinov, too, recalled that, for the conceptualization of this new line of investigation, "the introduction of a suitable physiological terminology was a
first necessity."[78]

   We should consider, then, the choice of one term that reveals much about
the dynamics of Pavlov's transition. To replace *psychic secretion,* he chose the
term *uslovnyi refleks,* commonly translated as "conditioned reflex."[79] *Uslovnyi
refleks,* however, can be translated as either "condition*ed* reflex" or "condition*al* reflex." Significantly, in the French-language abstract of the report in
which Tolochinov first used this term—an abstract edited by Pavlov—
uslovnyi refleks is translated not as *le réflexe conditionné* but as *le réflexe condi-
tionnel.*[80]

   What exactly did Pavlov mean by *conditional reflex* in the years that concern us here? Why did he use this term to replace Snarskii's *association* or *habitual reflex* and Tolochinov's *reflex at a distance*? Recall Tolochinov's account that, having discovered the phenomenon of extinction, he suggested a specifically physiological terminology for psychic secretion: "I proposed that the phenomenon of salivation during irritation of the dogs at a distance by foodstuffs be considered *a reflex at a distance,* which was accepted by prof. *I. P. Pavlov,* who termed it a *conditional reflex* as distinct from the *unconditional* reflex received during the direct irritation by edible and inedible substances of the mucous membrane of the roof of the mouth."[81] For Tolochinov, *reflex at a distance* captured what he considered essential about this phenomenon: it was a determined, reflexive phenomenon elicited, like those from the knee and the eye, without physical contact. Like some of Tolochinov's experiments, this term also suggests he was thinking in terms of an analogy from Newtonian physics, in which gravity was "action at a distance."[82]

According to Orbeli, who worked in the laboratory from 1901 to 1917, Pavlov used the term *conditional reflexes* "in part because their very inclusion as reflexes then had for him a conditional character."[83] This fits Pavlov's common use of the word *uslovyni,* "conditional," which he employed as a synonym for "tentative" or "hypothetical."[84]

For Pavlov, I think, *conditional reflex* reflected not only whatever ontological reservations he may have had but also, much more importantly, the test this potential new line of investigation had to pass in order to qualify as "good physiology." To make this point, I first explore Pavlov's basic distinction between "unconditional" and "conditional" reflexes.

In his first public statement on this subject, to the Fourteenth International Congress of Physiologists in 1903, Pavlov rejected Tolochinov's notion that the difference between simple physiological reflexes and psychic secretion was that the former resulted from the organism's direct contact with an exciter and the latter from action "at a distance." In "psychic experiments," Pavlov argued, an object irritated the animal's salivary glands by acting on various bodily surfaces—the nose, the eyes, or the ear—"by means of the environment (the air, the ether) in which both the organism and the irritating substance are located." This was a form of direct contact and resembled many "simple physiological reflexes" that were transmitted through the same organs.[85]

The essential difference between conditional and unconditional reflexes, then, lay elsewhere. "In the physiological case, the activity of the salivary glands is linked with the same qualities of the object upon which the effect of the saliva is directed." These qualities irritated the specific nerve endings in the roof of the mouth. So, "in physiological experiments the animal is irritated by the essential, unconditional qualities of the subject, by those related to the physiological role of saliva." This was an *unconditional reflex*. In psychic experiments, on the other hand, the animal was irritated by "the qualities of external objects that are inessential or even completely accidental in relation to the work of the salivary glands." For example, the visual, auditory, and olfactory properties of meat could elicit psychic secretion even though these properties had no "business relation" to the work of the glands. Indeed, in psychic experiments the salivary glands could be irritated by "absolutely everything in the surroundings," including the dish in which the substance was presented, the attendant who brought it into the room, the noises he made, and so forth. "So, in psychic experiments the connection of objects irritating the salivary glands becomes altogether distant and subtle."[86] This was the *conditional reflex*.

Both the promise and the peril of research on psychic secretion resided in the apparent "conditionality" of the relationship between stimulus and response.[87] On the one hand—and this was Pavlov's gut feeling—this conditionality perhaps represented the animal's complex but determined adaptation to the subtlest change in its conditions: to changing signals about available food or an approaching predator. By means of "the distant and even accidental characteristics of objects the animal seeks food, avoids enemies, and so forth."[88] On the other hand this conditionality might represent either the indeterminacy of the idiosyncratic psyche or a determinacy inaccessible to physiological methods. In either case, conditionality would deprive experiments on this subject of the determinedness that was the sine qua non of "good physiology." As Pavlov put the central question in 1903 (answering it, perhaps, with a bit more conviction that he actually felt):

> The center of gravity in our subject lies, then, in this: is it possible to include all this apparent chaos of relations within certain bounds, to make these phenomena constant, to discover their rules and mechanism? It seems to me that the several examples which I shall now present give me the right to respond to these questions with a categorical "yes" and to find at the basis of all psychic experiments always the very same special reflex as a fundamental and most common mechanism. True, our experiment in physiological form always gives one and the very same result, excluding, of course, any extraordinary conditions—this is an unconditional reflex; the basic characteristic of the psychic experiment, on the other hand, is its inconstancy, its apparent capriciousness. Nevertheless, the result of the psychic experiment also recurs, otherwise we could not even speak about it. Consequently, the entire matter resides only in the great number of conditions influencing the result of the psychic experiment as compared with the physiological experiment. This will be, then, a conditional reflex.[89]

Pavlov saw the conditional reflex as a suitable subject for physiological research only if it could be revealed, in the final analysis, as a fully determined conditioned reflex. As an experimentalist and laboratory manager, he defined the question operationally: "To what extent can *pravil'nye* results be acquired in the laboratory?" This is what made Tolochinov's discovery of extinction so important to Pavlov: it represented the first case in which conditional reflexes behaved in a quantifiably repeatable, orderly fashion. After his report in Helsingfors, Tolochinov conducted a number of trials (ending in April 1903) that gradually reinforced Pavlov's intuition that research on conditional reflexes could reveal "firm lawfulness," "constantly recurring facts." For example, conditional reflexes diminished and disappeared if the conditional stimulus was repeated without repetition of the unconditional stimulus on which it was based, it was renewed by a strong unconditional stimulus, and it became stronger when a stimulus excited several sensory organs rather than just one. Babkin's research of 1903–4 further buttressed Pavlov's growing confidence that the conditional reflex would, with sufficient research, prove to be fully determined.

This ability to generate *pravil'nye* results was the consistent theme of Pavlov's initial speeches on the new line of investigation. In Madrid (1903), as we have seen, he answered with an emphatic yes his own question about whether the chaos of psychic experiments could be rendered *pravil'nyi*, and he enumerated the "main conditions guaranteeing [the experiments'] success, that is, their constancy."[90] In his second public communication, "On the Psychic Secretion of the Salivary Glands" (1904), he commented extensively on both the complexity of conditional reflexes and the several "constant relationships" that indicated the subject could "be studied with complete success." By this time, Babkin's research had begun and Pavlov could claim that "the results obtained in the laboratory by one worker were easily reproduced by others [*sic*] on new dogs. It was clear that the path chosen for the study of complex nervous phenomena was successful."[91] In his third statement, on accepting the Nobel Prize in 1904, he emphasized that "the observed relations between the external phenomena and variations in the secretory work" of the salivary glands during psychic experiments are "lawful (*zakonomernye*), since they can be repeated as often as one wishes, like ordinary physiological phenomena, and can be systematized in a definite manner."[92] Babkin emphasized this same point in the laboratory's first doctoral dissertation devoted to conditional reflexes: "The constancy of phenomena [and] the ease with which they can be reproduced . . . leave no doubt that the salivary glands of the dog are a very convenient object for investigation of several functions of the large hemispheres of the brain."[93]

After Tolochinov's and Babkin's experiments had established a few basic, repeatable patterns, Pavlov could address the conditional reflex in basically the same manner as he had addressed digestive physiology. Feeding the same dog the same quantity of the same food in two different experiments had *never*, after all, yielded exactly the same secretory results. The differences were explained by reference to the dog's personality, mood, and so forth, and varying results were contained within the characteristic secretory curves. Similarly, differing results in two apparently identical experiments with conditional reflexes could be contained by invoking an increasing number of laws and the numerous uncontrolled variables that, as Pavlov, like Bernard, knew, always existed in any complex animal machine. As in the construction and explanation of the characteristic secretory curves, interpretive moments played an essential role in Pavlov's transformation of the conditional reflex into a conditioned reflex.

Confidence in the ability of experiments on the conditional reflex to generate relatively regular, quantifiable results was also critical to Pavlov's concerns as manager of a large laboratory. Regardless of its scientific promise, the new line of investigation was only feasible if it could be pursued by the physiologically untrained physicians who performed the vast majority of experiments in his laboratory. Pavlov could not sit on the bench beside each of these coworkers, who usually numbered about twelve at a time. To supervise their work and interpret their results adequately—to exercise "quality control"—the chief required that experiments be of relatively simple design and that results be expressible *quantitatively*. This was not just part of Pavlov's notion of good physiology; it also provided a simple language in which his coworkers could gather results and communicate them to the chief for final interpretation. Pavlov "trusted" numbers both as a reflection of physiological reality and as a managerial tool.[94] By 1903–4 he was confident that the investigation of conditional reflexes could generate reasonably regular, repeatable numbers, as had investigations of digestive physiology over the previous fifteen years, and so meet the scientific and managerial criteria for a new focus of laboratory research.

Reinterpreting an Old Experiment, Reconsidering an Old Truth

In February 1903, Pavlov's praktikant Nikolai Geiman reported to the Society of Russian Physicians that the mouth of a dog responded to both chemical and mechanical excitation. On the basis of acute experiments, Geiman reported that mechanical irritants, including rocks, sand, and other inedible substances, elicited salivation through a simple reflex transmitted from the tongue, lips, and cheek through the lingual and glossopharyngeal nerves to the salivary glands.

These findings constituted a fundamental change, both in the interpretation of Vul'fson's experiments with stones and sand in 1896 and in Pavlov's long-standing rejection of mechanical exciters of the digestive glands. For Vul'fson and Pavlov in 1897, as we have seen, the production of a plentiful, mucin-rich saliva when sand was placed in a dog's mouth and no saliva at all when stones were placed there demonstrated that the salivary response reflected an acute judgment by the psyche, which determined that the stones were to be ejected from the mouth and the sand washed out. The chemical composition of the two substances was identical, and the possibility that they exercised a mechanical effect was so contrary to laboratory doctrine that it was not taken seriously. "What other than an evaluation of particular circumstances, and so the psychic moment," Vul'fson had asked rhetorically, "could determine this difference in salivation?"[95] As we have seen, this position was part of a more general opposition to hypothesized mechanical exciters, which Pavlov considered emblematic of the outmoded mechanistic view of the digestive system.

Now, however, the laboratory view was changing, at least with respect to the salivary glands. According to the emerging new line of investigation, every conditional reflex was based on an unconditional reflex, and mechanical exciters could well serve as just such an unconditional stimulus. This is not to say that Geiman's specific experimental results did not play some role, but they clearly underdetermined the conclusions drawn from them (as had the results of Vul'fson's previous experiments on this same issue). Pavlov had sometimes disregarded the results of acute experiments (dismissing their results as the artifacts of trauma to the experimental animal), but he chose not to do so in this case.

The importance here of the interpretive moment emerges clearly if we compare Geiman's report of February 1903 with the data and argumentation in his doctoral thesis, completed the following year. According to the thesis, Geiman's acute experiments encountered a number of problems that interfered with normal salivary secretion. Tracheotomized dogs with salivary fistulas were curarized (to paralyze the motor system and so keep the animal still) and kept alive through artificial respiration. Their mouths were propped open and, in most experiments, the dogs were placed on their belly. If the dog moved during the experiment, it received an additional injection of curare. The curare sometimes clearly affected salivation, even ending it completely—in which case the dog received an injection of pilocarpine to increase the ex-

citability of the glands. These injected chemicals clearly interfered with normal salivation, so Geiman devised another technique to render his animals
motionless: he severed the spine of three experimental dogs, hoping in this
way to prevent movement while avoiding chemicals that interfered with normal salivation. The resultant trauma caused salivation to increase for several
hours and then to cease completely. According to Geiman, only with his thirty-
ninth experiment was he able to immobilize the dog without clearly disrupting normal salivation. His thesis makes clear that this experiment was conducted in March 1903—that is, *after* his report to the Society of Russian Physicians, and so after Pavlov had committed himself publicly to the new position on mechanical excitation of the salivary glands.[96]

The discussion of Geiman's results at the meeting serves to highlight the interpretive choices behind that new position. This is especially striking because
of Bekhterev's aggressive questioning. Well-acquainted with the work of the
Pavlov laboratory and keenly aware of the shift in perspective reflected in
Geiman's report, Bekhterev was clearly determined to wrest from Pavlov an admission of previous error. Pavlov spoke first, summarizing Geiman's results in
a manner calculated to deemphasize the change in the laboratory's basic view.

> Thanks to a series of investigations in our laboratory there have accumu-
> lated many facts [testifying to] subtle adaptiveness. So, there arose a second
> question about the nature of this adaptiveness. It was proposed that here the
> irritation of the ends of the sensory nerves, based on specific conditions of
> irritation, plays a role. This view was most easily verified with the salivary
> glands, with regard to which there had already been established two facts:
> first, that the dryer the food, the more the saliva; and second, that edible,
> tasty food elicits a thick, mucin-rich saliva, while substances that are not
> tasty, [are] repulsive, elicit a watery [saliva]. Clearly, these phenomena are
> entirely purposive, but what is the nature of these purposive relations?
> [Geiman's] investigation confirms that it all resides in the specific irritabil-
> ity of the nervous endings. With the contact of unpleasant substances, ap-
> parently, the chemical endings of the nerves are irritated, but food sub-
> stances irritate salivation mainly by a mechanical path, and in a purely
> mechanical manner there is acquired a thick saliva. If these facts are con-
> firmed by further investigations, everything will be reduced to simple re-
> flexive relations.[97]

Pavlov thus emphasized the continuity between previous views and those expressed in Geiman's report: the praktikant was merely elaborating the laboratory's long-standing position on the purposiveness of salivary secretion.
Bekhterev was not going to let Pavlov get away with this so easily.

Bekhterev: "It seems to me that your recent investigations do not entirely accord with the data in Snarskii's work, since there much attention is given to various psychic phenomena."

Pavlov: "Not in the least. In these experiments we are investigating the simplest reflex, but this does not exclude the possible existence of a more complex reflex, a psychic reflex. One must distinguish between these two forms of reflex. There is a simple salivary reflex from the mucous membrane of the mouth, but a complex one from the eyes and olfactory organ. In certain conditions [the complex reflex] occurs, but in others it does not."

Bekhterev: "It seems to me that in the present work, for some reason, you are especially emphasizing the mechanical moment."

Pavlov: "The fact remains as before, only its interpretation has changed. We already knew from Vul'fson's experiments that the secretion of saliva is linked to the dryness of food. We drew the conclusion that it was specifically this dryness that influenced [salivation], without investigating more closely the causes of this phenomenon. But, you know, [dryness] can influence [salivation] either chemically or physically. Actual investigations make clear that here one must seek this influence, apparently, in mechanical causes, but the form of mechanical irritation, undoubtedly, varies—and so its consequences vary as well. . . ."

Bekhterev (pressing for an admission of previous error): "The experiments with stones [and sand] always seemed unconvincing to me."

Pavlov: "No, they all the same have their significance. Sand drives saliva, but stones do not. What does this mean? Clearly, the sand clings to the mucous membrane and must be lubricated for removal; this is absolutely unnecessary for removal of the stones. So, this experiment indicates great purposiveness, indicating that even mechanical influences must be differentiated." (This was not, of course, Pavlov's original interpretation of this experiment.)

Bekhterev: "I also wanted to say that the experiment with the stones does not correspond to irritation with food, since here there is another mechanical action." (Bekhterev is criticizing here the very form of experiments in which the Pavlov laboratory evaluated the possible effect of the mechanical influence of food.)

Pavlov (reacting to this general challenge): "But there is no hint whatsoever in the stomach of mechanical sensitivity."

Bekhterev (pressing the attack): "But here also mechanical influences have enormous significance, since the mechanical activity of the stomach, like a millstone, is indispensable for a *pravil'nyi* digestive process. This is why the question occurs to me: are the data of previous authors regarding these mechanical influences really entirely excluded and refuted? It is not clear how this influence is expressed, but it is indubitable in such a complex organ as the stomach with its motor function."

At this point, Bekhterev mentioned related investigations in his own laboratory, which allowed Pavlov to change the subject by attacking Bekhterev's methodology. Nikolai Kravkov, professor of pharmacology at the Military-Medical Academy supported Bekhterev's position, noting that "in the question of the action of substances, especially food, on the mucous membrane it is extraordinarily difficult to distinguish between mechanical and chemical irritation." Geiman and Pavlov then summarized their experimental evidence for the laboratory's new position on the importance of mechanical exciters (for the salivary glands, but not for the gastric glands).[98] Their reinterpretation of experimental results now provided the unconditional reflex on which the more complex system of conditional reflexes was based.

Pavlov's Tale

Pavlov himself related several times the story of his transition to research on conditional reflexes, each time with the story line first presented in a speech of 1906 at a London ceremony honoring T. H. Huxley.[99] Especially interesting is one aspect of this tale, presented here in the version Pavlov offered in the preface to *Twenty-five Years of Objective Study of the Higher Nervous Activity (Behavior) of Animals* (1923).

> I began to investigate the question of this [psychic] secretion with my collaborators, Drs. Vul'fson and Snarskii. While Vul'fson collected new and important material regarding the details of the psychic excitation of the salivary glands, Snarskii undertook an analysis of the internal mechanism of this excitation from the subjective point of view; that is, considering the imagined internal world of the dogs (upon whom our experiments were conducted) by analogy with our own thoughts, feelings, and desires. There then occurred an event unprecedented in the laboratory. We differed sharply from each other in our interpretations of this world and could not by any further experiments come to agreement on any general conclusion, despite the laboratory's consistent practice by which new experiments undertaken by mutual agreement usually resolved any disagreements and arguments.
>
> Dr. Snarskii held to his subjective explanation of the phenomena, but I, struck by the fantastic nature and barrenness for science of such an approach to the problem, began to seek another exit from this difficult position. After persistent deliberation, after a difficult intellectual struggle, I de-

cided, finally, in the face of the so-called psychic excitation, to remain in the role of a pure physiologist, that is, of an objective external observer and experimenter, dealing exclusively with external phenomena and their relations. For implementation of this decision I also began with a new coworker, Dr. I. F. Tolochinov, and there subsequently followed twenty years of work with the participation of many tens of my dear coworkers.[100]

According to Pavlov, a distant influence from his youth gave him the courage to address psychological phenomena "objectively."

> I think that . . . the most important impetus for my decision, although at the time an unconscious one, was the influence, from the long distant years of my youth, of the talented brochure of Ivan Mikhailovich Sechenov, the father of Russian physiology, entitled *Reflexes of the Brain* (1863). You know, the influence of an idea that is powerful by virtue of its novelty and truthfulness to reality, especially in one's younger years, is so profound, so enduring, and, one must add again, often concealed. In this brochure, a brilliant attempt was made—a truly extraordinary attempt for that time (of course theoretically, in the form of a physiological scheme) to represent our subjective world in a purely physiological manner.[101]

Notice that Pavlov's version of his conflict with Snarskii corresponds neither to the content of Snarskii's thesis nor to Pavlov's actual position on psychic secretion in the years immediately before and after Snarskii's research. Pavlov and Snarskii clearly differed about something, and perhaps their disagreement indeed concerned differing estimations of the fruitfulness of contemporary psychology.[102] As we have seen, however, Snarskii did not distinguish himself by "holding to a subjective explanation of the phenomena." He was the first laboratory coworker to insist that psychic secretion was "an association" or "habitual reflex" and that "the consciousness of the dog plays no important role."[103] Furthermore, he developed this idea in a polemic against Vul'fson's view that the psyche actively chooses and judges, a view that Pavlov had enthusiastically endorsed and continued to propound in his lectures through at least fall 1902.[104] As for the "unconscious" influence of Sechenov's "Reflexes of the Brain," this tract, published in 1863 and unmentioned by Pavlov until his tale of 1906, was cited—for the first time in any laboratory publication—by Snarskii himself in his doctoral thesis. It is quite possible that Snarskii's invocation of Sechenov's essay awakened Pavlov's youthful enthusiasm for it, in which case, here again, Pavlov was being strangely reticent about his praktikant's role.

Assuming that my account of the transition to research on conditional reflexes is correct, why would Pavlov have told such a tale? This is especially puzzling since Pavlov was usually scrupulous—even generous—in crediting co-workers for their contributions. For example, although Pavlov was critical of Tolochinov's failure to write up his results and deeply disturbed by "the mixture of fact and fantasy" in his former collaborator's articles of 1912–13, he always acknowledged Tolochinov's importance—and privately, even his "priority"—in the initial research on conditional reflexes.[105]

I can suggest two reasons. First, in his tale Pavlov cast himself as a committed struggler for the scientific worldview in the spirit of Darwin, T. H. Huxley, and other such heroes. Snarskii-the-subjective-psychologist served him well here as a villain. Second, Pavlov's tale established a reputable physiological paternity for a line of research that he regarded, in Savich's phrase, as a vulnerable "child." Dismissed by many as speculative and ridiculed by others as "spitting science," the study of conditional reflexes was a risky endeavor for a basically conservative man who treasured the respect of his colleagues.[106] In Pavlov's tale, this line of investigation was born through a combination of Vul'fson's and Tolochinov's experiments, the conceptual influence of Sechenov, "the father of Russian physiology," and Pavlov's own courage and faith in the scientific worldview. It was, in other words, respectably modern and objective; it was "good physiology," untainted by influences from psychology, a discipline that had been associated in Pavlov's formative intellectual years with barren, reactionary metaphysics. In this same spirit, as Pavlov embraced his new line of investigation with growing enthusiasm and confidence he redefined it in increasingly physiological terms. In his annual reports of 1903–5 he termed it "the study of questions of experimental psychology on animals"; in 1906 he renamed it "the objective investigation of the higher divisions of the central nervous system"; and in 1907 he finally settled on "the investigation of the activity of the large hemispheres and sense organs."[107]

Pavlov's tale, then, was a part of this changing lexicon, this redefinition and legitimation of his new endeavor. In this sense, it was a final step in his transition to research on conditional reflexes.

In a science that has been cultivated so thoroughly as has
physiology, one could hardly expect that a single person
could make so many important contributions as has Pavlov.

—KARL MÖRNER, memo to the Nobel Committee (1902)

Throughout the 1890s and into the early twentieth century, Ivan Pavlov's physiology factory fairly hummed with
activity. But what exactly did it produce?

For one thing, of course, it produced a substantial number of knowledge claims about the digestive system. These
included relatively simple facts (e.g., the pepsin content of
gastric acid), broader physiological generalizations (the important role of the vagus and the psyche), theoretical insights (the specific excitability of the glands), and sweeping
metaphorical statements (the digestive system is a chemical
factory). The laboratory's extensive experience with intact
animals also enabled its members to offer advice to physicians about the treatment of various digestive disorders in
human patients.

These knowledge claims were formulated and communicated in a constant output of literary products, processed
differently for various audiences. The least processed of
these were the doctoral dissertations, which were edited by
the chief for a few readers. A thesis often contained contradictory data and interpretations, confessions about experimental difficulties, and other impurities absent in more refined literary products. Next came the praktikanty's public

reports and published articles. These were tightly edited, sharply focused, self-consciously public statements that projected the laboratory's confident voice to outside audiences. Many of the contradictions and complexities contained in the dissertations were omitted (although these sometimes reemerged in public discussions). Finally, the most highly processed literary products—in which vision and data meshed most grandly and smoothly—were the chief's synthetic publications, particularly his *Lectures on the Work of the Main Digestive Glands.*

Literary products served also to advertise the laboratory's methods, techniques, and dog-technologies. Ranging from the Mett method for determining proteolytic power of a digestive juice to the Pavlov sac (isolated stomach), these products attracted a number of western scientists to St. Petersburg. Enhancing production in other laboratories, establishing the chief's presence in scientific discourses distant from his own area of expertise, and fortifying the scientific status of physiology in general, such products provided a stable source of Pavlov's authority even when some of his knowledge claims were called into question.

Perhaps most important in this respect were the various types of surgically altered dogs, impressive embodiments of laboratory achievements. Pavlov displayed several of these animals at the All-Russian Hygiene Exhibit in 1893 to impress the general public with the power of experimental physiology. They also served in public lectures, not only for scientific-pedagogical purposes but more broadly for, as Pavlov put it, "convincing the usually so stubborn public of the correctness and obvious usefulness of experiments on animals."[1] Exhibited proudly to the laboratory's visitors and in a 1904 photo album celebrating Pavlov's achievements, laboratory dogs also made an impressive gift to a valued colleague.

These dogs, in turn, enabled the physiology factory to produce large quantities of pure digestive juices, which themselves proved a quite popular item among scientists and clinicians. Many Russian and western investigators requested samples of these gastric and pancreatic secretions to facilitate their research on the physiological chemistry of digestion and the process of protein absorption.

Pure gastric juice created a sensation in the international medical market. Pavlov's "small gastric juice factory" (as he proudly called it), which swung into production in 1898, bottled the gastric juice drawn from esophagotomized dogs by sham-feeding and sold it as a remedy for dyspepsia. By 1904, this enterprise was selling over three thousand flagons a year, increasing the laboratory budget by over 65 percent. Perhaps more important, Pavlov's "natural gastric juice of the dog" provided a dramatic demonstration of the clinical value of experimental physiology, considerably enhancing the chief's reputation among physicians and physiologists in Russia and abroad.

The laboratory also produced about a hundred alumni. Just as they had when praktikanty, these alumni qualitatively extended Pavlov's reach. Armed with a doctoral degree, they often rose to influential positions in Russia's medical establishment. About one-half acquired professorial positions in clinical medicine (often combining these with a clinical position in a hospital); others assumed posts in the state medical bureaucracy and in a wide range of military and civilian institutions throughout the Empire. Few became physiologists, although this began to change at the turn of the century. Even alumni who attained only modest professional heights enhanced Pavlov's reputation simply by making their way, in the course of their everyday lives, into innumerable milieus inaccessible to the chief for many reasons, including the sheer limitations on the time of any single person. Former praktikanty lived throughout the Empire, treating and chatting with patients, attending meetings, delivering and commenting on papers, recommending the laboratory's home remedy for dyspepsia, and, apparently quite often, regaling acquaintances with tales of their investigative experiences in St. Petersburg. Favored alumni continued to perform important tasks for the laboratory; several traveled abroad on study leaves, teaching Pavlovian techniques and otherwise extending the chief's European contacts. Traveling alumni were especially important to Pavlov given the rarity of his own forays beyond Russia's borders, and they generated significant return traffic to St. Petersburg.

The broad appeal of these various products made the laboratory the hub of an expanding network and, in a sense, a product itself. Visitors came from Russia and abroad to marvel at the operation of this modern physiological laboratory and to master its techniques. Impressed by what they had read or seen, a number of scientists solicited the chief's expert advice in laboratory design.

This simple inventory highlights an important feature of factory production: the efficient generation of sheer numbers of diversified products. The great quantity of experiments and data—over which only the chief had total access and control—allowed Pavlov to mobilize them selectively for his purposes. The sheer quantity of praktikanty and lines of investigation afforded him great flexibility and a "panoramic view," allowing him to move among related projects at will, to note interesting similarities and differences among them, and to initiate new ones as he saw fit. The sheer number of knowledge claims and literary products—and their "diversification"—appealed to various audiences and vitiated the importance of a challenge to any single claim.

The sheer quantity of alumni amplified Pavlov's voice and extended his reach both in Russia and abroad.

This brings us to the laboratory's final product: Pavlov himself. The talented but undisciplined procrastinator who labored erratically during the 1870s and 1880s became part of the purposive, precise, and regular operation of his physiology factory. No longer did he work by inspiration or stroll along the Neva River on weekdays, dreaming of future accomplishments. Every moment was accounted for, and those who sought unscheduled counsel could usually obtain it only, literally, on the run.

Pavlov himself became a product in yet another important sense. As chief, his name became associated with the entire array of products generated by his physiology factory and so with a broad and impressive set of achievements that seemed (and indeed was) beyond the powers of any single individual. This composite Pavlov was much more than the sum of his parts. He acquired symbolic status—as a precise and ingenious experimenter, a synthetic thinker, and the embodiment of a modern, clinically relevant, laboratory-based physiology. He became, one might say, a Claude Bernard for his time, a Bernard for the emerging era of factory science.

With one notable exception, these products were not exchanged for money. Pavlov usually took payment in the currency of authority, reputation, professional position, and expansion of his network. These, in turn, provided capital for the continual refueling and expansion of his physiology factory. This, for Pavlov, was the most important material reward for his scientific labors. (It was of distinctly secondary importance, though still much appreciated, that the currency of scientific success proved readily convertible into salary increases and, eventually, great material comfort.)[2]

In Part III I explore the fate of laboratory products from three different perspectives. Chapter 8 is devoted to the history of a single product, gastric juice—the laboratory's most important direct contribution to clinical medicine. In Chapter 9 I examine Pavlov's rise to international renown as the spokesman for and embodiment of the laboratory's product line. Here, too, I explore developments in turn-of-the-century physiology and physiological chemistry that, just as Pavlov attained a "European reputation," subjected some of his key knowledge claims to serious criticism and signaled a shift in physiological discourse on digestion away from his beloved organ physiology. Finally, in Chapter 10, I address the response to laboratory products by one important audience: the Karolinska Institute's Nobel Committee.

Chapter 8

GASTRIC JUICE
FOR SALE

Appetite is juice. To restore a person's appetite means to give him a large portion of good juice at the beginning of a meal.

—IVAN PAVLOV, *Lectures on the Work of the Main Digestive Glands* (1897)

Appetite was not only central to Pavlov's analysis of the digestive system, it was also both the productive force and the product of what the chief proudly called his "small gastric juice factory."[1] This laboratory enterprise thus captured and deployed the dual nature of appetite according to the laboratory view in the 1890s. Appetite was both the psychic impulse that generated large quantities of pepsin-rich gastric juice and the "igniter juice" itself, necessary to the efficient processing of foodstuffs in the stomach. Those unhappy individuals who suffered from a lack of appetite and so from various digestive ailments could be effectively treated by providing "a large portion of good juice at the beginning of a meal." The Pavlov laboratory's most successful contribution to the clinic was the production and sale of "appetite juice"—"the natural gastric juice of the dog"—as a remedy for dyspepsia.

This enterprise significantly increased Pavlov's laboratory budget, but the intangible profits were greater still. Natural gastric juice flowed not just from Pavlov's dogs but also from his fundamental scientific insights. By selling the juice to clinicians he was also marketing other products with a much higher price tag and a much longer shelf-life: laboratory-based scientific medicine in general and his own physiological enterprise in particular. The notion of a scientific medicine had gained a great deal of ground in the last decades of the nineteenth century, though its heroes were not Bernard's experimental physiologists but rather bacteriologists such as Louis Pasteur and Robert Koch. Pavlov's "natural gastric juice of the dog" was a down payment on behalf of the Bernardian vision of a physiologically based scientific medicine—the first specifically therapeutic product that his laboratory could offer the medical community.[2]

The development and success of the small gastric juice factory resulted not simply from Pavlov's ideas and skills but from his laboratory system as a whole. Like the laboratory's less tangible products, this medical innovation developed through the fusion of the chief's scientific-managerial vision with the institutional context in which he worked—in particular, with the nature of his laboratory's workforce.

Physiologists, Physicians, and the Pepsin Market

Pavlov recognized immediately that his success with Ekaterina Shumova-Simanovskaia in obtaining a plentiful psychic secretion by sham-feeding esophagotomized dogs was significant not only for the scientific arguments it enabled them to make but also for the physical product of these experiments. In their article "The Innervation of the Gastric Glands of the Dog" (1890), the authors were primarily concerned with making several scientific points—specifically, about the secretory role of the vagus nerves and of appetite; but they also noted that sham-feeding esophagotomized dogs was a reliable means of producing "a great mass" of pure gastric juice. "Now with the aid of our experiment one can replace, so to speak, the chemical extraction of pepsin with the physiological [extraction], which has the advantage of making available an almost pure solution of pepsin without the admixture of other substances."[3]

Generations of physiologists and clinicians had sought precisely that. Johann Eberle's discovery in the early 1830s that an extract of mammalian gastric mucosa coagulated egg white sparked a search for the active agent. In 1835 Johannes Müller and his pupil Theodor Schwann published an article on the artificial digestion of egg white in which they concluded that the active principle in gastric juice was an organic ferment, which in 1836 Schwann named *pepsin*.[4] In subsequent decades, those who investigated the properties of pepsin were hampered by the difficulty of obtaining sufficient quantities in pure form. Many pigs, goats, and cows were slaughtered in largely unsatisfactory attempts to isolate and gather pepsin from the stomach's mucous membrane and from rennet bags. The procedure developed by Pavlov and Shu-mova-Simanovskaia offered an unprecedented opportunity for the systematic chemical analysis of gastric juice in general and pepsin in particular, an op-portunity that Shumova-Simanovskaia subsequently exploited in the Imperial Institute's Chemistry Division in the early 1890s.[5]

By 1890 pepsin had also become a well-established remedy for a widespread nineteenth-century ailment—dyspepsia. The literature on the causes of and cures for this ill-defined malady was vast. Sergei Luk'ianov, director of the Institute and head of its Pathology Division, observed that "dyspeptic phenom-ena, as the expression of disturbances of gastric processes with respect to se-cretion, absorption, motility, and sensitivity, are distinguished by their great variety."[6] These varied symptoms, according to another observer, included generalized distress or discomfort, nausea, vomiting, vertigo, hemicrania (mi-graine), nervous phenomena, palpitation of the heart and cardiac pain, im-pairment of sexual function, malnutrition, and "muscular weakness, or a feel-ing of worthlessness, especially in the morning."[7] Physicians attributed dys-pepsia to various combinations of nervous, chemical, and emotional causes and often described the "dyspeptic" personality as hypochondriacal and self-indulgent.[8]

A landmark in the clinical treatment of dyspepsia was the proposal by the eminent French physician Lucien Corvisart in 1855 that dyspepsia resulted from a pepsin insufficiency in the stomach and so was best treated by the in-gestion of a pepsin-rich substance. Commercial (artificial) pepsins produced by various means soon flooded the medical marketplace. In the late 1850s a lawsuit against a purportedly impotent pepsin concoction led the Société de pharmacie de Paris to investigate the array of available *poudres nutrimentives.* The commission reported in 1865 that these commercial pepsins, although very popular among French physicians, were of uneven and undependable quality. For example, one product sold as "German pepsin" and obtained from the stomach of slaughtered pigs was rich in active ferment but could hardly be endorsed for medical use owing to its "reprehensible odor," which testified to the presence of putrefying animal matter. The commission recommended an "official pepsin" prepared by rinsing the rennet bag of a slaughtered goat and refining the residue, and this procedure entered the French pharmacopoeia.[9] In subsequent decades, scientists, physicians, and entrepreneurs in Europe and the United States developed over twenty methods for preparing commercial pepsin.[10] Russia's pharmacopoeia recommended a native variation, *pepsinum rossicum,* prepared by passing the mucous membrane of a freshly slaughtered animal through a hydraulic press, freezing the residue, cleaning it by dialysis,

and freezing it again "with a strong draft of dry air."[11] By the 1890s a number
of pharmaceutical companies were responding to the ever-increasing demand
by selling commercial preparations, which were administered in combination
with hydrochloric acid (the other active component of gastric juice). St. Petersburg's pharmacies sold four such preparations produced by Merck alone.

A Spurned Offering from Laboratory to Clinic (1893–97)

When the military physician P. N. Konovalov arrived at the laboratory in 1892,
Pavlov assigned him a dissertation topic designed to explore the feasibility of
replacing commercial, artificial pepsins with natural gastric juice. Were a dog's
feeding habits, weight, and health disrupted by using sham-feeding to drain it
of gastric juice? How much juice could be obtained, and how frequently? What
was the chemical nature of this juice, and how long did it remain stable? How
did it compare with the commercial preparations sold in St. Petersburg as a
remedy for dyspepsia? Pavlov also asked Konovalov to compare the therapeutic efficacy of natural gastric juice and commercial products, but the praktikant was unable to do so "due to lack of time." This latter issue, clearly, was
considered the least pressing.[12]

For Konovalov, this research always remained a dramatic high point of his
professional career. Associating himself proudly with the early history of natural gastric juice, he recalled some forty years later that "during the public defense of my dissertation at the Military-Medical Academy, one of my examin-
ers, Professor of Physiological Chemistry A. Ia. Danilevskii, told me: 'Having
worked all my life on the secretions of various glands of the living organism,
I could never imagine that one could obtain gastric juice of such ideal purity;
when your teacher I. P. Pavlov showed me a glass tube of the juice you obtained
I thought he was joking and showing me a tube of distilled water.'"[13]

In his dissertation, "Commercial Pepsins in Comparison to Natural Gastric
Juice" (1893), Konovalov described the technique he developed for drawing 150
to 300 cubic centimeters of gastric juice from a dog daily with no ill effects. He
also analyzed the properties of the juice and argued that, by comparison, the
thirteen commercial pepsins sold in St. Petersburg's pharmacies were "extremely crude and faulty."[14]

Despite the laboratory's attachment to dogs as experimental subjects,
Konovalov also attempted to acquire gastric juice from the most popular animal among producers of commercial pepsin—the pig. Pigs survived the
esophagotomy well, but they proved temperamentally unsuited to the job.
"Obtaining the juice from them is absolutely impossible. As soon as the pig is

placed in the stand it begins to squeal frenziedly; terror paralyzes its extremities and, apparently, temporarily halts the functioning of the gastric glands completely, so not a drop of gastric juice is acquired. The pig turns out to be a very nervous and sensitive animal, and so is extremely unsuitable for physiological experiments requiring calm."[15]

Why test pigs at all? Discussing his adventures with these noncompliant creatures, Konovalov noted that dog juice would suffice "for physiological goals," but from a therapeutic perspective it raised the problem of "squeamishness" (p. 6). Russians did not eat dog meat, and potential customers might find ingestion of a dog's bodily fluids aesthetically displeasing. Russians did eat pork, however, and so would perhaps respond more enthusiastically to pig juice. Furthermore, dogs were meat-eaters and therefore expensive to feed in St. Petersburg, inevitably driving up the cost of their gastric juice. These two concerns—squeamishness and expense—may also explain Konovalov's consideration of goats and calves, neither of which proved suitable since the esophagotomy excised their cud, condemning them to a quick (and uneconomical) death.

Whatever their disadvantages, dogs proved most cooperative producers. Three or four weeks after implantation of an esophagotomy and gastric fistula, a dog was ready for work. Upkeep was minimal: one needed to clean the animal's wounds three times a day (since digestive fluids dripped constantly through the aperture of the esophagotomy and around the fistula) and feed it solid food through the fistula and milk and water through a cannula.

After denying the dogs food for about twelve hours, Konovalov placed them on a wooden stand and fastened them with leather straps to an overhead wooden beam. Unlike the pigs, Konovalov's two dogs stood "calmly," making extraction of their gastric juice most "convenient" (p. 8). Konovalov then opened the gastric fistula to drain the dog's stomach of any mucus and undigested food. The fistula was corked and a thin glass tube inserted through the cork. Konovalov then waited another few minutes for the occasional "volitional secretion" to pass. He then began slowly to feed the dog small pieces of meat. "The dog eats with great enthusiasm," he reported, "and the swallowed pieces, moistened with copious saliva, fall through the opening in the digestive tract [from the esophagotomy] and out into a vessel" (p. 9). Five to ten minutes later, transparent drops of "reflexive gastric juice" appeared in the glass tube. Sometimes the juice contained bits of mucus, which Konovalov removed with pincers or a glass filter. He usually collected 200 to 300 cubic centimeters of juice in a one-hour session, setting it aside in five-minute portions for later analysis. He concluded the session by feeding the dog a generous meal of two pounds of meat and a half pound of black bread (inserted through the fistula) and two bottles of milk (through a cannula).[16]

Konovalov's dogs usually remained nameless.[17] Unlike Druzhok, Zhuchka, and other experimental animals, their personalities (aside from a healthy appetite) were largely irrelevant to their laboratory task. The experimenter was interested only in their ability to perform *chernaia rabota*, "hard labor," in a factory regimen that made taxing physical demands on their bodies.

For several months Konovalov tested the productive capacity of his dogs under various regimens. He discontinued experiments with his first dog after only twelve sessions, perhaps because it produced a mere 160 cubic centimeters of juice in an average session, compared with the second dog's 235 cubic centimeters. From late October 1892 through mid-March 1893, he tested this second dog's reaction to an increasingly demanding schedule, initially collecting the juice every three or four days, then every other day, and finally, daily. In forty-five sessions the dog produced 10,606 cubic centimeters of gastric juice. Its production per hour increased constantly until Konovalov moved to daily collections, when yield per hour declined (p. 12).

Both dogs, Konovalov observed, were quite happy in their work. Throughout the experiments, they "looked like entirely healthy, normal animals. Both always greeted me and the attendant joyfully; upon being released from the kennel they themselves rushed to the room where I collected juice and leapt on the table and into the stand. This was of course facilitated, aside from their habituation to the experimental setting, by the fact that sham-feeding obviously brought the animal great satisfaction, since during actual feeding through the fistula its taste sensations were but little satisfied" (pp. 11–12). Moreover, while providing enormous quantities of gastric juice both dogs actually gained weight, and did so in direct proportion to the frequency with which the juice was collected (perhaps because they were fed much more sumptuously by Konovalov than in the kennel). Clearly, Konovalov concluded, the gastric glands were capable of producing such a massive quantity of juice that the daily loss of hundreds of cubic centimeters did not interfere with digestion. The glands adapted to the demands made on them (pp. 11, 14).

Konovalov devoted considerable space in his thesis to a description of this gastric juice and its chemical properties (pp. 16–41). He analyzed these properties with the help of Marcel Nencki, chief of the Institute's Chemistry Division, and he used them to establish a norm for later comparison with commercial substitutes. The juice was "an entirely transparent, colorless liquid without any scent, or with the same light scent of a fresh solution of hydrochloric acid; very acidic to the taste, it not only is not repugnant but is even pleasant" (p. 32). The gastric juice was composed largely of hydrochloric acid, which gave it an important bactericidal function. About 0.5 percent of the juice consisted of solids. When boiled or frozen, the juice yielded a white sediment—this was the ferment, pepsin, responsible for gastric juice's proteolytic power. The juice's acidity fluctuated in the course of secretion but had a mean value of 0.544 percent. Its proteolytic power varied with the acidity and was usually between 7.4 and 9 millimeters (as measured by the Mett method). Pepsin was most effective—its proteolytic power highest—when the acidity was 0.2 percent. Preserved at room temperature, gastric juice lost its proteolytic power, and did so rapidly after the second month. In sum, "the purity of reflexive gastric juice, its taste, its enormous proteolytic power and, finally, its antimicrobial qualities give it every right to attract the serious attention of therapeutists" (p. 41).[18]

Konovalov then analyzed the commercial pepsins competing in the Russian market. The demand for these was constantly increasing, but the products remained largely unregulated. Their ingredients were varied and mysterious.[19] "Almost every factory and pharmacy that prepares pepsin for sale varies it in its own way," wrote Konovalov, "varying additives, taste, scent and so forth; moreover, most of these manipulations are entirely unknown to the public and physicians. The greater number of methods are unsatisfactory. The general, heretofore inherent problem with all these methods is that, essentially, nobody knows the substance which they are attempting to obtain . . . The second great insufficiency characteristic of all these means is the crudeness of approach, of chemical and physical agents which can hardly extract the ferment . . . without changing its essential physicochemical qualities" (p. 49). These commercial products varied widely in their fermenting power, and sometimes lacked it completely. Their therapeutic efficacy was inevitably inconsistent, and "it fell to physicians to sort out what to accept and what to reject among those offered for sale, which are usually accompanied by advertisements that promise much" (pp. 49–50). Physician-investigators in England, France, Germany, and Russia had, on occasion, systematically examined commercial pepsins, but these studies suffered from a superficial understanding of natural pepsin itself. Furthermore, since various commercial pepsins constantly appeared and disappeared from the market, such studies were quickly outdated (pp. 50–51). (Konovalov cited more than half a dozen such studies in the 1880s alone.)

Konovalov found all the commercial pepsins he tested to be foul, useless, and potentially dangerous (p. 62). When natural gastric juice proteolyzed albumins, the resulting fluid was "pure, transparent, colorless, scentless"; commercial pepsins, on the other hand, produced a turbid, yellowish, opalescent liquid "with a heavy, unpleasant, and even fetid scent" (p. 59). Natural gastric juice did not rot, but the impurities in artificial varieties produced the unmistakable odor of putrefaction by the second day of use. The proteolytic power of commercial pepsins varied greatly but was uniformly much lower than that of natural gastric juice. The usual prescribed dosage of ten grams for an adult contained "too little ferment to have any effect." Larger doses were likely to produce dangerous "peptotoxins" in the unhappy patient's stomach, so permeated were these products by "unnecessary admixtures in a state of rotting degeneration" (p. 63). By virtue of "its purity, anti-putrefaction qualities, taste, appearance, and enormous proteolytic capacity," the natural gastric juice of the dog was a vastly superior product (p. 62).

Confessing that he had lacked time to test systematically the therapeutic efficacy of natural gastric juice, Konovalov nevertheless pronounced it "a new, useful, and very powerful medicinal substance" (p. 65). He could himself testify to its effectiveness against dyspepsia. Having ingested it in doses of 30 cubic centimeters, he reported "not only the absence of any unpleasant sensations, but, taking it after a meal, I always received a sensation of special lightness in the stomach area, an absence of that sensation of unpleasant heaviness during digestion that is characteristic of the chronic, low degree dyspepsia that I have" (p. 63). Nor was this his experience alone. "In just the same way, my comrade physicians in the laboratory and those who visited the Institute drank the juice, in part from curiosity, in part with the medicinal goal of easing a sense of heaviness during digestion, which was always quickly attained. Then, for some time the juice was tested in the treatment of children's dyspepsia in one of the children's hospitals; a result worthy of great attention was immediately acquired, about which there will probably be a report in time" (p. 63).[20]

Konovalov also noted several earlier claims that gastric juice relieved suffering in patients with cancerous tumors, cleared up soft chancres, and dissolved pellicles in diphtherial patients. These positive findings—from an earlier era in which large quantities of pure gastric juice had been impossible to obtain—promised a bright therapeutic future under new conditions of production.[21]

Gastric juice therapy, however, did not catch on. Pavlov proudly exhibited both the juice and the dogs that produced it at St. Petersburg's All-Russian Hygiene Exhibition in 1893, but his laboratory did not produce the juice regularly or market it aggressively. Pavlov's attitude seems to have been that his laboratory had provided the clinic with a gift and a clear scientific demonstration of its value; it was for clinicians themselves to make use of it.

Konovalov's report to the Society of Russian Physicians probably did inspire the first reported clinical use of natural gastric juice by a physician outside the Pavlov laboratory. A. Troianov was a leading St. Petersburg physician, senior surgeon at the city's Obukhovskaia Hospital, and a regular participant at meetings of the Society of Russian Physicians. In 1893 he was treating a patient who suffered from a number of symptoms of poor digestion, including a sensation "like a flow in the stomach, accompanied by unusual heat and rumbling in the intestines." Troianov first treated him with bouillon, milk, thin cereal, and *kissel* (boiled, thickened, and sweetened fruit juice or puree). When the patient began to suffer from heartburn, diarrhea, and gassiness on eating, his physician concluded that the problem lay with "a disturbance of the chemism of the gastric juice" and "decided to try the gastric juice of the dog." Troianov turned to the Imperial Institute of Experimental Medicine and, "thanks to the kindness of professor I. P. Pavlov," received half a liter of juice. After the patient had ingested a tablespoon of juice twice a day for two weeks, "we received a good result": his appetite improved, his heartburn diminished, and his diarrhea and edema disappeared.[22]

In 1895, Russia's medical press reported that a French physician, Frémont, had developed a surgical procedure—a form of isolated stomach based on the Thiry intestinal fistula—that enabled him to obtain about 800 cubic centimeters of gastric juice from a dog, and that Frémont was highly enthusiastic about the juice's clinical potential.[23] Pavlov defended his priority in a "Historical Note on the Secretory Work of the Stomach" (1896), which appeared (in both French and Russian) in the Institute's *Archive of Biological Sciences*. He reminded readers that "the basic results flowing from doctor Frémont's experiments were established long ago by me, together with my coworkers."[24] Citing his article of 1890 with Shumova-Simanovskaia and the dissertations of two *praktikanty*, Pavlov pronounced it "understandable that doctor Frémont acquired from his dogs the very same hundreds of cubic centimeters of pure juice that we, too, acquired." Frémont's work, Pavlov continued, was actually much inferior in both its scientific and its clinical import to the earlier findings in his own laboratory. The isolated stomach developed "by me together with doctor Khizhin" was superior to Frémont's technique since it allowed food to come into direct contact with the stomach's mucous membrane. As for the acquisition of gastric juice for clinical use, the combination of fistula and esophagotomy used in Pavlov's laboratory was an "incomparably easier operation" than Frémont's isolated-stomach procedure and yielded juice with an acidity level much closer to that of "pure" juice. And "our dogs live many years in flourishing health"; could Frémont say the same? Pavlov added that "the juice of our dogs has for the past three or four years been used many times by various Petersburg physicians for disturbances of gastric digestion and, judging by their oral testimonies, with good use." Unfortunately, the pioneering work in this direction by Konovalov ("in a dissertation completed in my laboratory") remained entirely unknown outside Russia.[25]

Frémont's threat to Pavlov's scientific priority, and his increasing success in convincing French physicians of the merits of gastric juice therapy, was obviously much on Pavlov's mind in 1897, as is evident from *Lectures on the Work of the Main Digestive Glands.* Here Pavlov noted that his method allowed one to acquire juice from a dog "almost as milk is acquired from cows," converting the animal into "an inexhaustible manufactory of the purest product." He reminded Russian physicians of Konovalov's demonstration that the artificial solutions of pepsin and hydrochloric acid they used "cannot even think of competing" with the effectiveness of natural gastric juice. Pavlov's frustration with the unresponsiveness of Russian physicians is evident in the next passage in *Lectures.*

> That it is obtained from dogs can hardly be a serious obstacle to the use and distribution of the gastric juice of dogs as a pharmaceutical preparation. Numerous trials in the laboratory upon ourselves testified to its usefulness rather than to any harm. Its taste is not at all unpleasant, no worse than a solution of hydrochloric acid of corresponding strength. In view of prejudice [against the juice of dogs] it is entirely possible to acquire juice in this manner from animals that people eat. I cannot refrain on this occasion from expressing regret that this project, which at least deserves a serious trial, is not progressing in Russia, although I have many times brought it to the attention of comrade physicians . . . As of last year the pure gastric juice of the dog, obtained by doctor Frémont from an isolated stomach on the principle of Thiry's well-known intestinal fistula, is recommended abroad as a therapeutic agent for various illnesses of the digestive canal. Will it prove the case that this product long known to us will be more successful under a foreign flag?[26]

This digression from Pavlov's narrative, ten pages into the first chapter of *Lectures,* is interesting for several reasons. First, of course, Pavlov is clearly concerned to establish his own scientific priority. Second, he attributes the failure of natural gastric juice to catch on to "prejudice" against ingesting the juice of a dog. This was indeed a problem, one that Pavlov exacerbated by stressing proudly the "natural" origins of his juice and that Frémont vitiated by obscuring those origins with a well-chosen brand-name—gasterine.[27] Third, Pavlov was clearly not primarily concerned here with turning a profit for his laboratory: he suggested that others could draw natural gastric juice from another animal (perhaps a pig) that Russians ate and so, presumably, would ingest more enthusiastically, and he offered technical assistance to those who might attempt to do so. Finally, Pavlov simultaneously invoked the authority of French medicine (which was already using gastric juice as a remedy) and appealed to Russians' patriotic pride (this remedy was "long known to us").

As of late 1897, then, Pavlov thought he had offered compelling laboratory evidence in favor of his natural gastric juice, but Russian physicians made little use of it. For one thing, Pavlov spoke the language of the laboratory physiologist rather than that of the clinician. For another, he had not advertised his product or made it generally accessible. Commercial pepsins were sold freely in St. Petersburg's pharmacies, but a physician could acquire natural gastric juice only by making arrangements with Pavlov himself. Finally, whatever the limitations of commercial pepsins, these did not include surmounting the cultural prejudice against ingesting the juice of a dog.

From the Laboratory to the Medical Market (1898–1901)

The confluence of several circumstances at the turn of the century transformed the natural gastric juice of the dog from a scientific innovation into a well-known clinical remedy. Certainly, the publication of Pavlov's Lectures and Luk'ianov's textbook of pathology—each of which extolled the virtues of gastric juice treatment—provided added publicity. Three other factors seem to have been even more important: (1) the clinical background, interests, and practice of Pavlov's praktikanty, many of whom became effective advocates of gastric juice therapy in the medical community; (2) the Institute's practical orientation and budgetary problems, both of which put Prince Ol'denburgskii constantly on the alert for potential sources of new income; and (3) the vogue of gastric juice therapy in France and Germany, which enabled Pavlov and others to use both the prestige of western medicine and an appeal to Russian patriotism in marketing this product to Russian clinicians.

For Pavlov, the laboratory's tasks were largely limited to developing a method for obtaining the juice, analyzing its qualities, and establishing scientifically its advantages over artificial products. His praktikanty, however, had a different perspective. Having come to his laboratory with a medical degree, they almost always returned to medical practice afterwards. Many of them became active proponents of scientific medicine and the beneficiaries of its prestige. They thus provided a conveyor belt of influence from the laboratory into the clinical community and sought to persuade their medical colleagues in a language somewhat different from the chief's.

For example, in 1898 Abram Virshubskii, an experienced physician who was working in the laboratory on the influence of fats on digestion, contributed to the widely read clinical journal *Gazeta Meditsina* (Medical Gazette) a two-part article entitled "The Old and the New in the Sphere of the Secretory Function of the Stomach."[28] Here Virshubskii not only outlined the "new doctrine" on digestion that had emerged from Pavlov's laboratory but also engaged clinical experiences and literature to make the connection between laboratory and clinic. "The fundamental reform in the physiology of the stomach," he argued, "will inevitably be reflected in the pathology and therapy of this organ."[29] His central example of the contribution of physiology to a new scientific therapeutics was natural gastric juice.

According to Virshubskii, the new doctrine scientifically explained and confirmed some accepted clinical practices (e.g., the use of milk to treat hyperacidity and hypersecretion) but, more importantly, entailed a new approach to the diagnosis and treatment of gastric problems. Engaging the (primarily German) clinical literature, he outlined the clinical perspective emerging from the work of Pavlov's laboratory. First, the secretory functions of the gastric glands were "indispensable" to the digestion of food and its movement along the digestive tract, and the acid content of gastric juice also stimulated pancreatic secretion. Second, disturbances of this secretory function were manifested in a great number of apparently "independent illnesses" ranging from acidic dyspepsia to diarrhea to "reflexive suffering" in other organs. And third, clinicians frequently failed to correctly diagnose these disturbances and treat them appropriately.[30]

This brought Virshubskii to "the question of the use of gastric juice," which pitted the new doctrine against old prejudices and blind empiricism. Noting that the therapeutic advantages of gastric juice had been mentioned twice previously in print—but "without consequences"—he suggested that physicians' failure to adopt the new remedy reflected their false belief that a solution of pepsin and hydrochloric acid was equally effective.[31] Again Virshubskii engaged the clinical literature, using the reports of Russian, German, and French physicians on their results with artificial pepsins to demonstrate that this view "cannot withstand the most indulgent critique." Using Konovalov's laboratory findings, Virshubskii explained scientifically the disappointing clinical results with commercial products. Not only were these products low in proteolytic power and permeated by various harmful admixtures, but given that they were administered with a random amount of hydrochloric acid, they might actually inhibit digestion (since the optimal acidity level for gastric juice was 0.2 percent). This, Virshubskii argued, explained what one noted Russian clinician admitted to be the "capricious" results obtained by administering hydrochloric acid to patients suffering from gastritis. "Consequently," he concluded, "given the contemporary state of our clinical information" the results of treatment with artificial pepsin and hydrochloric acid "are dependent on blind chance."[32]

Virshubskii's clinical evidence for the efficacy of natural gastric juice was, however, revealingly sketchy. Having devoted about nine printed pages to an explication of the new doctrine and three to debunking the use of artificial pepsin and hydrochloric acid, only on his last page did he turn to the positive clinical results of his preferred remedy. Gastric juice, he explained, had demonstrable bactericidal powers, which would perhaps prove useful in treating diarrhea in children.[33] Because the juice was a "natural product, at one with the organism" and could be easily obtained in great quantity and ideal purity through Pavlov's method, it should certainly win itself a place in the medical armamentarium.[34] As for clinical experiences, he offered only the splendid results achieved on himself and other praktikanty: "I personally have been drinking it for three months, taking 20–25 cc. once or twice a day after eating, and am indebted to it for the disappearance of continual diarrhea and meteorism. Others drank 40 cc. at a time with good results in the sense of the easing of the subjective symptoms of dyspepsia."[35] "Mutatis mutandum," concluded Virshubskii. "The radical reform of gastric physiology is inevitably reflected in the pathology and therapeutics of this organ."[36]

As the 1890s wore on, natural gastric juice acquired an important constituency by becoming a home remedy in Pavlov's laboratory. One clinician who visited in 1900 observed that the advocates of gastric juice therapy included not only Pavlov and Luk'ianov, "the best representatives of the physiology and pathology of digestion," but also "many of Pavlov's students" who had become its "fervent champions." Pavlov himself referred in his *Lectures* to the convincing results obtained by "numerous trials in the laboratory upon ourselves." Publications by such praktikanty as Konovalov (1893), Virshubskii (1898), and V. N. Boldyrev (1907) also mention their own happy experiences, and those of their coworkers, with the laboratory's favorite remedy.[37] Needless to say, these "trials" transpired in an atmosphere conducive to positive results. Several coworkers advocated gastric juice treatment in the medical press, but, probably more important, they no doubt spoke up on its behalf during informal and formal gatherings of their medical colleagues (as the laboratory's visitor found in 1900). By the turn of the century, the sheer number of physicians who passed through the laboratory, became enamored of its home remedy, and subsequently returned to medical practice and administration created a vanguard constituency for this contribution to scientific medicine.

The leaders of the Institute were always alert to potential sources of funds, favorable publicity, and concrete medical contributions, so they also developed an interest in promoting gastric juice therapy. At a meeting of the Institute's governing body on October 5, 1898, the director, Sergei Luk'ianov, announced that "in view of the successful application of natural gastric juice" Prince Ol'denburgskii thought it "necessary to organize the regular preparation of this juice for sale." Pavlov had clearly been consulted beforehand, and he expressed his willingness to produce the juice in 200 cubic centimeter portions for sale at 50 kopecks per flagon through the Institute's financial office.[38] In his report to the Tsar at year's end, Ol'denburgskii explained this decision as a result of "heightened demand for this preparation."[39]

This heightened demand was actually quite meager (see Table 3 below), but it sufficed to tax Pavlov's limited production capacity. It seems likely that Virshubskii's article, the first to make an extended argument on behalf of natural gastric juice in terms (and in a journal) calculated to influence a clinical audience, led a few physicians to make their way to Pavlov's laboratory and request a flagon or two (just as Troianov had done in 1893 after publication of Konovalov's report). If so, this would have put Pavlov in an awkward position: His entreaties to Russian physicians notwithstanding, he was not prepared to provide gastric juice to all comers. He had other uses for his dogs, and it was hardly an efficient use of his resources to have an esophagotomized dog standing by just in case a physician should stop by. Furthermore, an outsider's arrival would disrupt the chief's rigid laboratory regimen and require him to pull a coworker away from his research in order to draw the gastric juice. Pavlov would not have failed, however, to see this "heightened demand" as an opportunity to expand his operations, enhance his prestige at the Institute, and provide a service to physicians. He may well have turned to either Luk'ianov or Ol'denburgskii with the suggestion that the gastric juice operation be put on a firmer footing.

The Institute Council reached the decision to begin regular production of natural gastric juice amid a controversy regarding the Institute's anti-plague serum. Several flagons of this preparation had been taken to the Pasteur Institute for analysis by Emile Roux and by A. A. Vladimirov, chief of the Institute's Epizootology Division, both of whom pronounced it effective. Their report was attacked vigorously, however, by several Council members, including Pavlov, who complained that the experimental evidence was too scanty to support this conclusion. The Institute had invested considerable sums in the anti-plague serum, which was a pet project of Prince Ol'denburgskii, the subject of many proud words in his annual reports to the Tsar, and the source of both good publicity and much-needed revenue. (Both Prince and Princess Ol'denburgskii seem generally to have taken an adventurous approach toward the use of Institute serums.)[40] Luk'ianov had adopted a compromise position, agreeing that more experimental investigations were required but also insisting that the serum had "some" preventive and therapeutic value.[41] Himself a proponent of Pavlov's remedy for dyspepsia, and also preoccupied with the Institute's financial problems, Luk'ianov probably invoked Prince Ol'denburgskii's authority to forestall a similar debate about natural gastric juice.

The Institute Council decided on November 10, 1898, to advertise its new medical product in the official government journal *Pravitel'stvennyi Vestnik* (Herald of the State), the conservative newspaper *Novoe Vremia* (New Times), and the leading medical journal *Vrach* (The Physician). During the last two months of the year, a mere ten flagons were sold (adding a grand total of 5 rubles to Institute coffers). Yet a meeting of the Institute's Council on January 7, 1899, with Pavlov in attendance, further resolved that, in view of the numerous people who were coming to the Institute to obtain gastric juice, the juice would be dispensed only through pharmacies or with a physician's prescription.[42] This "heightened demand" produced sales of 114 flagons in 1899 and 262 in 1900.

In 1901, a qualitative increase in demand led to the establishment of a full-fledged factory-style laboratory operation. Luk'ianov reported to the Institute Council that the "significant increase" in demand had caused frequent delays in satisfying the requests of pharmacies and physicians. Clearly, as in 1898, Pavlov and Luk'ianov had conferred before the Council meeting, but this time they had not reached an accord on all relevant issues. Pavlov responded to Luk'ianov's report by insisting that his staff could not produce enough gastric juice to meet the increased demand without disrupting his laboratory's scientific research. He suggested that an assistant be hired and paid 30 rubles a month from the Institute's general coffers. Luk'ianov then raised a contentious issue: like other Institute preparations, Pavlov's natural gastric juice should come with instructions for its proper use. The director had raised this in an earlier meeting in January but had dropped it when Pavlov insisted that the necessary information was already available in the scientific literature. Luk'ianov now raised it again, clearly as a negotiating point. Pavlov again resisted (the Council protocols record "an exchange of opinions"), and the Council finally agreed to ground rules for a much-expanded operation: Pavlov would be allowed to hire an assistant at 30 rubles a month to prepare the gastric juice under his direction. This assistant's salary would be approved monthly by the Council and could be terminated if juice sales declined. At Luk'ianov's suggestion, the price of a flagon of gastric juice was raised to 80 kopecks to cover this added expense. Pavlov would write instructions for the use of this product and submit them for the Council's approval.[43] I discuss these instructions and Pavlov's reticence about writing them below, but first we need to consider one question: why did the demand for gastric juice increase so sharply, from 262 flagons in 1900 to 864 flagons in the first nine months of 1901?

The answer, I think, lies in the rave reports about the efficacy of gastric juice therapy in the French medical press in 1900 and 1901. Between January 1900 and June 1901, French physicians published about a dozen articles with glowing reports in *Bulletin de Société Médicale des Hôpitaux* and *Le Bulletin Médical*. Their general tone was captured by the eminent clinician Henri Huchard's pronouncement that natural gastric juice was as powerful a remedy for the treatment of gastric illnesses as was foxglove for disturbances of the heart.[44] These reports elicited published discussions celebrating the therapeutic benefits of gastric juice in treating various digestive ailments.

These French physicians were using Frémont's gasterine, the product that had prompted Pavlov's defensive "Historical Note" of 1896 and the patriotic warning in *Lectures* that this "product long known to us will be more successful under a foreign flag." Gasterine had caught on much more quickly with French physicians than had Pavlov's natural gastric juice with Russian physicians.[45] At this time there were close ties between the Russian and French medical communities, and Russia's leading medical journal, *Vrach*, regularly reported on French developments.

In February 1900, a certain D.J. contributed an item to *Vrach* in which he drew attention to a report made by the French physician Le Gendre to the Parisian Société Médicale des Hôpitaux earlier that year. Describing Le Gendre's reported clinical successes with Frémont's gasterine, D.J. noted, "Gasterine is Frémont's name" (to conceal from patients the nature of the substance administered to them) "for *gastric juice* obtained *from the isolated stomach of the dog* (strange that there is no mention at all of prof. I. P. Pavlov, to whom we are indebted for this means of obtaining gastric juice)."[46] Doctors Frémont and Le Gendre had reported on thirteen cases in which gasterine (which D.J.

loyally insisted on calling "gastric juice of the dog") had cured all manner of digestive complaints, failing only with cancer sufferers.

These developments corresponded perfectly with the rhetorical tactic by which Pavlov, in both his "Historical Note" and *Lectures,* positioned his "natural gastric juice" vis-à-vis Frémont's "gasterine." Pavlov's nostrum benefited both from the authority of French medicine (which had demonstrated its clinical efficacy) and from patriotic reaction against what was portrayed as the vaguely dishonest exploitation of a Russian discovery by French physicians.

French developments were clearly much on the mind of the Khar'kov physician A. A. Finkel'shtein, who arrived at Pavlov's laboratory in 1900 to learn about the production and use of natural gastric juice. Pavlov arranged an "extraordinarily kind reception" and instructed his visitor on the techniques of gastric juice production. Finkel'shtein then operated on his own dogs in the faculty clinic of Khar'kov University's medical school and used the gastric juice thus obtained to treat twenty-two patients with various stomach ailments.[47] His article in *Vrach,* "Treatment with Natural Gastric Juice," began and concluded with references to the nationalist aspects of this remedy. In his introduction he wrote pointedly, "Our illustrious physiologist prof. I. P. Pavlov, having illuminated so much about digestion in a new light, has expressed, incidentally, regret that the use of the gastric juice of dogs as a therapeutic agent, which merits a thorough trial, has not caught on here in Russia" (p. 963). In his concluding paragraph, Finkel'shtein "quoted" Pavlov's patriotic plea in *Lectures* (quoted earlier in this chapter), adding several words in a perhaps unconscious indication of his own patriotic preoccupations: "Will it prove the case that this product long known to us *and produced in a Russian laboratory* will be more successful under a foreign flag?" (the emphasized words are Finkel'shtein's, absent from Pavlov's text). He ended with the ominous observation that "efforts in this direction have already been made in recent times (Le Gendre, Frémont, Leyden)" (p. 963).

Finkel'shtein endorsed the now traditional arguments in favor of using natural gastric juice rather than a solution of pepsin and hydrochloric acid, adding his own observations that natural gastric juice "is pleasant in sight and scent" and that most pepsin solutions "even diminish the proteolytic power of gastric juice in man" (p. 963). More important, he detailed the encouraging results obtained in eight patients suffering from "catarrh of the stomach," two with stomach cancer, nine with typhus, and three with miscellaneous gastric problems. Natural gastric juice had not, of course, restored the "chemism" in the stomach of cancer patients, but it did relieve their pain on eating, improve

their appetite, and vitiate several other symptoms. For those suffering from catarrh, it had proven a "powerful healing agent," relieving symptoms and restoring digestive functions. "The most striking data was acquired by detailed investigation of the patients' gastric juice: both its secretion and chemism
changed sharply for the better" (p. 964). Finkel'shtein reported equally dramatic improvements in anemic patients. Knowing that anemia was frequently
accompanied by a diminution of gastric secretion and a revulsion for food, he
had administered natural gastric juice to two anemic patients. His hopes were
"justified brilliantly" when their appetite returned, improving the quality of
both their gastric juice and their blood (p. 965). Seven of nine typhus patients
also responded positively, much more so than to the hydrochloric acid with
which they had earlier been treated: natural gastric juice stimulated their appetite, cured their constipation, relieved symptoms in their intestines and nervous system, and may even have shortened the course of their illness. In sum,
the gastric juice of dogs had proven "a new, powerful, therapeutic agent"
(p. 965). As a product of "the Russian laboratory," it deserved the attention of
Russian physicians who cared about their patients and their country.

Factory Production

In a small room on the ground floor of the laboratory, five young, large dogs,
weighing sixty to seventy pounds and selected for their voracious appetite,
stood on a long table harnessed to the wooden cross-beam directly above their
heads. Each was equipped with an esophagotomy and fistula, and each faced
a short wooden stand tilted to display a large bowl of minced meat (Figure 24).

During the period of their employment—about five or six years on the
average—these "factory dogs" (*fabrichnye sobaki*) were fed sumptuously to
compensate for the extraordinary demands made on them.[48] Every other day
they were called on to produce about one or two quarts of gastric juice for collection, enough, one knowledgeable coworker wrote, to treat two or three patients.[49] They constantly lost a great deal of fluid as their saliva ran out through
the aperture of their esophagotomy, and so, at midday, they were fed large
quantities of water through their fistula. Their weight was monitored constantly; if it declined, they were pulled off the line. The dogs were fortified daily
with a breakfast and dinner each consisting of about two pounds of meat and
a large bowl of oatmeal. It is not surprising, then, that one assistant, who could
not afford to indulge himself so generously, was discovered eating the dogs'
supplies.[50]

The amount of meat necessary to sustain factory dogs drove up the price

FIGURE 24. *Pavlov's "small gastric juice factory." Courtesy of Academy of Sciences Archive, St. Petersburg branch*

of their labor, but the continued exploration of alternatives had come to naught. In theory, pigs and cats could also generate the pepsin-rich psychic secretion necessary to ignite a patient's digestive process, but the chief was prejudiced, both personally and institutionally, in favor of dogs (and against cats). As we have seen, Konovalov had demonstrated that pigs were "too nervous" for the job. One former praktikant, Nikolai Riazantsev, had explored the possibility of using the juice of cats (1894) and bulls (1898). The former had proven too small and too inefficient as producers. The latter generated a relatively inexpensive and plentiful product, but the gastric juice of a bull failed, predictably, to proteolize albumins and so was unfit for human use.[51]

So it was dogs that, every other day, were put to work (Figure 25). Having been denied food since the previous evening, they bounded hungrily to their places in the small gastric juice factory. The assistant washed out their stomachs with water, attached each dog to a harness, corked the fistula, and ran a glass tube (containing some glass wadding to catch any mucus) through the cork to a large collecting bottle. Only then did he or she place before the dogs the basin filled with chopped meat. "The dogs were fed raw minced meat, which, because of the esophagotomy, fell out of the upper end of the gullet into a dish. The animal bent its head, took the same meat in its mouth and swallowed, and again the meat fell into the dish. The dog could never know

FIGURE 25. *Boldyrev's sketch of a "factory dog" at work. Food consumed by the dog excites its appetite but falls out of the aperture of the esophagotomy (C) and so never reaches the stomach. The resulting gastric secretion (or "appetite juice") flows out through a fistula (E) into a receptacle. After heating and filtering, the juice was bottled and sold as a remedy for dyspepsia. From V. N. Boldyrev, "Natural'nyi zheludochnyi sok, kak lechebnoe sredstvo, i sposob ego dobyvaniia,"* Russkii Vrach, *1907, no. 5: 156*

the reason why it might continue eating indefinitely and yet become more and more hungry . . . Large, hungry dogs could produce up to 1,000 cc. of gastric juice at one session. Gastric juice was collected from an animal every other day so as to allow the animal's organism to replenish the water, salts, and proteins lost with the secretion."[52]

As Pavlov delighted in pointing out, the gastric juice factory was powered by appetite. In his lectures at the Military-Medical Academy, he proudly demonstrated the operation of this physiological perpetual-motion machine.

Yesterday somebody asked me: why does the dog not refuse to eat? I must tell you that I have had hundreds of such dogs and not one behaved as your comrade suggested . . . They eat in the most marvelous fashion until their jaws become tired. Well, of course, this is a sham meal, a physiological *perpetuum mobile* . . . Does it guess that it is being deceived? That is for the dog to know. What, after all, stimulates us to eat? Necessity. But the dog has this very same necessity, even more so, because it eats and eats and is not satisfied. You have heard from history that in dissolute times Roman gentlemen would feast and then take purgatives and again begin to eat. And this is people—but the dog does as God commands. With its operation it needs no purgatives. Eat as much as you wish![53]

Pavlov concluded this lecture by offering the class an opportunity to sample the gastric juice produced by the dog before them. "Taste it. It is very acidic in taste."[54]

The dogs usually labored for about two or three hours a session, during which time they each produced about a quart and a half of juice. The assistant then passed the juice through filter paper and heated Chamberland-Pasteur tubes composed of unglazed porcelain. This was calculated not only to eliminate bacteria and improve the scent but also, as one coworker revealed, to catch any "worm" eggs—"if," he hastened to add, "the dog has worms and if one allows that these could somehow penetrate to the stomach, which, by the way, has never been seen in the laboratory."[55]

Management of this operation entailed "great responsibility." By 1904 the gastric juice factory was adding thousands of rubles to laboratory coffers, and quality control was especially essential in the face of the cultural prejudice against ingesting a dog's bodily fluids. On days when the factory was in operation, Pavlov made it his "first duty" on arriving at the laboratory "to check the thermostat upon which the control bottles stood in order to make sure that there was no fungal infection."[56] The first assistant, hired after the Institute Council meeting of October 1901, was a female student, one Filatova.[57] As profits swelled in subsequent years, Pavlov placed the operation under the supervision of his most trusted assistant, Evgenii Ganike.

These profits made a substantial difference in the life of Pavlov's laboratory and in his clout at the Institute. As we saw in Chapter 1, the Institute was constantly strapped for funds and the budgets of its divisions remained essentially flat from 1891 to 1914—except for the monies earned by the sale of various serums. The impact of the gastric juice proceeds is clear from the data in Table 3.

By 1904 juice proceeds had increased the laboratory budget by about 70 percent and by the First World War by more than 500 percent. Earlier, both Ganike and Pavlov had routinely reached into their own pockets to pay for necessary supplies when money ran short.[58] Now Pavlov had considerable discretionary funds at his disposal and was able to purchase whatever animals and equipment he desired, to expand his laboratory's permanent staff and working hours, to raise Ganike's salary, and even to provide Ganike with his own assis-

Table 3
Gastric Juice for Sale, 1898–1904

YEAR	JUICE INCOME (RUBLES)	SALES (FLAGONS)	LAB BUDGET FROM INSTITUTE OF EXPERIMENTAL MEDICINE (RUBLES)	TOTAL FUNDS (RUBLES)
1897	—	—	3,400	3,400
1898	5	—	3,400	3,405
1899	66	114	3,400	3,466
1900	71	262	3,400	3,471
1901	374	1,091	3,400	1,818
1902	930	1,191	3,800	4,730
1903	2,321	2,358	3,400	5,721
1904	2,570	3,275	3,400	5,970

Source: Figures compiled from the annual budgetary reports of Pavlov's laboratory, in Tsentral'nyi Gosudarstvennyi Istoricheskii Arkhiv Sankt-Peterburga 2282.

Note: The relationship between the volume of sales and income changes inexplicably in the years 1902–4, perhaps reflecting changes in pricing or in the percentage of sales income received by the laboratory. Beginning in 1903, Pavlov supplied physicians and institutions with free samples and bulk orders, making them, in effect, distribution centers.

tant.[59] (It was perhaps a sign of the Physiology Division's prosperity that in 1903 it became the first to submit *typed* annual reports to Prince Ol'denburgskii.) Gastric juice funds also inevitably increased Pavlov's stature at the Institute. For example, in January 1903 he requested a raise, to be paid from the special fund at Prince Ol'denburgskii's disposal to augment the salaries of selected division chiefs. The much-esteemed head of the Chemistry Division, Marcel Nencki, had received 4,000 rubles annually from this fund until his death the previous year; Sergei Vinogradskii, head of the Division of General Microbiology, received 1,000 rubles. A yearly sum equal to Vinogradskii's, Pavlov wrote, would "entirely satisfy" him; and he added by way of justification, "In conclusion, I permit myself to share with Your Excellency my happiness that the budget of the Physiological laboratory of the Institute has increased by almost 1,000 rubles thanks to the ever increasing sale of gastric juice."[60] This argument proved persuasive.

The growing wealth of Pavlov's laboratory resulted not only from the juice sales themselves but also from his successful negotiation of Institute politics. Other Institute divisions that produced serums and medicines for sale were permitted to keep only a small percentage of the proceeds, most of which entered the Institute's general coffers. The Epizootological Division, for example, kept only 10 percent of the profits from its sale of malein and tuberculin. Pavlov's Physiology Division, however, received *100 percent* of the gastric juice proceeds. Recall that in 1901 Pavlov agreed to produce gastric juice if the Institute provided 30 rubles a month so that he could hire a helper and thus avoid disrupting the scientific work of his laboratory. In 1902, when Nencki died, Pavlov took advantage of the situation to plead poverty and object to the Chemistry Division's privileged budgetary status (Chemistry received 4,900 rubles a year, 1,500 more than Physiology—which received 1,250 more than General Pathology). Pavlov was partially mollified by a one-time transfer of 150 rubles from the Chemistry Division's budget and by another one-time payment of 50 rubles from the Division of General Pathology. Much more important, the Institute Council agreed that "in view of the new difficulties for the Physiology Division" all expenditures for the production of gastric juice would be reimbursed monthly from the Institute's general coffers. The Physiology Division would reimburse the Institute at the end of the year, however, for any net losses from gastric juice production.[61]

The Institute, then, made the initial investment in the gastric juice factory. Pavlov received 400 rubles a year to pay an assistant's salary of 360 rubles and to cover other costs. When the operation began to turn a profit, Pavlov assumed this 400 ruble fixed cost and successfully resisted attempts to appropriate the profits for the Institute's general needs. Boris Babkin, a coworker at the time, later recalled Pavlov's argument for this special arrangement: "From time to time the Institute administrators sought to deprive Pavlov of this income from the sale of gastric juice, which they wished to apply to the general needs of the Institute. Their efforts always called forth stormy protests from Pavlov, and he invariably defended the right of his laboratory to use funds gained from the sale of a product obtained by his own method. Of course, he himself did not receive a penny from this undertaking, and, as far as I know, he never patented his method of procuring gastric juice."[62] Here again, Pavlov's close relationship with Ol'denburgskii no doubt worked in his favor. (Babkin's observation about Pavlov's failure to patent his procedure is worth pondering, and I do so below.)

The Difficult Link to the Clinic

How exactly were physicians supposed to use natural gastric juice? The simple answer, from Pavlov's standpoint, was as follows: "We know for scientific reasons, and from our own experience, that it works. Discover the clinical details for yourselves." As noted earlier, in 1898 Pavlov had refused to write instructions for the use of his remedy, insisting that the necessary information was "available in the scientific literature." He agreed to do so only in 1901 in return for the Institute's subsidy of his operation. Those instructions, reproduced in full below, provided little guidance for clinicians.

Natural gastric juice is obtained at the Physiological Division of the Imperial Institute of Experimental Medicine from dogs operated upon by the method of the Member of the Institute I. P. Pavlov. The operation is described in the article by I. P. Pavlov and E. O. Shumova-Simanovskaia, "The innervation of the gastric glands in the dog" (*Vrach*, 1890, No. 41, p. 929).

Before obtaining the juice the dog's stomach is washed completely clean and it is then allowed to swallow a piece of meat, which, nevertheless, does not reach the stomach but rather falls out to the surface through a fistula in the esophagus. Due to the food's excitation of the appetite there begins a very strong flow of gastric juice through the gastric fistula.

The juice obtained in this manner is first aerated with the goal of eliminating, as much as possible, the unpleasant specific scent which is absent only in rare cases, and then the juice is filtered through heated tubes made of unglazed porcelain to eliminate all microorganisms and accidental admixtures.

The juice is poured into sterilized glass receptacles of 200 cubic centimeters each. Gastric juice should be used only at a physician's instructions. The medium dose for an adult: 50–200 cubic cm. a day, in several doses, taken with food. For a child the dose should be correspondingly less.

Of the reports in the scientific literature about the use of gastric juice, the following can be mentioned: 1. A. Finkel'shtein, "Treatment with natural gastric juice" (*Vrach*, 1900, No. 32, p. 963); 2. Frémont ("Gazette des hôpitaux," 1895, No. 58); 3. Le Gendre (*Vrach*, 1900, No. 6, p. 179); 4. Lannois (*Vrach*, 1900, No. 7, p. 211); 5. A. M. Virshubskii, "The Old and the New in the Sphere of the Secretory Function of the Stomach" (*Meditsina*, 1898, No. 26, p. 8).

It is necessary to keep in mind that the juice spoils over time. At room temperature the juice can be maintained without marked weakening for only one month; but on ice it can be preserved for about one year, and therefore this is the preferred means of preserving it.

Kept on ice, the juice sometimes produces a sediment which dissolves if the juice is returned for a while to room temperature. This sediment is composed of pure pepsin and does not at all mean the preparation has spoiled. Juice that has grown dark at room temperature is not fit for use. For understandable reasons, juice that has spoiled under any conditions should be withdrawn from use.

A 200 cubic cm. vial of juice costs 80 kopecks. Juice cannot be returned unless there are reasons to recognize that the spoilage occurred for reasons dependent on the Institute. Uncorked vials are not accepted back.

Prof. I. Pavlov
St. Petersburg
November 26, 1901[63]

Here the authority for gastric juice therapy rests almost entirely on the scientific literature and the hygienic quality of laboratory procedures, with an added implicit appeal to the reputation of French medicine. The illnesses or symptoms amenable to gastric juice treatment are not even mentioned. The recommended "medium dose," to say nothing of the "correspondingly less" dose for children—both without regard to illness or symptoms—represented, at best, rough guesswork. (To take some of the sources cited in the "instructions": Le Gendre treated indigestion with 90 to 150 grams a day; Finkel'shtein treated catarrh of the stomach with 25 grams each time his patient complained of pain; and Virshubskii reported no clinical experiences or dosages.)[64]

Pavlov's coworker Alexander Sokolov candidly admitted the roughness of these instructions in July 1901 when one physician, I. K. Konarzhevskii, requested ten flagons of gastric juice and precise instructions for its use. "Instructions for the use of natural gastric juice and the means of its application at the bedside must be developed by the clinic and practical medicine," replied Sokolov. "We cannot devote ourselves specifically to the elaboration of this question and therefore cannot provide the exact instructions that you need. Published information on the use of natural gastric juice for therapeutic goals is still very scanty. You, therefore, have no alternative other than to yourself observe this remedy, guided by the conclusions obtained by experiment and published in an entire series of works from our laboratory."[65] Armed with "the good wishes of the comrade and the purely theoretical character of the information," Konarzhevskii used gastric juice to treat twenty patients at the ambulatory clinic of the private hospital where he worked.

The results were most encouraging. "Natural gastric juice," Konarzhevskii concluded, "should in the near future occupy a conspicuous place as a therapeutic agent in the treatment of chronic illnesses of the stomach, including cancer of that organ." Taken shortly before a meal, it most "energetically restored the appetite," much more effectively than did hydrochloric acid and artificial pepsin. The restoration of appetite was not the result of "psychic influence" (i.e., a placebo effect), since most of the patients did not know what they were taking or why. Natural gastric juice reversed the decline in "general nutrition" associated with chronic catarrh of the stomach, curing it completely in about three weeks; it produced an aversion to liquor in alcoholics and was an effective palliative for those suffering from cancer of the stomach. It also cured "neurosis of the stomach" and inflammation of the stomach's mucous membrane, and it halted unhealthy fermentative processes. Konarzhevskii concluded, as did many Russian commentators on this subject, on a patriotic note: "It is extremely desirable that science be indebted to Russian clinical and practical medicine for the decisive resolution of the question of evidence for the use of natural gastric juice and the means of its application, just as science is entirely indebted to the Russian school of experimental medicine for the development of the method for its acquisition from dogs" (p. 26).

Intended to publicize the therapeutic efficacy of natural gastric juice, Konarzhevskii's clinical accounts also reveal the difficulties engendered by the problems of expense and squeamishness. Perhaps apprehensive about his patients' reactions, Konarzhevskii did not generally inform them of the nature of "the acidic medicine" they were ingesting. Some, however, discovered the truth one way or another—perhaps he told the first few patients and the reaction convinced him not to inform subsequent ones; perhaps he informed only his well-educated clientele. One female patient who suffered from anemia, "globus hystericus," and "neurosis of the stomach" responded "very reluctantly, with great skepticism and revulsion" to the proposal that she ingest the juice of a dog. Another sufferer from "neurosis of the stomach" proved an inconveniently knowledgeable man: "Although much revolted by the very idea that he would have to take natural gastric juice—as a naturalist by education, he was already familiar with the means by which it was obtained—when the neurosis of the stomach began to torture him badly our patient began to take a liquor glass-full twice a day before eating. He easily adjusted to this in several days, finding it even rather tasty and saying that, however skeptical he might have been about gastric juice as a therapeutic agent, he recognized its usefulness for himself, since after taking it for several days he undoubtedly felt a significant relief from the torturous, extremely unpleasant 'revolutions' of his stomach" (p. 18). The patient's appetite improved markedly and he was still gratefully taking his medicine when Konarzhevskii completed his article.

Konarzhevskii's most detailed accounts concerned six female patients suffering from chronic catarrh and chlorosis. Three were "rural girls," two were peasants, and one was an urban woman. The rural girls apparently did not know what they were taking. They were well aware of its cost, however. Instructed to ingest 30 to 43 cubic centimeters three times a day, they attempted to economize by watering down the gastric juice to a 50/50 solution. Only on the twelfth day of their treatment did they begin to notice an improvement in their condition. The two peasant women remarked from the first day that "this acidic medicine" was restoring their appetite; by the fourth day they were eating without pain. They ended their treatment after two weeks (pp. 20–22).

Unlike the peasants and like the male intellectual, the urban woman was well aware of the nature of the nostrum: she "initially ingested natural gastric juice very reluctantly, with great revulsion; but I [Konarzhevskii] then considered it an urgent necessity at every meal. During the first two or three days she ate little and without appetite, just as she had before taking the gastric juice; but from the fourth day, that is, after the ingestion of an entire 200 cc. flagon, it was clear to all that her appetite was enhanced and that the symptoms of chronic catarrh had noticeably subsided. She began to eat much and with clear satisfaction, commenting that she was experiencing dog-like proclivities in the form of a craving for food" (p. 21). Possessing a good sense of humor, she was clearly expressing her discomfort with the nature of the cure.

Konarzhevskii worked in a private hospital where his patients possessed at least a modicum of discretionary funds. We can only estimate the cost of their treatments, but clearly they were rather expensive. Assuming that the rural girls who watered down their gastric juice terminated their treatment at the first sign of improvement, they each purchased about five flagons at a cost of 4 rubles. The two peasant women, who ingested full-strength gastric juice for two weeks, each spent 8 rubles. The urban woman, who followed the standard regimen for three weeks and then ingested reduced portions of juice during a fourth week, probably spent about 14 rubles. For the price of their treatments, the rural girls could have purchased forty pounds of apples and the peasant women eighty pounds of meat on the hoof; the urban woman could have paid one week's rent on a nice two-bedroom apartment in St. Petersburg.[66]

In their public comments, Pavlov's coworkers discussed several obstacles to the sale of natural gastric juice, but its expense took pride of place. Perhaps because they didn't want to exacerbate the problem, none mentioned patients' aesthetic resistance to ingesting dog juice. Babkin later concluded that the chief obstacle was the "inconvenience of its administration": "Even though twice diluted with water, it still had an unpleasantly sour taste. The juice had to be taken in large quantities, being sucked through a glass tube so that its acid would not damage the teeth."[67] This observation was no doubt correct for the relatively small group that could afford Pavlov's nostrum. Boldyrev's diagnosis in 1907, however, was closer to the mark as a general market analysis: unfortunately, he observed, the high cost of feeding a dog in St. Petersburg drove the price up well beyond that affordable in the public clinics of rural Russia and rendered it impractical for "practice among the poor."[68]

The steadily increasing sales from 1902 to 1914, then, almost certainly reflected purchases by physicians, pharmacies, and medical institutions catering to St. Petersburg's wealthy and, especially, to its growing middle classes. Unlike Germany, Russia lacked national health insurance, and few peasants or workers—together comprising over 90 percent of the population—could afford to indulge in such an expensive cure for their stomach ailments. Natural gastric juice, then, never became a truly mass phenomenon, but its sales grew with St. Petersburg's middle classes, reaching fifteen thousand flagons annually by the eve of the First World War.[69]

Selling Scientific Medicine

Why did Pavlov not patent his natural gastric juice? This was, after all, a lucrative product of his scientific discoveries, and one with demonstrable potential in both Russian and foreign markets. There are a number of possible explanations, including Pavlov's distaste for commercial ventures that he considered inappropriate to the life and values of the true scientist.[70]

The interesting point, though, is that, far from patenting this product, Pavlov and his coworkers actively encouraged others to produce it themselves. Konovalov's dissertation (1893) and *Lectures* itself (1897) contained the basic information necessary for aspiring creators of a gastric juice factory. When Frémont and other westerners began producing their own juice, Pavlov objected only to their failure to acknowledge his scientific priority and to the possibility, hurtful to his patriotic pride, that Russians would eventually purchase gasterine rather than Russian-made natural gastric juice. We have seen that Finkel'shtein and Konarzhevskii received guided tours of the gastric juice operation and were apprised of all necessary technical information to reproduce what they saw there. Furthermore, in 1907 two of Pavlov's coworkers, Boldyrev and I. S. Tsitovich, published articles in *Russkii Vrach* specifically designed to aid their countrymen in just such an endeavor. As Boldyrev explained, the juice was expensive to produce in St. Petersburg, an obstacle to its use that "is easily overcome, since the operations that transform the dog into a producer of natural gastric juice, as is evident from their description [in his article], are so simple that they can be performed by any physician with even a little mastery over the knife." He provided readers not just with precise instructions for the care and feeding of the dogs but even with the address of the medical supply company in St. Petersburg where the equipment necessary for sterilizing the juice could be purchased by catalogue for a mere 3 rubles and 25 kopecks.[71] However important the financial benefits of his operation, these clearly did not exhaust Pavlov's motivations for spreading the gospel of gastric juice therapy.

For Pavlov and his praktikanty, the distribution of natural gastric juice and associated technologies was a means of selling physiologically based scientific medicine in general and publicizing the achievements of Pavlov's laboratory in particular. For precisely this reason, perhaps, Pavlov insisted on marketing his product as "the natural gastric juice of the dog," despite the risk of causing revulsion in potential customers. By such nomenclature he emphasized that this medical product was the creation of experimental physiology, the result of understanding and manipulating the laws governing living organisms.

Gastric juice therapy fit perfectly the rhetorical tactics by which Pavlov sought to convince physicians of the importance of experimental physiology. Throughout the 1890s, Pavlov emphasized the necessarily symbiotic relationship of laboratory and clinic and offered a physician-friendly physiology that was respectful of traditional clinical wisdom, which he delighted in defending against the criticism of mechanistic physiologists.[72] Within this rhetorical context, and given the failure of physiology to provide the same dramatic medical breakthroughs as had bacteriology, Pavlov's specific examples of physiology's contribution to medical practice were almost always cases in which the laboratory had *confirmed and refined* accepted clinical practices. For example, physicians had always stressed the importance of the patient's individuality—and Pavlov's studies of the digestive role of the capricious psyche had confirmed this; physicians had always prescribed milk in cases of weak digestion—and Pavlov's discovery of milk's richness in chemical exciters had revealed the sound physiological principles behind this wise practice.

Gastric juice treatment embodied this same relationship between laboratory and clinic. Through his synthetic physiology, Pavlov had confirmed and explained the traditional medical emphasis on appetite and associated practices. He had debunked the debunkers, refuting the crude laboratory experiments used to convince physicians that their attention to appetite was unscientific. He had vindicated the wisdom of those physicians who, although the concept of appetite had disappeared from recent textbooks, continued to employ therapeutic measures, such as the use of bitters, to promote a healthy appetite.[73] The "natural gastric juice of a dog" represented physiology's refinement of an established clinical practice—the replacement of unreliable artificial pepsins with a superior natural product—and so embodied a partnership between laboratory and clinic.[74]

Gastric juice therapy thus became part of Pavlov's physician-friendly physiology, a physiology that self-consciously defended traditional medical practices against the criticism of crude and mechanistic experimentalists, that restored appetite to the central place to which empirical medicine had earlier assigned it, that reinforced physicians' authority over the regimen of their patients, and that, as Claude Bernard had promised decades ago, provided therapeutically useful scientific knowledge.

Conclusion

The success of the "small gastric juice factory" resulted not just from Pavlov's ideas and skills but also from the very nature of his laboratory enterprise. Pavlov had developed the basic scientific ideas and surgical skills for the gastric juice operation by 1890, but it was the institutional context of the Institute and the nature of his laboratory system that transformed these into a successful enterprise. That context encouraged division chiefs to develop and market medical remedies in order to secure favorable publicity and much-needed funds. The incorporation of a large number of physicians into laboratory life provided the skilled hands that turned appetite and a specific dog-technology into a medical innovation and provided a vanguard constituency for that innovation within the medical community. These physicians spoke the language of clinical medicine more fluently than did the chief. They proved effective salesmen, recommending "the natural gastric juice of the dog"—this emblem of the value of their own scientific expertise—in articles and daily interactions with their peers.

By any reasonable historical definition, gastric juice treatment provides an example of laboratory-based scientific medicine. Based on the path-breaking research of a leading experimental physiologist, publicized under the banner of contemporary science and buttressed by the experiences of numerous physicians in several different countries, the natural gastric juice of the dog enhanced Pavlov's reputation among physiologists and physicians alike. I suspect that gastric juice therapy was typical of the relatively small services that physiologists rendered to the clinic in this period, services that were far less dramatic than those rendered by bacteriology and much less likely to attract the attention of historians looking for contributions that we might recognize as scientific today, but nevertheless provided good reason to believe in the clinical promise of experimental physiology.

Chapter 9

HAIL TO
THE CHIEF

I will even venture the opinion that Pavlov [is] . . . as great as
Claude Bernard.

—MAURICE PALÉOLOGUE, *An Ambassador's Memoirs* (1925)

That France's last ambassador to the tsarist court would,
in his private diary (September 25, 1916), equate Ivan Pavlov
with Claude Bernard speaks volumes about the Russian's
reputation. Paléologue's assessment, moreover, echoed that
of more authoritative commentators. Even before Pavlov
became the first physiologist to win a Nobel Prize, his
French colleagues were comparing him to Bernard and
Charles Brown-Séquard, while Germans invoked the names
of Carl Ludwig and Rudolf Heidenhain.

Paléologue and his scientific contemporaries could each,
no doubt, have pointed to one or more of Pavlov's particular
contributions, but when paying homage to the Russian physiologist they usually emphasized the great number and range
of his achievements. Theoretical generalizations and esoteric
facts, new techniques for measuring secretions and surgically
altered animals, endless supplies of digestive fluids for investigative and therapeutic purposes—these and other laboratory products formed the "material substrate" of Pavlov's
reputation.

The composite Pavlov who emerged, however, was much
greater than the sum of his parts. As the single man associated with these many varied achievements, and as an effec-

tive spokesman for their broader significance, Pavlov himself became a powerful symbol of a modern, clinically relevant, laboratory-based physiology that was simultaneously precise and synthetic, a physiology that truly brought living organisms, in all their complexity, within the purview of modern science.

The historical development of this composite Pavlov followed a relatively simple chronology: From 1891 to 1897 Pavlov concentrated on the Russian medical audience, rising to prominence in his native country. In 1898 the German translation of *Lectures on the Work of the Main Digestive Glands* effectively "breached the frontiers" (in the words of one contemporary), and the synergistic effect of laboratory products and effective marketing brought Pavlov a "European reputation." This status was both reflected and qualitatively enhanced by the Nobel Prize he received in December 1904.

The next part of the story both complicates this picture and underlines the importance of factory production. In the first years of the twentieth century, as Pavlov was enjoying his growing reputation, scientific developments raised serious questions about some of his most important knowledge claims, his interpretive practices, and even the "normalcy" of one of his key dog-technologies. The increasing importance of physiological chemistry, the rise of a new humoralism, and insider criticisms by a disaffected former coworker suggested the limits of Pavlov's style of organ physiology, exposed the tendentiousness of some of his interpretations, and compromised the sweeping vision of a precise and purposive digestive factory that Pavlov had offered just a few years before. These subjects remained contentious for decades. Pavlov's reputation, however, did not rest on the resolution of any one or even several such issues. As the representative of the entire product line of his physiology factory, and the symbol of all that this product line had come to represent, he was praised as a great physiologist even by those who rejected the knowledge claims he held most dear.

Capturing the Russian Medical Market, 1891–97

In the early and mid-1890s, Pavlov used the steady stream of products generated by his laboratory to occupy a strategic position as the leading Russian physiologist in St. Petersburg's medical community and as the exemplar of a mutually respectful relationship between laboratory physiology and the medical practitioner. The principal market for these products was the Society of Russian Physicians, where praktikanty and chief alike delivered the great bulk of their scientific reports. This choice suited both Pavlov's scientific vision and his institutional circumstances, and it did much to shape, in turn, the rhetoric he later brought to a broader audience in *Lectures.*

Russia's physiological community was small and scattered, but six leading practitioners worked at its epicenter in St. Petersburg. The most prestigious position, at the Academy of Sciences, was occupied by the aging F. V. Ovsiannikov, who concentrated on histology, embryology, and neurophysiology (and whose intervention on behalf of V. N. Velikii had cost Pavlov the position at Tomsk University in 1889). The professor of physiology at St. Petersburg University was the neurophysiologist N. E. Vvedenskii, who had narrowly defeated Pavlov for that university post in 1889. Four scientists pursued physiological research in various departments of the Military-Medical Academy: Ivan Tarkhanov, professor of physiology until his retirement in 1895; K. N. Ustimovich, who presided over a small physiological laboratory in the Veterinary Department; V. M. Bekhterev, professor of psychiatry and neuropathology; and Pavlov, who headed the department of pharmacology until 1895, when he replaced Tarkhanov as professor of physiology.

Russian physiologists lacked both their own professional society and their own specialized journal. They published primarily in the country's various medical journals and, when seeking a more specialized and prestigious outlet, in such leading western scientific journals as Virchow's *Arkhiv,* Pflüger's *Arkhiv, Centralblatt für Physiologie,* and (less frequently) *Archives de Physiologie.* (During the First World War, when German outlets were closed and patriotic feelings ran high, Pavlov and Vvedenskii would join two other colleagues to found Russia's Physiological Society and its *Fiziologicheskii Zhurnal.*)

The Russian market for physiology expanded qualitatively with the professionalization of Russian medicine in the years after the Crimean War, particularly with the founding in the 1880s of what quickly became the medical community's two leading institutions: the journal *Vrach* and the Society of Russian Physicians.[1] The Society's vibrant St. Petersburg branch included about 150 of the capital's most eminent physicians, professors of clinical medicine, and medical administrators, a number of whom gathered twice a month from September through May to hear and discuss brief reports. The proceedings of the St. Petersburg Society were widely published, both in its *Trudy Obshchestva Russkikh Vrachei* (Works of the Society of Russian Physicians) and in other Russian medical journals (which sent their own reporters to Society meetings).

Pavlov was St. Petersburg's only physiologist to make the Society the major outlet for his laboratory's literary products. Tarkhanov and his relatively few coworkers published most of their scientific work in French and German scientific journals, and they reported only infrequently to the Society. Ovsiannikov and Ustimovich appeared at Society meetings rarely if at all, and Vvedenskii was not even a member. Bekhterev and his coworkers published extensively abroad and concentrated their domestic efforts on a specialized segment of the Russian medical market—psychiatrists and neuropathologists.[2]

The nature of Pavlov's laboratory, his institutional base, and his style of physiology made the Society of Russian Physicians an ideal audience. As the physiology factory moved into full operation, it was capable of generating about six reports a year, researched and delivered by practicing physicians and frequently concerning the vital processes in intact animals. Pavlov used the occasions of these reports to propagate his vision of a physiologically based medicine and to gather potentially useful advice from his audience. Furthermore, the Imperial Institute of Experimental Medicine strongly encouraged outreach to the broader medical community, so Pavlov's participation in the Society enhanced his institutional position (and his praktikanty's publications in the Society's *Works* added heft and "good numbers" to his annual reports). The praktikanty's reports also redounded to their own professional advantage, helping them make connections in the capital as they prepared to return to medical practice. The chief routinely used a successful report to advance a praktikant's candidacy for membership in the selective Society itself. (Leon Orbeli later recalled that some influential members resented the influx of Jewish candidates from Pavlov's laboratory and that only the chief's threat to resign from the Society reversed the rejection of one Jewish nominee.)[3]

The reliable supply of reports from Pavlov's laboratory also sustained the Society of Russian Physicians, which often experienced difficulty finding speakers for its semi-monthly meetings. Pavlov could always be depended on to provide the necessary reports, and this constant supply, in turn, influenced the shape and tone of Society discourse and Pavlov's status within it.

Rarely did any investigator take the podium at Society meetings more than once a year, but between 1891 and 1904 Pavlov and his coworkers delivered about ninety reports. Carefully edited by the chief, these reports were tightly focused ten-minute statements on some aspect of the praktikant's doctoral research. Delivered by physicians and buttressed by impressive experimental data, they conveyed the range, methodologies, fundamental conclusions, and therapeutic promise of the laboratory's activities. Furthermore, their sheer volume, their interlocking position along Pavlov's lines of investigation, and the chief's periodic synthetic presentations made each of these reports part of an ongoing discourse. Frequently, a challenging question to a praktikant-reporter was answered by reference to an earlier report to the Society or to on-going laboratory research. These questions were usually handled by Pavlov himself, who often rose at the conclusion of a praktikant's report to summarize its significance and, almost always, to deal with any objections.

The sheer quantity of these occasions created a role for Pavlov: he became, as he once put it, "the voice of contemporary times," the experimental physiologist explaining to practicing physicians the nature and value of laboratory research.[4] His status as the authoritative voice of laboratory science within St. Petersburg's leading medical organization was reflected in his election as the Society's vice-president in September 1893, a post he held until assuming the presidency in 1907.

Mounting the podium regularly at Society meetings, Pavlov became increasingly expert at marketing his laboratory's products to an audience composed "primarily of representatives of practical medicine."[5] The rhetoric and specific examples he would use so effectively in *Lectures* to describe the relationship between laboratory physiology and medical practice (both in general and with respect to his laboratory's specific products) were developed, enriched, and polished during these constant exchanges at Society meetings.

In his remarks at these meetings Pavlov always insisted on the special advantages of laboratory research in analyzing complex phenomena that were difficult or impossible to untangle at the bedside, but he was also (usually) careful to acknowledge the unique perspective conferred by physicians' daily experiences. His effectiveness in making the connection between laboratory and clinic, and doing so in a manner attractive to medical practitioners, is clear from the frequency with which physicians responded to a report from the Pavlov laboratory by referring to their own clinical experiences.[6]

The dynamics of these exchanges is evident in the discussions of Khizhin's, Lobasov's, Volkovich's, and Val'ter's reports. When Pavel Khizhin reported to the Society in September 1894, explaining the key role of appetite in digestion, one physician asked Pavlov his opinion of force-feeding patients. The chief did not condemn that practice—noting that even in the absence of appetite, a patient required nourishment—and took the occasion to tie his laboratory findings to long-standing clinical practices. "It is well-known to all physicians that appetite is an important thing, and it is therefore a constant concern of the physician . . . and this is because it is linked with ignitory secretion. Appetite is the first portion of gastric juice, with which digestion begins."[7]

The harmony between laboratory results and clinical traditions broke down, however, over another issue. Physicians in the audience insisted that the mechanical excitation of the stomach by food contributed to gastric secretion, a position that contradicted Khizhin's report and undermined Pavlov's claim that the isolated sac faithfully mirrored secretion in the intact stomach. On this issue Pavlov was adamant, insisting that here clinical experience was misleading. "I invite you to my laboratory," he responded. "You can tickle the stomach [of a laboratory dog] as much as you please and will not receive even a drop of juice."[8]

One physician related his experiences with "starving patients with cancerous constrictions of the digestive tract" to explain his skepticism about Pavlov's argument. These patients had a "fearsome appetite" and "think only about food," yet when the physician opened their stomachs he found "not a drop of gastric juice." After placing two tablespoons of food in the patients' stomachs, he observed "the most enormous secretion of juice, entire cups of it." Pavlov professed himself "profoundly certain that there is an error here" since the results of laboratory experiments were clear and consistent. "If you take a dog with a gastric fistula and perform on it an esophagotomy . . . and then force the dog to fast for two days, and then approach it with a piece of meat, there will begin an ignitory secretion, and one will receive a mass of juice. I would want personally to see your trials for gastric juice in the patients you describe."[9]

Ivan Lobasov's report the following year prompted one physician in the audience to discuss his own experiences with the responses of dyspeptic patients to meat. Pavlov replied that the laboratory had this "in view," and he offered some advice. "I would like to turn the attention of physicians to the fact that fat is a normal inhibitor of gastric secretion, and when the physician is struggling with hypersecretion of the stomach and at the same time wants to activate the pancreas, fat is a wonderful means to do so. I think that this can also be useful with an ulcer."[10]

After Andrei Volkovich's report to the Society in March 1898, Pavlov complained that physicians had paid too little attention to his laboratory's findings and that one recent physician-investigator, "based on clinical material, expresses the conviction that there are no differences in the gastric juice elicited by various types of food." "Such a conclusion," he asserted, "is explicable only by the unsuitability of the clinical method for resolution of such questions." Pavlov also noted that one of Volkovich's dogs had developed an ulcer, "the symptoms of which corresponded entirely to those observed among patients in the clinic." Drawing upon laboratory experiences with this dog, he offered physicians some advice about diagnosing similar cases.[11]

When Anton Val'ter reported to the Society about the distinctive pancreatic responses to various foods, the potential clinical import of his research led several members of the audience to press him on some sensitive details. Why, asked one professor of clinical medicine, did Val'ter compare amounts of various foods with equivalent nitrogenous content rather than with equivalent amounts of carbohydrates or fats? "You know, I feed a patient, not so that he will absorb nitrogen, but to give him strength—which means I have in mind carbohydrates and fats . . . I see that if I want to introduce nitrogen, it is best to feed [the patient] milk, but what should I do if I want to enhance nutrition?" Val'ter responded that this would be addressed in future experiments, but that he had begun with amounts equivalent in nitrogen because "nitrogen is the main component part of food." His questioner, however, pressed the point, noting that "in that case, you are comparing incomparable things," since the caloric content differed in the tested amounts of milk and bread. "From a clinical point of view, it would be more interesting to determine what it is most advantageous to feed [the patient] in order to [efficiently] introduce nutrients." At this point, Pavlov took the floor. The questioner's point would be addressed more precisely in future experiments, he said, but Val'ter's "approximate calculation of calories" already indicated that the most advantageous food for that purpose was milk. Another physician suggested that pediatricians would be especially interested in laboratory tests of mother's milk, which perhaps was more easily digested than cow's milk; another added that a comparison of the digestibility of the milk of various animals would also be useful.[12]

V. N. Sirotinin, professor of clinical medicine at the Military-Medical Academy, raised another point. Val'ter had demonstrated the "surprisingly rational" nature of pancreatic secretion, but was this necessarily the result of nervous reflexes? What happened when this reflex was destroyed experimentally? Val'ter responded that it was impossible to test this experimentally, since "these reflexive paths have not been elucidated anatomically and therefore cannot be destroyed by vivisection." Another professor objected that these paths could surely be found in the spinal cord. Pavlov again rescued his floundering praktikant. Severing the spinal cord was too "crude" an operation to yield definitive results, he explained. The laboratory did, however, have supporting data of another sort: unlike dogs with a Pavlov sac, dogs with a Heidenhain sac manifested no "fluctuations in secretion." Clearly, then, the severing of the vagus nerves in the Heidenhain sac destroyed a "subtle [nervous] reflex." (Here, of course, Pavlov was using an observation concerning gastric secretion to answer a question about pancreatic secretion.)

Two other physicians—again, having in mind specific issues about the treatment of patients—pressed Val'ter about other details: how big were the food particles fed to his dogs, and what was the temperature of the food and the feeding time? Pavlov again stepped in, insisting on the experimenter's need first to identify basic patterns and only afterwards to confront complicating variables. "In all our experiments," he explained, "we constantly took bread as bread, meat as meat, and milk as milk, since each type of food has its own character. There then arises the question: why does this occur, what depends upon the food, what upon its volume, what upon various qualities of firm materials? This already belongs to the province of analysis. For us, it was most necessary to first characterize the work [of the glands] with each food, and then there will be analysis."[13] Another clinician then related some experiences that accorded with Val'ter's (and Khizhin's) conclusions, and the Society president, Popov, engaged in an agreeable exchange with Val'ter about the importance of a patient's (or dog's) individuality. Popov concluded the session by thanking Pavlov for the "valuable data" generated by his laboratory and expressing his hope that research on the questions discussed would continue.

When laboratory findings seemed to contradict clinical experiences, Pavlov always insisted on the reliability of his results, but he sought, wherever possible, to reconcile the two. By both insisting on the precision of controlled laboratory experiments and conceding that these experiments could not reproduce the complexities that the physician encountered in a patient, he upheld the conclusions of his laboratory while leaving room for the sometimes contrasting experiences of physicians. For example, when one praktikant's report on the passage of food from the stomach to the intestines encountered objections based on medical practice, Pavlov responded, "I generally find that one cannot compare clinical observations of chronic illnesses with a physiological experiment: these [observations] cannot be refuted by the data of an experiment. A mass of other influences enter into [clinical observations], and the physician cannot eliminate these . . . In the physiological experiment all these peripheral influences are eliminated, the phenomena are simplified, and so the results stand out in greater relief." The physicians in the audience were able to agree that the laboratory had isolated one of several mechanisms in the passage of food from the stomach to the intestines and therefore the apparently conflicting experiences of laboratory investigators and clinical observers could be reconciled.[14] Similarly, Pavlov once noted that "from ancient times" physicians had used alkalies to treat gastric disorders in the belief that they excited the secretion of gastric juice. This, his laboratory had discovered, was erroneous. Still, Pavlov took empirical medical wisdom on faith. Clinical observations, after all, were "incomparably broader than experimental observation";

it would therefore be "unjust" to conclude that alkalies had no therapeutic benefits. It was simply "premature" to expect the laboratory to explain everything that worked in practice.[15]

Pavlov did, however, insist on the fundamental "analogy" between laboratory findings with dogs and clinical realities with human patients, and sometimes moved easily between the two realms. For example, he noted that the laboratory's finding that cold foods disturbed gastric secretion was confirmed by the frequent illness of peasants during harvest time (when they were too rushed to prepare hot meals).[16] When one physician drew upon his experiences with patients to criticize a praktikant's finding that fat inhibited the movement of food through the digestive tract, Pavlov responded, "I think that the fact related by the report is beyond doubt and that its meaning is clear. I cannot allow that there is a fundamental difference between a human and a dog in this regard. If doctor Akimov-Perets did not succeed in finding in people the inhibitory influence of fat upon the movement of food, then one can explain this only by the fact that his thought was not directed toward this. If he would now repeat his experiments, then, probably, he would find in people the very same as the reporter [i.e., the praktikant] found in dogs."[17] When another clinician reported experiments on patients that confirmed the laboratory's conclusions about the role of appetite, the chief noted approvingly, "Your experiments demonstrate in the best possible manner that if one repeats and verifies any investigation, one must perform the experiment under the very same conditions . . . Clinicians have already attempted to verify your conclusions, but always depart from the conditions of the experiment, . . . having a preformed idea, and, of course, do not acquire your results. Your service consists in the fact that you have entered completely into the spirit of an experimenter . . . and so you have received entirely the same [results] as did the experimenters."[18]

Pavlov repeatedly invoked his own experiences and attitude as exemplary of the fruitful relationship between laboratory and clinic. On assuming the vice-presidency of the Society of Russian Physicians in 1893, he applauded its practice of dividing the top posts between representatives of "practical and theoretical laboratory medicine." This policy was "the best guarantee of the vitality and purposiveness of our Society" and embodied the "mutually beneficial union" between laboratory and clinic.[19]

Similarly, he portrayed his relationship with Sergei Botkin as one between exemplary representatives of the laboratory and the clinic. Recall that Botkin, Russia's most eminent physician until his death in 1889, had in 1878 become Pavlov's patron by selecting him to manage the Botkin laboratory at the Military-Medical Academy. The choice of Pavlov rather than a clinician as speaker on the fifth anniversary of Botkin's death was itself symbolically meaningful. Throughout his speech, Pavlov portrayed Botkin as the ideal "thinking physician," possessing a broad clinical perspective, cognizant of his lack of physiological expertise, and respectful of the laboratory's potential contribution to medicine. In this 1894 speech he recalled an instructive episode from the previous meeting of the Society.

> My much-esteemed comrade, professor Tarkhanov, reporting on his animal experiments, expressed a somewhat new view about sleep. At the end of his report, with many reservations and labeling his own words almost audacious, he permitted himself to propose the existence of a special sleep center [in the brain]. Then a member of the audience clearly recalled that 20 years earlier the late Sergei Petrovich Botkin, at one of his lectures analyzing cases of pathological sleep, bravely dwelled on his own view that these cases were best understood if one acknowledged the existence of a special sleep center. I find this case instructive and truly characteristic of the relationship of the physiologist and physician to certain issues. Therefore, it seems to me desirable that physiologists be better acquainted with the clinic and especially with clinical casuistry. How many cases one can find where clinical observations led to the discovery of new physiological facts![20]

The exchange Pavlov was describing here had occurred just two weeks previously, so many listeners surely remembered that the "member of the audience" was Pavlov himself.[21] In this anecdote, then, Pavlov both flattered physicians—crediting their great representative, Botkin, with the initial insight—and advertised himself as a physiologist respectful of the physician's unique perspective. (It was no doubt a bonus that he accomplished this at Tarkhanov's expense.)

Pavlov closed this address by invoking his own (idealized) relationship with Botkin as an exemplary one between physiologist and physician: "I had the pleasure of entering into special relations with the late Sergei Petrovich [Botkin]. I was a *laborant* in his clinical laboratory. I well remember now and will long recall those cases when I came to him with laboratory results. Sergei Petrovich did not like to engage in physiological critiques, but, through his extraordinary perspicacity, he immediately found confirmation of the facts brought to him [by Pavlov], and these [facts] also elucidated for him murky sides of clinical observations, . . . opening new perspectives for the posing of new questions, for [the discovery of] new factors. I naturally hope that our exchange after my report today will have the very same fruitful character."[22] His wish was granted. Clinicians in the audience shared their experiences with various remedies and solicited Pavlov's opinion. Sirotinin, professor of clinical medicine at the Military-Medical Academy, termed Pavlov's findings "a joyous phenomenon" for all physicians, since "they reconcile therapeutic, pharmacological, and experimental results, which have long contradicted one another, for example, with respect to bitters."[23] Pavlov later incorporated this information and the very tone of this exchange into *Lectures*.

In his addresses to the Society of Russian Physicians, Pavlov developed and honed the rhetoric on the laboratory-clinic relationship that he would use so effectively in *Lectures*. His speech on assuming the vice-presidency in 1893 both asserted the necessity of the laboratory to modern medicine and recognized the uniqueness of the practitioner's perspective. On the one hand, "the practical physician is called upon to fix a machine that nobody understands as is necessary. Take the example of a watch craftsman. If he takes it upon himself to fix a watch, he knows how it is built and, of course, his activity is entirely purposeful and precise. The very same is demanded of the physician—to fix that which is broken—but to do so in a machine about which there is no complete information. Therefore I understand the ideal aspiration of the practical physician toward that which provides this knowledge; I understand why practical medicine at present holds tightly to the theoretical, to the laboratory."[24]

On the other hand, the "theoretical" medicine generated in the laboratory required the broad experience and textured perspective of practical medicine if it was to avoid the dangers of rigid, one-sided thinking. "It is clear," said Pavlov, "that practical medicine, as knowledge existing from the time that man first fell ill, knowledge that has been collected always under the powerful pressure of the instinct of health and life—that such knowledge must have uncovered immense material and does so every moment. I understand, then, the interest of theoretical physicians dedicating themselves to the laboratory who, in order to avoid clichédness (*shablonnost'*) and broaden their worldview, turn to practical medicine's profound and broad supply of observations."[25]

Addressing "the relationship between physiology and medicine in questions of digestion" in his 1894 speech commemorating the fifth anniversary of Botkin's death, Pavlov again emphasized the symbiotic relationship between laboratory and clinic and positioned himself as an intermediary between the two. The physician was "the mechanic of the human body" and like any mechanic required precise knowledge of the mechanism he was repairing. There was, Pavlov recognized, a nonscientific component to medical practice, yet the "talent and art of separate physicians" were but "fleeting phenomena" that perished "together with individuals." Practitioners' contribution to "the science of the human organism" would prove more enduring, since that science would best provide mankind with "its greatest happiness—health and life." As a component of science, medicine consisted of "the collection of data about the phenomena occurring in the human organism under extremely varied conditions, both external and internal, in the majority of cases arising on their own and only in part resulting from the interference of the physician in the course of the vital process." Physiology, on the other hand, was the investigation of phenomena "produced almost entirely by the investigators themselves," and consequently, "observation, as the basic method of medicine, is transformed in the hands of physiology into experiment."[26]

This replacement of observation by experiment was "an enormous advantage" for the physiologist as investigator, but Pavlov hastened to assure his audience that he was not "a true theoretician . . . prepared to view all practical medicine as applied physiology." The "thinking physician," too, possessed important advantages. For one thing, "all ailing mankind" was in the "laboratory of the physician," who therefore was heir to generations of empirical knowledge and constantly encountered phenomena produced not by man's "weak hand" but by "powerful life and nature." Not surprisingly, then, "the physiological horizon of thinking physicians" was sometimes "broader and freer than that of physiologists themselves." Furthermore, physiology consisted largely of limited, analytical truths obtained by breaking the organism into its parts; the physician, on the other hand, dealt "with synthesis, with the whole of life." This constituted both a great strength and a source of errors for the "thinking physician." It also made the practical use of physiological knowledge a difficult, challenging task.[27]

The fusion of this synthetic dimension with the precision of laboratory physiology was, as we have seen, a hallmark of Pavlov's scientific vision. Speaking in 1899 on the tenth anniversary of Botkin's death—and two years after the publication of *Lectures*—he claimed some important practical success in doing so and thus took a slightly more aggressive position on physiology's status in scientific medicine. Entitled "The Contemporary Unification in Experiment of the Main Aspects of Medicine in the Case of Digestion," his speech drew upon his own preliminary successes in "experimental pathology and therapeutics" to proclaim the dawn of a new age in medicine.[28] "Victorious experiment," he announced, "is, under our very eyes, now extending its power to both pathology and therapy . . . It seems to me that the most important success of contemporary medicine consists in this—that it has acquired the possibility at the present time of being elaborated, in its main aspects, experimentally.[29] Physiology, then, was increasingly able to incorporate the synthetic dimension that had earlier been the exclusive province of practical medicine.

Pavlov referred here to his laboratory's efforts to treat the afflictions of experimental dogs.[30] The practical results, he conceded, were still "trivial," but they justified the confident hope that "experimental, laboratory therapeutics" would soon "indicate to the clinic—efficiently and with complete competence—a purposive mode of action" against specific illnesses.[31]

The most impressive of these results concerned the survival of a dog with a double vagotomy. Physiologists had long been aware that this operation condemned an animal to death, and Moritz Schiff had recently proposed that a vagotomized animal perished from the disruption of its digestive system. In a dramatic presentation to the Society of Russian Physicians in 1896, Pavlov stirred his audience (and become a bête noire among Russian antivivisectionists) by displaying a vagotomized dog that he had kept alive for many months.[32] He explained that, by first identifying the chain of events that led from a vagotomy to fatal digestive disturbances, he had been able to intervene effectively and preserve the animal's health. "Here is a clear example of an entirely laboratory-based, rational therapy of a serious, fatal disturbance of the organism," in other words, an example of experimental therapeutics.[33]

Pavlov's other examples involved much less dramatic successes in treating the various digestive problems of his laboratory dogs. For example, he claimed some success in using alkalies to treat dogs with hypersecretion. As noted earlier, physicians had long treated digestive problems with alkalies in the belief that they excited gastric secretion, a belief that Pavlov had refuted. Yet he had nevertheless urged physicians not to abandon this practice, since it represented generations of empirical experience. Now, in 1899, he claimed that his laboratory had justified long-standing medical wisdom: alkalies actually inhibited gastric secretion and so were useful in the treatment of hypersecretion. Preliminary experiments also indicated that hypersecretion might be "easily" treated by varying an animal's diet.[34]

For Pavlov in 1899, the physician's authority rested primarily on mastery of the clinical art. As in 1894, he closed his speech with some inspirational words about his model "thinking physician," Sergei Botkin, a man of "magical" clinical abilities who appreciated the unique virtues of the laboratory.

> Was this not a clinician with a stunning ability to unravel the riddle of illnesses and to find the best means against them! His charm among patients truly carried a magical character: he would often cure with only a word, with only one visit. How many times did one hear from his student-clinicians the sad admission that the very same prescriptions in, apparently, the very same cases, that were ineffective for them produced miracles in the hands of their teacher. It would seem that the soul of this great clinician would be com-

pletely satisfied . . . but, nevertheless, his profound intellect, not seduced by his success, sought the key to the great puzzle: what is an ill person and how to help him through the laboratory, through animal experiment. Under my very eyes, he sent tens of his students into the laboratory. And this clinician's great appreciation of experiment comprises, in my conviction, Sergei Petrovich [Botkin's] glory no less than does the clinical activity so well-known throughout Russia.[35]

As Pavlov dominated the Society of Russian Physicians, his other laboratory products reinforced his position as the leading physiologist among St. Petersburg's physicians. The laboratory's various publications were reviewed appreciatively by physician V. F. Orlovskii in his annual summary of developments in digestive physiology and their practical significance.[36] The "natural gastric juice of the dog" found a growing market among the capital's practitioners, and as laboratory alumni occupied positions throughout the Empire they spread the word about laboratory achievements—occasionally, in articles that applied their scientific expertise to issues related to their subsequent careers, ranging from the treatment of various ailments to the scientific feeding of cattle.[37]

Professional Successes in Russia

By 1897 Pavlov had become Russia's institutionally most powerful physiologist, and he enjoyed the various kinds of recognition that marked a successful Russian scientist in his prime.

Laboratory products made solid if undramatic headway in Russia's small physiological community. The production of pure digestive juices facilitated work in the burgeoning area of physiological chemistry by the physiologist A. Ia. Danilevskii and by two chemists at the Institute, Marcel Nencki and Ekaterina Shumova-Simanovskaia. Beginning in 1902, E. S. London, head of the Institute's Division of General Pathology, used a dog with a Pavlov isolated sac (created by Pavlov's assistant A. P. Sokolov) in his pioneering studies of protein absorption.[38] As of 1904, however, the laboratory had launched only a handful of alumni on a career in physiology. Pavlov's favorite, Val'ter, had died in a train accident in 1902, prompting the chief's lament that "there are no physiologists now, and they are needed."[39] Pavlov ignored here two former praktikanty who, although practicing physiologists, brought him no solace: Alexander Samoilov had abandoned digestive studies for electrophysiology, and Lev Popel'skii, as we shall see below, was hardly conducting himself as an ideal alumnus. (A cohort of future physiologists—Babkin, Boldyrev, Orbeli, and Savich—had not yet completed their doctoral dissertations and left the laboratory.)[40]

Pavlov was keenly disappointed by the response of Russian physiologists to *Lectures*. His wife, no doubt echoing his assessment, recalled that the book "had no success at home in Russia."[41] Yet Pavlov's dissatisfaction seems more revealing of his high expectations than of his work's actual reception. *Lectures* inspired only a few reviews—not one by a senior physiologist—and certainly did not transform physiology, or even digestive physiology, in Russia. The paucity of reviews, however, reflected the absence of specialized physiological journals and the rarity of book reviews in Russia's medical journals. The silence of such leading physiologists as Danilevskii, Ovsiannikov, Sechenov, Tarkhanov, and Vvedenskii was at least partly explicable by their primary interest in other aspects of their discipline. Furthermore, *Lectures* offered little that was new to the many Russian physicians and physiologists who regularly read the *Works* of the Society of Russian Physicians, where both the praktikanty's reports and the chief's synthetic statements had previously appeared.

The few reviews that did appear were positive, praising Pavlov for his synthetic view of the digestive system, his contributions to methodology and technique, and the usefulness of *Lectures* for medical practice. The reviewer for *Russkii Arkhiv Patologii, Klinicheskoi Meditsiny i Bakteriologii* (Russian Archive of Pathology, Clinical Medicine, and Bacteriology) praised *Lectures* as a "very important contribution to the scientific-physiological literature," although he did note that the book was not "a separate independent investigation of digestive physiology." "Professor I. P. Pavlov's work is, nevertheless, highly valuable," he wrote, "since the author presents in lively form, in the form of lectures, everything accomplished over the past ten years on the physiology of the stomach and pancreatic gland by an entire series of investigators in the laboratory he directs. The important scientific significance of such a 'summary' of a decade of laboratory work, in which the initiative, direction, and guidance of work, rested entirely, of course, on the director of the laboratory, is not subject to any doubt." The reviewer emphasized Pavlov's important methodological innovations, which had enabled him to provide "an entirely precise response to the main questions of digestive physiology." He concluded on a strong positive note: *Lectures* would "fundamentally change the reader's view of the activity of the digestive organs" and would become a "handbook for every physician."[42]

The same journal carried a much longer essay, "The Old and the New in the Field of Digestion," by S. S. Salazkin, who had worked briefly in Pavlov's laboratory in 1893. Salazkin, who had embarked on a successful career in physiological chemistry, described the history of digestive physiology in purely Pavlovian terms. He touched every point dear to his former chief, emphasizing Pavlov's methodological contributions, his demonstration of the central role of appetite and nervous control, his refutation of previous views about the role of mechanical exciters, his proof of the precise, regular, and "enormously adaptive" operation of the digestive glands, and the clinical importance of Pavlov's achievements. "The works of the Pavlov school," Salazkin concluded, "have played a very prominent role, having fundamentally altered . . . not a few former views and having shed light on many areas in which darkness had earlier reigned."[43] Salazkin's reference to the "Pavlov school" would prove typical of a transformation that took place as *Lectures* was reviewed: the praktikanty, whom Pavlov had always called "coworkers" (*sotrudniki*), became instead his "pupils" and part of his "school."

The leaders of Russian physiology wrote little or nothing about Pavlov's work in these years. Tarkhanov's assessment of Pavlov's scientific contributions, in an article of 1898 for the Brokgauz and Efron encyclopedia, offered faint praise indeed: "I. P. Pavlov's great significance consists in his introduction and perfection of a method to obtain various digestive juices in pure form."[44] S. I. Chir'ev, professor of physiology at Kiev University, wrote thirty pages on digestive physiology in his textbook *Human Physiology* (1902) without once mentioning his St. Petersburg colleague.[45]

Nor was Pavlov among the scientists most familiar to the Russian public. Unlike such popular figures as Ivan Sechenov and Dmitrii Mendeleev, he was neither a preeminent scientist nor a frequent writer and lecturer to lay audiences. When in 1901 a columnist for the popular weekly *Niva* (The Cornfield) reviewed the scientific legacy of the past century, he mentioned three Russian physiologists: Sechenov, Tsion, and Vvedenskii.[46]

Pavlov did, however, enjoy considerable professional success in these years. In 1897 he was promoted to full professor at the Military-Medical Academy, and his laboratory there, though never a match for that at the Institute, was fundamentally renovated. In 1900 he shared with Bekhterev the Academy of Science's prestigious von Baer prize for scientific research. A committee chaired by Ovsiannikov concluded that "the experiments of professor Pavlov, thanks to the new methods he has introduced, have yielded such brilliant results by their precision that his name is inextricably linked with all the most significant data in the sphere of digestion."[47] Significantly, the Academy's committee also noted that Pavlov's *Lectures* had been translated into German, a sign of the "European reputation" much prized by Russia's scientific community. In 1901 Pavlov was elected to corresponding membership in the Academy of Sciences, and so had good reason to believe that he might succeed Ovsiannikov as its physiologist, and so acquire there his third laboratory (this did occur, in 1907). There also began a trickle of honorary memberships: in the Ekaterinoslav Medical Society (1900), the Society of Russian Physicians (1901), the Moscow Society of Naturalists (1902), the St. Petersburg Medical Society (1903), and the Society of Odessa Physicians (1903). Finally, as we shall see in Chapter 10, in 1901 Pavlov's colleagues at the Institute and the Military-Medical Academy nominated him for a Nobel Prize. At least in the latter case, Pavlov was almost certainly aware that they had done so.

What, then, was the source of Pavlov's disappointment in the Russian response to *Lectures?* What could he have wanted from his colleagues that they had not bestowed? Perhaps the answer, in part, is this: they had not recognized him as the preeminent physiologist in the country and they could not bestow upon him the prerequisite for that status: a European reputation.

Attaining a European Reputation, 1898–1904

Until the publication of the German edition of *Lectures,* Pavlov did little to publicize his laboratory products or his special contribution as laboratory chief in the West. In the years 1891–97 he published only infrequently in western journals and eschewed all international conferences. His few publications in western languages concerned either the innervation of the digestive glands or topics far from his laboratory's main lines of investigation, and they lacked entirely the synthetic quality of his published addresses to the Society of Russian Physicians. During these same years, edited versions of key doctoral dissertations—including those by Sanotskii, Dolinskii, Khizhin, and Lobasov—appeared in the French edition of the Institute's *Archive of the Biological Sciences.* Though not widely available, this journal did make some of the laboratory's key knowledge claims accessible to western specialists. Pavlov himself, then, was not cited substantially more frequently, or in greater detail, than were his praktikanty.

This state of affairs is clear in Edward Schäfer's authoritative *Text-Book of Physiology,* published in 1898 as an English-language successor to L. Hermann's *Handbuch der Physiologie* of the previous decade. In his contributions to this multi-authored volume, J. S. Edkins cited the two Pavlov articles that had been published in German, one on vagus control of the pancreas, the other on vagus control of the gastric glands and the importance of appetite. Drawing upon the publications of Pavlov and a number of praktikanty—Kudrevetskii,

Dolinskii, and Popel'skii—Edkins noted that "the later experiments of Pawlow and the St. Petersburg school have greatly amplified our knowledge of the nervous influence" on the pancreas.[48] Pavlov's scientific contribution received no special emphasis here. In his discussion of gastric secretion, Edkins devoted more space to Khizhin's research than to that of any other member of "the St. Petersburg school," crediting him with the important discovery that the course of secretion differs for various foods.[49]

The German edition of *Lectures,* then, was Pavlov's first effort to bring before a western audience the full range of laboratory products, his own synthetic vision of the digestive system, and his role as laboratory chief. In September 1897, even before the Russian edition had appeared, he wrote the preface to the German translation: "These [lectures] present a connected and complete review of everything that has previously been scattered among a dozen separate articles. Several of these articles are written only in Russian, others were published in the form of dissertations and reports to meetings, and so remained entirely unknown to the scientific world of other countries. This book contains that which, in its sum, was not accessible to foreign specialists and to the contemporary literature."[50]

The German edition owed much to Pavlov's Institute colleague Marcel Nencki, who used his extensive European contacts to facilitate Pavlov's negotiations with the German publisher, and to Pavlov's favorite praktikant of the time, Anton Val'ter. Val'ter had completed his path-breaking thesis on pancreatic secretion in 1897 and received a grant from the Military-Medical Academy to work in Ewald Hering's laboratory of experimental pathology. By this time he had also begun his German translation of Pavlov's masterwork. Perhaps in part as a result of his affection for and familiarity with the chief and his laboratory, Val'ter succeeded in giving this translation the same authoritative, informal, and lively tone as the Russian original.[51] The German edition appeared in 1898, followed in 1901 by a French edition and in 1902 by an English edition, translated from Val'ter's German (and including Pavlov's 1899 address to the Society of Russian Physicians).

Lectures was reviewed quickly, widely, and with uniform enthusiasm by physicians, physiologists, and other laboratory scientists in Western Europe and the United States. Reviewers did not fasten upon any single knowledge claim, but rather commented consistently on the work's great range, synthetic quality, and relevance to medical practice. A brief summary of some typical responses illustrates both the separate appeal of the different elements of *Lectures* and the symbolic value that Pavlov himself acquired as the single figure identified with them all.

In his review of *Lectures* for *Zeitschrift für Physikalische Chemie,* the eminent physical chemist Wilhelm Ostwald emphasized the exemplary relationship between "the Pavlov school" and the chief's synthetic work. Noting Pavlov's "lively language," "brilliantly conducted experiments with animals," and "numerous new explanations," Ostwald emphasized that *Lectures* resulted from "the general work of the laboratory" and thus embodied the great progress over recent decades in the very nature of scientific work. Liebig's laboratory had been the first example of the "organization of such collaborative work of master and comrades, where the latter are themselves on the path to becoming masters themselves." "This scientific-practical method" was destined to reshape world science, and, wrote Ostwald, "we can hail the present work as a desirable and fortunate sign of this success." (Ostwald clearly knew little about Pavlov's laboratory beyond what he had read in *Lectures;* as we have seen, the great majority of praktikanty were not, in fact, "on the path to becoming masters.") Ostwald had a developed interest in the nature of scientific creativity, and he was unique in identifying Pavlov's laboratory system as the most significant feature of *Lectures,* but this theme, and the routine transformation of Pavlov's coworkers into his "pupils" and his "school," marked other reviews as well.[52]

Many reviewers noted the great sweep and synthetic quality of Pavlov's investigations, which recalled the triumphs of an earlier era. J. Boas, a specialist on the treatment of digestive ailments and editor of *Archiv für Verdauungs-Krankheiten,* pronounced Pavlov the successor to "the great era of Bernard, Ludwig, Heidenhain and their schools." Hermann Munk, professor of physiology at Berlin University, concluded that Pavlov's "enormous work" was comparable only to that of "Beaumont and Blondlot, and, recently, Heidenhain." The anonymous reviewer for the *British Medical Journal* pronounced Pavlov's investigations "not unworthy to be compared with the inimitable researches of Claude Bernard."[53]

Reviewers were uniformly sympathetic to Pavlov's argument for the specific excitability of the glands and the purposive nature of their work, and several noted his discovery of "definite periodic laws" for gastric and pancreatic secretion. Interestingly, however, not a single reviewer mentioned, let alone endorsed, his factory metaphor.[54]

Reviewers consistently mentioned that Pavlov's work would prove "indispensable" to both physiologists and clinicians. Munk urged "all physiologists, clinicians, and physicians . . . interested in questions of the secretion of the digestive juices and their dependence on the nervous system to study this instructive work." Boas predicted that "everybody working in the sphere of digestive illnesses will acquire from this book plentiful new ideas and impulses." The reviewer for the *Lancet* urged *Lectures* on "all who practice medicine and who desire to have intelligent reasons for recommending to their patients systems of diet adapted to their particular derangements of digestion."[55] Pavlov's rhetoric about the relationship between laboratory and clinic clearly pleased his reviewers, who often quoted approvingly his exhortation that "it is only by an active interchange of opinion between the physiologist and the physician that the common goal of physiological science and of medical art will be quickly and surely reached."[56] Among the specific knowledge claims held by reviewers to have practical medical significance were Pavlov's demonstration of the importance of appetite, the response of the gastric glands to Liebig's meat extract, the inhibiting effect of fat and alkalies upon the glands, the role of hydrochloric acid as an exciter of the pancreas, the influences of various foodstuffs on the glands, and the key secretory role of the vagus and sympathetic nervous system.

Medical reviewers did not, of course, accept all Pavlov's practical recommendations. Boas disputed Pavlov's faith in bitters, and Theodor Rosenheim, a specialist in the treatment of digestive disorders, cautioned that Pavlov's results had yet to be verified through clinical experience with humans.[57] Yet these reviewers were uniformly supportive of the Russian physiologist's attempt to link his laboratory findings to clinical concerns, and their criticisms lacked entirely the defensive glee that sometimes accompanied a practitioner's discovery of weaknesses in the arguments of laboratory scientists. Pavlov's rhetoric—his engagement of clinicians in what he presented as a discourse among partners—clearly played an important role in this positive response to his work.

In his short review in *Science,* the U.S. biochemist Lafayette Mendel united the various elements of *Lectures* noted by other reviewers. "Among the comparatively recent contributions to physiological literature," he claimed, "no book has exerted a more stimulating influence" than *Lectures.* Emphasizing the broad range of Pavlov's contributions, he noted that the highly original work of this "brilliant Russian investigator" had "aroused the interest of both physiologists and physicians; and the work has already served in fulfillment of the author's hope, to further physiological science by promoting a more active interchange of ideas between the practitioner and the laboratory worker." The research of "Pavlov's school" was also interesting "from the general biological point of view" for its "demonstration of the purposeful character of secretion into the alimentary canal. Quantitatively and qualitatively the work of the glands varies with the character of the substances upon which they exert their action at different times." Mendel also mentioned a number of Pavlov's narrower knowledge claims and his development of new experimental methods. These new methods, particularly those for acquiring pure digestive juices, had acquired an importance independent of Pavlov's scientific conclusions. As Mendel noted with respect to his own field, "With pure digestive juices made thus readily available, it is not surprising to find interest in the study of their composition renewed." For Mendel, as for Ostwald, Pavlov's work and laboratory embodied larger developments in science, demonstrating the truth of Michael Foster's observation that "the heart of physiology is in the laboratory. It is this which sends the life-blood through its frame; and in respect to this, perhaps, more than anything else, has the progress of the past years been striking."[58]

Pavlov's various knowledge claims rapidly made their way into the specialized physiological literature and into textbooks. One important example is Johns Hopkins University physiologist W. H. Howell's *American Text-Book of Physiology* (1896, 1900). Howell had read the German edition of *Lectures* and structured his discussion of the mechanism of pancreatic and gastric secretion according to that work. He related approvingly Pavlov's experimental proofs for the role of the vagus nerve in pancreatic and gastric secretion, the failure of mechanical irritation to excite gastric secretion, the important role of the psyche, and the two stages of gastric secretion. His response to Pavlov's argument for precise, purposive secretion was respectful but cautious: Pavlov's findings were "suggestive" and, if borne out, would demonstrate "an adaptation whose mechanism is very obscure."[59]

Howell's textbook illustrates dramatically the manner in which *Lectures* established the chief's intellectual credit for the work performed by his praktikanty. Compare one section from the first edition of Howell's text, published in 1896, before the publication of *Lectures*, with the same section in the second edition of 1900. Howell is discussing the nervous mechanisms of gastric secretion.

First edition, 1896

> Some notable experiments recently made by Pawlow and Khigine [Khizhin]
> ... have, however, thrown some light upon this difficult problem ... Khigine has made similar experiments, but altered the operation so that the isolated fundic sac retained its normal nerve supply, which in Heidenhain's operations was apparently injured. The results which he obtained are much more complete than any hitherto reported. He was able in the first place to determine the effect of various diets ...

Second edition, 1900

> The notable experiments recently made by Pawlow and his pupils . . . have, however, thrown some light upon this difficult problem . . . This operation has since been modified by Pawlow in such a way that the isolated fundic sac retains its normal nerve supply . . . Pawlow has been led by his interesting experiments to give a different explanation of the normal mechanism of secretion . . . On a given diet the secretion will assume certain characteristics, and Pawlow is convinced . . .[60]

The knowledge content of these passages is identical, but in the post-*Lectures* version Pavlov emerges as laboratory chief and head of a school—and gains the intellectual credit for the isolated-sac operation and the knowledge claims that flowed from it.

Pavlov's knowledge claims also permeated Robert Tigerstedt's *Lehrbuch der Physiologie des Menschen* (1898), which one leading physiologist characterized as "the standard text-book of German [medical] students."[61] When Pavlov sent Tigerstedt a copy of *Lectures* in 1898, the Finnish physiologist replied that "as you have perhaps seen from the first volume of my textbook on physiology, I make use of your investigations of secretion and the qualities of digestive juices . . . The summary of results on this subject in your lectures is extremely important. I am profoundly convinced that it will be received with great gratitude by other specialists."[62]

Pavlov edited the Russian translation of Tigerstedt's textbook in 1900, adding a preface that praised its "outstanding qualities." No doubt with the author's blessing, Pavlov drew upon recent developments in his laboratory to make some additions to Tigerstedt's treatment of several aspects of digestive physiology. His explanation exudes confidence in his growing international authority.

> As editor, I decided generally not to add notes. For what purpose? The author advances more than sufficient facts. To oppose my own opinion to the guiding opinion of the author seems inappropriate—for this is his textbook and not mine. And there are many other opinions besides mine. I permitted myself the single exception of the section on the work of the digestive glands—and this, I think, with good foundation. This sphere has for about ten years been cultivated almost exclusively by my laboratory—and here my opinion, my choice of facts, might well turn out to be closer to the truth than that of anybody else. Nevertheless, even here I limited myself to but a few [comments], since our facts have already been taken into account by the author.[63]

Tigerstedt was apparently pleased by the Russian edition of his book, and the increasingly warm relationship between the two physiologists would soon prove a great boon to Pavlov.[64]

The positive response to *Lectures* also spurred other developments that enhanced Pavlov's international reputation. First, because his knowledge claims and dog-technologies were relevant to clinical practice and clinically oriented investigations, several western physician-investigators took these as the basis for their own research. Perhaps the first to do so was Franz Riegel, a clinician and laboratory investigator in Giessen with numerous publications on the diagnosis and treatment of circulatory and digestive ailments. Based on Pavlov's demonstration of the importance of the vagus nerve, Riegel and his coworkers explored the usefulness of atropine (which paralyzes nervous activity) and various medicaments for regulating gastric secretion. Riegel's coworker Walther Clemm used a dog with a Pavlov sac to explore the secretory response to sugar, a subject about which clinicians had long held differing views. Riegel and two other coworkers also experimented on patients to determine the secretory effect of chewing, and they confirmed Pavlov's view of the importance of appetite.[65] Another German physician-investigator, Professor Heinrich Schüle of Freiburg, addressed this same subject. Using a cannula to insert food directly into a patient's stomach (as had Khizhin with Druzhok), he concluded that appetite played little part in normal human digestion.[66] Others reached precisely the opposite conclusion.[67] This question remained contentious, but the dispute itself served to incorporate Pavlov's specific knowledge claim and his dog-technologies into medical discourse.

Lectures also served to advertise Pavlov's "numerous ingenious experimental methods" and his laboratory design, generating a demand for these products and transforming the St. Petersburg laboratory into the hub of an expanding network of contacts.[68] Many western scientists and physicians either initiated correspondence with Pavlov or traveled to St. Petersburg. Some correspondents requested literature on Pavlov's techniques, others solicited advice about specific methodologies, and still others sought his counsel on laboratory design. The last group included Johann Orth (a professor of pathological anatomy who was planning a laboratory for the study of experimental biology at the Virchow Institute in Berlin), Robert Emerson (a physiological chemist at Harvard University who was seeking a model for a floor devoted to operations on animals), and Francis Benedict (who was planning a laboratory for studies of metabolism at the new branch of the Carnegie Institute in Boston).[69] A number of other scientists—including Emil Abderhalden, H. J. Hamburger, Carl Lewin, F. Rollin, Paul Mayer, Robert Tigerstedt, and Karl

Mörner—wrote to Pavlov for advice about obtaining pure digestive juices and creating the dog-technologies necessary for their production.[70]

Creation of these dog-technologies, however, required surgical skill, suitable facilities, and the craft knowledge acquired only through experience. A number of western scientists, therefore, requested permission to come to St. Petersburg in order to study the necessary operations with the generally acknowledged master. By 1904 this group included Otto Cohnheim, W. Friedenthal, Walther Gross, Waldemar Koch, Hermann Munk, Johann Orth, Ernest Stadler, F. A. Steeksma, G. Stewart, Walther Straub, Armand Tschermak, Alois Velich, and Robert Webster.[71]

The demand for laboratory technologies did not depend on a scientist's agreement with, or even interest in, Pavlov's specific knowledge claims. This is clear from the sizable contingent of physiological chemists among the laboratory's correspondents and visitors. Abderhalden, Cohnheim, Emerson, Munk, and Webster all studied the biochemistry of ferments and proteins and the process of protein absorption. This scientific orientation, originating in the early decades of the nineteenth century, had begun to take off at about the same time as *Lectures* reached the West—and, as we shall see below, it was rapidly moving scientific discourse on digestion *away* from Pavlov's beloved organ physiology. These physiological chemists were indifferent to purposive secretory curves, the problem of appetite, or the innervation of the gastric glands, but they were very interested in learning how to create dog-technologies that would provide a reliable supply of pure digestive juices.[72]

The yearly contingent of foreigners in Pavlov's laboratory that is best represented in the archival record is that for 1902. Visitors during that year included the physiological chemist Cohnheim, the pathologist Gross, and the pharmacologist Straub from Heidelberg University; Walther Friedenthal, an assistant to the physiological chemist Munk at Berlin University; physiologist Tschermak of Vienna; and Dutch physician Steeksma. Each sought to master the creation of Pavlov's dog-technologies.[73]

After his stay in Pavlov's laboratory, Steeksma informed his former host that "I have begun to conduct the very same experiments as I conducted in your laboratory and intend to acquaint my colleagues in Holland with them at the April 1905 meetings of the Holland Congress of Physicians, where I will demonstrate several operated dogs." He also passed on a colleague's request for medical advice: This physician was treating a patient who, after an operation to remove a stone, had a fistula of the bile duct. The patient was now suffering from a softening of the bones, "exactly as that in dogs" in Pavlov's laboratory. Would Pavlov please tell him how to treat this condition?[74]

The pathologist Gross spent over four months in Pavlov's laboratory in early 1902, during which time he conducted work on gastric digestion along the laboratory's line of investigation. On returning to Heidelberg, he wrote an article based on this research and sent it to Pavlov for his inspection. "If you are agreed with the text of my article," he wrote to Pavlov in June 1902, "please be so kind as to send it to the journal that seems most appropriate. It seems to me that the order in which experiments are presented should be changed, some of them excluded, and data added from the articles of your pupils." Shortly thereafter, Gross became an assistant to Friedrich von Müller at the medical clinic of Munich University. With Müller's encouragement, in 1905 Gross requested a return trip to St. Petersburg in order to "master your surgical method on dogs and conduct work in your laboratory under your guidance."[75] He arrived in September of that year and continued research begun by another foreign visitor, B. Lonnqvist.[76] Like Pavlov's praktikanty, Gross reported on his results to the Society of Russian Physicians.[77]

The most eminent visitor of 1902 was Otto Cohnheim, who, according to one modern commentator, "inaugurated the modern phase in the study of protein absorption." Before his trip to St. Petersburg, Cohnheim had published a book on protein chemistry, and in 1901 he claimed to have identified an enzyme, erepsin, that prepared proteins for absorption into the blood. In 1902 he was investigating the origin of erepsin and its site of action, and, apparently, thought Pavlov's dog-technologies would facilitate this research.[78] After returning from St. Petersburg in late 1902, he wrote a letter to Pavlov "to again express my heartfelt thanks for your kindness . . . You greeted me with such kindness and so readily showed me your new and most interesting investigations that I will all my life recall with the greatest satisfaction the wonderful weeks that I was permitted to spend in your laboratory. I will soon begin to conduct here the operations that I studied [in St. Petersburg] . . . In Leipzig, I attempted to perform the Eck fistula operation, but without success, since the animal's loss of blood was so great." Five years later, Cohnheim informed Pavlov that "our Institute is completely adapted for surgical operations and I will diligently conduct the operations that I studied under you." In 1910 he sent Pavlov a number of his recent articles, "from the content of which you will see that I have become your diligent student."[79]

Another leading physiological chemist, Emil Abderhalden, wrote to Pavlov in October 1904 to request assistance in the creation and maintenance of a dog with a pancreatic fistula. Abderhalden was collaborating with H. E. Fischer (who had won the Nobel Prize in Chemistry two years earlier). "For the splitting of synthetic polypeptides we need the most active pancreatic ferment possible," he wrote. "I want to prepare a dog with a pancreatic fistula and use juice from the fistula. I would be very grateful if you would be so kind as to advise me, having such valuable investigative experience, how one can preserve the life of the operated-upon animal as long as possible, what nutrients it needs, and how best to care for it. It would also be very important for me to know how I can acquire the most active juice . . . If it would be useful for our experiments (with professor Fischer), I would not fear a long trip to visit your wonderful laboratory." Pavlov responded by sending both some digestive juices and his praktikant Boris Babkin to Berlin, where Babkin labored unsuccessfully to create the necessary dog-technology in Abderhalden's laboratory. Abderhalden wrote again in September 1905: "I must again ask of you the great kindness of sending us pancreatic, intestinal, and gastric juices so that we (prof. Fischer and I) can continue our work . . . It has been completely impossible here in Berlin to acquire a dog with a pancreatic fistula. Doctor Babkin is horrified by the operative facilities in the local Institute and has little hope for the operations. We can only hope that it will soon be possible to conduct the fistula implantation operation here. For now, we are wholly dependent on your kindness, for which we express our heartfelt thanks." Many years later, Abderhalden recalled that, even before Pavlov had rendered this technical assistance, the Russian physiologist had influenced him profoundly: "As a young student I was fascinated by your remarkable experiments on the secretion of digestive juices, and from that time you have been my teacher. I cannot find the words to express my debt to you."[80]

Pavlov's international reputation was reflected in these years by his election to a series of honorary memberships. Beginning with Italy's Anton Alzatel Society in 1898, by 1904 these had been conferred by scientific and medical societies in Uppsala, Stockholm, Helsingfors, Copenhagen, Berlin, Vienna, Paris, and New York. When Pavlov traveled to Paris in 1900 for the International Congress of Physicians—his first foreign trip in fifteen years—he did so as one of the Congress's honorary presidents.

The Manifestation Mondiale of 1904

The year 1904 marked the twenty-fifth anniversary of Pavlov's graduation from medical school and so, according to Russian academic tradition, offered an occasion for official celebration of his achievements. Pavlov disliked official festivities and torpedoed those planned by the Academy of Sciences and the Military-Medical Academy by threatening to leave St. Petersburg on the day they were scheduled. He did consent to a celebratory meeting at the Society of Russian Physicians, and his jubilee was also marked by congratulatory notes from numerous scientific and medical societies, a photo album celebrating his achievements, and a special volume of the Imperial Institute's journal devoted to reminiscences and scientific articles honoring his scientific achievements.[81]

As the Parisian physiologist J. P. Langlois observed, Pavlov's reputation had so completely "breached the frontiers" that "physiologists of the entire world" transformed his jubilee into a *manifestation mondiale* ("worldwide demonstration" of respect for his achievements).[82] Langlois had in mind here the celebratory volume of the Institute's *Archive of the Biological Sciences,* which was dedicated to Pavlov from his "colleagues, pupils, and admirers." This volume reflected the great range of his contributions to science and medicine, testified to his international reputation, and codified the existence of a "Pavlov school."

Russian contributors included Pavlov's friend D. A. Kamenskii, who described the chief's early days in the Botkin laboratory; the former praktikant I. Shirokikh, who wrote an article entitled "The Significance of I. P. Pavlov's Ideas for the Development of a Doctrine about the Feeding of Agricultural Animals"; and another former praktikant, V. V. Kudrevetskii, who reported his latest clinical experiences in the treatment of dyspepsia (and the use of "the natural gastric juice of the dog"). Two Russian professors of physiology, Danilevskii and Tarkhanov, also contributed articles, although neither cited Pavlov's work and Tarkhanov could manage only a two-page abstract. Four of Pavlov's coworkers and colleagues contributed an annotated bibliography of 139 "works of I. P. Pavlov and his pupils." The eleven contributions by westerners included a warm biographical sketch by Tigerstedt and scientific articles by five recent visitors to the laboratory (Cohnheim, Friedenthal, Langlois, Straub, and Tschermak), British physiologists William Bayliss and Edward Schäfer, French physiologists Marcel Gley and Jean Camus, and Swedish pharmacologist C. E. Santesson.[83]

Describing the reasons for this *manifestation mondiale* to readers of *La Presse Médicale,* Langlois portrayed Pavlov not merely as the discoverer of a dazzling array of scientific facts but as the very embodiment of modern physiology, a visionary, and even a bit of a wizard. Having visited Pavlov's laboratory in February 1904, Langlois explained that it was there, "among the numerous students that he directs and inspires, that one must see Pavlov." Pavlov resembled Langlois's "venerated master" Brown-Séquard by his physical and intellectual vivacity and his love for his collaborators. Also like Brown-Séquard, Pavlov had, in earlier years, shared his own modest lodgings with his experimental animals. That, however, was a bygone era. Now, "the small room in a narrow alley has been singularly transformed: Pavlov's laboratory, constructed according to its master's instructions, can be considered a model of its genre." His greatness could be fully appreciated only by seeing the laboratory in operation and by reviewing all the works produced there by the master and his pupils. These investigations were "guided in their every step by therapeutic concerns" and united by a single "directing idea, the same philosophical conception": "to demonstrate the adaptation of all the organs to accomplish the best for the entire organism in its vital processes" (p. 187).

Lovingly describing the surgical facilities of Pavlov's Physiology Division— "a milieu that is identical to the best hospital's operating room"—Langlois observed that the dogs, like Pavlov's coworkers, were enraptured by their master: "All these dogs, with their double or triple fistulas, have a particularly gay air, and welcome the arrival of their master with expressions of joy. The dogs who are manufacturing gastric juice, pancreatic juice, and saliva, suspended by a double strap under their belly, interrupt their abundant 'sham meal' to cast their gaze at Pavlov and to request his habitual caress" (pp. 186–87).

Langlois's Pavlov, then, is considerably more than the sum of his laboratory's products. He is the beloved master of a modern laboratory and everything within it—not merely nerves and glands, but coworkers and dogs. He collects both the juices and the affection of his experimental animals, who, despite the rattling of their sophisticated technological apparatus, retain recognizably pet-like characteristics. As a visionary and the master of a modern laboratory system that produced so many diverse products, Pavlov thus embodied an appealing synthesis of some troublesome polarities: efficient production and scientific imagination, laboratory physiology and medical practice, precise science and the complexities of living creatures. "The Russian master," Langlois concluded, "occupies the very pinnacle in the domain of experimental medicine" (p. 187).

Criticism from Within, Complexity from Below

Even as Pavlov enjoyed his growing reputation in the first years of the twentieth century, cracks began to appear in the intellectual edifice he had constructed.

Three related developments combined to cast doubt on some of his central knowledge claims, to devalue and even discredit one of his key dog-technologies, and to shift discourse on the digestive glands away from organ physiology and his approach to it. First, growing attention to the interaction and biochemistry of ferments cast doubt on the conclusions Pavlov had drawn from the regularities he perceived at the organ level and signaled an important shift in discourse toward issues that were resolvable only at a sub-organ level.[84] Second, Pavlov's former coworker Lev Popel'skii published several articles subjecting Pavlov's concept of purposiveness, and the quantitative data used to support it, to an "insider" criticism. Finally, Bayliss and Starling's discovery of secretin undermined Pavlov's nervist portrayal of pancreatic secretion and generated a growing interest in the humoral mechanisms of the digestive apparatus. All three of these developments resulted in part from discoveries in Pavlov's laboratory.

Pavlov's reputation remained intact, but these developments did suggest that the scientific vision and style that had brought him to the pinnacle of international science would no longer be the source of cutting-edge developments in digestive physiology.

THE "FERMENT OF FERMENTS" AND BIOCHEMICAL COMPLICATIONS

Before the publication of *Lectures,* Pavlov's laboratory had devoted little attention to the lowly intestines.[85] In 1897, however, the chief assigned N. P. Shepoval'nikov to investigate intestinal secretions, and two years later the praktikant reported a stunning (and disturbing) fact: the addition of intestinal juice to pancreatic juice raised the strength of all three pancreatic ferments and transformed the albuminous ferment from a zymogenic (i.e., inactive) form into an active form, trypsin. Describing Shepoval'nikov's finding in an address to the Society of Russian Physicians in 1899, Pavlov named the newly discovered agent *enterokinase,* from the Greek for "intestinal" and "move" or "excite."[86] For Pavlov, the discovery of enterokinase was exciting because this "ferment of ferments" represented a further mechanism for the precise and purposive work of the digestive glands. "Initial experiments," he announced, "give us good reason to expect that the joint chemical activity of three fluids—pancreatic juice, bile and intestinal juice—provides the broadest scope for the most precise adaptations to various objects of digestion."[87]

This discovery, however, also had another, disturbing implication. It raised the possibility that dog-technologies with Pavlov's pancreatic fistula were "abnormal" and that the data generated by them were flawed. In dogs with a Pavlov fistula, pancreatic juice flowed out over a piece of the duodenum and so might well be affected by enterokinase. This, in turn, raised the possibility that the fermenting power of the pancreatic juice analyzed by Vasil'ev and Val'ter had been selectively—and even idiosyncratically—raised by the "ferment of ferments." Because enterokinase acted most decisively on one of the three pancreatic ferments—and because the laboratory had not been aware that it was dealing with ferments that were in some combination of zymogenic and active states—measurements of the "precise, purposive" response of that gland to various foodstuffs may well have been skewed. Furthermore, the rate of flow (which influenced the proportion of the pancreatic juice flowing out of the fistula that had made contact with the duodenum), and even the elapsed time between secretion and the measurement of ferment content, may well have influenced the experimental data. At a meeting of the Society of Russian Physicians, Pavlov conceded that, with the benefit of hindsight, there had been signs of this possibility. "There actually were several indications of these different states of the albuminous ferment. For example, it was clear that while sitting in the thermostat [after secretion from the animal] the juice became stronger, that is, it digested more egg white. But this issue was not further elaborated until now."[88]

Even after Shepoval'nikov's discovery, however, the laboratory continued to rely on its standard pancreatic fistula, the chief having redeemed it with a physiological argument: "It was thought that this small piece [of duodenum] does not secrete enterokinase, since it is subjected to a strong irritation, which, according to laboratory experiments, inhibits the secretion of kinase."[89] So, in 1900 Pavlov assigned another praktikant, I. I. Lintvarev, to use the standard fistula in experiments designed to incorporate the discovery of enterokinase into laboratory doctrine on the adaptiveness of pancreatic secretion.[90]

The Pavlov laboratory, however, did not monopolize discussions of this question. Twenty-five years earlier, Heidenhain had obtained from the pancreas a substance that did not itself possess proteolytic power but from which he could acquire an active ferment. He had named this substance *zymogen*, a term that by the late 1890s was commonly used for this entire class of substances (sometimes referred to as *precursors* or *proferments*). By this time, a number of investigators were studying the physiological and chemical processes that converted digestive zymogens to active enzymes—transforming, for example, the pepsinogen secreted by the gastric glands into pepsin, and the trypsinogen secreted by the pancreas into trypsin.[91]

In 1902 two physiologists at the Pasteur Institute, C. Delezenne and A. Frouin, built upon the discovery of enterokinase to deliver a "severe blow" to the Pavlov laboratory view.[92] Using a catheter to obtain pancreatic juice that had not passed over the duodenum, the French physiologists concluded that the juice "does not possess its own digestive action vis-à-vis albumin." Under normal physiological conditions, then, pancreatic juice possessed no proteolytic power until it combined with the enterokinase secreted by the duode-

num. Yet Pavlov and his coworkers had claimed to measure the varying proteolytic power of pancreatic secretion in response to various foods—and had
built their characteristic curves upon just such data. "The positive results offered by Pavlov and his pupils," Delezenne and Frouin explained, "must be attributed to the intervention of the intestinal juice secreted by the fragment of
[duodenal] mucous membrane" that supported the orifice of the pancreatic
duct in dogs with a pancreatic fistula.[93] In a second report delivered the following year, they concluded that Pavlov's evidence for the precise, purposive
adaptation of pancreatic secretion was, at least with regard to variations in
proteolytic power, an artifact of experimental error. Their new data, they
wrote, would "completely change the classical notions about the proteolytic
power of pancreatic secretion that resulted from the work of Heidenhain, and
from the numerous investigations of Pavlov and his pupils. The experiments
of Pavlov's school that were apparently so suggestive about the adaptive variations of pancreatic secretion lose their significance."[94] By this time, Delezenne and Frouin could cite two other investigators, one German and one Russian (see below), who had independently reached the same conclusion.

Pavlov assigned Boris Babkin to check the French physiologists' results.
Babkin raised questions about the normalcy of Delezenne and Frouin's dog-
technology and about some of their specific conclusions, but he confirmed
their critique of the pancreatic fistula and the results based upon it.[95] Perhaps
the best indication of Pavlov's state of mind regarding this question is that,
rather than assigning Babkin to continue his work on the pancreas, he instead
assigned him to write the laboratory's first doctoral thesis on conditional reflexes.

These developments proved only the beginning of increasingly complex
discussions about the physiology and chemistry of digestive secretions, discussions in which some members of Pavlov's laboratory group, including
Babkin and Savich, would participate actively. The question of the adaptiveness of pancreatic secretions was not settled and would prove contentious for
decades to come. Yet this challenge did undermine one of Pavlov's key knowledge claims and signaled a far-reaching change in the very nature of scientific
discourse on digestion.

That discourse was spiraling away from organ physiology toward studies of
intermediary metabolism. Fueled by the increasing intellectual and institutional power of physiological chemistry, the center of gravity in digestive physiology was, then, moving away from Pavlov's aesthetic preferences and intellectual strengths.[96] Pavlov did not enjoy technical supremacy in this emerging
discourse, nor did he speak with special authority. This is clearly evident in the

publications from his laboratory that did address these issues: in marked contrast to those of earlier years, these publications contained ubiquitous citations to data and interpretations generated by other laboratories. As we shall see in Chapter 10, these forays into unfamiliar territory hardly enjoyed unmixed success.

THE RENEGADE POPEL'SKII

Sometime in 1902–3, Pavlov's former coworker Lev Popel'skii visited the laboratory for a talk with the chief. Their encounter ended with a vintage explosion of Pavlov's temper. Babkin witnessed the aftermath: "After a stormy interview in the laboratory at the Institute of Experimental Medicine, Pavlov broke off relations with Popielski completely and refused to shake hands with him on parting. It happened that I was working in the laboratory at the time. The scene took place in Pavlov's study on the second floor. We heard loud shouts and then down the small winding staircase leading to our room came Popielski [Popel'skii], almost falling downstairs in his hurry, and red as a lobster. He quickly put on his coat and went out, without saying good-bye to anyone. After this there was dead silence and it was a long time before Pavlov appeared."[97]

For Babkin, this incident illustrated Pavlov's integrity, his unwillingness to "compromise in matters relating to scientific truth." The chief was "incensed beyond words," not because "of any personal considerations of Pavlov's or because his pride was hurt by Popielski's criticisms," but "solely because Popielski had had the temerity to criticize Pavlov's laboratory in print, after working there for four years, and that he did so without performing a single new experiment to prove his words. Pavlov would not have objected if Popielski had based his criticism on experimental evidence, instead of which he had looked for small discrepancies in the figures or illogical presentation of facts in the M.D. theses of some of Pavlov's students dealing with the adaptation of ferments to food."[98] The notion of "experimental criticism" to which Babkin referred here was indeed part of Pavlov's views on good science. Pavlov agreed with Bernard that, since a "negative fact" did not refute a "positive fact," physiologists should ground their criticisms in experiments that limited or refuted the conclusions of another scientist. Popel'skii's criticisms, however, did in fact draw upon his own experiments. In any case, it seems most unlikely that Pavlov's wrath was ignited by this epistemological nicety.

A much more likely source of Pavlov's fury was Popel'skii's "disloyalty" (to which Babkin alludes). This was not simply a matter of personal allegiances.

Unlike most consumers of laboratory products, Popel'skii was an insider. Having written his doctoral thesis in the laboratory, he had studied closely the dissertations of Pavlov's coworkers and was intimately familiar with laboratory procedures, and was therefore able to criticize the laboratory's conclusions from a unique perspective.

Lev Popel'skii belonged to the small group of coworkers that, at the turn of the century, seemed to comprise the nucleus of a genuine "Pavlov school." He differed from the other members of this cohort, however—and from other laboratory coworkers as well—in two important respects. First, as a graduate of the mathematics division of St. Petersburg University, he could engage in quantitative reasoning with some comfort and authority. Second, he had decided on a career in physiology and begun to acquire expertise in that discipline before working with Pavlov. As Tarkhanov's assistant in the Physiology Department of the Military-Medical Academy from the early 1890s through 1895, Popel'skii had begun research on the influence of cocaine, ether, and alcohol on nervous excitability. When Pavlov replaced Tarkhanov in 1895, Popel'skii remained as assistant to the new chief and conducted doctoral research under Pavlov's guidance.

Popel'skii's doctoral thesis (1896) attests to both his goodwill toward the chief and his intellectual independence. He thanked Pavlov for choosing the theme of his thesis and for "the fundamental physiological education which I received under his guidance," and he consistently noted the dependence of his work on the methodological contributions of "my most-esteemed teacher."[99] His review of the literature, however, departed from standard form by failing to emphasize the contributions of Pavlov's laboratory (although he gave due credit to its important achievements) and by the ubiquitous citations of work produced in other, mostly French and German, laboratories. This review also lacked the standard explication of the importance of methodology in general and the methodological contributions of Pavlov's laboratory in particular. Popel'skii's thesis also contained a considerable amount of first-person-singular language that clearly expressed the author's self-image as not simply a pair of skilled hands but a physiologist in his own right. Finally, Popel'skii did not employ Pavlov's favorite words for describing physiological processes. In particular—and this was a harbinger of things to come—the word *purposiveness* does not appear once. One of the propositions appended to Popel'skii's thesis may well have reflected his problem with this concept so dear to the chief: "The question of the specific exciters of the digestive glands requires further, serious exploration."[100]

Like Pavlov, Popel'skii was a convinced nervist. In his thesis, *On the Secre-*

tory-Inhibitory Nerves of the Pancreas (1896), he argued that two sets of nerves—one excitatory, the other inhibitory—connected the vagus nerves to the pancreas. Using acute experiments, he demonstrated that stimulation of the vagus could inhibit the pancreatic secretion that was otherwise elicited by the presence of hydrochloric acid in the duodenum. For both Popel'skii and Pavlov, this "antagonism" between excitatory and inhibitory nerves characterized the nervous control of bodily organs.[101] Popel'skii also provided a nervist explanation for a previously troubling phenomenon: dogs with severed vagus and sympathetic nervous connections still responded with a pancreatic secretion to the presence of hydrochloric acid in the duodenum. Popel'skii attributed this to hypothesized local nerve centers that presumably connected the duodenum with the pancreas. This conclusion proved to be the high-water mark for nervist explanations of pancreatic secretion; it appeared in *Pflügers Archive* shortly before Bayliss and Starling's discovery of an alternative humoral mechanism.[102]

After defending his doctoral thesis, Popel'skii was appointed *privatdozent* in physiology at the Military-Medical Academy, where he continued his research on digestive physiology. As the Academy approached the hundredth anniversary of its founding, Popel'skii was commissioned to write a history of its Physiology Department.[103] Published in 1899, his account was appropriately complimentary toward the four major figures of the Department's modern era—Sechenov, Tsion, Tarkhanov, and Pavlov; but the latter could hardly have been pleased by the relatively faint praise he received. Popel'skii found suitable superlatives to describe the contributions of Sechenov ("the glory and pride of the Academy"), Tsion ("a first-class scientist"), and Tarkhanov ("indisputably one of the leading contemporary physiologists"). Describing Pavlov's work, he could muster only the following: "The scientific activity of prof. I. Pavlov was expressed in part in his own investigations, published in various Russian and foreign specialized publications, and in part in his guidance of the work of up to 40 beginning authors, largely on topics provided by the professor."[104] In what was, at the very least, an impolitic oversight, Popel'skii failed to mention *Lectures on the Work of the Main Digestive Glands* or even to include it in his list of Pavlov's scholarly works.

In 1899 Popel'skii was bitterly disappointed by his failure to secure an Academy stipend to study abroad. According to Babkin, he "attached the blame for this misfortune to Pavlov" and took his revenge in print. "He began writing critical articles in a somewhat unpleasant vein, which appeared in the *Russian Physician* . . . in 1901 and 1902. In them he disparaged the work that was being published at that time from Pavlov's laboratory concerning the adaptation of ferments of the pancreatic juice to different kinds of food. These criticisms had no experimental basis and were purely speculative."[105] Pavlov did have difficulty at this time securing coveted European travel grants for his favorites, in large part because of his ongoing feud with Viktor Pashutin, the powerful president of the Military-Medical Academy. As we have seen, however, there were distinct signs of strained relations between Pavlov and Popel'skii before the latter's failed candidacy, strains that were perhaps related to Popel'skii's prior physiological training, his independent cast of mind, and his continued loyalty to Tarkhanov (who had always viewed Pavlov's scientific style as dogmatic). In any case, Popel'skii's subsequent criticism of Pavlov's work was rooted in his independent expertise in mathematics and physiology and, as we have seen, was foreshadowed in his dissertation (three years before his failed candidacy of 1899).

In 1901–3, while serving as a military physician in Moscow, Popel'skii published a series of articles criticizing Pavlov's notion of "purposiveness" and the quantitative data on which it was based. Using the laboratory facilities at the Moscow Military Hospital, he also employed Pavlovian techniques and dog-technologies, particularly the improved pancreatic fistula, to advance an alternative view of the differing glandular responses to various foods.[106]

In sharp contrast to outside reviewers of the knowledge claims generated by Pavlov's laboratory, Popel'skii analyzed in detail the relationship between the experimental data generated by praktikanty and the conclusions based on them. "The facts," he argued, "cannot be considered convincing," and he repeatedly noted "a certain subjectivism in interpretation."[107] When confronted with complex phenomena, Popel'skii observed, scientists frequently become "the victim of enthusiasm for some preconceived idea." Pavlov's enthusiasm for his vague notion of "purposiveness" had proven so powerful that his theory "cannot be considered strictly scientific."[108]

For Popel'skii, the facts generated by Pavlov and his praktikanty themselves clearly refuted the laboratory view that the glands secreted purposively. For example, the pepsin in gastric juice acted only on albumins, so what purpose was served by the gastric secretion elicited by alcohol, hydrochloric acid, starch, and bread? Conversely, what purpose was served by what Pavlov admitted to be the "puzzling" failure of egg white to elicit any secretion whatsoever?

Popel'skii's critique focused, however, on his specialty—the pancreatic gland. He argued that Pavlov simultaneously insisted on two incompatible propositions: (1) the amount and ferment content of pancreatic secretion elicited by particular food followed the stereotypical courses identified in Val'ter's thesis, and (2) the secretory pattern of the pancreatic gland adapted over time to an animal's diet. If the latter claim was true, how could any one animal produce the "stereotypical" secretory pattern identified in the first proposition? Pavlov's investigative assumptions were also, in Popel'skii's view, biochemically unsupportable. "In Vasil'ev's, Iablonskii's, and Lintvarev's experiments," wrote Popel'skii, "professor I. P. Pavlov maintained dogs on a milk-bread diet, expecting that the albumin ferment [would] disappear from pancreatic juice and that the gland [would] come to produce a heightened amount of starch ferment, in correspondence to the food's starch content. With meals of meat, I. P. Pavlov expected that the starch ferment would disappear, and that the gland [would] produce only albumin ferment . . . A simple chemical analysis of these substances (a meal of milk-bread consists of albumins, fats, and starch) . . . indicates that professor I. P. Pavlov's attempt must turn out unsuccessfully" (pp. 1242–43). Since Pavlov had *succeeded* in demonstrating this biochemical impossibility, Popel'skii suggested that the secret of this "success" must reside in the interpretation of data.

He proceeded to analyze in detail the data provided by the doctoral dissertations that sought to "justify" Pavlov's doctrine of the purposiveness of pancreatic secretion—specifically, those of Vasil'ev, Iablonskii, Lintvarev, and Val'ter. He pointed to significant differences in the results of identical experiments and to what he perceived as a consistent pattern of ignoring differences within a category (say, experiments on the secretory response to feedings of meat) while emphasizing lesser differences between categories (say, between the secretory responses to meat and bread) (pp. 1242–43). He adduced numerous cases in which the experimental data seemed to contradict Pavlov's notion of purposiveness. For example, contrary to Pavlov's theory, a diet of milk and bread elicited a greater amount of albuminous ferment than did a diet of meat (p. 1244). Similarly, the pancreatic juice poured upon fat contained both albuminous and starch ferment, and this had no discernible purpose. For Popel'skii, the experimental data actually fit a pattern that was precisely opposite to that predicted by Pavlov's doctrine: "with small exceptions, in [Val'ter's] experiments there is visible a parallelism between the [rise and fall] of the proteolytic power of all three ferments" (p. 1245). In other words, if a meal elicited a high content of one ferment, it tended to elicit a high content of the other two ferments as well. (See the relevant curves from *Lectures*, reproduced in Figure 21 in Chapter 6.) There was no evidence, then, for the specificity—let alone the purposive specificity—of pancreatic responses to the three foods.

Popel'skii argued, further, that the very nature of the digestive process rendered Pavlov's theory highly improbable.

Ingested food is subjected in the stomach to a fundamental physicochemical reprocessing, as a result of which: 1) it is transformed into a homogeneous gruelly mass; 2) it acquires an acidic reaction; 3) albumins turn into peptones; 4) starch, due to [the ferments in] the saliva, [is] partially transformed into sugar; 5) fat, judging from Volhard's investigations, in all probability is also subjected to certain alterations. If one considers all this, it is understandable that there is no basis to speak about a *specific* action of the component parts of food—of fat, starch, and albumins in their natural state—on the pancreas. Furthermore, the acid in gastric juice obliterates the differences between types of food: acid is such an energetic exciter of the pancreas that the question of other exciters falls to the background.[109]

Reinterpreting Pavlov's data, and offering the results of his own experiments,[110] Popel'skii proposed a sweeping alternative to Pavlov's explanation of the fluctuations in the course of pancreatic secretion. Rejecting Pavlov's belief in "a variety of specialized irritants" and nerve endings, Popel'skii argued that physical and chemical exciters all acted on the same nerve. The amount and proteolytic power of the pancreatic juice depended on the amount and strength of the exciters.[111] As for Pavlov's insistence that the amount and proteolytic power of pancreatic juices fluctuated independently (i.e., that some substances elicited a great amount of weak juice, others a small amount of strong juice), Popel'skii suggested that juice that seemed weak in proteolytic power was actually strong in "proferment" and that this juice became fully "kinetic" only in the course of seven to ten hours.[112]

> Our conclusion that the qualities of pancreatic juice are determined by the force and amount of the irritant is very important for physiology, since it eliminates prof. I. P. Pavlov's complicated hypothesis of the existence in the mucous membrane of the digestive canal of an entire mass of varied nervous endings, capable of becoming active under the influence of strictly defined, specific exciters.
>
> In our view, at the basis of the work of the glands lies a simple nervous apparatus (*one type of nervous ending* of the centripetal nerves, a reflexive center from which there emanate secretory nerves), that, under the influence of *the changing qualities* of the irritant, of its strength and amount, can condition the production of juices with the most varied qualities.[113]

Popel'skii first advanced these arguments in Russian—in a "preliminary communication" in *Vrach* in 1901 and a detailed exposition in *Russkii Vrach* in 1902–3.[114] That, no doubt, would have sufficed to infuriate the chief. One can imagine Pavlov's reaction to the republication of Popel'skii's preliminary communication in *Deutsche Medicinische Wochenschrift* (November 1902) and to the appearance of a second article, which detailed Popel'skii's experiments undermining Val'ter's thesis, in *Centralblatt für Physiologie* (May 1903).[115]

Shortly after writing these articles, Popel'skii left Russia for the University of Lwow, where he became an eminent professor of pharmacology and director of the Research Institute of Experimental Physio-Pharmacology.[116] A renowned European scientist, he would always be recalled with distaste by Pavlov and his coworkers. Not only had Popel'skii violated the integrity of "we, the laboratory," but, as we shall see in Chapter 10, his timing had been most unpropitious.

THE HUMORALIST CHALLENGE

The "humoralist tide" that in the 1890s had largely recast scientists' perception of immunological processes reached the digestive system in 1902 with William Bayliss and Ernest Starling's discovery of secretin.[117]

The British physiologists had been following up Dolinskii's discovery of 1893 in Pavlov's laboratory that the introduction of hydrochloric acid into the duodenum elicited a pancreatic secretion. As we have seen, Popel'skii had explained this phenomenon as resulting from hypothesized nervous mechanisms linking the duodenum with the pancreas. According to Bayliss and Starling, further experiments by the French physiologist Emile Wertheimer had convinced them that this secretion might be caused by the same "local reflexes" as those responsible for the intestinal movements (peristalsis) they had been studying. They anesthetized a dog, isolated a loop of its duodenum, tied it at both ends, and destroyed its nervous connections, creating a loop connected to the rest of the body only by its blood vessels. A friend who was present later recalled what followed: "On the introduction of some weak HCl [hydrochloric acid] into the duodenum, secretion from the pancreas occurred and continued for some minutes. After this had subsided a few cubic centimeters of acid were introduced into the denervated loop of duodenum. To our surprise a similarly marked secretion was produced. I remember Starling saying 'Then it must be a chemical reflex.' Rapidly cutting off a further piece of jejunum [duodenum] he rubbed its mucous membrane with sand and weak HCl, filtered and injected it into the jugular vein of the animal. After a few moments the pancreas responded by a much greater secretion than had occurred before. It was a great afternoon."[118]

Bayliss and Starling concluded that the active substance, which they named *secretin*, was produced by the action of the hydrochloric acid on a precursor stored in the mucous membrane of the duodenum. Traveling through the blood, secretin excited pancreatic secretion.[119]

They quickly incorporated their discovery into a broad vision of the regulation of bodily processes, a vision that assigned much the same role to humors as Pavlov assigned to nerves. Also like Pavlov, they were soon able to mobilize the research of other investigators. In 1904, F. A. Bainbridge, who was writing his M.D. thesis at the University of London under Bayliss and Starling's supervision, announced a humoralist explanation of the adaptation of pancreatic secretion to various foods. The following year, J. S. Edkins, working in the physiological laboratory at St. Bartholomew's Hospital in London, announced his discovery of gastrin, a humoral mechanism of gastric secretion.[120]

In Bayliss and Starling's Croonian Lecture to the Royal Society in March 1904, and in Starling's Croonian Lecture the following year, they rejected, of course, Pavlov's nervist portrayal of gastric and pancreatic secretion, but they also confirmed Pavlov's status as the founder of modern digestive physiology by testing their own views against the basic framework he had established.[121] Most important, they adopted Pavlov's basic view that the digestive glands adapted sensitively to the demands placed on them, and they explored the manner in which humoral mechanisms might accomplish this. Judging from their published texts, this generated problems that soon led them to follow Popel'skii's lead in seriously questioning some of the data and knowledge claims generated by Pavlov's laboratory.[122]

In their original communication, Bayliss and Starling were careful to allow for the possible role of nervous mechanisms. Alluding to Pavlov's claim that the vagus was the secretory nerve of the pancreas, they wrote that "we have in no way stated that our experiments disproved the existence of secretory nerves to the pancreas. So far, however, as our own experiments go we have been unable to obtain secretory effect from the vagus in the neck." They did insist, however, that the second stage of pancreatic secretion was purely humoral; it "does not depend on a nervous reflex, and occurs when all the nervous connections of the intestine are destroyed."[123] By 1905 their position had hardened, and they suggested that all cases of presumably nervous secretion were due to experimental error: "We are inclined to believe that the chemical mechanism is the only one involved in the secretion of pancreatic juice and that in all Pawlow's experiments where secretion was excited by the stimulation of nerves such as the vagus or splanchnics the effect on the pancreas was really a secondary one due to movements of the stomach arising as a result of the nerve stimulation and squeezing some of its acid contents into the first part of the small intestine . . . We ourselves have never been able by nervous means to obtain secretion of pancreatic juice provided that we excluded all possibility of entry of acid into the upper part of the small intestine."[124]

In his lecture of 1905, Starling accepted Pavlov's demonstration of the nervous character of the first, psychic phase of gastric secretion, but drawing on Edkins's research he rejected any role for the nerves in the second phase: "Pawlow has shown that this second stage occurs even when both vagus nerves are divided and the possibility is at once suggested that this stage is not due to nervous processes at all, but is determined by some chemical mechanism similar to that which we have studied in the case of the pancreas. Pawlow is apparently of the opinion that the secondary secretion is due to local reflexes in the wall of the viscus, but recent experiments by Edkins have shown that it is unnecessary to invoke the aid of any such obscure mechanisms." Analyzing Edkins's results by analogy with their own, Bayliss and Starling concluded that "certain constituents of the food have a special influence in promoting the secretion of gastric juice." "These elements can by their action on the pyloric mucous membrane give rise to the production of gastric secretin [i.e., Edkins's gastrin]."[125]

A problem emerged, however, when Bayliss and Starling sought to account for the fine adaptation of the glands to various meals—as demonstrated by Pavlov and his coworkers—by the action of *a single* chemical messenger for each gland. According to their view, a precursor to secretin (prosecretin) was stored in the mucous membrane of the duodenum until acid triggered the production and release of secretin itself. Traveling through the blood to the pancreas, secretin caused this gland to release its stored ferments. (They hypothesized an analogous mechanism for gastrin and gastric secretion.) Here, however, was the problem: how could this single agent account for the differences in amount and proteolytic power that Pavlov and his coworkers identified for specific foods, and for the ostensibly independent fluctuations of the amount and proteolytic power of pancreatic secretion? How, in other words, could it generate Pavlov's stereotypical secretory curves? We have seen that Pavlov handled this problem by hypothesizing the existence, first, of separate secretory and trophic nerves (the former governing the production of fluids, the latter the production of ferment bodies); and, second, of a variety of different nervous receptors that responded selectively to the extractive substances in various foods. Bayliss and Starling's humoral agents, however, did not afford them such an explanation. Secretin and gastrin were unitary substances elicited by undifferentiated exciters (secretin by the acid in the duodenum, gastrin by peptogenous substances in the stomach), and they triggered the release of the *entire* stock of stored ferments in the pancreas and gastric glands.

They may well have been encouraged by Popel'skii's articles (cited in both Croonian Lectures) to solve this problem by questioning the data and interpretation in key Pavlovian doctoral dissertations. "A striking feature . . . of the pancreas," they wrote, "is its alleged power of adapting its secretion to the nature of the food taken in by the animal. It has been stated by Pawlow that according as the food consists chiefly of proteids, carbohydrates, or fats, so do we find a relative preponderance of the ferments acting respectively on each of these three classes of foods. The evidence on which this statement is based, although lending to it considerable support, is not absolutely convincing."[126]

Following Delezenne and Frouin, Bayliss and Starling pointed out, first, that the discovery of enterokinase invalidated the data collected by the Pavlov laboratory on the proteolytic power of pancreatic ferments. It was "impossible to say what proportion of the trypsinogen of the juice secreted in these experiments had been converted into trypsin by the small amount of intestinal mucous membrane at the mouth of the duct."[127]

Furthermore—and here they echoed Popel'skii—Val'ter's "results do not entirely bear out his contentions." "Although milk contains no starch, it evokes the secretion of a large amount of amylopsin and . . . meat causes a secretion of more steapsin than does milk, although this latter contains much more fat than the meat diet." Val'ter and other representatives of the Pavlov laboratory "regard the adaptation as determined by the stimulation of special nerve endings in the mucous membrane by each constituent of the food, a conclusion hardly borne out by the results just quoted." Finally, "Another disturbing factor of these experiments is the large variation in total quantity of juice secreted with different food-stuffs" (an objection first raised, again, by Popel'skii).[128]

Bayliss and Starling's "chemical correlation" of digestive functions differed fundamentally from Pavlov's precise, nerve-governed digestive factory. For the British physiologists, neither the pancreas nor the stomach nor the intestines responded "purposively and precisely" to the requirements for digesting a particular food. Rather, each gland was stimulated by a single chemical agent to release its stock of ferments. This process, however, was coordinated by chemical reflexes. First, the peptogenous substances created in the stomach by the contact between psychic secretion and food led to the production of gastrin, which released the stomach's stock of ferments. Next, the acid produced by gastric digestion led to the production of secretin, which excited *both* the release of pancreatic ferments (and proferments) and the secretion of bile. Bile salts, in turn, "doubled or tripled" the strength of the fat-splitting and starch-splitting ferments in pancreatic juice. So, by exciting both glands simultaneously, secretin served to coordinate the timing and volume of their secretory activity. Bayliss and Starling predicted the discovery of a similar mechanism— also "of a chemical nature"—that synchronized the activity of the pancreas and the duodenum. In this way, the trypsinogen produced by the pancreas would be converted to trypsin by the enterokinase secreted by the duodenum.[129] For these physiologists, then, the efficient functioning of the digestive system was due not to nervous mechanisms that fine-tuned the secretions of each individual gland but rather to chemical mechanisms that coordinated the gross activity of these glands as a whole.

The British physiologists concluded with an explanation of the fluctuations in pancreatic secretion that more closely resembled Popel'skii's than Pavlov's.

> The quantity of juice secreted will depend on the amount of secretin turned into the circulation, and this, in its turn, on the amount of acid entering the duodenum from the stomach. The amount of juice will, therefore, be measured by the stay and resistance to digestion of the substance in the stomach rather than to any direct nervous or other influence of the duodenal contents on the pancreas. A repetition of Walther's experiments by Popielski . . . , working independently, has in fact led the latter to deny altogether the adaptation of the pancreatic juice to the nature of the food. Popielski concludes from his experiments that variations in the juice depend only on the intensity and duration of the stimulus, the intensity of the stimulus determining the amount of enzymes, whilst its duration determines the total quantity of juice.[130]

In their discussion of the salivary and gastric glands, Bayliss and Starling conceded a greater role to nervous mechanisms, but here, too, they rejected Pavlov's notion that specific nervous exciters accounted for any fine adaptation of secretion to specific foodstuffs. They agreed with Pavlov that any specificity of the salivary response to particular foods resulted from "the psychic reproduction of previous actual stimulation of peripheral sense organs." As for the gastric glands, these were probably governed by "a chemical mechanism or reflex . . . associated with, and probably subordinated to, a nervous reflex mechanism." Any specific secretory response to particular foods could, however, be explained without recourse to specific nervous exciters: "As a rule the more indigestible the foodstuff the longer will it remain in the stomach; the greater, therefore, will be the secretion of acid gastric juice."[131]

There followed decades of experimental investigations and interpretive conflicts about the role of nerves, humors, and the psyche in digestive processes. Experiment was piled on experiment, interpretation on interpretation, and methodological quarrel on methodological quarrel. This fascinating chapter in the history of experimental physiology, which seems to involve distinctly different national orientations, awaits a systematic historical analysis.[132]

Epilogue: One Critic's Concluding Assessment

In his *Principles of General Physiology* (1915), William Bayliss introduced his discussion of the secretion of digestive juices with an imposing, almost full-page photograph of Ivan Pavlov. "The modes in which the secretion of these juices is set going," he explained, "have been worked out almost entirely by Pavlov and his coworkers and will be found described in his book [*Lectures*]."[133]

Bayliss had not surrendered his earlier criticisms of Pavlov's conclusions, and his description of pancreatic and gastric secretion fully incorporated the humoralist corrections to *Lectures*. (For example, "Whether the vagus takes any part in the normal process [of pancreatic secretion] we have seen to be doubtful.") Still, much remained from the Russian's now classic work: for example, the ineffectiveness of mechanical excitation of the stomach, the role of the vagus in gastric secretion, and the important role of appetite. ("When, therefore, Macbeth wishes for his guests that 'good digestion' may 'wait on appetite,' he is merely expressing a physiological fact.") Bayliss also referred his readers to Pavlov's instructions for the creation of salivary and pancreatic fistulas, and he lauded "the ingenious operation" by which the Russian had created "a greatly improved form" of the Heidenhain sac.[134]

Even Pavlov's British critic, then, considered him a father figure in the field. The Russian physiologist's reputation did not rest on any single knowledge claim and had not been significantly tarnished by the suggestions of experimental errors and tendentious interpretation. Bayliss did not even mention Pavlov's characteristic secretory curves, which would have fit only with great difficulty into his own conception of digestive mechanisms. This, too, mattered little. Pavlov had transformed digestive physiology through his laboratory's many varied products and had become the very symbol of physiology's status as a modern, precise, medically relevant experimental science. As for the digestive system itself, Pavlov had replaced the earlier image of a mousetrap-like mechanism that responded bluntly to a wide variety of irritants with a new image of a complex and coordinated system that responded subtly to specific exciters. Bayliss and Pavlov, and the successors to each, might disagree about the relative importance of various exciters, the degree of their specificity, and the mechanism of their action, but these disagreements themselves transpired within the basic framework established by the visionary factory physiologist of Lopukhinskaya Street.

Chapter 10

THE NOBEL PRIZE

Pavlov is the soul and the leader even in the research that his workers and students in the laboratory carry out.

— KARL MÖRNER, memo to Nobel Committee (1903)

On December 10, 1904, Ivan Petrovich Pavlov stepped forward from his place on the platform of Sweden's Royal Musical Academy's Great Hall to accept the Nobel Prize in Physiology or Medicine from King Oscar II. He was the first physiologist and the first Russian to win that award.

The Nobel Prize—the thrilling climax to Pavlov's jubilee year—resulted from the same features of his scientific vision and laboratory system as had his renown in previous years. The great number and range of Pavlov's knowledge claims, his synthesis of these claims in a compelling description of a precise and integrated digestive system, the appeal of his laboratory's various products to scientists and physicians, and his own symbolic appeal as an embodiment of modern laboratory physiology and its clinical promise—all are evident in the deliberations of the Nobel Committee. So, too, are the criticisms raised in the early twentieth century against Pavlov's conclusions.

Pavlov's Nobel Prize was not, of course, inevitable. Various contingencies—for example, the changing identity of the Nobel Committee's five members and its group dynamics, to say nothing of the merits of Pavlov's competitors—played an important role in the decision-making process.

This is clear from Pavlov's being nominated four times (1901–4) before winning the award in 1904. Yet the imprint of Pavlov's laboratory system and scientific vision remains clearly visible in the history of his candidacy.[1]

Alfred Nobel's will, and the statutes created to implement it, specified that, beginning in 1901, the medical faculty of the Karolinska Medical-Surgical Institute would award a prize to the person "who shall have made the most important discovery within the domain of physiology or medicine." Nominations were solicited from individuals in a wide range of scientific and medical institutions throughout the world.

In 1901, three of the four letters nominating Pavlov were postmarked in St. Petersburg. Two were written on the letterhead of the Imperial Institute of Experimental Medicine: both its director, Sergei Luk'ianov, and the chief of its Chemistry Division, Marcel Nencki, nominated their institution's physiologist. A third nominating letter was signed by thirty of Pavlov's colleagues at the Military-Medical Academy.

The circumstances surrounding the Academy's nomination underline the extent to which it was an expression of institutional and national pride. Such a collective letter could have been organized only by—or, at the very least, with the active participation of—the Academy's strong-willed president, Viktor Pashutin. A well-known pathologist, Pashutin had, like Pavlov, worked in Botkin's, Ludwig's, and Heidenhain's laboratories and was a strong proponent of scientific medicine. He had earlier lent important support to Pavlov's candidacy for Academy positions in pharmacology (1890) and physiology (1895). Since that time, however, the two had engaged in a bitter ongoing quarrel about what Pavlov perceived as the president's dictatorial ways. In fact, one month after sending the Academy's letter to Stockholm, Pashutin collapsed of a heart attack and died at a faculty meeting—immediately after a heated exchange with the nominee.[2] No record exists of the history of the nominating letter, but it seems clear that Pashutin's appreciation of Pavlov's contributions and his own strong desire to advance the interests of his Academy and Russian science trumped his dislike for Pavlov. Whatever his faults, Pavlov was the only member of the Academy who possessed both a European reputation and accomplishments that fit the ambiguous criteria for a Prize in "physiology or medicine."

Had Pavlov read the Academy's nominating letter he would certainly have been less disappointed with the Russian response to *Lectures*.

For almost fifteen years Mr. Pavlov has studied in his laboratory the fundamental questions of the physiology and pathology of the digestive glands.

These systematic researches have provided science with an entire series of new methods, ingeniously conceived, which, as applied by their creator, have led to the discovery of a number of extremely important facts regarding the physiology and chemistry of the digestive glands and their secretions. Guided by the correct principle that the study of physiological phenomena should be conducted upon subjects that are as normal as possible, Mr. Pavlov has conducted his experiments upon animals that have earlier been operated upon and which are in a normal physiological condition. The ensemble of these experiments, the number and fertility of which grows every year, has provided the author with vast material for a profound analysis of the mechanisms and laws governing the work of the digestive glands, and subsequently for a synthetic conception of their work under diverse real conditions. The incontestable importance of Mr. Pavlov's work for pure physiology and its application to practical medicine serve as sufficient proof of the justice of our proposition.[3]

The fourth nominating letter attested to Pavlov's international reputation in the wake of the German edition of *Lectures*. The Johns Hopkins University physiologist William Howell wrote, with the Committee's selection criteria in mind, that "Professor Pavlov's work upon the physiology of digestion and secretion seems to me the most important contribution to physiology in recent years that can be traced solely or mainly to a single individual."[4]

Forty-one candidates were nominated for the Nobel Prize in Physiology or Medicine in 1901; Pavlov was one of twenty deemed worthy of further investigation. This task was passed to the sole physiologist on the Committee, the Karolinska Institute's professor of physiology J. E. Johansson. It was probably Johansson who decided to enlist the aid of his former coworker Robert Tigerstedt and to visit St. Petersburg in order to witness Pavlov's experiments firsthand. Neither the use of an outside expert nor the site visit was unusual in the early years of the Committee's deliberations, when its experts were called upon, if possible, to evaluate independently the truthfulness of a nominee's knowledge claims.[5]

Tigerstedt, as we have seen, was already acquainted with the nominee, who had just edited the Russian edition of Tigerstedt's textbook on human physiology (see Chapter 9). Johansson also had a previous connection with Pavlov, though both men were probably unaware of it. After meeting Alfred Nobel in October 1890, the Swedish physiologist had, at Nobel's request, conducted in Nobel's laboratory at Sevran "a certain number of tests connected with blood transfusion in which [Nobel] took a special interest."[6] (I described in Chapter 3 Nobel's identical proposal to the Imperial Institute and Pavlov's subsequent transfusion experiments.) At Nobel's urging, then, both physiologists had experimented briefly with blood transfusion in the early 1890s.

Johansson had subsequently corresponded with Nobel and played an important role in discussions about the meaning of two key terms in Nobel's will. The philanthropist had specified that five separate awards were to be made to those who "*during the preceding year,* shall have conferred the greatest benefit on mankind," and that one prize was for "the person who shall have made the most important discovery *within the domain of physiology or medicine.*" On deliberation, "preceding year" was interpreted to mean "recently," and "the domain of physiology or medicine" was interpreted to encompass "all the theoretical as well as the practical medical sciences."[7]

Having informed Pavlov of their visit beforehand, Johansson and Tigerstedt arrived in St. Petersburg on June 8, 1901. The nominee had prepared "two dozen dogs upon which various operations had been performed," and for the next ten days, from 10 A.M. until 6 or 7 P.M., he used them to demonstrate his "most important results."[8] Pavlov treated his visitors to an abbreviated live version of *Lectures* (which, in turn, closely resembled his lectures on digestive physiology to the Military-Medical Academy).

The two visiting physiologists were most favorably impressed. Their extensive joint report of July 1901, and Tigerstedt's lengthy addition to it, became part of the Nobel Committee's permanent record, framing discussions of Pavlov's achievements during all four years of his candidacy. For some combination of reasons—their appreciation of Pavlov's scientific contributions and, perhaps, shared disciplinary allegiances and personal relationships that developed during their ten days together in St. Petersburg—Johansson and Tigerstedt became Pavlov's committed advocates.

Invoking their firsthand observations, Johansson and Tigerstedt testified to the veracity of proposition after proposition advanced in *Lectures* on the salivary, gastric, and pancreatic glands, and in Pavlov's other publications on the survival of vagotomized dogs, the role of enterokinase, and so forth. They witnessed Pavlov's creation of an isolated sac ("one of the most delicate [operations] in contemporary physiology") and his demonstration that this sac faithfully mirrored the activity of the intact portion of the stomach. They described approvingly Pavlov's experimental demonstration of the powerlessness of mechanical irritation to excite the gastric glands, the governing role of nervous mechanisms, and the important role of the psyche in salivary, gastric, and pancreatic secretion. Pavlov apparently did not demonstrate the existence of his precise characteristic secretory curves, but the visiting physiologists did report that on operated-upon animals, "one could observe that the gastric acid after ingestion of meat differs from that after the intake of bread. Therefore, gastric secretion has a different course for different substances."[9]

In the conclusion to their report, Johansson and Tigerstedt attested to the truthfulness of Pavlov's scientific findings. "Summarizing our observations, which, as should be evident from [this report] extended to most of the areas of prof. Pavlov's work—in which a rigorous scientific precision has been observed—we wish to emphasize that it is our certain conviction that the evidence presented to us by prof. Pavlov must be considered to be fully and positively established—it can be presented at any time, naturally under the presumption that the experiments are conducted with the necessary precision and care."[10] Johansson and Tigerstedt's submission of an extensive list of Pavlov's achievements is especially interesting in the light of the Committee's charge to "call attention to the special aspect of the [nominee's] discovery which [the expert] considers decisive."[11] Committed advocates of Pavlov's candidacy, they did not identify any single decisive "discovery." Rather, they indicated the great *number* and *range* of Pavlov's contributions to his field.

A final comment in their report indicates that an important reservation about Pavlov's candidacy had already been raised: to what extent were Pavlov's works really Pavlov's? "We also feel that we should mention," wrote Johansson and Tigerstedt, "that during our stay in prof. Pavlov's laboratory we reached the conclusion that all the works issued from it, whether or not they carry prof. Pavlov's name, to a substantial degree constitute his intellectual property, as he has not only carried out all the operations on the animals used in the experiments, but has also been the leader and organizer with regard to the planning, development, and implementation of the special investigations."[12]

Tigerstedt was apparently dissatisfied with the case he and Johansson had made, and shortly thereafter he submitted his individual appraisal of the candidate's achievements. He began with the issue of intellectual credit. Pavlov's work, he conceded, indeed represented a synthesis of "a large number of specialized dissertations" written by other people. Yet Pavlov himself had "awakened" these researchers' interest in the subject, and their dissertations were permeated by the chief's "guiding idea." These works "must therefore, to a substantial degree, be seen to constitute one single man's intellectual property, although this man, despite his clear work capability and his great endurance, would not by himself have been capable of collecting the colossal amount of material which has been observed and that [is] put forth in these works."[13] The practice of "coworkership," then, should not "constitute any hindrance to attributing these texts to prof. Pavlov."

The "guiding idea" uniting all these works, Tigerstedt explained, was that "the digestive organs, through a many-sided and extremely subtle regulation, cooperate for the resolution of their task" (pp. 9–10). This idea, in turn, encompassed two insights of general significance. First, Pavlov's analysis of the role of the psyche provided "an extremely obvious example of how the activity of organs that definitely are not under the influence of our will can still be rather closely dependent on our mental state—and we have thereby received a new intimation of the close dependence in which mind and body stand in relation to one another." Second, and of no less "general physiological interest," Pavlov had demonstrated that the principle of specific excitability extended beyond the sensory organs to the inner organs as well.

Tigerstedt again reviewed Pavlov's many specific knowledge claims. These, he attested, were "as correct as facts in physiology can be" and had been "tested so many times on different individuals that any doubt concerning the correctness of the observations must be excluded" (p. 3). Reviewing Pavlov's methodological contributions, he apparently thought it necessary to respond to the objection that these were not original. Conceding that other physiologists had previously devised fistulas and even an isolated sac, he noted that Pavlov's improvement of these operations, and his unprecedented success in caring for the experimental animals thus created, had enabled him to avoid the errors of his predecessors and so to accomplish something substantially new (pp. 1–2).

Tigerstedt's and Johansson's arguments proved persuasive for a key member of the committee—Karl Mörner, rector of the Karolinska Institute and its professor of chemistry and pharmacy. In a short note of July 30, 1901, Mörner pronounced himself satisfied that "prof. Pavlov's work regarding the glands of the digestive canal is of the nature and importance" worthy of the Nobel Prize in Physiology or Medicine.[14]

Pavlov, however, was but one of four candidates who had passed through the initial stages of the selection process. He joined Emil von Behring, renowned for his development of serum therapy, particularly against diphtheria; Ronald Ross, for his work on malaria; and Niels Finsen, for his development of light therapy against various diseases. The archival record does not include even the briefest summary of Committee discussions, but a memo from Tigerstedt, written after one Committee session of September 1901, indicates that Pavlov's relative paucity of publications in his own name and the related problem of intellectual property continued to weaken his candidacy.[15] It was, Tigerstedt insisted, "totally incorrect" to regard Pavlov's Lectures as "a kind of compilation of the experimental dissertations upon which they are based." Quite to the contrary, both the theses and the chief's synthetic work represented "the contributions of the Pavlovian school" and should be considered jointly in evaluating the candidate's achievements.[16]

In August 1901 a rumor circulated throughout the international scientific community that Pavlov and Finsen would share the Nobel Prize in Physiology or Medicine. Printed as fact in the conservative newspaper *Novoe Vremia*, in *Vrach*, and in the Military-Medical Academy's *Izvestiia*, the news also elicited a congratulatory letter from Pavlov's proud mentor, Ilya Tsion, who was then living in Paris. An abashed Pavlov could only reply that "as for the prize, the press has confused something—it has probably still not been awarded, but your happiness for your pupil of long ago is for me a very great reward."[17]

Three weeks later, Pavlov finished third in the Committee's balloting and fourth in the final deliberations. The majority of the Committee voted to award the prize jointly to Ross and Finsen, with a minority favoring a joint award to Ross and Pavlov. Behring's candidacy suffered from the investigators' conclusion that "both the fundamental discovery and the proof of its practical value are so old that, while admitting that in other respects they fully deserve a prize, we cannot now recommend them for the honor."[18] The Faculty Collegium of the Karolinska Institute, however, had the last word, and it overruled the divided Committee. The prize finally went to Behring "for his work on serum therapy, especially its application against diphtheria, by which he has opened a new road in the domain of medical science and thereby placed in the hands of the physician a victorious weapon against illness and death."[19]

The decision-making process of 1901 did, however, produce enduring and ultimately critical successes for Pavlov. For one thing, the glowing reports by Johansson and Tigerstedt became the basic documents for future evaluations of his candidacy. Even more important, the two physiologists became his determined champions. Each did what he could to advance Pavlov's candidacy in subsequent years.

Tigerstedt's visit to Pavlov's laboratory proved the beginning of a long friendship. The Finnish physiologist's first letter to Pavlov after his return to Helsingfors was addressed to "Respected Colleague," but by September 1902 he had adopted the salutation "Dear Friend."[20] Perhaps one element in their budding friendship was Pavlov's generosity regarding a laboratory product that had especially intrigued his colleague. Tigerstedt was interested in research on digestive fluids and apparently had expressed enthusiasm about Pavlov's method for acquiring pure gastric juice. Pavlov responded by offering to send him the necessary dog-technology. Tigerstedt eventually demurred at "great self-sacrifice," professing concern that he would be unable to care properly for the animal in the constricted setting of his laboratory. He did, however, request and receive "a little natural gastric juice." (Several years later, when Tigerstedt embarked on a study of "animal fluids," he again requested a sample of digestive juices, since "conditions in my present laboratory do not permit us to conduct the operation for the implantation of fistulas.")[21] By October 1902 Mörner, too, had received a sample of gastric juice from the nominee.[22]

Tigerstedt waited less than one week after the Nobel Committee had settled the 1901 Prize on Behring to nominate Pavlov for the succeeding year.[23] He also advised Pavlov that he and his coworkers should participate in force at the upcoming Congress of Northern Naturalists and Physicians in Helsingfors, which would be attended by some of the Nobel judges.[24] Babkin later recalled that the chief "mobilized his whole laboratory that spring in order to present as many papers as possible at the Congress," knowing that it "would be attended by the Swedish scientists."[25]

The laboratory was indeed well-represented at Helsingfors, where its representatives delivered eight separate reports. These addressed a wide range of topics, which reflected the changing nature of scientific discourse on digestive secretion. Only two reporters—Babkin and A. P. Sokolov—developed the laboratory's standard lines of investigation.[26] A third, Ivan Tolochinov, delivered the first public account of conditional reflexes and the laboratory's changing view of the psyche (see Chapter 7). V. V. Savich presented two reports: one on enterokinase and a second that reaffirmed nervous control over the pancreas in the wake of Bayliss and Starling's discovery of secretin. Anton Val'ter also addressed the humoral challenge, presenting experimental evidence to indicate that the blunt exciters of acid and secretin could not explain the specificity of pancreatic secretion. As Karl Mörner (one of the Nobel Committee members attending the conference) noted in his report, two other presentations, one by Evgenii Ganike and another coauthored by Pavlov and his praktikant S. V. Parashchuk, "touched upon territory that Pavlov and his disciples had earlier avoided—specifically, questions of a purely chemical nature." Ganike addressed the chemical nature of pepsin, and Pavlov advanced a bold chemical proposition: that pepsin and rennet were one and the same substance.[27]

The two Nobel Committee members present—Mörner and Oskar Medin, professor of pediatrics at the Karolinska Institute—were unimpressed by the forays into physiological chemistry. Medin noted that Ganike's report "did not contain anything new."[28] Mörner added that Ganike had merely repeated earlier experiments conducted by Cornelis Pekalharing in Utrecht. (He observed, however, that Pekalharing had relied on Pavlov's method for acquiring pure gastric juice and that this redounded to the chief's credit.)[29] Mörner also raised a number of criticisms about Pavlov and Parashchuk's report (including one concerning Pavlov's interpretation of quantitative data), and he observed that the chemist Emil Hammersten had introduced Pavlov's talk by offering several important reasons to reject the Russian's conclusions. Medin offered the blunt assessment, "I was not convinced."[30]

Nevertheless, Mörner concluded his report to the Committee on a strong positive note. He endorsed Johansson and Tigerstedt's earlier summary of Pavlov's scientific contributions, added that the Russian's methodological innovations had "paved the road" for future advances, and emphasized the broader significance of three of Pavlov's accomplishments: his proofs for specific excitability and the role of the psyche and his elucidation of nervous mechanisms that controlled the glands. "In a science that has been cultivated so thoroughly as has physiology," wrote Mörner, "one could hardly expect that one person could make so many important contributions as has Pavlov." Taken together, these contributions constituted a "thorough transformation" of digestive physiology and were "fully deserving of a Nobel Prize."[31]

In its final deliberations, the Committee quickly identified the same three leading candidates as it had the previous year—Finsen, Ross, and Pavlov. Johansson was not on the Committee this year, so Pavlov lacked a strong advocate. The five members split between two basic positions. Emil Holmgren (professor of histology) argued—and Mörner agreed—that Ross should be ranked first and Pavlov second. In Holmgren's view, Ross's work best combined theoretical and practical contributions. Pavlov was "an extraordinarily skillful and talented scientist," but his work had not conferred the same "benefit upon mankind" as had Ross's. Conversely, Finsen had devised a "very beneficial method of treatment" but was an undistinguished scientist. Holmgren suggested, then, that the prize be divided between Ross and Pavlov, a decision that would signal the Committee's appreciation of both practical and theoretical contributions. Medin, Ernst Almquist (professor of hygiene), and Carl Sundberg (professor of pathological anatomy) adopted another position: all three leading candidates fully met Nobel's criterion of conferring "the greatest benefit upon mankind." Finsen's and Ross's contributions were of more immediate practical benefit, yet Pavlov's achievements, too, showed signs of medical usefulness. Finsen's and Ross's discoveries, however, were more "original" and theoretically exciting than Pavlov's. Considering Ross's work to be more scientifically sophisticated than Finsen's, this group ranked Ross first, Finsen second, and Pavlov third and recommended that Ross alone receive the prize. All five committee members, then, ranked Ross first, and the majority voted to award him an undivided Nobel Prize. The Faculty Collegium approved this decision.[32]

Pavlov's candidacy proved substantially weaker the following year. He was nominated in 1903 by five individuals—including Tigerstedt and Johansson—and both the chief investigator of his candidacy and the Committee's composition remained unchanged. But criticism of some of Pavlov's key knowledge claims over the past two years now entered the Committee's deliberations.[33] In his report Mörner concluded that, although Pavlov's many achievements still made him worthy of the honor, "I believe that it would not be opportune to award Pavlov this year's Nobel Prize."[34]

Mörner mentioned "the emergence of vague questions" about four aspects of Pavlov's research (p. 14). First, Pavlov's foray into physiological chemistry— his identification of pepsin with rennet—had proven unconvincing. Some scientists supported Pavlov's position, others opposed it; the issue had not yet been "definitively resolved" (pp. 13–14). Second, and "more important," was Bayliss and Starling's discovery of a humoral mechanism of pancreatic secretion, which cast doubt on Pavlov's broader notions of nervous control and specific excitability. Mörner listed a number of important scientists who agreed with Bayliss and Starling, but concluded, again, that this issue remained unresolved (p. 13). Third, alluding to the research of Delezenne and Frouin, Mörner noted that Pavlov's investigations on pancreatic secretion had been conducted before scientists had fully appreciated the difference between zymogenic and active ferments, and this realization had created "an opening" for criticism of Pavlov's conclusion about the specific and purposive secretion of the pancreas (pp. 13–14).[35]

Mörner devoted most attention, however, to the disturbing criticisms raised by "one of Pavlov's former disciples," Lev Popel'skii.[36] These went to the very heart of Pavlov's claim about the "precise adaptation" of pancreatic secretion. Mörner reported that Popel'skii's argument was grounded in "criticism of some of the details of works conducted by some of Pavlov's students," and cogently summarized Popel'skii's critique of Vasil'ev's, Iablonskii's, Val'ter's, and Lintvarev's theses. For example, he noted that "Popel'skii says that Lintvarev selected certain data that agreed with the [laboratory's] preconceived view" and added that "Popel'skii cites the [complete] data, with a reference to the page of the dissertation in which they appear."[37] Popel'skii's critique of the concept of precise and purposive secretion extended also to Pavlov's analysis of gastric and salivary secretion—and here, again, Mörner provided some detail. He also mentioned that Popel'skii had himself conducted experiments on pancreatic secretion, and he described for the Committee Popel'skii's alternative explanation of secretory patterns in the main glands.[38]

Pavlov had not replied to Popel'skii's critique, and Mörner predicted that the future would bring "both attack and defense." Pavlov would certainly prove able to refute some of the criticisms, but Popel'skii had established that "Pavlov is certainly guilty of one-sidedness in his consistent finding that secretory activity is governed by purposiveness." Here, argued Mörner, Pavlov's claims clearly outstripped his evidence, and even the existing factual foundations "may not be fully reliable."[39]

Both in his lengthy report and in his remarks at the Nobel Committee's meeting of September 23, 1903, Mörner made clear that he remained impressed by the candidate's achievements. Recent criticisms concerned "only a part" of Pavlov's results, and the more Mörner pondered the Russian's corpus as a whole, "the more important it seems."[40] He also agreed with Tigerstedt and Johansson that Pavlov deserved intellectual credit for the work of his praktikanty, since he remained "the soul and the leader even in the research that his workers and students in the laboratory" carried out.[41] Still, during the Committee's deliberations Mörner reiterated his position that, since "certain aspects of Pavlov's work are now under debate," it would be "inopportune" to award him the prize.

Nobody on the Committee took issue with this conclusion, and the competition narrowed to Finsen and Koch. Finsen carried four votes out of five and became the winner of the 1903 Nobel Prize for Physiology or Medicine.[42]

By 1904, then, Pavlov was the sole finalist of 1901 who had not yet won the prize. He again received multiple nominations, including one from Vinsens Czerny, director of the surgical clinic at Heidelberg University, and another submitted jointly by Johansson and C. G. Santesson, professor of pharmacology at the Karolinska Institute.[43] The Swedish scientists' unusually lengthy nominating letter signaled the beginning of a determined campaign to finally gain the prize for Pavlov. They apparently identified four weaknesses in his candidacy. The first was the issue of intellectual property, which the nominators dealt with by referring to Tigerstedt and Johansson's conclusion of 1901 that "the credit for the achievements in [the laboratory's] works belongs to Pavlov himself." The second weakness was the recent objections to Pavlov's work reported by Mörner in 1903; these, they argued, were relatively minor when viewed against the background of Pavlov's many undeniable contributions, both methodologically and to scientific knowledge itself. Third, in a reference to Pavlov's notion of "purposiveness," they conceded that the Russian physiologist "had given his doctrine a somewhat teleological formulation that might appear strange to some modern scientists," but argued this should "not have any importance."[44]

Santesson and Johansson devoted most space to the fourth point: the objection that Pavlov's contributions, however scientifically important, did not have sufficient practical import to satisfy Nobel's directives. They pointed out that A. F. Hornborg—in experiments, we should note, that were conducted under Tigerstedt's supervision—had recently confirmed the importance of appetite to gastric secretion in a young boy.[45] Theoretical works such as Pavlov's, they argued, "only slowly" entered the practical realm. Pavlov's research, however, was clearly oriented toward practical goals. Drawing upon the eighth chapter of *Lectures,* they contended that it was surely not "too audacious to predict" that Pavlov's research would, for example, facilitate the rational organization of the diet and the treatment of digestive diseases. This was especially probable in view of the "transforming influence" of his methodological contributions on research in pathology and pharmacodynamics. They closed with the argument that Nobel had intended to reward not only practical achievements but also "more theoretical works." Johansson's well-known relationship to Nobel no doubt lent great authority to this interpretation of the philanthropist's intent.[46]

It was Pavlov's great fortune that Johansson replaced Holmgren on the Nobel Committee that year and, as the Committee's sole physiologist, was assigned to investigate his own nominee. Mörner's report of 1903 had constituted a reluctant brief for Pavlov's critics; Johansson's of 1904 presented a whole-hearted and powerful argument for his admirers.

Johansson's basic rhetorical tactic was to compare, gland by gland, the state of digestive physiology before and after Pavlov's work, to enumerate the number and range of the candidate's scientific contributions, and so to portray as unimportant any doubts concerning a few of them. Having visited Pavlov's laboratory in 1901, Johansson reminded the Committee, he and Tigerstedt had witnessed "with our own eyes" the experiments underlying Pavlov's knowledge claims.[47]

Briefly reviewing the scant and contradictory knowledge about glandular mechanisms before Pavlov's research, Johansson enumerated Pavlov's major discoveries (unapologetically identifying many of these findings with the dissertations written by praktikanty). Concerning the gastric glands, Pavlov's main contributions were his demonstrations that the mucous membrane of the stomach did not respond to mechanical irritation, that appetite—mediated by the vagus—played an important secretory role, that the gastric glands could be excited only by specific substances, that fat inhibits gastric secretion, and that the secretions elicited by different foods differed distinctively in their amount and proteolytic power (Johansson here referred to Khizhin's and Lobasov's dissertations) (p. 4). Concerning the pancreatic gland, Pavlov had demonstrated the governing role of the vagus and sympathetic nerves, the specific secretory roles of hydrochloric acid and fat, the characteristic nature of the secretory response to fat and various foods, the adaptation of pancreatic secretion over time to particular diets, and the role of enterokinase (Johansson here mentioned the dissertations of Kudrevetskii, Popel'skii, Dolinskii, Damaskin, Val'ter, Vasil'ev, Iablonskii, and Lintvarev) (p. 9). Johansson provided similar lists of Pavlov's contributions regarding the salivary glands (three knowledge claims), the gall bladder (three knowledge claims), and the passage of food from the stomach to the duodenum (two knowledge claims) (pp. 12–13). These discoveries, united by Pavlov's principles of specific excitability and purposiveness, had yielded a fundamentally new view of the digestive canal, which had been revealed as a sensitive, interconnected, and adaptive system.

Viewed against this background, Johansson insisted, the criticisms raised recently against Pavlov's work were, at the most, trivial. Bayliss and Starling's discovery of a possible humoral mechanism of pancreatic secretion did not negate the importance of Pavlov's nervous mechanisms. "Pavlov, so far as I can see, assumed a reflex mechanism on good grounds. There is no decisive evidence against this view." Furthermore, "for the Nobel Committee, it should be sufficient to state that [Pavlov's] actual discovery of the two exciting substances [hydrochloric acid and fat] cannot be contradicted, that this discovery is of the greatest importance for physiology, and that it forms the point of departure for discussions of the importance of so-called secretin" (p. 9).

Johansson directed his main fire against Popel'skii's criticisms, which he treated with disdain. Like Mörner, Johansson did not have access to Popel'skii's detailed critique, which was available only in Russian. He simply ignored Popel'skii's criticism of Val'ter's thesis (this criticism had been important to Mörner the year before), dealing only with criticisms directed against Vasil'ev, Iablonskii, and Lintvarev. Johansson concluded that the basic patterns in their data indeed supported the laboratory's conclusions and that Popel'skii's criticism was "illegitimate." Furthermore, although Popel'skii wanted "to insinuate that some data [have] been hidden," the experimental protocols were fully reported in the dissertations and "one can fully follow the experiments" (pp. 9–12).[48] Again invoking his authority as an eyewitness, Johansson expressed his own opinion on one scientific issue in dispute: "My view on this question, which I base not only upon the above-mentioned experiments [as reported in the published literature] but also upon the observations that I made myself during my visit to Pavlov's laboratory with Tigerstedt, is that the pancreatic juice really does lose its proteolytic power when [the animal] is fed milk and bread for a long time. As for the other ferments, I believe that the experiments are noteworthy, but that further proof is required" (p. 11).

Johansson concluded that Pavlov, by virtue of his "pioneering operational and experimental methods" and his "comprehensive revision" of scientists' understanding of the digestive canal, was worthy of the Nobel Prize (p. 14).[49]

A postscript written on this report in a different hand attested to the effectiveness of Johansson's argument: "I agree to the above stated. K. A. H. Mörner." Johansson had convinced Pavlov's half-hearted critic of 1903 to support the Russian's candidacy in 1904.

By the Committee's final session on September 24, the field had been winnowed to two finalists: Pavlov and Koch. The major weaknesses in Koch's candidacy were that his "fundamental, epoch-making discoveries" regarding cholera and typhoid had occurred long ago and that Committee members considered his later contributions either derivative or untested. Koch's fervent advocate on the Committee, Ernst Almquist, presented a strong argument for his favorite. Koch and Pavlov had each begun his research at about the same time, in the late 1870s, and each had "delivered one contribution after another to the solution of a complex and comprehensive question." In each case, no single one of these contributions was "really worthy of a Nobel Prize," but together they constituted "a great advance for science." Koch had "established the foundations for the rational struggle against epidemics," and although not one of his "beautiful discoveries after 1890" deserved a prize, if his work was regarded as a whole—in the same manner as Johansson was urging the Committee to view Pavlov's—it certainly did. "I find Koch's work to be of greater importance," Almquist concluded, "and give him my vote."[50]

The other Committee members had decided otherwise. The perfunctoriness of their remarks testifies to the fact that the die was cast. Citing Johansson's report and "Prof. Mörner's concurrence," Sundberg voted for Pavlov, explaining that his "discoveries, and especially the general acknowledgement of these discoveries, are so contemporary . . . that they coincide with the spirit of [Nobel's] testament." Johansson voted without comment. Medin briefly alluded to Johansson's and Mörner's position and to the lack of any "pioneering discovery" in Koch's recent work, and he, too, voted for Pavlov. Mörner commented briefly about the datedness of Koch's most original work and cast a final vote for the Russian physiologist. The Committee's recommendation, by a vote of four to one, was endorsed by the Faculty Collegium on October 20, 1904.

On October 21, Mörner sent the following letter to Pavlov.

> Most Respected Colleague!
> I have the honor and the pleasure to inform you that the Collegium of Professors of the Royal Karolinska Medical-Surgical Institute yesterday (20/X) decided to award you this year's Nobel Prize in Physiology and Medicine for your work on the physiology of digestion.
> The financial value of a Nobel Prize is exactly 140,858 kroner and 51 ere (or 198,000 francs).
> The decision of the Collegium of Professors will be published in newspapers on 10 December 1904.
> It would be preferable if you refrained from a premature disclosure of this news.
> We invite you to come to Stockholm for a personal audience on 10 December. Do you intend to deliver a presentation, and if so in which language: German, English, or French?
> If you agree to deliver a presentation I will take all the necessary measures.
> With profound respect, your devoted
>> Rector of the Royal Karolinska Medical-Surgical
>> Institute, Professor, Doctor of Medicine,
>>> Graf Karl Mörner.[51]

In her memoirs, Serafima Pavlova recalled that her husband was "absolutely stunned, so unexpected was this for him . . . He had never thought that his work might be valued so highly, especially since [his] book . . . had enjoyed no success in Russia. Having myself always considered I. P. [Pavlov's] work brilliant, I was delighted that it was finally being properly appreciated. I. P. was not happy with my attitude and said 'You have created for yourself an idol and now enjoy kneeling before it. There is nothing special about my work. It consists wholly of the logical development of thought on the basis of conclusions from facts.'"[52] He was clearly enjoying his role as the great man of science.

Tigerstedt invited the Pavlovs to spend a week with his family in Helsingfors on the way to Stockholm, and so, having "ordered an evening coat for I. P. and two dresses and a nice fur coat" for Serafima, they departed for a week of friendly evenings and formal receptions. The Pavlovs were feted no less generously in Stockholm, where they began what proved to be enduring relationships with the Mörners and with Sir William Ramsay, who shared that year's Nobel Prize in Chemistry.[53] There they also met Emmanuel Nobel, who years earlier had negotiated Pavlov's first "Nobel prize"—the 10,000 ruble gift of 1893 that had financed the construction of his new laboratory and so made possible the second Prize of 1904 (see Chapter 3).

The ceremonies of December 10–12, 1904, were a triumph for Pavlov and his country. The speeches hailing the Nobel laureates each ended with a climactic passage in the prize-winner's native language. Mörner had learned enough Russian to properly conclude his review of Pavlov's many contributions to digestive physiology, "which have accomplished a revolution and comprised an epoch in the history of that sphere of knowledge."[54] King Oscar II, who formally awarded the gold Nobel medallion on December 10, startled Pavlov by greeting him in Russian: "*Kak Vashe zdorov'e, kak Vy pozhivaete?*" (How is your health, how are you?). Emmanuel Nobel later confided to the Pavlovs, however, that the Russian Nobelist's democratic attire had made a bad impression on the king: "I fear your Pavlov," he reportedly told Nobel, "he wears no orders and is probably a socialist."[55]

On December 12, Pavlov addressed the Swedish Academy of Science in halting German. He devoted the first two-thirds of his presentation to a review of the digestive system as a purposive "series of chemical laboratories" governed by specific excitability and the omnipresent psyche. He then gave an intensely delivered account of his laboratory's recent research on psychic secretion, to which he now referred as a conditional reflex. He concluded with these words:

> Essentially, only one thing in life is of real interest to us—our psychical experience. Its mechanism, however, was and still is shrouded in profound obscurity. All human resources—art, religion, literature, philosophy, and the historical sciences—all have joined in the attempt to throw light upon this darkness. But humanity has at its disposal yet another powerful resource— natural science with its strict objective methods. This science, as we all know, is achieving gigantic successes every day. The facts and conceptions that I have advanced at the end of this lecture constitute one of many attempts to study the mechanism of the higher vital processes in the dog . . . through the consistent application of a purely natural scientific manner of thinking.[56]

With the ceremonies behind them, the Pavlovs attended yet more receptions, enjoyed a performance of *Eugene Onegin*, and paid a memorable visit to Stockholm's zoological garden. Finally, "exhausted by the endless festivities . . . we happily left for home."[57]

The Nobel laureate returned to a celebration of his "International Nobel Prize" at Prince Ol'denburgskii's palace, to much-enlivened festivities at the Society of Russian Physicians in honor of his twenty-fifth jubilee, to a laboratory that was more attractive than ever to Russian coworkers and foreign visitors, and to the three decades of research on conditional reflexes that would transform him from an internationally famous physiologist into a cultural symbol of twentieth-century science.[58]

EPILOGUE

By 1904 Pavlov's physiology factory was well-established, successful, and humming with activity. The beleaguered workshop physiologist of 1888 who had lamented that "my time and strength are not spent as productively as they could be" now commanded a laboratory system that, by harnessing skilled hands to his recast Bernardian vision, had transformed digestive physiology and brought him international renown.

Yet the Nobel Prize marked only the midpoint of Pavlov's career and that of his enterprise. The Pavlov of conditional reflexes and salivating dogs would emerge over the next three decades in a time of unprecedented historical change marked by two revolutions, a civil war, and a world war; the massive expansion and transformation of Russia's scientific community; the end of imperial rule and the consolidation of the Soviet state, first by Lenin and then by Stalin.

What, then, became of the physiology factory? It is not surprising that these world-shaking events—to say nothing of the chief's attention to a new scientific subject—affected Pavlov's enterprise. Much more striking, however, is that these events elicited only slight variations on familiar themes: the chief's scientific-managerial vision and the operation of his laboratory system remained fundamentally the same in old times and new. A brief look at subsequent developments serves to highlight the physiology factory's essential features.

As Russia's preeminent physiologist after receiving the Nobel Prize, Pavlov expanded and improved his laboratory facilities in the last decade of imperial rule. He acquired a more spacious laboratory at the Military-Medical Academy, and his election to the Academy of Sciences brought a third one under his wing. A new patron emerged with the growing vitality of Russian capitalism: the Moscow merchants' Ledentsov Society for the Support of the Experimental Sciences and Their Practical Application, which financed a large fortress-like addition to Pavlov's laboratory at the Imperial Institute of Experimental Medicine. The chief personally designed this building, the so-called Towers of Silence, for maximum control over the environment of animals during experiments on conditional reflexes, and it became the site of numerous technical innovations to that end. In 1916 Pavlov's longtime assistant Evgenii Ganike became chief of the Institute's new Physico-Physiological Division, which served mainly as a workshop for the development of techniques and technologies to facilitate Pavlov's investigations. Coworkers and foreign visitors flooded Pavlov's laboratories as never before.

The good times came to an abrupt end in August 1914. The First World War, the Bolshevik seizure of power in 1917, and the civil war that followed disrupted laboratory work at a critical juncture. Pavlov's coworkers left for the front, depriving his factory of the skilled hands necessary to its operation. Openly hostile to the Bolsheviks, depressed by his inability to continue his research under deteriorating material conditions, and outraged by communist suggestions that scientific laboratories should be run by a council of its workers, Pavlov reluctantly considered emigration. He finally decided to remain in Russia, in part because no western country was prepared to provide the seventy-year-old scientist with the large laboratory and staff he required, and in part because Lenin promised to do just that. Lenin was motivated by his conviction that Russia's scientific community was essential to socialist construction and by the potential propaganda value of a Nobelist prospering in revolutionary Russia. Having established by 1921 his special status with the Soviet state, Pavlov turned it to good advantage in subsequent years, openly criticizing communist policies while enjoying the privileges of those at the apex of the Soviet star system in science.[1]

Under both Lenin and Stalin, the state granted Pavlov a virtual *carte blanche*. He and his favored coworkers were allowed to travel abroad with relative freedom; his laboratories were continually expanded and renovated; and, in the late 1920s, construction began on his own science city in Koltushi, a small village just outside Leningrad (as St. Petersburg was renamed in 1924). Here, at the Institute of Experimental Genetics of Higher Nervous Activity, Pavlov and his coworkers studied conditional reflexes in a wide variety of organisms, most notably the anthropoid apes Roza and Rafael, and prepared for what Pavlov envisioned as a major study—with eugenic goals—of the role of heredity and environment in determining an organism's "nervous type." Renowned as the "World Capital of Conditional Reflexes," Koltushi became a scientific showplace and a hub of Pavlov's ever-expanding international network. By this time, Pavlov was coordinating the work of about forty coworkers a year in four separate facilities. His basic managerial approach remained unchanged: he continued to use his coworkers for the gathering of experimental data which he himself interpreted.

Yet the qualitative expansion of his laboratory enterprise and the new circumstances of Soviet science complicated this task and posed new managerial challenges. For one thing, however energetically he raced from laboratory to laboratory, it was now impossible for Pavlov to have more than the most casual contact at the bench with the great majority of his coworkers. He seems to have adjusted to this by expanding middle management in his laboratories and also by convening the so-called Pavlovian Wednesdays, weekly gatherings of his coworkers to hear reports on their experiments and discuss interpretive possibilities. (Time permitted only a few coworkers to report, however, and this honor became a point of intense competition.) Attempting to exert quality control, Pavlov often instructed coworkers in different laboratories to conduct the same experiments so that he could compare their results. These results, however, like those from earlier experiments on digestive physiology, inevitably varied, requiring the chief to make the interpretive choices inherent to his style of physiological research.

The changing nature of the scientific workforce seems also to have had consequences for the social-cognitive dynamics in Pavlov's laboratories. With the massive expansion of Soviet science, Pavlov now employed not just physiologically untrained physicians but, increasingly, coworkers who were themselves preparing for, and sometimes had already embarked upon, a career in physiology. Were there now fewer merely "skilled hands" and more Boldyrevs, Popel'skiis, Snarskiis, and Tolochinovs? It seems, indeed, that it became much more common for important questions to remain open in the laboratory, with the more frequent coexistence of different opinions—about, for example, the nature of inhibition—than had earlier been the case.

As in earlier years, some coworkers brought particular skills and perspectives that exercised a special influence on laboratory work. An important example in later years was Pavlov's lover and longtime coworker Maria Petrova, who convinced him to launch several new lines of investigation, including one in which chief and coworker collaborated closely to study the effect of castration on conditional reflexes. (Pavlov's attachment to Petrova also influenced his managerial style in a more mundane way, sometimes depriving other coworkers of his attention altogether.)

Soviet power challenged Pavlov's managerial traditions in a number of ways. For one thing, by the late 1920s his new workforce included a number of Communist Party members. Pavlov remained the unquestioned final authority in the laboratory, but the Communist cell created another interpretive center there. The relationship of Pavlov's research to broader ideological questions—to philosophical issues about dialectical materialism, free will, and the nature of cognition—gave Party members reason to question the interpretive authority of their bourgeois chief.[2] Communist coworkers sometimes met to debate the relationship between laboratory work and dialectical materialism, and on occasion a coworker urged their ideas on Pavlov. The broader politicization of science in Soviet society also generated challenges to Pavlov's authority, yet his special status allowed him to successfully resist purges, plans, and other Party policies that radically reduced the autonomy of other scientists in the 1930s.[3] Only after he had passed from the scene in 1936 did the politics of Stalinist Russia sweep unimpeded through his laboratories, condemning some of his most valued coworkers to the gulag.

Pavlov himself was always very proud of his laboratory operation and its many contributions to Russian science and world physiology; yet, in his last days he seems to have regretted one disadvantage of his managerial style. Ruminating about a possible successor, he lamented the fact that the "thinking people" who had passed through his laboratory had largely gone their own way, while those who remained in the fold lacked the qualities to replace him as chief.

Just as Pavlov preserved with minor variations the basic system of production in his laboratories, so did he maintain his basic approach to physiology. Research turned from digestive processes to "higher nervous activity," and the formerly idiosyncratic psyche now became the fully determined object of laboratory investigations. Yet Pavlov's scientific vision remained unchanged.

The later research, too, was conceived and approached as a form of organ physiology (the physiology of the brain and sensory organs), and at its heart lay Pavlov's adaptationist faith. It relied on chronic experiments with intact dogs fitted with a salivary fistula and proceeded simultaneously along clearly identifiable lines of investigation. As in the 1890s, these different lines of inquiry—for example, on the nature of inhibition, the laws of excitation, and the characteristics of various "nervous types"—influenced each other and provided Pavlov with a panoramic view of his subject. Just as Pavlov's earlier research was conducted with an eye toward the clinic, generating an "experimental pathology of digestion," medical advice, and gastric juice for sale, so did his later work feature an "experimental pathology of higher nervous processes," the diagnosis of psychiatric patients, and advice about treating them with caffeine and bromides. Just as research on digestive physiology was structured by a metaphor rooted in Russian political economy of the day—the digestive system as a complex chemical factory—so was research on higher nervous activity increasingly informed by a metaphorical relationship between excitation (which Pavlov equated with freedom) and inhibition (which he equated with discipline), a metaphor that acquired force and meaning from Pavlov's life experience, particularly his response to the Bolshevik seizure of power and its aftermath.[4]

The fundamental nature of the interpretive moments in Pavlov's style of "physiological thinking" also remained unchanged. As in earlier years, these were rooted in the dynamic tension at the heart of his scientific vision—in his determination both to encompass the complete animal whole *and* to attain the precise, repeatable, determined results that were the hallmark of any true science. In the years 1904–36 he pursued this same dual goal in investigations of the brain. This, of course, proved an immeasurably more difficult and complex task. Pavlov had required only about six years to convert laboratory experiments into a synthetic monograph on digestion; for his new subject he would need twenty-five years.

Even then, the chief confessed to his coworkers that he was tormented by a "beast of doubt": uncertainty about the success of his great quest. Confident, even aggressive, in public, Pavlov was constantly haunted by this "beast" through at least the mid-1920s. His longtime coworker Vladimir Savich observed in 1924 that the chief continually looked for basic reassurance in the results of new experiments: "Even many, many years later, when the new method had already clearly recommended itself, I. P. [Pavlov] would often say: 'Look, this new fact entirely justifies our approach, we could hardly be greatly mistaken.'"[5] In 1926, speaking privately to a group of coworkers at a celebration of the twenty-fifth anniversary of this research, Pavlov oscillated between self-congratulation and confession.

> I am unfortunately burdened by nature with two qualities. Perhaps they are both objectively good, but one of them is very burdensome for me. On the one hand, I am enthusiastic and surrender myself to my work with great passion; but together with this I am always weighed down by doubts. The smallest obstacle disturbs my balance and I am tortured until I find an explanation, until new facts bring me again into balance.

I must thank you for all your work, for the mass of collected facts—for having superbly subdued this beast of doubt. And now, when the book is appearing in which I give the conclusions of our 25 years of work—now, I hope, this beast will retreat from me. And my greatest gratitude for liberating me from torment is to you.[6]

Pavlov's "beast of doubt" was nourished by the very nature of the interpretive process in this research. This process was, as we have seen, essentially identical to that in his research on digestive physiology: in each case, the task was to discern the contours of a fully determined process amid a mass of highly variable data. In each case, he proceeded by invoking uncontrolled variables, hypothesized structures, and tentative laws to contain varied data within a framework that owed much to an intuited pattern.

In the later research, however, the distance between the data and the phenomena under investigation was considerably greater, and that space was bridged by piling interpretation upon interpretation. We must keep in mind that, for Pavlov, the conditional reflex comprised not so much a *discovery* as a *method* for experiments on higher nervous activity. In these experiments, Pavlov and his coworkers relied almost entirely on one source of data: saliva drops. As in the earlier research on digestive physiology, the amount of secretion was presumed to reflect with precision underlying physiological processes, and all good experimental data were presumed to be explicable in terms of Bernardian determinism.

Using the quantitative patterns of salivation during tens of thousands of experiments, Pavlov built an edifice of interpretive judgments about higher nervous processes, constructing a complex picture of brain activity that included excitation, inhibition, the inhibition of inhibition, generalization, radiation, various "nervous types," and so forth. However ingenious the experimental design and conceptual apparatus, this interpretive process offered very little firm ground, certainly nothing even remotely resembling the satisfyingly crisp conclusions to Claude Bernard's experimental trials as described in his *Introduction to the Study of Experimental Medicine*.

Pavlov uttered his only recorded public allusion to his private "beast of doubt" in a short speech on accepting an honorary doctorate from the University of Paris in 1925. It is not surprising that he flattered his French audience by praising a French physiologist, but it is, I think, revealing that he acknowledged his debt to Bernard specifically for the moral support offered by his writings in dealing with the traumas of the interpretive moment. Pronouncing Bernard "the original inspiration of my physiological activity" and praising "the force and compelling clarity of his thought," Pavlov confided: "The experiments so masterfully conducted by Claude Bernard served as an example for me during my works on the determination of biological phenomena, and directed and supported me in the hours of often painful doubts during my recent investigations of the physiology of the large hemispheres. In this field, still an unclear one for physiology, there is a great variety and complexity of the conditions in which nervous activity . . . takes place, and therefore the search for and determination of the active agents can here elicit justifiable anxiety."[7]

Pavlov's research on higher nervous activity, like his earlier work on digestion, gained broader symbolic significance from the tension at the heart of his scientific vision. In the years 1891–1904 his effort both to encompass the complex animal whole and to attain precise, repeatable, *pravil'nye* results had helped make him a symbol of a modern, clinically relevant, experimental physiology. In the years 1904–36 that same effort, now directed at the organ that most defines what it means to be human, made him an icon of twentieth-century science. Pavlov's status transcended that of a mere Nobelist. He and his salivating dogs entered the common culture of the twentieth century as a symbol of the hope—and, for many, the fear—that science might enable us to understand and perhaps even to control human nature.

APPENDIXES

Appendix A. Praktikanty and the Knowledge Claims in Lectures

This chart provides some indication of the great range of knowledge claims in *Lectures* and of the dependence of Pavlov's work on the experiments conducted by his praktikanty. Page numbers are from the first English edition of *Lectures* (W. H. Thompson, *The Work of the Digestive Glands* [London: Charles Griffin and Co., 1902]) and refer only to pages on which praktikanty are mentioned by name. This, of course, understates the dependence of *Lectures* on their experiments.

PRAKTIKANT (YEAR OF DISSERTATION/ PUBLICATION)	PAGE(S) IN LECTURES	RESEARCH TECHNIQUE, TOPIC, OR RESULT
Lecture 1: A General Overview of the Subject—Methodology		
Kuvshinskii (1888)[a]	6	Techniques for constructing a pancreatic fistula
Vasil'ev (1893)	7	Proper diet for dogs with a pancreatic fistula
Iablonskii (1894)	7	Proper diet for dogs with a pancreatic fistula
Shumova-Simanovskaia (unpublished)[b]	9	Technique for constructing an esophagotomy
Konovalov (1893)	11	Qualities of natural gastric juice
Khizhin (1894)	12	Creation and qualities of isolated sac

PRAKTIKANT (YEAR OF DISSERTATION/ PUBLICATION)	PAGE(S) IN LECTURES	RESEARCH TECHNIQUE, TOPIC, OR RESULT
Lecture 2: The Work of the Glands during Digestion		
Khizhin (1894)	21	The amount of gastric secretion is proportional to the quantity of the ingested food
	22	Under identical experimental conditions, the course of gastric secretion is identical from one trial to another
	33, 35, 37	Meat, bread, and milk elicit characteristic gastric secretory responses in terms of amount and proteolytic power
Val'ter (1896)[c]	27	Technique for estimating strength of proteolytic and amylolytic pancreatic ferments
	29, 38, 40	Meat, bread, and milk elicit characteristic patterns of secretion of the pancreas's fat-splitting, amylolytic, and tryptic ferments
Glinskii (1895)	27	Technique for measuring proteolytic power
Ketcher (1890)	31	Relationship of acidity and volume in gastric secretion
Vasil'ev (1893)	41	Pancreatic secretion adapts to a long-term change in diet
Iablonskii (1894)	41	Pancreatic secretion adapts to a long-term change in diet
Vasil'ev (1893)	41	Pancreatic secretion adapts to a long-term change in diet

PRAKTIKANT (YEAR OF DISSERTATION/ PUBLICATION)	PAGE(S) IN LECTURES	RESEARCH TECHNIQUE, TOPIC, OR RESULT

Lecture 3: The Efferent Nerves of the Gastric and Pancreatic Glands

Iurgens (1892)	51	Vagus control of gastric secretion
Ushakov (1894)	54	Vagus control of gastric secretion
Shumova-Simanovskaia (unpublished)	51, 54	Important role of appetite as evident in sham-feeding experiments
Sanotskii (1892)	55	Vagus control of gastric secretion
Kudrevetskii (1890)	59	Sympathetic nervous control of salivary secretion
Popel'skii (1896)	60	Vagus control of pancreatic secretion Sympathetic nervous inhibition of pancreatic secretion

Lecture 4: A General Scheme of the Complete Innervative Apparatus . . .

Glinskii (1895)	66–68	Technique for constructing salivary fistulas Salivary response varies according to moisture of food
Vul'fson (1897)	69	Response of parotid salivary gland to meat and bread
Sanotskii (1892)	72	Teasing excites gastric secretion

Lecture 5: The Place and Significance of Psychic (or Appetite) Juice in the Entire Work of the Stomach . . .

Khizhin (1894)	77	Appetite as first exciter of gastric secretion
Kotliar (unpublished)	78	Psychic secretion increases amount and proteolytic power of gastric secretion

PRAKTIKANT (YEAR OF DISSERTATION/ PUBLICATION)	PAGE(S) IN LECTURES	RESEARCH TECHNIQUE, TOPIC, OR RESULT
Lobasov (1896)	78–83	Psychic secretion in large stomach and isolated sac is identical No secretory effect of mechanical stimulation of stomach Total gastric secretion equals psychic secretion plus nervous-chemical secretion
Sanotskii (1892)	84	Psychic secretion does not account for all gastric secretion in digestive act

Lecture 6: The Chemical Exciters of the Innervative Apparatus of the Gastric Glands . . .

Khizhin (1894)	94	Water excites gastric secretion
Sanotskii (1892)	94	Water excites gastric secretion
Iurgens (1892)	94	Psychic secretion depends on vagus nerves
Khizhin (1894)	95	Meat ash, chloride of sodium, and hydrochloric acid do not excite gastric secretion; sodium bicarbonate inhibits gastric secretion
Khizhin (1894)	96	Egg white in the stomach does not excite gastric secretion
Riazantsev (1896)	96	Egg white in the stomach does not excite gastric secretion
Lobasov (1896)	97	Peptone does not excite gastric secretion (this corrects error by Pavlov and Khizhin, as explained on p. 96 of *Lectures*) Liebig's meat extract excites gastric secretion
Khizhin (1894)	97	Starch and fat do not excite gastric secretion

PRAKTIKANT (YEAR OF DISSERTATION/ PUBLICATION)	PAGE(S) IN LECTURES	RESEARCH TECHNIQUE, TOPIC, OR RESULT
Lobasov (1896)	99–106	Psychic juice converts bread and egg albumin into secretory exciter more powerful than those foods Starch heightens the secretory response to meat proteins Fat lessens the amount and proteolytic power of the secretory response to meat Fat lessens the amount and proteolytic power of the secretory response to milk
Khizhin (1894)	101–6	Psychic juice converts egg albumin into secretory exciter more powerful than this food—it is an "igniter" Moistening of meat and desiccation of bread do not influence gastric secretory response to these substances Fat lessens the secretory response to meat The Pavlov isolated sac mirrors the activity of the larger stomach

Lecture 7: Normal Exciters of the Innervative Apparatus of the Pancreatic Gland . . .

Bekker (1893)	113	Salts and alkalies have opposite effect on pancreatic secretion
Dolinskii (1894)	113	Acids excite pancreatic secretion
Shirokikh (1894)	115	Pepper and mustard have no effect on pancreatic secretion
Popel'skii (1896)	117	The hydrochloric acid in gastric secretion excites pancreatic secretion from the intestine, not from the stomach

PRAKTIKANT (YEAR OF DISSERTATION/ PUBLICATION)	PAGE(S) IN LECTURES	RESEARCH TECHNIQUE, TOPIC, OR RESULT
	118	This excitation (listed above) is reflexive in nature
Val'ter (1896)	119	The volume of pancreatic secretion is proportional to its acidity
	120	Starch may exercise an influence on the trophic component of pancreatic secretion
Dolinskii (1894)	121	Fats excite pancreatic secretion
Damaskin (1896)	121	Fats excite pancreatic secretion
	125	Water excites pancreatic secretion
Kuvshinskii (1888)	124	Psychic excitation results in pancreatic secretion (although this may be the indirect result of gastric secretion)
Bekker (1893)	126	Alkaline salts inhibit pancreatic secretion

Lecture 8: Physiological Data, Human Instinct, and Medical Empiricism

Riazantsev (1896)	143	Milk is an efficient source of nitrogen for the body
Val'ter (1896)	144	Fibrin is a powerful exciter of pancreatic secretion
Khizhin (1894)	147	Alkalies do not excite gastric secretion

[a]Kuvshinskii conducted his research in the Botkin laboratory, but he chose his subject at Pavlov's suggestion and pursued it under Pavlov's guidance.

[b]Pavlov and Shumova-Simanovskaia collaborated in 1889-90, just before Pavlov's appointment at the Institute.

[c]Pavlov cites Val'ter's 1896 report to the Society of Russian Physicians, since Val'ter's thesis was not completed when *Lectures* went to press.

Appendix B. Pavlov's Choices among Khizhin's Characteristic Secretory Curves

Pavlov had at his disposal data for the gastric secretion elicited by one quantity of bread, one quantity of milk, and three quantities of meat (a limitation due to experimental difficulties arising from Druzhok's dislike of milk and avidity for bread; see Chapters 4 and 5). In his thesis, Khizhin presented characteristic secretory curves for all three quantities of meat (Figure B.1). In *Lectures,* Pavlov made an interesting choice among Khizhin's curves, enshrining that for 200 grams as the "meat curve" (Figure B.2).

Pavlov freely conceded that he did not know what specific substances in meat (or any other foodstuff) excited nervous-chemical secretion, so he had some latitude in choosing the quantity of meat to compare with the data for bread and milk. In his discussion of pancreatic secretion, however, he justified the comparison of 100 grams of meat with 250 grams of bread and 600 cubic centimeters of milk by their equivalent nitrogen content.[1] Val'ter discussed the choice of quantities at length and provided calculations for nitrogenous content that justified the comparison of these three quantities. A second reason for choosing these quantities, Val'ter explained, was that they permitted an easy comparison with Khizhin's results for the gastric glands.[2] Why, then, did Pavlov, in *Lectures,* use a characteristic secretory curve constructed on the data for *200* grams rather than 100 grams of meat?

There are, I think, two plausible explanations, each of which touches on interesting aspects of Pavlov's presentation of data.

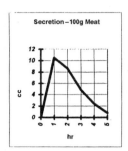

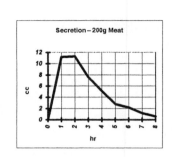

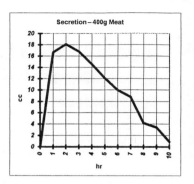

FIGURE B.1. *Khizhin's three curves (see fig. 13a–c) for the amount of gastric secretion with a meal of various amounts of meat. From P. P. Khizhin,* Otdelitel'naia rabota zheludka sobaki, Military-Medical Academy Doctoral Dissertation Series *(St. Petersburg, 1894), appendix*

First, Pavlov sometimes used average data and sometimes used the results of individual trials as an "ideal type." We have seen that in the first set of curves presented in *Lectures,* Pavlov demonstrated the precise stereotypicity of gastric secretion by using the data from two experiments with 100 grams of meat. To do so, he chose the two experiments of five with results that most closely resembled each other. These results, however, did not correspond to the average data for these five trials. Had he used the average data from trials with 100 grams of meat to make his characteristic "meat curve," this would have presented the reader with two different curves—each ostensibly stereotypical and

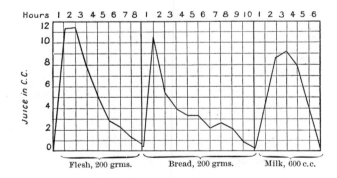

FIGURE B.2. *Pavlov's characteristic curve (fig. 22) for the amount of gastric secretion with a meal of meat, bread, and milk. I. P. Pavlov,* Lektsii o rabote glavnykh pishchevaritel'nykh zhelez *(1897), in* Polnoe sobranie sochinenii *(Moscow: USSR Academy of Sciences, 1951), 2, pt. 2: 57, 64; W. H. Thompson,* The Work of the Digestive Glands *(London: Charles Griffin and Co., 1902).*

precise—for trials with 100 grams of meat. As we have seen, this is precisely what occurred in Pavlov's portrayal of the characteristic curve for the pancreatic response to milk. (For that gland, however, Pavlov had at his disposal data for only one quantity of each foodstuff, so he could not avoid this problem.) By using the results of two concrete trials with 100 grams of meat to demonstrate the "astonishing precision" of the glands, and the average results of trials with 200 grams of meat to construct his characteristic curve, Pavlov avoided having to present two different curves again.

The second possible explanation is that Pavlov based his meat curve on results with 200 grams because the curve for 100 grams too closely resembled the bread curve. As we saw in Chapter 5, there was an inherently subjective dimension to judgments about the relative similarity or difference between two curves, and such judgments changed with the emergence and hardening of the factory metaphor in 1894–97. For Pavlov in 1897, the fundamental difference between the meat curve and the bread curve was that the bread curve peaked sharply in the first hour and was of relatively long duration. Had he used Khizhin's curve for 100 grams of meat, this latter difference would have disappeared: both the bread and meat curves would have showed a sharp peak in the first hour. So, why not use Khizhin's curve for 400 grams of meat? First, according to Pavlov's calculations, the nitrogenous content of 400 grams of meat was substantially larger than that for the quantities of bread and milk that he used for his characteristic curves. Second, the total amount of secretion elicited by 400 grams of meat was so much higher than that for these other substances that it dwarfed any other differences. Finally, the shape of this meat curve would then acquire some close resemblances to that of the milk curve. As in all cases of such choices, Pavlov had a ready scientific explanation for the problems that he was dealing with rhetorically. The differences among the curves elicited by differing quantities of meat were easily explicable. Meat always elicited a strong psychic secretion and so a sharp rise in secretion in the first hour after feeding. Larger quantities of meat, however, contained more extractive substances, and so generated a greater secretion in the second, nervous-chemical phase. This drove up the quantity (and proteolytic power) of secretion from the second hour onwards, changing the shape of the secretory curve.

Notes

1. I. P. Pavlov, *Lektsii o rabote glavnykh pishchevaritel'nykh zhelez*, in *Polnoe sobranie sochinenii* (Moscow: USSR Academy of Sciences, 1951) 2, pt. 2: 60; W. H.

Thompson, *The Work of the Digestive Glands* (London: Charles Griffin and Co., 1902), 39; and Val'ter's and Pavlov's comments at the Society of Russian Physicians, recorded in A. A. Val'ter, "Rabota podzheludochnoi zhelezy pri miase, khlebe, moloke i pri vlivanii kisloty," *Trudy obshchestva russkikh vrachei,* September 1896, pp. 38, 39.

    2. A. A. Val'ter, *Otdelitel'naia rabota podzheludochnoi zhelezy* (Military-Medical Academy Doctoral Dissertation Series, St. Petersburg, 1897), 40–41.

NOTES

Works frequently cited in the notes are identified by the following abbreviations:

Academy diss.   Dissertation in the Military-Medical Academy Doctoral
Dissertation Series, St. Petersburg

Babkin, *Pavlov*   Boris Babkin's manuscript biography of Pavlov, held at the
(MS)   archive of McGill University (a longer version of Babkin's
Pavlov: A Biography [Chicago: University of Chicago Press,
1949]); the manuscript is paginated separately for each chapter

Heidenhain,   R. Heidenhain [Geidengain], *Rukovodstvo k fiziologii*, vol. 5,
Fiziologiia   pt. 1 of *Fiziologiia otdelitel'nykh protsessov* (St. Petersburg,
1886)

Pavlov, *Lectures*   Authorized English translation of Pavlov's *Lektsii o rabote*
(trans.)   *glavnykh pishchevaritel'nykh zhelez* by W. H. Thompson, *The*
Work of the Digestive Glands (London: Charles Griffin and
Co., 1902)

Pavlov, *Lektsii*   Ivan Pavlov, *Lektsii o rabote glavnykh pishchevaritel'nykh zhelez*
(Lectures on the Work of the Main Digestive Glands) [1897],
in *PSS*, 2, pt. 2

PSS   I. P. Pavlov, *Polnoe sobranie sochinenii*, 6 vols. (Moscow: USSR
Academy of Sciences, 1951–52)

RGIA   Rossiiskii Gosudarstvennyi Istoricheskii Arkhiv, St. Petersburg

SPF ARAN   Sankt-Peterburgskii filial Arkhiva Rossiiskoi Akademii Nauk,
St. Petersburg

TsGIA SPb   Tsentral'nyi Gosudarstvennyi Istoricheskii Arkhiv Sankt-
Peterburga, St. Petersburg

The location of Russian archival materials is identified as, for example, SPF ARAN 259.1a.3: 12, meaning the item is located in SPF ARAN, collection (*fond*) 259, inventory (*opis'*) 1a, file (*delo*) 3, page (*list*) 12. Where pages of archival papers are numbered only on the front side, the reverse side is indicated by *ob;* for example, 54ob means the reverse side ("obverse") of page (*list*) 54.

The journal *Vrach* was renamed *Russkii Vrach* in 1901.

All English translations are my own, unless otherwise noted.

Introduction

1. This was the physiologist Robert Tigerstedt's characterization of the "totally incorrect" opinion of some members of the Nobel Prize Committee. It appears in his memo of September 8, 1901, "P. M. angaende prof. J. P. Pawlowa arbeten," in P. M. Försändelser och Betänkanden 1901, Nobel Archives, Karolinska Institute, Stockholm.

2. Among the many works that treat science as social production are Bruno Latour and Steve Woolgar, *Laboratory Life: The Social Construction of Scientific Facts* (Beverly Hills: Sage, 1979); Karin Knorr-Cetina, *The Manufacture of Knowledge: An Essay on the Constructivist and Contextual Nature of Science* (Oxford: Pergamon Press, 1981); Steven Shapin and Simon Schaffer, *Leviathan and the Air-Pump: Hobbes, Boyle, and the Experimental Life* (Princeton: Princeton University Press, 1985); Adele Clarke and Joan Fujimura, eds., *The Right Tools for the Job: At Work in Twentieth-Century Life Sciences* (Princeton: Princeton University Press, 1992); Steven Shapin, *A Social History of Truth: Civility and Science in Seventeenth-Century England* (Chicago: University of Chicago Press, 1994); and Robert Kohler, *Lords of the Fly: Drosophila Genetics and the Experimental Life* (Chicago: University of Chicago Press, 1994). I read Kohler's splendid book closely while thinking about Pavlov's laboratory and acknowledge the benefit of his many insights. Shapin provides a close analysis of the division of labor and work process in Boyle's workshop laboratory in *Social History of Truth*, 355–407.

3. For example, paging through *Nature*, one finds the following (this is only a partial list): "The Schorlemmer Memorial Laboratory," 52 (May 16, 1895): 63–64; Alfred Earl, "Tonbridge School Laboratories," 52 (May 23, 1895): 88–89; William H. Welch, "The Evolution of Modern Scientific Laboratories," 54 (May 28, 1896): 87–90; "The Davy-Faraday Research Laboratory," 54 (July 2, 1896): 200–201; "The New Research Laboratory of the Royal College of Physicians of Edinburgh," 55 (November 26, 1896): 88–89; "The New Laboratories at Guy's Hospital," 56 (June 3, 1897): 105–6; "The New Physical Laboratory of the Owens College, Manchester," 58 (October 27, 1898): 621–22; "National Physical Laboratory," 63 (January 24, 1901): 300–302; "Some Scientific Centres: The Leipzig Chemical Laboratory," 64 (June 6, 1901): 127–29; [F.H.N.] "Some Scientific Centres: The Laboratory of Wilhelm Ostwald," 64 (August 29, 1901): 428–30; "Some Scientific Centres: The Lab-

oratory of Henri Moissan," 55 (January 16, 1902): 252–54; "Some Scientific Centres: The Chemical Laboratory of the Royal Institutions," 66 (September 11, 1902): 460–62; "Opening of the Johnston Laboratories for Medical Research in the University College, Liverpool," 68 (May 14, 1903): 43–44; "Some Scientific Centres: VI. The Cavendish Laboratory," 69 (December 10, 1903): 128–32; "New Buildings of the University of Liverpool," 71 (November 17, 1904): 63–65; and M. Foster, "Some Scientific Centres: VII. The Physiological Research Laboratory of the University of London," 71 (March 9, 1905): 441–44.

4. Welch, "Evolution of Modern Scientific Laboratories," 88.

5. For the differing scientific-managerial styles in six large chemistry laboratories, see Joseph Fruton, *Contrasts in Scientific Style: Research Groups in the Chemical and Biochemical Sciences* (Philadelphia: American Philosophical Society, 1990). No similar work exists for large physiological laboratories, but reminiscences and the observations of historians point to important differences among them. See, for example, Gerald L. Geison, *Michael Foster and the Cambridge School of Physiology: The Scientific Enterprise in Late Victorian Society* (Princeton: Princeton University Press, 1978); Frederic L. Holmes, "The Formation of the Munich School of Metabolism," in *The Investigative Enterprise: Experimental Physiology in Nineteenth-Century Medicine*, ed. William Coleman and Frederic L. Holmes (Berkeley: University of California Press, 1988), 179–210; Simon Flexner and James Thomas Flexner, *William Henry Welch and the Heroic Age of American Medicine* (Baltimore: Johns Hopkins University Press, 1993), 84–86; Robert Frank Jr., "American Physiologists in German Laboratories, 1865–1914," in *Physiology in the American Context, 1850–1940*, ed. Gerald Geison (Bethesda, Md.: American Philosophical Society, 1987), 11–46, esp. 27–38; Merriley Borell, "Instruments and an Independent Physiology: The Harvard Physiological Laboratory, 1871–1906," in Geison, *Physiology in the American Context*, 293–321; and E. M. Tansey, "The Wellcome Physiological Research Laboratories 1894–1904: The Home Office, Pharmaceutical Firms, and Animal Experiments," *Medical History* 33 (1989): 1–41.

For the contrasting styles of Pavlov and his longtime rival, neurologist V. M. Bekhterev, see Boris Babkin, *Pavlov: A Biography* (Chicago: University of Chicago Press, 1949), 67–76, 80–82, 115–29. Babkin described these laboratories more fully in his original manuscript (*Pavlov* [MS]). Robert Frank Jr. of the University of California, Los Angeles, is currently writing a book on biomedical laboratories in Germany and the United States, and I am indebted to him for sharing his extensive knowledge of this subject in several conversations.

The comparison of laboratory and factory has a long history. For two quite different reflections, see Max Weber, "Science as a Vocation" [1919], in *From Max Weber: Essays in Sociology*, trans. and ed. H. H. Gerth and C. Wright Mills (New York: Oxford University Press, 1946), 135; and Bruno Latour, "The Costly Ghastly Kitchen," in *The Laboratory Revolution in Medicine*, ed. Andrew Cunningham and Perry Williams (Cambridge: Cambridge University Press, 1992), 295–303, esp. 299.

My thanks to Keith Barbera for bringing Weber's comments to my attention. W. Bruce Fye was, to my knowledge, the first to refer to a physiological laboratory as a *factory,* in his "Carl Ludwig and the Leipzig Physiological Institute: 'A Factory of New Knowledge,'" *Circulation* 74, no. 5 (1986): 923.

6. My comments about Voit are based on Holmes, "Formation of the Munich School"; those about Foster are based on Geison, *Michael Foster.* For information on Ludwig, see Robert Frank Jr., "American Physiologists in German Laboratories, 1865–1914," in Geison, *Physiology in the American Context,* 11–46, esp. 27–38; Heinz Schröer, *Carl Ludwig* (Stuttgart: Wissenschaftliche Verlagsgesellschaft, 1967); Timothy Lenoir, "Science for the Clinic: Science Policy and the Formation of Carl Ludwig's Institute in Leipzig," in Coleman and Holmes, *Investigative Enterprise,* 139–78; Carl Rothschuh, *History of Physiology,* ed. and trans. Guenter B. Risse (Huntington, N.Y.: R. E. Krieger, 1973); and Warren Lombard, "Life and Work of Carl Ludwig," *Science,* September 15, 1916, pp. 363–75.

7. Raphael Samuel, "Workshop of the World: Steam Power and Hand Technology in Mid-Victorian Britain," *History Workshop* 3 (spring 1977): 52. My thanks to Harry Marks for drawing my attention to this article. The standard text on the emergence of the factory system remains David S. Landes, *The Unbound Prometheus: Technological Change and Industrial Development in Western Europe from 1750 to the Present* (Cambridge: Cambridge University Press, 1969). For an introduction to this same revolution in science, see Robert W. Smith, "Large-Scale Scientific Enterprise," in *Encyclopedia of the United States in the Twentieth Century,* ed. Stanley Kutler (New York: Charles Scribner's Sons, 1996), 2: 739–65.

8. The first article in this series was A. Gray, "Famous Scientific Workshops: I. Lord Kelvin's Laboratory in the University of Glasgow," *Nature* 55 (March 25, 1897): 486–92.

9. On the dynamics of experiment, see, for example, Frederic L. Holmes's fine-grained studies of three scientists and his discussions of scientific creativity in *Claude Bernard and Animal Chemistry: The Emergence of a Scientist* (Cambridge, Mass.: Harvard University Press, 1974), *Lavoisier and the Chemistry of Life: An Exploration of Scientific Creativity* (Madison: University of Wisconsin Press, 1985), and *Hans Krebs,* 2 vols. (New York: Oxford University Press, 1991, 1993). See also Gerald Geison, *The Private Science of Louis Pasteur* (Princeton: Princeton University Press, 1995); Gerald Holton, "Subelectrons, Presuppositions, and the Millikan-Ehrenhaft Dispute," in *The Scientific Imagination: Case Studies* (Cambridge: Cambridge University Press, 1978), 25–83; François Jacob, *The Statue Within: An Autobiography* (New York: Basic Books, 1988); Peter Medawar, *Memoir of a Thinking Radish: An Autobiography* (Oxford: Oxford University Press, 1986); Hans-Jorg Rheinberger, *Toward a History of Epistemic Things: Synthesizing Proteins in the Test Tube* (Stanford: Stanford University Press, 1997); Peter Galison, *Image and Logic: A Material Culture of Microphysics* (Chicago: University of Chicago Press, 1997); Lawrence Principe, *The Aspiring Adept: Robert Boyle and His Alchemical Quest*

(Princeton: Princeton University Press, 1998); Shapin and Schaffer, *Leviathan and the Air-Pump;* Latour and Woolgar, *Laboratory Life;* Kohler, *Lords of the Fly;* and Andrew Pickering, *The Mangle of Practice: Time, Agency, and Science* (Chicago: University of Chicago Press, 1995).

10. Karl Marx, *Capital,* vol. 1 [1867], online transcription of the English edition of 1887, available from http://csf.Colorado.edu/cgi-bin/mfs/28/csf/web/psn/marx/Archive/1867-C1/Part3/ch07.htm?79. With respect to factory production, the full passage is as follows: "Intelligence in production expands in only one direction, because it vanishes in many others. What is lost by the detail labourer is concentrated in the capital that employs them and the labourer is brought face-to-face with the intellectual potencies of the material process of production as the property of another, as a ruling power . . . The separation of the intellectual powers of production from the manual labour and the conversion of these powers into the might of capital over labour, is . . . finally completed by modern industry erected on the foundation of machinery." From http://csf.Colorado.edu/cgi-bin/mfs/28/csf/web/psn/marx/Archive/1867-C1/Part4/ch14.htm?908.

11. Weber, "Science as a Vocation," 135–36. According to Gerth and Wright Mills, this article originated as a speech in 1918 and was first published in 1919.

Chapter 1. The Prince and His Palace

1. Contemporary data indicated that European Russia regularly led Europe in infant mortality (roughly 50% mortality for children under five) and in the incidence of almost every serious infectious disease (e.g., smallpox, scarlatina, diphtheria, typhus). G. Khopin and F. Erisman, "Meditsina i narodnoe zdravie v Rossii," in *Rossiia* [1898], ed. F. A. Brokgauz and I. A. Efron (Leningrad: Lenizdat, 1991), 224–26.

2. On the early history of the Imperial Institute of Experimental Medicine, see A. P. Salomon, "Imperatorskii Institut Eksperimental'noi Meditsiny v S.-Peterburge," *Arkhiv Biologicheskikh Nauk* 1 (1892): 3–22; N. A. Kharauzov, "K istorii Instituta Eksperimental'noi Meditsiny AMN SSSR," in *Ezhegodnik Instituta Eksperimental'noi Meditsiny za 1956* (Leningrad: Akademiia Meditsinskikh Nauk, 1957), 625–28; N. M. Gureeva, "K istorii organizatsii Instituta Eksperimental'noi Meditsiny," in *Ezhegodnik Instituta Eksperimental'noi Meditsiny za 1956,* 628–37; N. A. Chebysheva "Nauchno-organizatsionnaia rol' I. P. Pavlova v Institute Eksperimental'noi Meditsiny v 1891–1916 gg. (po materialam arkhiva IEM)," in *Ezhegodnik Instituta Eksperimental'noi Meditsiny za 1956,* 637–51; and Iu. P. Golikov and K. A. Lange, "Stanovlenie pervogo v Rossii issledovatel'skogo uchrezhdeniia v oblasti biologii i meditsiny," in *Pervyi v Rossii issledovatel'skii tsentr v oblasti biologii i meditsiny* (Leningrad: Nauka, 1990), 7–75.

3. P. Kh. Grebel'skii and A. B. Mirvis, *Dom Romanovykh,* 2d ed. (St. Petersburg, 1992), 224.

4. See Adele Lindenmeyr, *Poverty Is Not a Vice: Charity, Society, and the State in Imperial Russia* (Princeton: Princeton University Press, 1996). Note that Russian family names have male and female forms. So, it is Ivan *Pavlov* and Prince *Ol'denburgskii*, but Serafima *Pavlova* and Princess *Ol'denburgskaia*.

5. A grand duchess (*velikaia kniagina*) was a woman married to a son of the royal family who was not a tsar but merely a grand duke (*velikii kniaz'*). Elena Pavlovna's husband was the Grand Duke Mikhail, son of Tsar Paul I.

6. On the Sisters of Mercy and Russian Red Cross, see John F. Hutchinson, *Champions of Charity: War and the Rise of the Red Cross* (Boulder, Colo.: Westview Press, 1996).

7. E. A. Annenkova and Iu. P. Golikov, *Russkie Ol'denburgskie i ikh dvortsy* (St. Petersburg: Almaz, 1997), 130.

8. P. P. Khizhin, *Istoricheskii ocherk deiatel'nosti Ramonskoi lechebnitsy ee Imperatorskoi Vysochestva Printsessa E. M. Ol'denburgskoi so vremeni osnovaniia lechebnitsy* (Voronezh, 1893).

9. *Novoe Vremia*, January 7 [January 19], 1893, no. 6056.

10. Babkin, *Pavlov* (MS), chap. 7, p. 2.

11. See *Novoe Vremia*, January 7 [January 19], 1893, no. 6056, and January 8 [January 20], 1893, no. 6057: 3. On the Prince's honorary citizenship, see *Novosti i Birzhevaia gazeta*, January 16, 1904, p. 3.

12. S. Iu. Vitte [Witte], *Vospominaniia* (Moscow, 1960), 2: 564.

13. Ibid.

14. A. A. Mosolov, *Pri dvore poslednego imperatora: zapiski nachal'nika kantseliarii ministra dvora* [1937] (St. Petersburg: Nauka, 1992), 145.

15. Babkin, *Pavlov* (MS), chap. 7, p. 5.

16. Adrien Loir, *Le Prince Alexandre D'Oldenbourg, initiator de l'alliance Franco-Russe, et ses relations avec Pasteur* (Le Havre: Havre Eclair, 1933), 8.

17. See *Vrach*, 1901, no. 45: 1369–70; and TsGIA SPb 2282.1.145: 590b. The Institute's "light treatment station," Russia's first, was based on the research and techniques of Nils Finsen, who won the Nobel Prize in Physiology or Medicine in 1903 (see my Chapter 10). Established in 1899, the station offered treatment for a variety of ailments, including skin cancer. Prince Ol'denburgskii's vibrating table was tested in the clinic of neurologist-psychiatrist V. M. Bekhterev (who hoped to open a psychoneurological section at the Prince's Institute), where it was credited with a "positive therapeutic effect on patients with various nervous problems, both functional and organic, and also on one patient from the psychiatric ward." See *Obozrenie psikhiatrii, nevrologii, i eksperimental'noi psikhologii* 1 (1896): 122.

18. Gerald L. Geison devotes two fine chapters to Pasteur's development of a rabies vaccine in his *The Private Science of Louis Pasteur* (Princeton: Princeton University Press, 1995). For a different account, see Patrice Debré, *Louis Pasteur,* trans. Elborg Forster (Baltimore: Johns Hopkins University Press, 1998).

19. V. G. Ushakov, "Antirabicheskoe otdelenie," in *Materialy k istorii Vserossi-iskogo Instituta Eksperimental'noi Meditsiny* (Moscow: Medgiz, 1941), 112.

20. From N. A. Kruglevskii's account in his letter of February 28, 1886, in *Vrach,* 1886, no. 10: 193. For the logic behind such serial passages, see Geison, *Private Science of Louis Pasteur,* 184–85.

21. Ibid.; N. S. Stolygo, "K istorii otkrytiia pervykh pasterovskikh stantsii v Rossii," in *Iz istorii meditsiny* (Riga, 1959), 2: 165–70; Ushakov, "Antirabicheskoe otdelenie," 113. According to Kruglevskii, on February 13, 1886, he and Gel'man injected a rabbit with a rabies emulsion that had been passed through five generations of rabbits. The rabbit took sick ten days later and died on February 25.

22. *Vrach,* 1886, no. 10: 195.

23. Ibid. The fact that *Vrach* mentioned only Ol'denburgskii's rabies station here (others had been established almost simultaneously in Moscow, Odessa, and other cities) perhaps reflects the Prince's influence. For a brief account of the Odessa station, see John Hutchinson, "Tsarist Russia and the Bacteriological Revolution," *Journal of the History of Medicine and Allied Sciences* 40, no. 4 (1985): 423–24.

24. Two physicians from Khar'kov canceled their trip due to Pasteur's "dubious reply." See RGIA 733.149.863, as cited in the notes of V. L. Merkulov, held in the Fundamental'naia Biblioteka Voenno-Meditsinskoi Akademii. N. F. Gamaleia, of the Odessa bacteriological station, took with him to Paris nineteen Russians bitten by wolves. On Gamaleia's trip, see Ushakov, "Antirabicheskoe otdelenie," 114. *Vrach* reported on April 3, 1886, that Pasteur had so far treated 493 French patients, 32 Algerian, 20 English, 19 Italian, 18 Russian, 13 Austro-Hungarian, 8 North American, 7 Belgian, 4 German, 4 Spanish, 1 Swiss, 1 Greek, and 1 Brazilian. On April 10 it reported that Russians were now the largest foreign contingent: since the beginning of the month, 67 Russians bitten by dogs and 38 bitten by wolves had made their way to Pasteur.

25. Loir brought another inoculated rabbit to St. Petersburg, performed further inoculations there, and worked closely with Gel'man at the anti-rabies station. For his account of this trip, see Loir, *Le Prince Alexandre D'Oldenbourg,* 2–13. Perdrix's papers at the archive of the Pasteur Institute include a file "Mission en Russie" with some correspondence from his trip. In late July, Perdrix informed Pasteur that the St. Petersburg anti-rabies station had begun treating patients. On Pasteur's instructions, Perdrix then traveled to the Ol'denburgskii estate in Ramon, where, together with Khizhin, he examined cattle ill with anthrax and inoculated seven animals against the disease. He then inspected an anti-rabies facility in Moscow and paid a final visit to St. Petersburg before departing for Paris in September. See also Stolygo, "K istorii," 168; Ushakov, "Antirabicheskoe otdelenie," 114, 119–20; V. I. Likhachev, *S.-Peterburgskaia stantsiia predupreditel'nogo lecheniia vodoboiazni po sposobu Pastera* (St. Petersburg, 1887); and Golikov and Lange, "Stanovlenie," 10–11.

26. Pasteur referred to these anti-rabies facilities as *instituts Pasteur* in his letter to Dr. Chautemps of July 22, 1886. See Pasteur Vallery-Radot, ed., *Correspondance de Pasteur, 1840–1895* (Paris: Flammarion, 1951), 4: 76. Pasteur used the existence of these facilities to defend himself against charges that he was secretive about his rabies treatment.

27. Gel'man and Shperk also traveled to Ramon in 1887 to open an anti-rabies station and a syphilis division at the hospital on the Ol'denburgskii estate. The anti-rabies station remained open for only six months owing to the dearth of patients. See Khizhin, *Istoricheskii ocherk*, 23–24.

28. On Mechnikov's life and thought, see Daniel P. Todes, "Mechnikov, Darwinism, and the Phagocytic Theory," in *Darwin without Malthus: The Struggle for Existence in Russian Evolutionary Thought* (New York: Oxford University Press, 1989), 82–103; and Alfred Tauber and Leon Chernyak, *Metchnikoff and the Origins of Immunology: From Metaphor to Theory* (New York: Oxford University Press, 1991).

29. Gamaleia's letter is cited according to Kharauzov, "K istorii Instituta," 631. Mechnikov discussed this proposal and his rejection of it in "Rasskaz o tom, kak i pochemu ia poselilsia za granitsei" [1909], reprinted in *Stranitsy vospominanii* (Moscow: Akademiia Nauk SSSR, 1946). For Pasteur's delight at Mechnikov's decision to remain in Paris, see his letter of August 21, 1890, to Emile Roux, in Vallery-Radot, *Correspondance de Pasteur,* 4: 310–11. See also Hutchinson, "Tsarist Russia and the Bacteriological Revolution."

30. Ol'denburgskii mentioned this stipulation in a letter of November 3, 1888, informing the Ministry of Justice of the Tsar's decision. The letter is reproduced in Kharauzov, "K istorii Instituta," 632.

31. RGIA 515.29.1485. Ol'denburgskii drew considerably from these funds in 1889 and 1890, clearly using them as seed money for his new Institute. The 200,000 rubles, however, represented much less than half the costs of construction.

32. Paul Weindling, "Scientific Elites and Laboratory Organisation in *Fin de Siècle* Paris and Berlin: The Pasteur Institute and Robert Koch's Institute for Infectious Diseases Compared," in *The Laboratory Revolution in Medicine,* ed. Andrew Cunningham and Perry Williams (Cambridge: Cambridge University Press, 1992), 170–88.

33. The garden soon gave rise to the Imperial Botanical Garden. In the eighteenth and nineteenth centuries, Russia's leading botanists lived on the island, oversaw the garden, and delivered lectures there. See M. I. Pyliaev, *Zabytoe proshloe okrestnostei Peterburga* [1889] (St. Petersburg: Liga, 1994), 38.

34. *Vrach,* 1890, no. 39: 901. Anrep had also participated in the work of the St. George's Commune, as had Kruglevskii, S. M. Luk'ianov (who served as Institute director from 1894 to 1902), V. A. Manassein (editor of *Vrach*), and Pavlov.

35. *Vrach,* 1890, no. 47: 1083.

36. Pavlov delivered these lectures from 1878 until the closing of the *fel'dsher* courses in 1882. Gureeva writes that P. P. Khizhin, chief physician at Princess Ol'denburgskaia's free hospital in Ramon, worked under Pavlov in Botkin's laboratory (Gureeva, "K istorii organizatsii," 633). If so, this connection may also have played a role in Pavlov's appointment to the Institute's organizing committee.

37. Koch himself was at the time parlaying tuberculin into his own institute. Unfortunately, there is no satisfactory historical account of the tuberculin episode in Western Europe or Russia. It is clear, however, that the great enthusiasm that greeted Koch's announcement passed quickly. Rudolf Virchow demonstrated tuberculin's toxicity in 1891, and favorable clinical reports quickly yielded to negative ones. Still, clinical testing of tuberculin continued at least into the early 1900s. See, for example, Magda Whitrow, "Wagner-Jauregg and Fever Therapy," *Medical History* 34 (1990): 300–303.

38. *Vrach,* 1890, no. 45: 1037.

39. *Bol'nichnaia gazeta Botkina* 1, no. 44–45 (1890): 1079. According to the Russian medical press at the time, tuberculosis came in various forms: lupus (*volchanka*), tuberculosis of the bones, joints, and glands (*bugorchatka*), and tuberculosis of the lungs (*chakhotka*). According to Anrep, tuberculin had as yet produced no positive results for patients suffering from tuberculosis of the lungs.

40. November 15 and 22 reports in *Vrach,* 1890, no. 46: 1061, and no. 47: 1083.

41. Tsar Alexander III to Prince A. P. Ol'denburgskii, December 6, 1890; copy in TsGIA SPb 2232.1.1: 1–2.

42. Richard S. Wortman, *Scenarios of Power: Myth and Ceremony in Russian Monarchy* (Princeton: Princeton University Press, 2000), 2: 159–306.

43. For a brief account of the short-lived use of tuberculin to treat lupus, see Brian Potter, "The History of the Disease Called Lupus," *Journal of the History of Medicine and Allied Sciences* 48 (1993): 81–82.

44. "Biulleten' o lechenii liupoznykh bol'nykh po sposobu prof. R. Koch'a," *Vrach,* 1890, no. 48: 1099–1100. The second report on this subject was published two weeks later, in *Vrach,* 1890, no. 50: 1144. The signatories of the first report were Shperk, V. V. Kudrevetskii, D. A. Kamenskii, and Khizhin. Curiously, Khizhin did not sign the second report.

45. D. A. Kamenskii, "Moe znakomstvo s Ivanom Petrovichem," in *I. P. Pavlov v vospominaniiakh sovremennikov,* ed. E. M. Kreps (Leningrad: Nauka, 1967), 104. In this source, the extract in the text is one long paragraph. According to Ushakov, Professor L. V. Popov of the Military-Medical Academy served as consultant, and physicians B. V. Verkhovskii, V. V. Kudrevetskii, and N. Ia. Ketcher supervised the patients. See V. G. Ushakov, "Iz istorii VIEM," in *Materialy k istorii VIEM* (Moscow, 1941), 1: 196. Gureeva writes that five physicians (Kudrevetskii, Verkhovskii, Kamenskii, Khizhin, and Ketcher), all of whom had formerly worked under Pavlov in Botkin's laboratory, were invited to participate in the tuberculin study. See

Gureeva, "K istorii organizatsii," 633. With the exception of Kudrevetskii, these names do not appear among the signatories of the report. I cannot reconcile these differing accounts from available sources.

Prince Ol'denburgskii was apparently competing for Russian priority with at least two other ventures. In St. Petersburg's Obukhovskaia Hospital, a doctor Kalmeir began on November 27, 1890, to treat tuberculosis with tuberculin acquired directly from Berlin. See *Bol'nichnaia gazeta Botkina* 1, no. 46 (1890): 1110. In early November the St. Petersburg Duma dispatched the physician N. I. Sokolov to Berlin to become acquainted with tuberculin. See *Vrach*, 1890, no. 46: 1061. Sokolov began giving treatments at the Aleksandrovskaia Barachnaia Hospital in the last week of November, and on December 13 he reported positive results to a meeting of the Society of Russian Physicians. See N. I. Sokolov, "Nabliudeniia nad lecheniem chakhotochnykh bol'nykh Kokhovskoi zhidkost'iu," *Bol'nichnaia gazeta Botkina* 1, no. 49–50 (1890): 1154–66.

46. Tuberculin studies played an important part in the early life of the Institute, and the Prince subsequently dated the beginning of its investigatory activities to these investigations of Koch's remedy. See TsGIA SPb 2282.1.31: 2. The clinical investigation of tuberculin was also one of the two practical tasks included in the Institute committee's report to the Ministry of Internal Affairs on the work of the projected Institute. See ibid., 1.1: 15. The examination of tuberculin played an important part in the joint work of several of the new Institute's sections in 1891 and 1892, just as its production (for diagnostic purposes) became a mainstay of the Institute's activities until the First World War.

On the basis of her archival studies, Gureeva concludes that Pavlov participated in these studies; in February 1891 he took supplies of the serum from the Institute for use in the clinic of the Military-Medical Academy. He also supervised the work of an early coworker (and his first assistant), V. N. Massen, on the pharmacology of tuberculin. Massen's article "Neskol'ko eksperimental'nykh dannykh k voprosu o tuberculine" (Some Experimental Data on the Question of Tuberculin) was one of the first produced in Pavlov's laboratory at the Institute of Experimental Medicine. In it Massen acknowledges that his investigation was conducted "at the suggestion of Prof. I. P. Pavlov." Massen concluded that tuberculin was "not an energetic pharmacological substance." See *Trudy obshchestva russkikh vrachei* 57 (1890–1891): 16–19. The investigation of tuberculin was the principal task of the Institute's Epizootological Division in 1892. See TsGIA SPb 2282.1.31: 40b.

47. Perhaps in deference to Shperk, the report was published as a "bulletin" rather than a conclusive study of tuberculin's effects.

48. In early 1894, Ol'denburgskii recalled that Shperk had been his "closest assistant in the initial and very difficult organization of the Institute." See TsGIA SPb 2282.1.56: 83.

49. *Vrach*, 1890, no. 48: 1109. Tuberculin became the subject of much political infighting with the state apparatus and the medical community. *Vrach* reported in

December 1890 on a discussion of tuberculin within the Medical Council of the Ministry of Internal Affairs (of which Anrep was scholarly secretary). This apparently stormy meeting ended in "compromise." The free sale of tuberculin was not permitted, but rather, in view of its potential importance, it was made a property of the state. The Ministry of Internal Affairs would purchase Koch's remedy and use it only after its quality had been verified in "special laboratories established for this purpose by the Ministry of Internal Affairs," and it would be released for use "in clinical institutions only under strict controls." This decision, published in the official *Pravitel'stvennyi Vestnik* of November 30, 1890, was reprinted in *Vrach*, 1890, no. 50: 1148. The results of various clinical trials with tuberculin appeared in *Vrach* and other medical journals throughout the 1890s. *Vrach* editorialized in 1898 (no. 1: 16) that the results were so consistently negative that it was past time to consider this issue closed.

50. The item on this "rumor" in *Vrach* (1890, no. 50: 1148) expressed high regard for Pavlov and relief that he would not abandon his recently acquired assistant professorship in pharmacology at the Military-Medical Academy to assume the Institute directorship. This combination of sentiments speaks volumes about the attitude of the medical community toward the Prince and his Institute after the tuberculin episode.

Pavlov's refusal reflected both his aversion to administrative tasks and his cautiousness about the eventual success of the Institute. In her unpublished memoirs, Pavlov's wife recalled that although she was "sorely tempted by the director's house," her husband "refused without hesitation every administrative post offered by Prince Ol'denburgskii." See Serafima Vasil'evna Pavlova, *Vospominaniia*, in SPF ARAN 259.1.171: 353. Kamenskii noted that Pavlov found his dealings with Ol'denburgskii difficult and was uncertain about the eventual fate of the Institute. See Kamenskii, "Moe znakomstvo," 105.

51. Kamenskii, "Moe znakomstvo," 105. Anrep had also planned to appoint Danilevskii's brother to head the Chemistry Division. As we shall see shortly, the sinking of this candidacy had an even more profound impact on the subsequent history of the Institute. According to an unsigned biographical fragment found in Pavlov's personal papers, Prince Ol'denburgskii originally intended to appoint physiologist V. Ia. Danilevskii as director of the Imperial Institute, but was unable to do so because Danilevskii was Jewish. If this was the case, it is also quite possible that anti-Semitism played a role in the failure both of Danilevskii's candidacy for head of the Physiology Division and of his brother's (A. Ia. Danilevskii's) candidacy for head of the Chemistry Division. I have been unable to find information bearing on this unsubstantiated claim. This biographical fragment is found in a file of miscellaneous papers and notes in SPF ARAN 259.1.58: 64–65.

52. *Vrach*, 1890, no. 50: 1148.

53. Ibid. Charles Chamberland represented Pasteur, Eduard Pfuhl represented Koch, and Watson Cheyne represented Lister. For an enthusiastic review in the

popular press on the Institute and its founding ceremony, see *Niva*, 1891, no. 7: 156–60.

54. This formulation from the committee's original statutes remained unchanged in the approved document. See TsGIA SPb 2282.1.1: 36 and the published statutes in *Vremennyi ustav Imperatorskogo Instituta Eksperimental'noi Meditsiny* (St. Petersburg, 1894), 1.

55. TsGIA SPb 2282.1.1: 13–130b.

56. Ibid., 1.1: 130b.

57. *Vremennyi ustav*, 4–5.

58. The Council was composed of the heads (*zaveduiushchie*) of the Institute's scientific divisions—that is, of its full members (*deistvitel'nye chleny*)—and set "the general orientation of the Institute's scientific and practical activity." Under the director's leadership the Council met at least once a month from September through May to discuss and resolve issues of academic policy such as joint work among various divisions; founding, expanding, or closing existing divisions; permitting outsiders to work at the Institute; overseeing publication of the Institute's journal, *Arkhiv Biologicheskikh Nauk* (Archive of the Biological Sciences); and budgetary problems. The director was obligated to discuss the agenda of each Council meeting beforehand with Ol'denburgskii, to whom all Council decisions (together with the minutes of meetings) were submitted for final approval. The director also presided over the Management Committee, which was comprised of himself and three other members appointed by Ol'denburgskii for two-year terms and which oversaw financial and administrative matters (monitored, again, by the Prince). See *Vremennyi ustav*, 6–8.

59. TsGIA SPb 2282.1.1: 140b-15.

60. Ibid., 1.4: 9.

61. There was much joint work in the Institute's early years, but subsequently, as each division matured and developed its own orientation, collaboration acquired an episodic character dependent not on any general plan but rather on agreements "from below" among division heads. By the eve of the First World War, Prince Ol'denburgskii ascribed the Institute's successes to the *absence* of any general plan, which, he claimed, facilitated the free initiative of its division heads: "Each of the divisions presents in its traditions an independent whole, linked with the others only in an administrative relation. Such a setting liberates the individual strengths of the intellectual-spiritual aspirations of each of the division heads, and gives each laboratory its particular coloration, ranging from that of the abstract scholar, busy with the elaboration of abstract scientific questions, to that of the American scholar, elaborating questions of vital practical importance, linked with the demands of the moment and the day." Ibid., 1.396: 5.

62. Ibid., 1.1: 15. The eight scientific sections listed were: pathological anatomy with histology, physiology, experimental pathology, pharmacology with pharmacognosy, biological chemistry, microbiology, epizootology, and botany or phyto-

biology. In approving the Institute's statutes, the State Council agreed that the number and nature of the Institute's divisions would be determined by Ol'denburgskii and the Institute's Council in collaboration with the Ministry of Internal Affairs. Ibid., 1.1: 37.

63. Ibid., 1.1: 160b.

64. Ibid., 1.1: 80b.

65. Ibid., 1.1: 30b-4. Full professors in the universities received 3,000 rubles a year, professors in the Military-Medical Academy 2,872, and professors in the Grand Duchess Elena Pavlovna Clinical Institute up to 2,500. The director of the Military-Medical Academy received 4,559 rubles and university rectors 4,500.

66. Ibid., 1.1: 42. These salary figures were published in *Vremennyi ustav*, 17.

67. Ibid., 1.1: 42; published in *Vremennyi ustav*, 17–18. On the Table of Ranks in the late nineteenth century, see L. E. Shepelev, *Tituly, mundiry, ordena* (Leningrad: Nauka, 1991).

68. *Svod uzakonenii i rasporiazhenii pravitel'stva po vrachebnoi i sanitarnoi chasti v Imperii*, no. 1 (St. Petersburg, 1895–96), 119, 132. The special provision for women was probably a product of negotiations to recruit M. V. Nencki from abroad. Nencki brought with him to the Institute his assistant at the University of Berne, the female doctor of medicine N. O. Ziber-Shumova. When Nencki died in 1901, Ziber-Shumova succeeded him as chief of the Chemistry Division and served in that position until 1916.

69. *Vrach*, 1891, no. 24: 585. St. Petersburg's sanitary institutions received a bit more than 94,000 rubles in 1890 (those of Moscow received about 76,000). See F. A. Brokgauz and I. A. Efron, eds., *Rossiia: Entsiklopedicheskii slovar'* [1898] (Leningrad: Lenizdat, 1991), 221.

70. TsGIA SPb 2282.1.1: 53.

71. *Vremennyi ustav*, 11–12.

72. The Institute at this point had no funds at its disposal, but Pavlov and several unpaid coworkers (engaged in their own doctoral research) used its laboratory facilities. On May 27, 1891, Pavlov informed Shperk that doctors N. P. Iurgens, T. S. Shubenko, and A. S. Sanotskii were at work in his Physiology Division. See SPF ARAN 259.4.225. V. N. Massen, who was subsequently to become Pavlov's first assistant, soon joined them and at Pavlov's suggestion studied the pharmacological action of tuberculin.

73. According to one later commentator in *Vrach*, Ol'denburgskii's reluctance to choose Shperk as director turned out to be well founded. That journal's obituary for Shperk in 1894 noted that he had "justly acquired European renown" for his work on prostitutes and prostitution, but "unfortunately, in 1891 he exchanged this field of activity for another for which he was hardly well-prepared—and became director of the Institute of Experimental Medicine." *Vrach*, 1894, no. 18: 540.

74. Bacteriological research was, however, central not only to Vinogradskii's division of General Microbiology but also to the Chemistry Division, where Nencki

and his associates pursued research on the etiology of diphtheria, cholera, and cattle plague and, from 1895 to 1901, produced large quantities of anti-diphtherial serum. The Epizootology Division pursued bacteriological studies of cattle plague, rabies, and other animal illnesses, aside from engaging in practical activities against them. Within the Division of Pathology-Anatomy, a pathological-bacteriological section collected microbes, gave courses on "practical bacteriology," and produced various forms of bacteriological warfare against pests (e.g., the "bacteria of mouse typhus to exterminate field mice"). TsGIA SPb 2282.1.113: 630b.

This divisional structure changed little in subsequent years. In 1894, when Shperk died, his Syphilology Division was disbanded and the Institute's new director, S. M. Luk'ianov, became chief of a new division of General Pathology. (Shperk's assistant, Ganike, found a home in the Physiology Division, beginning a close working relationship with Pavlov that would last until the latter's death in 1936.) The increasing importance to the Institute of serum production was reflected in the addition of a laboratory for the preparation of therapeutic serums (1895), which subsequently became a full-fledged Division of Practical Hygiene (1902). By that time, the Institute had also opened a laboratory for the production of anti-plague preparations in the Alexander I Fort near the island of Kronstadt.

75. S. N. Vinogradskii to I. I. Mechnikov, November 19, 1891; cited in V. G. Ushakov, "Materialy k istorii 'Arkhiva Biologicheskikh Nauk,'" *Arkhiv Biologicheskikh Nauk* 61, no. 1 (1941): 7.

76. V. G. Ushakov, "Iz istorii VIEM," in *Materialy k istorii VIEM* (Moscow, 1941), 195.

77. Kamenskii, "Moe znakomstvo," 104–5.

78. Quoted by Vinogradskii's assistant V. L. Omelianskii in his "Sergei Nikolaevich Vinogradskii," *Arkhiv Biologicheskikh Nauk* 27, no. 1–3 (1927): 20–21.

79. In a letter to Mechnikov, Vinogradskii explained that only on returning to St. Petersburg had he learned he was to edit *Arkhiv*. He resisted, but "there is nobody but me for the following reasons: According to the Statutes, Institute publications are printed in Russian and French. Nencki knows neither language, the other members of the Council know French not at all." The job promised to be especially taxing, since each article would have to be edited in French and Russian and "Nencki and several foreigners working with him write only in German." Furthermore, "I fear that I will not succeed in giving this journal the slightest consistency of content, that it will not have the slightest scientific physiognomy. These fears are well-founded since my mandatory coworkers leave much to be desired. I will try to preserve as much power as possible in the choice of works, but time will tell how this turns out in practice. This right of choice, apparently, exists only for works from outside [the Institute]; those presented by Council members are subject to no control whatsoever." Vinogradskii's letter of November 19, 1891, is quoted in Ushakov, "Materialy k istorii," 6–7.

Whatever his reservations, Vinogradskii presented his plan for the journal to the Council on November 15, and in its first year *Arkhiv* was distributed to roughly two hundred Russians and two hundred foreigners. See TsGIA SPb 2282.1.31: 31. It was originally published in separate Russian and French editions, but these were eventually collapsed into one bilingual volume for financial reasons. Publication in French, of course, reflected the Institute's original orientation toward Paris and perhaps the perception that, although about an equal number of Russian physicians knew German and French, French remained the better-known language in Europe. See *Vrach,* 1894, no. 27: 778–79.

80. B. L. Isachenko, "Otdel obshchei mikrobiologii," in *Materialy k istorii VIEM,* 88.

81. See Marcel H. Bickel, *Marceli Nencki* (Bern: Hans Huber, 1972); and Wlodzimierz Niemierko, "Nencki, Marceli," *Dictionary of Scientific Biography* (New York: Scribner and Sons, 1977), 10: 22–23. Prince Ol'denburgskii probably needed special permission for Nencki to live in the Russian Empire.

82. Ushakov recalled that it was Pavlov's idea to recruit Nencki (Ushakov, "Iz istorii VIEM," 196). This gains credence from the fact that Pavlov had a close working relationship with E. O. Shumova-Simanovskaia, whose sister, N. O. Ziber-Shumova, was an assistant to Nencki in Bern and came with him to the Institute. *Vrach* objected to the Nencki appointment for the same reasons that it had warned against Pfuhl. Mentioning "rumors" that Nencki was being lured with a 6,000 ruble salary, an apartment, and "unlimited expenses" for his laboratory, *Vrach's* columnist concluded, "We profoundly respect Professor Nencki's services, but absolutely refuse to believe that a worthy candidate cannot be found among Russian scholars." See *Vrach,* 1891, no. 12: 322.

83. By November 1891, as noted above, the Institute's journal editor Vinogradskii was already concerned about the difficulties he would have translating the works of Nencki's entourage into French and Russian.

84. Ushakov, "Iz istorii VIEM," 196.

85. TsGIA SPb 2282.1.56: 80.

86. On the new general pathology building, see *Vrach,* 1896, no. 34: 960. Ol'denburgskii may have financed this with the balance of the credit acquired in 1889 for a "bacteriological station." On the effects of state funding, see TsGIA SPb 2282.1.56: 800b.

87. From Ol'denburgskii's report of 1893 in TsGIA SPb 2282.1.56: 80.

88. Prince A. P. Ol'denburgskii to S. Iu. Vitte, March 21, 1893, in RGIA 565.5.20018: 110–11.

89. TsGIA SPb 2282.1.56: 810b, 830b.

90. Ol'denburgskii justified the physical expansion of the Inoculations Division by an influx of a different kind: 486 people bitten by various animals arrived in 1893, an unprecedented number of them from the provinces, where, owing to the cholera epidemic, local authorities refused to treat them. In 1893 alone, 140 of these patients were housed in the Division's new fifteen-bed unit. Ibid., 1.56: 82.

91. On the success of the Institute's animal "factory," see ibid., 1.31: 170b-18. At a Council meeting of April 29, 1893, Pavlov, Nencki, and Uskov still complained of a "constant insufficiency of small animals" and asked that the administration allot them an annual quantity that they could draw upon weekly. See ibid., 1.46: 140b-15. Another example of the Institute's growing pains appears in the minutes of its Council meeting of April 29, 1893. Here Pavlov complained that necessary laboratory materials were provided only after a delay of one or two weeks. The head of administrative affairs explained that it was difficult to keep up with the divisions' demands since the Institute had only one driver for its carriage—and he had been fined three times for leaving his horse unattended while collecting supplies in St. Petersburg. The Council decided to hire a driver's assistant so purchases could be made daily without risking this added expense. See ibid., 1.46: 130b-14.

92. Ibid., 1.56: 810b.

93. In his letter to Witte, Ol'denburgskii explained the unexpected influx of praktikanty by "the timeliness and importance of the questions addressed by the Institute's program of activity, and no less by the very successful choice of chiefs of its scientific divisions, among whom are individuals with a European reputation." TsGIA SPb 565.5.200018: 110–11.

94. The phrase "detour through the lab" is Bruno Latour's, from his "The Costly Ghastly Kitchen," in *The Laboratory Revolution in Medicine*, ed. Andrew Cunningham and Perry Williams (Cambridge: Cambridge University Press, 1992), 297.

95. According to Robert Frank Jr., who is writing a monograph on German physiological laboratories of this era, the German word *praktikanten* referred to people engaged in practical laboratory exercises for the purposes of education. My thanks to him for discussing this with me. As we shall see in Chapter 3, the word *praktikanty* would acquire a somewhat different meaning in Pavlov's Physiology Division.

96. Nancy Mandelker Frieden, *Russian Physicians in an Era of Reform and Revolution, 1856–1905* (Princeton: Princeton University Press, 1981).

97. These laws are collected in *Svod uzakonenii.*

98. In *Russian Physicians,* Frieden quotes a spokesman for the profession on the "oversubscription" of medical schools (p. 127). She also provides a study of a cross section of physician salaries in 1902: the median annual salary was about 1,600 rubles (p. 336).

99. V. I. Pashutin, *Kratkii ocherk Imperatorskoi Voenno-Meditsinskoi Akademii za 100 let eia sushchestvovaniia* (St. Petersburg, 1898), 19. The development of "scientific positivism" was judged so important that censorship of even the most "pernicious materialist" scientific works was curtailed. See Daniel P. Todes, "Biological Psychology and the Tsarist Censor: The Dilemma of Scientific Development," *Bulletin of the History of Medicine* 58 (1984): 529–44.

100. I. D. Delianov, "On the Question of Attracting Students to Serious Scientific Work under the Guidance of Teachers, and Other Measures for the Preven-

tion of Disorders," in TsGIA SPb 14.1.6780: 100b. The political consideration is reflected in the title of this dossier.

101. In 1893–94, 126 physicians defended theses for the doctoral degree in medicine at the Military-Medical Academy—an unusually high number because of their service in campaigns against epidemics, particularly of cholera, which had prevented many physicians from completing their dissertations in the previous two years. See *Vrach*, 1894, no. 21: 618. In the 1894–95 academic year, thirty-nine military physicians alone were on service leave (*komandirovka*) at the Military-Medical Academy for "improvement in the medical sciences." See *Vrach*, 1894, no. 35: 978–79.

102. *Svod uzakonenii*, 113, 116.

103. Ibid., 114, 116.

104. Ibid., 121.

105. Ibid., 127–30. In her analysis of a cross section of physicians, Frieden found that about 13 percent were Jewish. See her *Russian Physicians*, 333.

106. *Vrach*, 1898, no. 18: 546.

107. *Vestnik Rossiiskago obshchestva pokrovitel'stva zhivotnykh* 8 (1898): 261–62.

108. *Imperatorskii Institut Eksperimental'noi Meditsiny (1890–1910)* (St. Petersburg, 1911), 55. These figures do not include praktikanty in the practical divisions.

109. Compiled from Ol'denburgskii's annual reports in TsGIA SPb 2282.1.113, 145, 162, 163, 221, and 222. I have omitted praktikanty in the practical divisions, so these division totals do not equal the totals in the previous chart. These figures provide only the vaguest indication of the comparative popularity of different divisions, since aspiring praktikanty were sometimes turned away for lack of space and some division heads (e.g., Vinogradskii in Microbiology) preferred a small laboratory group.

110. Pavlov also drafted the Institute's "rules for the production of expertise"— that is, rules for paid consultations with Institute specialists in chemistry, bacteriology, pathological anatomy, and veterinary medicine. His handwritten manuscript of 1894 is in SPF ARAN 259.4.228.

111. *Pravila dlia storonnikh lits, zhelaiushchikh rabotat' v uchrezhdeniiakh Imperatorskago Instituta Eksperimental'noi Meditsiny* (St. Petersburg, 1894), 1–5.

112. Ibid., 6–15.

113. In 1898, for example, the Institute's monthly oil bill from the Nobel Brothers Company averaged well over 1,200 rubles. See TsGIA SPb 2282.1.120: 26, 40, 590b, 800b, 112, 1160b, 132, 149, 160, and 1680b. In 1916 Ol'denburgskii noted that from 1895 to 1911 the Institute's annual expenditure on gas had risen from 3,189 to 7,033 rubles, and on heating supplies from 16,466 to 27,085 rubles. Ibid., 1.396: 153.

114. A. P. Ol'denburgskii, *Kratkii predvaritel'nyi otchet po Imperatorskomu Institutu Eksperimental'noi Meditsiny s 1890–1915* (Short Preliminary Report on the Imperial Institute of Experimental Medicine in the Years 1890–1915), in ibid., 1.396: 154–1540b.

115. Ibid., 56.1: 25. Ol'denburgskii's selection of Luk'ianov as director of the Institute met with the hearty approval of *Vrach* (1894, no. 19: 567). Luk'ianov served as director until 1902, when he became Vice-Minister of Public Enlightenment.

116. TsGIA SPb 2282.1.183: 3–40b. Pavlov, especially, took advantage of Nencki's death to protest against Chemistry's privileged budgetary position.

117. Omelianskii, "Sergei Nikolaevich Vinogradskii," 23.

118. In the years 1891–1904 the budget of the Chemistry Division declined from 6,400 to 4,900 rubles, that of the Microbiology Division from 2,800 to 1,700 rubles, the Pathology-Anatomy Division from 2,400 to 2,150 rubles, and the Epizootology Division from 3,200 to 2,500 rubles. The budget of the Syphilology Division was 3,200 rubles in 1891; this division was transformed into the General Pathology Division in 1895, and by 1904 its budget had declined to 2,500 rubles. The budget of Pavlov's Physiology Division rose slightly, from 3,200 rubles in 1891 to 3,400 in 1904. These data are compiled from Ol'denburgskii's summary report of 1916 on the history of the Institute (in TsGIA SPb 2282.1.396) and from the yearly reports and Council meetings (in TsGIA SPb 2282). There are minor contradictions between the information in these sources, but these do not change the general picture of declining budgets.

119. Ibid., 1.396: 155.

120. Ibid., 1.113: 36.

121. Ibid., 1.162: 610b, and 1.183: 36. In the wake of the Revolution of 1905 the Institute received its first sizable state grant specifically for medical research: 40,000 rubles to investigate the etiology of syphilis and malaria. See ibid., 1.221: 20b.

122. For example, ibid., 1.56: 820b; *Vrach,* 1894, no. 42; *Vrach,* 1897, no. 24: 695.

123. TsGIA SPb 2282.1.145: 590b.

124. For Vinogradskii's gifts, see ibid., 1.163: 1250b, and 184: 800b; for the Nencki collection, see *Russkii Vrach,* 1902, no. 33: 1200, and no. 38: 1389; for the gift in Pavlov's honor, see TsGIA SPb 2282.1.238: 10.

125. TsGIA SPb 2282.56.1: 60–61. Emil von Behring and Shibasabūro Kitasato had first produced diphtheria antitoxin in sheep and goats. Roux announced his technique for using horses in the *Annales* of the Pasteur Institute in 1894. For other institutions involved in antitoxin production, see, for example, Jonathan Liebenau, "Paul Ehrlich as a Commercial Scientist and Research Administrator," *Medical History* 34 (1990): 65–78; and E. M. Tansey, "The Wellcome Physiological Research Laboratories 1894–1904: The Home Office, Pharmaceutical Firms, and Animal Experiments," *Medical History* 33 (1989): 1–41.

126. Luk'ianov's letter and the response to it were discussed in *Vrach,* 1894, no. 42: 1173. Among the contributors to the diphtheria antitoxin project were Prince and Princess Ol'denburgskii (5,000 rubles).

127. *Vrach,* 1894, no. 44: 1228.

128. On Hoechst and German commercial production of diphtheria antitoxin, see Liebenau, "Paul Ehrlich," esp. 67–71.

129. *Vrach*, 1895, no. 1: 24. In 1898 the Institute itself produced 24,786 bottles of diphtheria antitoxin, acquiring another 7,588 from Hoechst in Berlin. See TsGIA SPb 2282.1.113: 35.

130. On the ups and downs of the Institute diphtheria antitoxin, see *Vrach*, 1894, no. 42: 1173, and no. 44: 1228; *Vrach*, 1897, no. 49: 1437, and no. 50: 1464.

131. "Extreme financial difficulties" was the assessment of the Institute's director, Podvysotskii, in 1906. TsGIA SPb 2282.1.251: 11.

132. *Vrach*, 1904, no. 15: 369; TsGIA SPb 2282.1.262: 16–17.

133. TsGIA SPb 2282.1.262: 17.

134. Ibid., 1.56: 810b, 820b.

135. For 1898, 1899, and 1903 see ibid., 1.113: 640b, 1.145: 600b, and 1.221: 150b.

136. For a sample of *Vrach*'s coverage of the Institute, see *Vrach*, 1892, no. 9: 377; 1894, no. 50: 1389; 1896, no. 34: 960; 1901, no. 2: 68; 1901, no. 45: 1397; and the items on the Institute's diphtheria antitoxin in *Vrach*, 1894, no. 42: 1173, and no. 44: 1228; *Vrach*, 1897, no. 49: 1437, and no. 50: 1464. For a list of the publications originating in Institute laboratories, see *Imperatorskii Institut Eksperimental'noi Meditsiny*, 4–38.

137. *Vrach*, 1893, no. 21: 677–78.

138. V. Verekunov, V. Statsenko, and A. Rutkovskii, *Pervaia vserossiiskaia gigienicheskaia vystavka v S. Peterburge: Kratkii ocherk* (St. Petersburg, 1894), 69–70.

139. For two such photographs, one with the Prince and one with the Princess Ol'denburgskii, see *Peterburgskaia zhizn'*, September 7, 1897, pp. 2122–23.

140. Hutchinson, *Champions of Charity*, 186–87.

141. TsGIA SPb 2282.56.1: 28; *Vrach*, 1895, no. 6: 170, and no. 8: 230.

142. From the V. L. Merkulov papers. Merkulov gives the archival location of this information as RGIA 1289.1.17 (1904): 1.

143. For the invitation list for 1900, see TsGIA SPb 2282.1.163: 5.

144. Ibid., 1.145: 3, and 1.162: 48.

145. Ibid., 1.251: 400b. This event of December 1906 is also mentioned in Vitte, *Vospominaniia*, 3: 383.

Chapter 2. The Visionary of Lopukhinskaya Street

1. For an excellent essay on the necessity of considering a scientist's investigatory purpose and ideas when analyzing disciplines and laboratories, see William Coleman's "The Cognitive Basis of the Discipline: Claude Bernard on Physiology," *Isis* 76 (1985): 49–70.

2. By *scientific vision* I mean the cluster of elements that, as Ian Hacking has pointed out, are denoted (with varying shades of meaning) by "words like weltanschauung or Holton's . . . 'themata' and 'thematic presuppositions,' or even A. C. Crombie's 'styles of scientific reasoning.' We have expectations about what the world is like and practices of reasoning about it. These govern our theories and

our interpretation of data alike." Hacking also comments on the stability of such notions: "experimenters do not change their ideal conceptions of the universe in the course of, or at any rate because of, experimental work. Such notions are not molded to fit into [the other elements of experimental practice]: they stand above them." See Ian Hacking, "The Self-vindication of the Laboratory Sciences," in *Science as Practice and Culture*, ed. Andrew Pickering (Chicago: University of Chicago Press, 1992), 50–51. The phrase *mangle of practice* is Pickering's; see his *The Mangle of Practice: Time, Agency, and Science* (Chicago: University of Chicago Press, 1995).

3. Compare, for example, Pavlov's comments on vivisection in "Zhivosechenie" (1893) (in *PSS*, 6: 23–24) with those in "Obshchaia tekhnika fiziologicheskikh opytov i vivisektsii" (1910) (in *PSS*, 6: 324). One longtime favorite anecdote about the nature of *physiological thinking* concerned Ludwig and Heidenhain. Pavlov repeated it in "Zhivosechenie" (1893), his eulogy to Heidenhain (1897), and his lectures on physiology to the students of the Military-Medical Academy from 1895 to 1924. (The anecdote is discussed later in this chapter.) Other permanent elements of Pavlov's rhetorical arsenal were his pleasure at repeating an old experiment and various variants of mechanistic imagery to describe the animal organism.

4. On the biological sciences and the political context of Russia in the 1860s, see James Allen Rogers, "Darwinism, Scientism, and Nihilism," *Russian Review* 19 (1960): 10–23; Alexander Vucinich, *Science in Russian Culture* (Stanford: Stanford University Press, 1970), vol. 2; Daniel P. Todes, "V. O. Kovalevskii: The Genesis, Content, and Reception of His Paleontological Work," *Studies in History of Biology* 2 (1978): 99–165; Daniel P. Todes, "From Radicalism to Scientific Convention: Biological Psychology in Russia from Sechenov to Pavlov" (doctoral diss., University of Pennsylvania, 1981); Daniel P. Todes, "Biological Psychology and the Tsarist Censor: The Dilemma of Scientific Development," *Bulletin of the History of Medicine* 58 (1984): 529–44; David Joravsky, *Russian Psychology: A Critical History* (Oxford: Basil Blackwell, 1989); and Loren Graham, *Science in Russia and the Soviet Union: A Short History* (Cambridge: Cambridge University Press, 1993).

5. According to S. A. Liozner-Kannabikh, Pavlov told her in 1932 that he had convinced a librarian to inform him when the latest issues of these journals arrived at the public library and to leave a window open for him to steal into the library at night. Pavlov recalled that he would read until dawn, then replace the volumes on the shelves before the authorities were any the wiser. S. A. Liozner-Kannabikh, "Iz vospominanii ob Ivane Petroviche Pavlove," *Zhurnal nevropatologii, i psikhiatrii imeni S. S. Korsakova* 53, no. 7 (1953): 580–82.

6. G. A. Dmitriev-Krymskii, *Biografiia Ivana Petrovicha Pavlova*, in SPF ARAN 259.1.146: 51–55. In writing this biography Dmitriev-Krymskii took liberties with the available material, but his account of this period in Pavlov's life was based on interviews with Pavlov and his wife.

7. Samuel Smiles, *Selections from Lives of the Engineers, with an Account of Their*

Principal Works, ed. Thomas Parke Hughes (Cambridge, Mass.: M.I.T. Press, 1966). The quotation is from Hughes's introductory essay (p. 2).

8. See Hughes's introduction in Smiles, *Lives,* on the gospel of work (p. 14) and on the engineer, nature, and progress (pp. 11–12).

9. Serafima Pavlova, *Vospominaniia,* in SPF ARAN 259.1.169.

10. Samuel Smiles, *Self-help, with Illustrations of Character, Conduct, and Perseverance* [1859] (Chicago: Belford, Clarke and Co., 1884), 416, 305, 144, and 150.

11. Sechenov's article "Reflexes of the Brain" was first published in 1863 as an article in the medical journal *Meditsinskii Vestnik* (the censor forbade its publication in *Sovremennik*) and was later republished as a book: *Refleksy golovnogo mozga* (St. Petersburg, 1866). For a discussion of this work, see Todes, "From Radicalism to Scientific Convention," 239–92; and Joravsky, *Russian Psychology,* 53–62, 92–103, 125–33. For the censorship history of Sechenov's work, see Todes, "Biological Psychology," 535–39.

12. G. H. Lewes, *The Physiology of Common Life* (Edinburgh: Blackburn and Sons, 1859). The comment on this book is from M. A. Antonovich, "Sovremennaia fiziologiia i filosofiia," *Sovremennik,* 1862, no. 2: 236.

13. Boris Babkin, *Pavlov: A Biography* (Chicago: University of Chicago Press, 1949), 214. Pavlov told this same story to another coworker, Iu. P. Frolov. See Iu. P. Frolov, *Chetvert' veka bliz Pavlova* (1948), manuscript 278/3375 in the archive of the Dom-Muzei I. P. Pavlova, in Ryazan.

14. The professors of physiology were I. M. Sechenov, I. F. Tsion, and I. R. Tarkhanov. Tarkhanov's successor was Pavlov, who taught physiology at the Military-Medical Academy from 1895 to 1924.

15. From the in-house *Journal of the Censorship Committee* for 1866, in RGIA 777.2.7: 2.

16. K. Bernar [Claude Bernard], *Vvedenie k izucheniiu opytnoi meditsiny,* trans. N. Strakhov (St. Petersburg, 1866). On Strakhov, see Linda Gerstein, *Nikolai Strakhov* (Cambridge, Mass.: Harvard University Press, 1971).

17. From Pavlov's remarks on accepting an honorary doctorate from the University of Paris in 1925, in I. P. Pavlov, *Neopublikovannye i maloizvestnye materialy I. P. Pavlova* (Leningrad: Nauka, 1975), 77. (Pavlov made this revelation to a Parisian audience likely to appreciate it, but as we shall see, his high regard for Bernard is evident throughout his corpus.) He was probably referring here to Strakhov's translation of *Introduction a l'étude de la médecine expérimentale* (1865). During Pavlov's seminary years, however, another volume of Bernard's lectures was also available in Russian: *Leçons sur la physiologie et la pathologie du système nerveux* (1858), translated as K. Bernar, *Lektsii fiziologii i patologii nervnoi sistemy,* 2 vols., trans. F. V. Ovsiannikov (St. Petersburg: N. Nekliudov, 1866–67).

Pavlov was perhaps drawing on his own memories when, in 1897, he wrote the following in his course of self-study for nonspecialists interested in anatomy and physiology: "Happily for the lay reader, there is an extraordinary teacher in physiology in the person of Claude Bernard, who possesses the rare ability to write with such unusual clarity and accessibility that he can be understood by any educated person; and to do so in a manner that preserves entirely the scientific character and depth of presentation. In him, one can say, the boundary between a scientific and a popular exposition disappears." Pavlov recommended that the layperson begin with three works by this "brilliant mind": the Russian translations of *Introduction à l'étude de la médecine expérimentale* (1865), *De la physiologie générale* (1872), and one chapter of *La science expérimentale* (1878). See *Programma chtenii dlia samoobrazovaniia*, 2d ed. (St. Petersburg: Pedagogicheskii muzei voennykh uchebnykh zavedenii, 1897), quoted in N. M. Gureeva, "Uchastie I. P. Pavlova v deiatel'nosti pedagogicheskogo myzeia voennykh uchebnykh zavedenii," *Fiziologicheskii zhurnal SSSR* 45, no. 9 (1959): 1159.

18. Pavlova, *Vospominaniia*.

19. This transformation was typical of the Russians who, inspired by the radical scientism of the 1860s, became professional scientists in subsequent decades. For two other examples, see Todes, "V. O. Kovalevskii"; and the chapter on I. I. Mechnikov in Daniel P. Todes, *Darwin without Malthus: The Struggle for Existence in Russian Evolutionary Thought* (New York: Oxford University Press, 1989), 82–103. Pavlov's "identity crisis" can be traced through his letters to his fiancée and is the subject of a chapter in my forthcoming biography of Pavlov.

20. From Pavlov's "journal" for Serafima Karchevskaia, *Popal'sia* (Trapped), which he renamed *Chudnye dela tvoi, Gospodi* (Wondrous Are Thy Works, Lord), July 4, 18, and 29, 1879, in SPF ARAN 259.2.1299: 7–8, 17–19, 22.

21. Pavlov, *Chudnye dela tvoi, Gospodi*, July 29, 1879, p. 22.

22. At about this time, Pavlov wrote in a scientific article, "The natural sciences are the best applied logic, where the correctness of intellectual processes is sanctioned by the acquisition of results that make it possible to predict phenomena in an indubitable, errorless fashion." See Pavlov, "O sosudistykh tsentrakh v spinnom mozgu" [1877], in *PSS*, 1: 35.

23. Pavlov, *Chudnye dela tvoi, Gospodi*, August 21, 1879, p. 31.

24. Ibid., 35–36.

25. Tsion is usually referred to in western literature by the name he later adopted in France, Elie de Cyon. In his youth, Tsion had also apparently been attracted to science by its link to radical politics and had traveled for a time in Lassallian socialist circles.

26. Ivan Pavlov, "Ivan Petrovich Pavlov" [1879], in *PSS*, 6: 442.

27. I. P. Pavlov to I. F. Tsion, September 13 [1897]; copy in SPF ARAN 259.7.167: 3–4; published in E. M. Kreps, ed., *Perepiska Pavlova* (Leningrad: Nauka, 1970), 56.

28. I. P. Pavlov, "Vstupitel'naia lektsiia po fiziologii" [1895], in *Neopublikovannye i maloizvestnye materialy I. P. Pavlova* (Leningrad, 1975), 11.

29. Frederic L. Holmes, "Physiology and Experimental Medicine in the Nineteenth Century," 74–75. My thanks to Larry Holmes for sharing this unpublished manuscript with me.

30. As Katkov later noted, Tsion was a "strong opponent of the materialist orientation" and one who rendered special service by combating materialism in that science, physiology, in which it had especially "penetrated minds and acquired strength." M. N. Katkov to K. P. Pobedonostsev, June 21, 1887, in *K. P. Pobedonostsev i ego korrespondenty*, vol. 1, pt. 3 (Moscow, 1923), 715.

31. Tsion, *Kurs fiziologii professora I. Tsiona* (St. Petersburg, 1874), 6–7.

32. Ibid., 7.

33. Ibid., 8.

34. Ibid. Tsion used the term *vivisectionist techniques* for what Pavlov would call *physiological surgery.*

35. Ibid., 10–15. See also Anson Rabinbach, *The Human Motor: Energy, Fatigue, and the Origins of Modernity* (Berkeley: University of California Press, 1992).

36. I. Tsion, *Kurs fiziologii (Lektsii chitannye v 1872/73 uchebnom godu v Imperatorskoi Mediko-Khirurgicheskoi Akademii)* (St. Petersburg, 1873), 1.

37. Tsion, *Kurs fiziologii (Lektsii)*, 15.

38. Tsion, *Kurs fiziologii professora I. Tsiona*, 8–9.

39. Lev Popel'skii, *Istoricheskii ocherk kafedry fiziologii v Imperatorskoi Voenno-Meditsinskoi Akademii za 100 let (1798–1898)* (St. Petersburg: Ministry of Internal Affairs, 1899), 80–82.

40. Elie de Cyon, *Methodik der Physiologishchen Experimente und vivisectionen*, Mit Atlas, 2 vols. (Giessen, 1876). For Pavlov's comment, see *PSS*, 6: 326. Pavlov refers to Claude Bernard's *Leçons de Physiologie Opératoire* (Paris: Baillière, 1879).

41. The abstract of the first collaborative work by Pavlov and V. N. Velikii is in *PSS*, 1: 27; the second, Pavlov's "Eksperimental'nye dannye k voprosu ob akkomodatsionnom mekhanizme krovesnosnykh sosudov" (1877), is reprinted in *PSS*, 2, pt. 1: 49–87. Pavlov's examination of the relevant literature on pancreatic secretion began with a work by Claude Bernard. Pavlov's notes from 1874 are preserved in SPF ARAN 259.1.28.

42. Owing to an attack of "neurosismus" in his freshman year and to the time he spent on his scientific research, Pavlov required five years to complete his university requirements and graduated in 1875.

43. Pavlov, "Ivan Petrovich Pavlov," 442. This autobiographical sketch was written for the publication celebrating the twenty-fifth anniversary of the graduation of Pavlov's medical school class from the Medical-Surgical Academy.

44. I. F. Tsion, *Serdtse i mozg* (St. Petersburg, 1873).

45. Ibid., 17; Claude Bernard, "Sur la physiologie du coeur et ses rapports avec le cerveau," in *Leçons sur les propriétés des tissus vivants* (Paris: Baillière, 1866). Bernard's speech had been published in Russian as *Fiziologiia serdtsa i otnoshenie ego k golovnomu mozgu*, trans. N. Sokov'eva (St. Petersburg: O. I. Bakst, 1867). Otniel E. Dror briefly and ably summarizes the fundamental similarity between Bernard's and Cyon's speeches in his "Creating the Emotional Body: Confusion, Possibilities, and Knowledge," in *An Emotional History of the United States*, ed. Peter N. Stearns and Jan Lewis (New York: New York University Press, 1998), 173–74.

For the subsequent history of emotions in the laboratory, see Dror's "The Affect of Experiment: The Turn to Emotions in Anglo-American Physiology, 1900–1940," *Isis* 90 (1999): 205–37.

46. Tsion, *Serdtse*, 2.

47. Some of these sentiments are echoed in Pavlov's description of the pleasures of the mature mind.

48. See, for example, Tsion, *Kurs fiziologii professora I. Tsiona*, 24–34.

49. Tsion, *Serdtse*, 7.

50. Bernard, too, of course, frequently used mechanistic imagery.

51. N. K. Mikhailovskii, "Strannye protivorechiia," *Otechestvennye Zapiski*, no. 7, 1874. For political reasons, the faculty of the Military-Medical Academy had voted to appoint another, much less qualified candidate as professor of physiology, but the Minister of War had overruled this decision and appointed Tsion.

52. From the state's extensive documentation of this incident, in Gosudarstvennyi Arkhiv Rossiiskoi Federatsii 109.254, chast' 1, pp. 12–13.

53. N. N. Strakhov to L. N. Tolstoy, in *Tolstovskii Muzei* (St. Petersburg, 1914), 2: 53.

54. Ivan Pavlov, "O russkom ume," in SPF ARAN 259.1a.4: 12. For a brief discussion of this speech, see Daniel P. Todes, "Pavlov and the Bolsheviks," *History and Philosophy of the Life Sciences* 17 (1995): 384–86. The Academy's administration did, in fact, interview a number of students in an effort to identify the ringleaders of the October 1874 protests and to develop tactics against them. I have not yet been able to determine whether Pavlov was among the students interviewed.

55. I. P. Pavlov and E. O. Shumova-Simanovskaia, "Otdelitel'nyi nerv zheludochnykh zhelez sobaki (predvaritel'noe soobshchenie)" [1889], in *PSS*, 2, pt. 1: 138–41, and "Innervatsiia zheludochnykh zhelez u sobaki" [1890], in *PSS*, 2, pt. 1: 175–99.

56. Ivan Pavlov to Serafima Karchevskaia, 7 [October 1880], in SPF ARAN 259.2.1300/1.

57. Another awardee was Pavlov's future colleague and rival, V. M. Bekhterev.

58. On the Ludwig laboratory, see Robert Frank Jr., "American Physiologists in German Laboratories, 1865–1914," in *Physiology in the American Context, 1850–1940*, ed. Gerald Geison (Bethesda, Md.: American Philosophical Society, 1987), 11–46, esp. 27–38. See also Heinz Schröer, *Carl Ludwig* (Stuttgart: Wissenschaftliche Verlagsgesellschaft, 1967); Timothy Lenoir, "Science for the Clinic: Science Policy and the Formation of Carl Ludwig's Institute in Leipzig," in *The Investigative Enterprise: Experimental Physiology in Nineteenth-Century Medicine*, ed. William Coleman and Frederic L. Holmes (Berkeley: University of California Press, 1988), 139–78; Carl Rothschuh, *History of Physiology*, ed. and trans. Guenter B. Risse (Huntington, N.Y.: R. E. Krieger, 1973); and Warren Lombard, "Life and Work of Carl Ludwig," *Science*, September 15, 1916, pp. 363–75. For a list of Ludwig's "students," see Schröer, *Carl Ludwig*, 287–93.

One does not have to deny the pedagogical dimension of laboratory activities at Leipzig to realize that a stay there does not meaningfully qualify somebody as Ludwig's "student." This label seems to owe much to an uncritical acceptance of the metaphor—originating with the concerned scientists themselves—of "laboratory as school." Robert Tigerstedt made precisely this same point with respect to the convention according to which Ludwig was labeled a "pupil" of Johannes Müller, even though Ludwig was "a finished physiologist when he first visited Berlin." Lombard paraphrases and endorses Tigerstedt's point in "Life and Work of Carl Ludwig," 364.

59. Pavlov's laboratory notebook from this trip is held in SPF ARAN 259.1.1.

60. Pavlov's extensive notes from his systematic reading of German physiological journals are preserved in SPF ARAN 259.1.20.

61. Pavlov's friend Ia. Ia. Stol'nikov had first constructed this apparatus in the Leipzig laboratory during an earlier trip. Pavlov used a model that had been improved according to Ludwig's suggestions. For references to this apparatus, see Pavlov's "Innervatsiia sily serdechnykh sokrashchenii" [1887], in *PSS*, 1: 418, and "Usilivaiushchii nerv serdtsa" [1888], in *PSS*, 1: 420–24.

62. This supports a larger point made by F. L. Holmes in his unpublished essay "Physiology and Experimental Medicine": historians' preoccupation with the differences between French and German physiological traditions has obscured the important differences among individuals in each country.

63. Pavlov, "Ivan Petrovich Pavlov," 443.

64. Ivan Pavlov, "Pamiati R. Geidengaina," in *PSS*, 6: 108.

65. W. Bruce Fye, "Carl Ludwig and the Leipzig Physiological Institute: 'A Factory of New Knowledge,'" *Circulation* 74, no. 5 (1986), 920–28.

66. Warren P. Lombard, "The Life and Work of Carl Ludwig," *Science*, September 15, 1916, pp. 368, 370.

67. Ibid., 368.

68. Ibid., 370.

69. N. Ia. Chistovich, "Iz vospominanii o rabote pod rukovodstvom Ivana Petrovicha Pavlova v 1886–87 g.g.," in *Sbornok, posviashchennyi 75-letiiu akadmika Ivana Petrovicha Pavlova*, ed. V. L. Omelianskii and L. A. Orbeli (Leningrad: Gosudarstvennoe Izdatel'stvo, 1924), 29–30. In his autobiographical sketch, Pavlov mentions "certain obstacles" in the laboratory—"mainly, of course, the paucity of resources." See Pavlov, "Ivan Petrovich Pavlov," 442.

70. Ivan Pavlov to Serafima Karchevskaia [Pavlova], 16 [March 1881] and July 10 [1882], in SPF ARAN 259.2.1300/2. (After her marriage to Pavlov on May 25, 1881, Serafima took the last name *Pavlova*.) In frustration, Pavlov instead operated on some rabbits in his apartment; their survival further convinced him that "such difficult operations cannot succeed in the hygienic conditions of our Laboratory." Although he much preferred to experiment on dogs, Pavlov briefly used rabbits because they bred rapidly and were not in such short supply.

71. Pavlov to Karchevskaia, 16 [March 1881].

72. Chistovich, "Iz vospominanii," 30.

73. Pavlov, "Ivan Petrovich Pavlov," 442.

74. I have taken this text from N. M. Gureeva and N. A Chebysheva, *Letopis' zhizni i deiatel'nosti I. P., Pavlova*, with commentary by V. L. Merkulov (Leningrad: Nauka, 1969), 31–32. It was first printed in *Krasnaia Tatariia*, September 18, 1949. See also S. M. Dionesov and V. P. Mikhailov, "O naznachenii I. P. Pavlova professorom Tomskogo universiteta," *Fiziologicheskii zhurnal*, 39 (1953), no. 3.

75. Vvedenskii became professor of physiology at St. Petersburg University and V. N. Velikii (Pavlov's collaborator in his first scientific report) gained the position at Tomsk. Both were supported by powerful patrons: Sechenov and Ovsiannikov, respectively.

76. Pavlov's appointment at the Military-Medical Academy was very controversial. His champions, physiologist I. R. Tarkhanov and pathologist V. V. Pashutin, argued that physiological training was the most important qualification for a scientific pharmacologist. Tarkhanov, who had earlier dismissed Pavlov's work as unoriginal (when Pavlov was seeking the position in physiology at St. Petersburg University), now praised it highly. Pashutin pointed out that his own physiological training had sufficed for him to pursue a successful career as professor of pathology. Pavlov's experience as manager of Botkin's laboratory was also adduced as evidence that he "will develop into an outstanding pharmacologist when he devotes his activity specially to this subject." Opponents argued that he was clearly unqualified for the post, and one faculty member insisted that the appointment was so clearly inappropriate as to breach Academy regulations. Seventeen faculty members cast their votes for Pavlov and five against. His competitor, the Academy's *privatdozent* in pharmacology at the time, received eleven votes for and eleven against—so Pavlov was appointed. Rossiiskii Gosudarstvennyi Voenno-Istoricheskii Arkhiv 316.40.1855 and 316.43.3926.

77. D. A. Kamenskii, "Moe znakomstvo s Ivanom Petrovichem," in *I. P. Pavlov v vospominaniiakh sovremennikov*, ed. E. M. Kreps (Leningrad: Nauka, 1967), 105.

78. Kamenskii, "Moe znakomstvo," 104. The coworker, V. N. Massen, published his "Neskol'ko eksperimental'nykh dannykh o tuberkuline" (Some Experimental Data on Tuberculin) in *Vrach* in 1891.

79. Kamenskii, "Moe znakomstvo," 104.

80. Ibid. Pavlov commented on his doubly good fortune of 1891 in his brief autobiography of 1904: "Finally, in my forty-first year of life, I received a professorship and my own laboratory, and now not even just one, but two jobs: as professor of pharmacology (and subsequently of physiology) in the Military-Medical Academy, and as chief of the Physiological Division of the Institute of Experimental Medicine. So, there suddenly were also sufficient financial means and a great possibility to do what I wanted in the laboratory. Until then, the constant necessity to pay for every experimental animal, and this with meager financial re-

sources in general, made itself felt in the scale of laboratory activity." See Pavlov, "Ivan Petrovich Pavlov," 443.

81. By labeling Pavlov's vision *Bernardian* I do not mean to oppose it to a sharply distinguished *German* or *Ludwigian* vision of physiology. I agree with Frederic L. Holmes that these traditions had more in common than is often supposed and that their relationship remains to be thoroughly explored. By *Bernardian*, rather, I mean simply the set of postulates and attitudes propounded by Claude Bernard and summarized later in this section of the text.

82. Claude Bernard, *An Introduction to the Study of Experimental Medicine* [1865], trans. Henry Copley Greene (New York: Dover Publications, 1957).

83. This was also the spirit of the guiding theme of the gold medal competition won by Pavlov and his collaborator in 1874. Probably formulated by Tsion, it read: "La médecine est la science la plus vaste, car elle les comprend toutes." The more narrowly defined general topic was the nerves governing the work of the pancreatic gland. See Gureeva and Chebysheva, *Letopis'*, 19.

84. William Coleman, "The Cognitive Basis of the Discipline: Claude Bernard on Physiology," *Isis* 76 (1985): 49–70; and Bruno Latour, "The Costly Ghastly Kitchen," in *The Laboratory Revolution in Medicine,* ed. Andrew Cunningham and Perry Williams (Cambridge: Cambridge University Press, 1992), 295–303.

85. Frederic Lawrence Holmes, *Claude Bernard and Animal Chemistry: The Emergence of a Scientist* (Cambridge, Mass.: Harvard University Press, 1974), 1–2; and Coleman, "Cognitive Basis of the Discipline." In his brief discussion of a passage in Bernard's *Introduction to the Study of Experimental Medicine,* Holmes puts it this way: "Here, however, he was not defining ultimate modes of explanation, but methods of experimental practice" (p. 2).

86. Ivan Pavlov, *Zapisi lektsii po fiziologii pishchevareniia* (1902–1903), in SPF ARAN 259.1.78. Pavlov explained these laws to his students as follows: "The constancy of matter lies in this, that matter does not disappear, but is only transformed, and its weight remains unchanged. Forces (electrical, mechanical, and others) are transformed one into the other and never disappear. Finally, the constancy of elements consists in the combination of simple elements, during which one never passes into another." Ibid.

87. Pavlov, "Zhivosechenie," 12.

88. Ivan Pavlov, "Remarks in the discussion of A. A. Val'ter's report 'Otdelitel'naia rabota podzheludochnoi zhelezy'" [1898], in *PSS,* 6: 113. In this public forum, Pavlov expressed himself somewhat cautiously, saying that Val'ter's findings "will be a marvelous weapon in the hands of the defenders of that theory." Similarly, in 1893 Pavlov commented about physiologists' success in keeping alive dogs with a double vagotomy: "One must not but see that this point represents a victory for natural science and the dignity of medicine, and proof of the correctness of the physicochemical view of life." See Pavlov, "O vyzhivanii sobak s pererezannymi bluzhdaiushchimi nervami" [1893], in *PSS,* 1: 536.

89. See, for example, Pavlov's "Nobelevskaia Rech'" [1904], in *PSS*, 2, pt. 2: 348, 354; and his comments during his lectures on physiology in 1911, in "Lektsii po fiziologii," in *PSS*, 5: 11, 12, 14, 68, 69, 87, and 144. In his Nobel speech (1904) Pavlov characterized the digestive system as a "chemical laboratory," and in his lectures of 1911 as "a series of chemical laboratories." In both cases he was using the Russian word *laboratoriia* in its original univocal sense to mean a factory. In Vladimir Dal', *Tolkovyi slovar' zhivogo velikorusskogo iazyka* [1882] (Moscow: Russkii Iazyk, 1990), 2: 231, *laboratoriia* is defined as "an institution for chemical or metallurgical works, for filling explosive shells, for the preparation of fireworks." I discuss the emergence and consequences of Pavlov's metaphor of digestive system as factory in Chapter 5. His estimation of the extent to which the physiologist—as opposed to the physician—was the "mechanic of the living organism" changed over time, as we shall see in Chapter 8.

90. Pavlov, "Lektsii po fiziologii," 12. Here, characteristically, what begins as a philosophical statement becomes an expression of the scientist's professional goals and methods.

91. I. P. Pavlov, "Eksperimental'naia terapiia kak novyi i chrezvychaino plodot-vornyi metod fiziologicheskikh issledovanii" [1900], in *PSS*, 1: 572–73. For an early expression of this sentiment, see Pavlov's "O sosudistykh tsentrakh v spinnom mozgu" [1877], in *PSS*, 1: 35–63. Here the task of science is "to predict phenomena in an indubitable, errorless fashion" (p. 35). For the prevalence and significance of the view that the physician was the mechanic of the human body, see John Harley Warner, "The Fall and Rise of Professional Mystery: Epistemology, Authority and the Emergence of Laboratory Medicine in Twentieth-Century America," in Cunningham and Williams, *Laboratory Revolution*, 310–41.

92. Ivan Pavlov, "Rech', proiznesennaia po povodu izbraniia tovarishchem predsedatelia obshchestva russkikh vrachei v S. Peterburge" [1893], in *PSS*, 6: 28.

93. For Pavlov on delicate machine, see "Eksperimental'naia terapii kak novyi i chrezvychaino plodotvornyi metod fiziologicheskikh issledovanii" [1900], in *PSS*, 1: 572; on closed machine, "Lektsii po fiziologii," 14; on chain of moments, "Zhivosechenie," 195.

94. Pavlov, "Sovremennoe ob"edinenie v eksperimente glavneishikh storon meditsiny na primere pishchevareniia" [1899], in *PSS*, 6: 258.

95. I have found only one place in Pavlov's corpus where he suggests that physiological processes might manifest properties absent in physical and chemical entities: "Aside from physicochemical, here there exist also their own [i.e., physiological] laws, very complex, which we still do not understand; these are hidden by a mass of details, subtleties, the sense of which is still not entirely clear to us." Even here, however, he soon adds: "This is how one must understand those cases in which physicochemical explanations turn out at present to be inapplicable. This means that we still have not gone to the end (*doshel chered*), that we still do not know everything." See Pavlov, "Lektsii po fiziologii," 257.

96. For example, ibid., 256–57.

97. A. F. Samoilov, "Obshchaia kharakteristika issledovatel'skogo oblika I. P. Pavlova" [a shortened version of his 1925 memoir], in *I. P. Pavlov v vospominaniiakh sovremennikov,* ed. E. M. Kreps (Leningrad: Nauka, 1967), 204.

98. Ibid.

99. Pavlov, "Pamiati R. Geidengaina," 107, 104.

100. Pavlov, *Lektsii,* 2, pt. 2: 173; *Lectures* (trans.), 130. Unless otherwise noted, all translations of *Lektsii* are my own. The last few sentences of this extract are badly mistranslated in *Lectures* (trans.), 130, giving them precisely the opposite meaning of the original Russian.

101. See Frederic L. Holmes, "Early Theories of Protein Metabolism," in *The Origins of Modern Biochemistry: A Retrospect on Proteins,* ed. P. R. Srinivasan, Joseph S. Fruton, and John T. Edsall (New York: New York Academy of Sciences, 1979), 171–87; Frederic Lawrence Holmes, *Between Biology and Medicine: The Formation of Intermediary Metabolism* (Berkeley: University of California Office for History of Science and Technology, 1992); and Holmes, *Claude Bernard,* 240, 260–61, 377–400. My thanks to Larry Holmes for alerting me to the importance of these early studies of intermediary metabolism.

102. Pavlov, *Lektsii,* 21; *Lectures* (trans.), 2–3.

103. Pavlov, "Eksperimental'naia terapiia," 574. This general perspective may explain one striking difference between Pavlov's experimental practices and those common among investigators of the intermediate metabolic processes involved in digestion: In their experiments, Tiedemann and Gmelin, Voit, and others used simple foods (such as purified albumin, fibrin, starch, and fat), and Voit's Munich school had developed refined methods for purifying and standardizing diets. In sharp contrast, Pavlov's operational categories were the relatively crude "meat, bread, and milk." That is, he analyzed the response of the digestive organs to foods that dogs (and humans) actually ate in the normal course of their lives. For Pavlov, the use of these foods was appropriate to an organ physiology oriented toward synthetic knowledge about specifically "physiological reality and empirical rules of dietetics"; the "analytical" knowledge generated by more chemically oriented investigators would eventually be incorporated into the broader generalizations generated by his own research. As Pavlov explained to a session of the Society of Russian Physicians in 1896, in response to a question about the physical features of the food ingested by dogs during experiments, "in all our experiments we always take bread for bread, meat for meat, and milk for milk, since each type of food has its own special character. Further along, there arises the question: why does this [response] occur, what depends on the food and its volume, and what on the varied quantities of hard substances? All this relates to the issue of analysis. We must first of all establish the characteristics of [digestive] work with each food—and then comes analysis." See *PSS,* 6: 76.

Pavlov did take courses in chemistry at St. Petersburg University, but his physiological training included little if any instruction on physiological chemistry. Over time, the uniqueness of his laboratory techniques served, it seems, increasingly to isolate him from developments in other laboratories, unless these directly challenged his own conclusions. There is no record of any exchange in which Pavlov was challenged to explain his use of complex foods in his experiments, nor did any reviewers of *Lectures on the Work of the Main Digestive Glands* raise this question. My thanks to Larry Holmes for raising this question.

104. Pavlov, "Zhivosechenie," 14.

105. Ibid., 21.

106. Ivan Pavlov and E. O. Simanovskaia, "Innervatsiia zheludochnykh zhelez u sobaki" [1890], in *PSS*, 2, pt. 1: 189.

107. Pavlov, "Zhivosechenie," 19.

108. As we have already seen, Pavlov complained to his wife in 1882 that the "nasty" laboratory facilities available to him prevented him from conducting the chronic experiments on rabbits that he considered necessary for his doctoral thesis.

109. Ivan Pavlov, "Eksperimental'nye dannye k voprosu ob akkomodatsionnom mekhanizme krovenosnykh sosudov" [1877], in *PSS*, 1: 29.

110. Ivan Pavlov, "O normal'nykh kolebaniiakh krovianogo davleniia u sobaki" [1879], in *PSS*, 1: 72, 78.

111. Ibid., 75, 77. Pavlov's attempt to deal experimentally with the differences among various dogs also led him to employ a method he called *physiological cutting*. Rather than severing the vagus to observe the effects, he would freeze it. This allowed him to conduct the same experiments repeatedly on the same dog. See Ivan Pavlov, "Materialy k innervatsii krovenosnoi sistemy" [1882], in *PSS*, 1: 69.

112. I. P. Pavlov and E. O. Shumova-Simanovskaia, "Innervatsia zheludochnykh zhelez" [1890], in *PSS*, 2, pt. 2: 195–96.

113. Pavlov, "Zhivosechenie," 24–25.

114. Ibid., 25–26. This article was published in Russian two years before Ludwig's death and was never translated.

115. Ibid., 26.

116. Pavlov, "Pamiati R. Geidengaina," 104, 105. For a discussion of the theories in question, see Timothy Lenoir, "Science for the Clinic: Science Policy and the Formation of Carl Ludwig's Institute in Leipzig," in *The Investigative Enterprise*, ed. William Coleman and Frederic L. Holmes (Berkeley: University of California Press, 1988), 151–55; and Holmes, "Physiology and Experimental Medicine," 63–65.

117. Pavlov, "Pamiati R. Geidengaina," 107.

118. Ibid., 108.

119. Pavlov, "Lektsii po fiziologii," 270.

120. The Russian terms are *tselesoobraznost'* (purposiveness) and *prisposoblennost'* (adaptiveness). Pavlov expressed his conviction about the purposiveness of

physiological processes most fully and openly from the mid-1890s forward, but this conviction is also apparent in his earlier works. For example, in their article of 1890 Pavlov and Shumova-Simanovskaia noted that the time lag between psychic excitation and gastric secretion was a "rather determined" (i.e., consistent) phenomenon and therefore must have "a definite goal and a precisely operating mechanism." Pavlov and Shumova-Simanovskaia, "Innervatsiia zheludochnykh zhelez," 180.

121. Pavlov, "Eksperimental'naia psikhologiia i psikhopatologiia na zhivotnykh" [1903], in *PSS*, 3, pt. 1: 25. Pavlov put it this way in his speech on accepting the Nobel Prize in 1904: "As is clear to everyone, the animal organism is an extremely complex system, consisting of an almost infinite series of parts, which are joined, both one with the other, and so in the form of a single complex with the natural surroundings and being in complete harmony (*ravnovesii*) with it. The balance of this system, as with every other, is the condition of its existence. Where we are unable to find purposive connections, this depends only on our lack of knowledge." See Pavlov, "Nobelevskaia Rech'," 353.

122. Pavlov, "Eksperimental'naia psikhologiia i psikhopatologiia," 25–26. Here Pavlov specifically invokes Darwin to defend his adaptationist reasoning. My thanks to Roger Smith for a stimulating discussion of this subject.

123. Daniel P. Todes, *Darwin without Malthus: The Struggle for Existence in Russian Evolutionary Thought* (New York: Oxford University Press, 1989).

124. Pavlov's comment, from a discussion at a meeting of the Society of Russian Physicians in 1899, in *PSS*, 6: 145.

125. Ivan Pavlov, *Tsentrobezhnye nervy serdtsa* (Academy diss., 1893), in *PSS*, 1: 197. Here Pavlov said nervism was Botkin's "great service" to physiology. As Pavlov was well aware, however, a loosely defined *nervism* could be identified not only in the work of such other Russian physiologists as Sechenov and Tsion but also in that of many European physiologists of the time. Perhaps Pavlov's compliment to Botkin was sincere, but perhaps it was an attempt to flatter his powerful patron. Pavlov identified a certain cooling in Botkin's attitude toward him at the time and speculated that it might have resulted from his failure to acknowledge Botkin sufficiently in the conclusion to his doctoral thesis. "Obviously, the man [Botkin] likes groveling, it is to his taste." See I. P. Pavlov to Serafima Pavlova, June 13 [1883], in SPF ARAN 259.2.1300/2.

126. See my discussions in Chapters 5 and 9; and Pavlov, *Lektsii*, 150. For a different view of Pavlov's nervism, see Joravsky, *Russian Psychology*, 143–48. By *humoral response* scientists meant one transmitted through the body by a chemical carried in body fluids (the blood) rather than by the nervous system.

127. Ivan Pavlov to Serafima Karchevskaia, Thursday 25 [September 1880], in SPF ARAN 259.2.1300/2.

128. Ivan Pavlov to Serafima Pavlova, February 29, 1888; March 3, 1888; March 9, 1888; March 12, 1888; and March 15, 1888; in SPF ARAN 259.2.1300/2.

129. V. V. Savich, "Ivan Petrovich Pavlov," in *Sbornik, posviashchennyi 75-letiiu akademika Ivana Petrovicha Pavlova*, ed. V. L. Omelianskii and L. A. Orbeli (Leningrad: Gosudarstvennoe Izdatel'stvo, 1924), 11–12. Savich's reference to Pavlov having "halted in midstream" refers to the chief's response to the discovery in 1894 by one praktikant, Ivan Dolinskii, that the hydrochloric acid secreted by the gastric glands served as an exciter of the pancreatic gland. Pavlov accepted this finding but did not pursue the possibility of a humoral mechanism. Rather, he assigned another praktikant, Lev Popel'skii, to explore a potential nervist explanation for this phenomenon. For Popel'skii's conclusions, see my discussion in Chapter 9; and Horace W. Davenport, *A History of Gastric Secretion and Digestion: Experimental Studies to 1975* (New York: Oxford University Press, 1992), 187–88. Davenport also notes that Dolinskii had actually rediscovered a phenomenon uncovered by François Leuret and Jean-Louis Lassaigne seventy years earlier.

130. For Pavlov on the language of facts, see his "Eksperimental'naia psikhologiia i psikhopatologiia," 23; on science as facts, "Lektsii po fiziologii," 11, 13, 37. In Pavlov's usage, a *theoretician* was a person distracted from the demands of real life by chimerical abstractions. See, for example, his address to the Society of Russian Physicians in 1894, in which he denies being a "theoretician"—that is, somebody "who is prepared to look at the entirety of practical medicine as the application of physiology." See Pavlov, "O vzaimnom otnoshenii fiziologii i meditsiny v voprosakh pishchevareniia. Chast' I" [1894], in *PSS*, 2, pt. 1: 244. Similarly, he commented in 1899 that pharmacology was properly a part of physiology that developed in close relation to medical practice, but it had instead pursued "its own theoretical goals without connection to practical medicine." See *PSS*, 2, pt. 2: 277. In his lectures to medical students, Pavlov did add that "after you have seen the facts, I will be able to turn your attention to theory." But, as became clear in his subsequent lectures, this statement was intended to emphasize the importance of facts rather than prepare students for any discussion of "theory"—which was, in any event, not forthcoming. See Pavlov, "Lektsii po fiziologii," 56.

131. Pavlov, *Lektsii*, 11. For his remarks elsewhere, see, for example, in 1894, "O vzaimnom otnoshenii," 249; in 1897, *Lektsii*, 11; in 1899, "Sovremennoe ob"edinenie," 248; in 1911, "Lektsii po fiziologii," 39, 57.

132. Pavlov, "Vystupleniia v preniikh po dokladu N. A. Kashereininovoi" [1906], in *PSS*, 6: 251–52.

133. Pavlov, "Lektsii po fiziologii," 44.

134. Bernard, *Introduction*, 143.

135. Pavlov, "Lektsii po fiziologii," 83.

136. Bernard, *Introduction*, 173–78. My thanks to Larry Holmes for discussing with me the passage in Bernard's work on induced diabetes. Pavlov made the same distinction as did Bernard between the significance of "negative" and "positive" experimental facts. See, for example, Pavlov's "O sosudistykh tsentrakh v spinnom mozgu" [1877], in *PSS*, 1: 62.

Chapter 3. The Laboratory System

1. Babkin, *Pavlov* (MS), chap. 7, p. 10.

2. Claude Bernard, *An Introduction to the Study of Experimental Medicine* [1865], trans. Henry Copley Greene (New York: Dover Publications, 1957), 21–23.

3. Bernard, *Introduction*, 23–24.

4. As David Joravsky has noted, many years later Pavlov's friend and colleague Nikolai Kol'tsov observed of the coworkers who explored conditional reflexes that "these ephemeral scholars were often only hands, for which their teacher provided the head." For Joravsky, "the important question is whether or to what degree Pavlov turned his assistants into factory hands, made them do science in the regimented way that stifles creativity regardless of voice level or other peculiarities of command and execution. He was a pioneer of the twentieth century's 'big science,' which raises that troubling question everywhere." See David Joravsky, *Russian Psychology: A Critical History* (Oxford: Basil Blackwell, 1989), 390. See also N. K. Kol'tsov, "Trud zhizni velikogo biologa," *Biologicheskii Zhurnal* 5, no. 3 (1936): 401.

5. I discuss Pavlov's metaphor of digestive system as complex chemical factory in Chapter 5. Although he never directly called his laboratory a factory, he did equate the animal's digestive system both with a factory and with a laboratory in a manner suggesting he saw these two as interchangeable.

6. Pavlov's laboratory budget for 1891 was 3,200 rubles. See TsGIA SPb 2282.1.396: 164. His nearest competitor was I. R. Tarkhanov at the Military-Medical Academy, whose annual budget was 600 rubles. See Lev Popel'skii, *Istoricheskii ocherk kafedry fiziologii v Imperatorskoi Voenno-Medtsinskoi Akademii za 100 let (1798–1898)* (St. Petersburg, 1899), 118.

7. I have identified praktikanty from Pavlov's yearly reports to Ol'denburgskii in TsGIA SPb 2282.1. This information is amplified by two very useful volumes: D. G. Kvasov and A. K. Fedorova-Grot, *Fiziologicheskaia Shkola I. P. Pavlova* (Leningrad: Nauka, 1967); and N. M. Gureeva and N. A. Chebysheva, eds., *Letopis' zhizni i deiatel'nosti akademika I. P. Pavlova*, commentary by V. L. Merkulov (Leningrad: Nauka, 1969).

8. In the context of physiological experiments, a fistula is an inserted or artificially created channel between an organ and the exterior of the body or between two organs or (as in the Eck fistula) between two blood vessels.

9. The new chemistry building was completed in 1892 and the new pathology-anatomy building in 1893.

10. My discussion of this episode is based on examination of the relevant archives. Some of this material has been published in V. S. Meshkunov and A. M. Blokh, "Al'fred Nobel' i Imperatorskii Institut Eksperimental'noi Meditsiny v Sankt-Peterburge," *Voprosy istorii estestvoznaniia i tekhniki*, 1994, no. 1: 121–28. These authors have also provided some historical context for Nobel's contribution, which I have incorporated into my discussion. They indicate, for example,

that Emmanuel Nobel became an honorary "member-coworker" of the Institute in 1892, but the events leading to this are unknown. Alfred Nobel's letter to his nephew was written in French; a Russian translation appears in ibid., 124–25.

11. Alfred Nobel to Emmanuel Nobel, June 21, 1893, in TsGIA SPb 2282.1.47: 1–3. On the Russian branch of the Nobel family, see Robert W. Tolf, *The Russian Rockefellers: The Saga of the Nobel Family and the Russian Oil Industry* (Stanford: Hoover Institution Press, 1976). Pavlov referred to this gift from Nobel in his speech on accepting the Nobel Prize in 1904, recalling that "Alfred Nobel displayed in his letter a lively interest in physiological experiments and proposed several very instructive experimental projects that touched upon the greatest physiological tasks—the question of the aging and death of organisms." See Pavlov, "Nobelevskaia rech', proiznesennaia 12 dekabria 1904 g. v Stokgol'me," in *PSS*, 2, pt. 2: 358.

12. Ivan Pavlov to Emmanuel Nobel, May 18, 1894, in TsGIA SPb 2282.1.47: 19. In his official report to the Tsar for 1894, Ol'denburgskii referred to the use of this money for the physiology building as Nobel's original intention.

13. E. A. Ganike, "Vospominaniia ob Ivane Petroviche Pavlove," in SPF ARAN 259.4.82. Pavlov's collaborative work with Nencki on urine may also have been encouraged by Alfred Nobel's interest in this subject. Nobel described his ideas about urine in a letter of March 31, 1893, to Emmanuel, who may have relayed them to Ol'denburgskii or Pavlov. This letter is published in Meshkunov and Blokh, "Al'fred Nobel'," 123–24.

14. I. P. Pavlov, "K khirurgicheskoi metodike issledovaniia sekretornykh iavlenii zheludka" [1894], in *PSS*, 2, pt. 1: 275–76.

15. In much the same way, Latour's Pasteur insisted that the simplifying precision of laboratory microbiology could reveal secrets of infectious disease that would always remain invisible to hygienists encountering the multifactorial complexities of illness outside the laboratory. See Bruno Latour, *The Pasteurization of France* (Cambridge, Mass.: Harvard University Press, 1988).

16. Pavlov, *Lektsii*, 37; Lectures (trans.), 18. Unless otherwise noted, all translations of *Lektsii* are my own. See also Pavlov, "On the Surgical Method of Investigation of the Secretory Phenomena of the Stomach" (K khirurgicheskoi metodike issledovaniia sekretornykh iavlenii zheludka) [1894] in *PSS*, 2, pt. 1: 275–81.

17. Pavlov, *Lektsii*, 38; *Lectures* (trans.), 18.

18. Bernard, *Introduction*, 14.

19. Pavlov, "K khirurgicheskoi metodike," 275.

20. Ibid.; Pavlov, *Lektsii*, 37; *Lectures* (trans.), 18.

21. Pavlov's "Vivisection" (Zhivosechenie) [1893], in *PSS*, 6: 9–27; "On the Surgical Method of Investigation of Secretory Phenomena"; "On the Mutual Relations of Physiology and Medicine in Questions of Digestion" [1894–95], in *PSS*, 2, pt. 1: 245–63 (pt. 1) and 264–74 (pt. 2); *Lectures on the Work of the Main Digestive Glands* (Lektsii o rabote glavnykh pishchevaritel'nykh zhelez) [1897], in *PSS*, 2, pt. 2: 11–

198; "The Contemporary Unification in Experiment of the Main Aspects of Medicine, as Exemplified by Digestion" (Sovremennoe ob"edinenie v eksperimente glavneishikh storon meditsiny na primere pishchevareniia) [1899], in *PSS*, 2, pt. 2: 247–84; "Physiological Surgery of the Digestive Canal" (Fiziologicheskaia khirurgiia pishchevaritel'nogo kanala) [1902], in *PSS*, 2, pt. 2: 285–334; "The Psychical Secretion of the Salivary Glands" (Complex Nervous Phenomena in the Work of the Salivary Glands) (O psikhicheskoi sekretsii sliunnykh zhelez [Slozhno-nervnye iavleniia v rabote sliunnykh zhelez]) [1904], in *PSS*, 3, pt. 1: 40–57.

22. The one woman, E. O. Shumova-Simanovskaia, was Pavlov's collaborator, friend, and benefactor from pre-Institute days.

23. A minority of the praktikanty, thirteen of ninety-nine in the years 1890–1904, came to the laboratory with a doctorate already in hand—they are not considered in this percentage. Twenty praktikanty worked in the laboratory for only one year; the great majority of these did not complete their doctoral theses. Information about this group would much enhance our understanding of laboratory dynamics but is unfortunately unavailable. Thirty praktikanty spent two years in the laboratory, twenty-nine spent three years, nine spent four years, and one, V. V. Savich, served as a semipermanent praktikant until becoming an assistant in Pavlov's physiological laboratory at the Military-Medical Academy.

24. Babkin, *Pavlov* (MS), chap. 7, p. 12.

25. D. A. Kamenskii, "Moe znakomstvo s Ivanom Petrovichem," in *I. P. Pavlov v vospominaniiakh sovremennikov*, ed. E. M. Kreps (Leningrad: Nauka, 1967), 105.

26. On Khizhin's contribution, see Chapter 4; on Snarskii's and Tolochinov's contributions, see Chapter 7. I enlarge on the term *dog-technology* later in the text.

27. *Vrach*, 1898, no. 7: 212.

28. These assistants, with their years of service, were V. N. Massen (1891–93), Iu. M. Iablonskii (1891–94), E. A. Ganike (1894–1936), E. A. Kotliar (1895), N. I. Damaskin (1895–98), and A. P. Sokolov (1899–1909). The "member-coworker" was G. A. Smirnov (1893–1934). Smirnov's duties were the same as those of the assistants, but he chose his research topics independently. In 1916 Ganike became head of the newly formed Physico-Physiological Division at the Institute, but this served largely as a workshop attached to Pavlov's factory.

29. D. A. Sokolov, *25 let bor'by: Vospominaniia vracha* (St. Petersburg, 1910), 77.

30. Babkin, *Pavlov* (MS), chap. 14, pp. 5–6.

31. Sokolov, *25 let bor'by*, 31; Serafima Vasil'evna Pavlova, *Vospominaniia*, in SPF ARAN 259.1.170: 505; Babkin, *Pavlov* (MS), chap. 7, p. 11.

32. Babkin, *Pavlov* (MS), chap. 10, p. 11.

33. In *Lords of the Fly: Drosophila Genetics and the Experimental Life* (Chicago: University of Chicago Press, 1994), Robert Kohler approaches laboratory *Drosophila* as both biological entities and technologies (see his general discussion on pp. 6–11). For analysis of another organism-technology, see Bonnie Clause, "The Wistar Rat as a Right Choice: Establishing Mammalian Standards and the Ideal of

a Standardized Mammal," *Journal of the History of Biology* 26 (1993): 329–49. My approach to this duality is somewhat different, no doubt in part because of the differences among fly, rat, and dog, the various laboratories, and the precise experimental uses to which these organism-technologies were put.

34. Bruno Latour, "The Costly Ghastly Kitchen," in *The Laboratory Revolution in Medicine,* ed. Andrew Cunningham and Perry Williams (Cambridge: Cambridge University Press, 1992), 299.

35. On "local knowledge," see Susan Leigh Star, "Scientific Work and Uncertainty," *Social Studies of Science* 15 (1985): 391–428; and Harry M. Marks's unpublished manuscript, "Local Knowledge: Experimental Communities and Experimental Practices, 1918–1950" (1988).

36. Pavlov, "Fiziologicheskaia khirurgiia pishchevaritel'nogo kanala" [1902], in *PSS,* 6: 286. This article was originally published in German.

37. Pavlov, *Lektsii,* 22; *Lectures* (trans.), 4. See also Pavlov, "Fiziologicheskaia khirurgiia," 289.

38. On the history of the gastric fistula and the isolated sac, see Horace W. Davenport, *A History of Gastric Secretion and Digestion* (New York: Oxford University Press, 1992), esp. 138–43.

39. In my article on Pavlov's laboratory for *Isis* ("Pavlov's Physiology Factory," *Isis* 88, no. 2 [1997]: 205–26), I wrote that these fistulas did not result in "any visible pathological symptoms" (p. 223). I have since discovered the observation by one praktikant that the gastric fistula sometimes caused hypersecretion: "In dogs with gastric fistulas there develops sometimes, especially if they are not cared for with special attention, an extraordinary lability of the secretory function of the stomach, almost a pathological one. Among these dogs, the influence of even insignificant conditions elicits a plentiful secretion of gastric juice that, by its dimensions, far exceeds the established norm for those conditions." See A. A. Val'ter, *Otdelitel'naia rabota podzheludochnoi zhelezy* (Academy diss., St. Petersburg, 1897), 110–11.

40. For the concession that the pancreatic fistula was "not ideal," see Pavlov, *Lektsii,* 27–28; *Lectures* (trans.), 8. For his assessment five years later, see Pavlov, "Fiziologicheskaia khirurgiia," 290, 309, 312, 313. Pavlov noted that one dog in five had a "favorable individual predisposition" that enabled it to survive the operation with relative ease.

41. Ivan Pavlov, "O vzaimnom otnoshenii fiziologii i meditsiny" [1894], in *PSS,* 2, pt. 1: 251.

42. On the "Heidenhain stomach," see Davenport, *History of Gastric Secretion,* 14, 140. Heidenhain did concede that if emotions affected gastric secretion, as was sometimes reported, this would be evidence for the importance of central nervous mechanisms. I discuss this briefly in Chapter 4. On Pavlov's modifications, see Pavlov, "K khirurgicheskoi metodike," 279. The isolated sac was sometimes referred to as the "Heidenhain-Pavlov sac." The difficulty of convincing Russian clin-

icians that the isolated sac reflected normal gastric secretion is evident in the published protocols of the discussion at the Society of Russian Physicians in 1894. See *Trudy obshchestva russkikh vrachei* 61 (September 1894): 42–46; an abridged version of this discussion is published in *PSS,* 6: 40–45.

43. Pavlov, *Lektsii,* 33, 147; *Lectures* (trans.), 13, 108.

44. Pavlov made this remark while invoking his experience with the disorders among his laboratory dogs as a source of authority in discussions of pathology. See I. P. Pavlov, "Laboratornye nabliudeniia nad patologicheskimi refleksami s briushnoi polosti" [1898], in *PSS,* 1: 553–54. On the deterioration of Druzhok's isolated sac, see A. N. Volkovich, *Fiziologiia i patologiia zheludochnykh zhelez* (Academy diss., Kronstadt, 1898), 41–42. The laboratory seems never to have explored the potentially subversive implications of Pavlov's and Volkovich's observations for conclusions based on Druzhok's "normalcy." Rather, Druzhok's illness, and that of Sultan, the second dog to receive an isolated sac, was used to launch a new line of investigation: the experimental pathology and therapeutics of digestion. See my brief discussion on this in Chapter 4.

45. Susan Abrams discusses this aspect of the laboratory dog in her unpublished manuscript "A Dog's Life: Conflict and Contradiction in Horsley Gantt's Pavlovian Laboratories" (1994). See also Michael Lynch, "Sacrifice and Transformation of the Animal Body into a Scientific Object: Laboratory Culture and Ritual Practice in the Neurosciences," *Social Studies of Science* 18 (1988), 265–89.

46. The acknowledged importance of the psyche was the primary source of these interpretive moments, but hardly the only one. As experience with various surgical operations increased, even dogs-as-technologies acquired a "personality" of sorts. For example, the size of the isolated stomach varied from dog to dog, requiring some mathematical recalculations to compare the secretory responses in two animals. Similarly, in later years, with a growing appreciation of the differences between the fundal and pyloric regions of the stomach, the location of the isolated sac acquired significance. See, for example, Ia. Kh. Zavriev (Abo-Zavaridze), *Fiziologiia i patologiia zheludochnykh zhelez sobaki* (Academy diss., St. Petersburg, 1900), 155.

47. Pavlov, *Lektsii,* 102, 104; *Lectures* (trans.), 73, 75; Pavlov, "Fiziologicheskaia khirurgiia," 304–5. The relationship between the presumably *pravil'nye* nervous-chemical mechanisms and the idiosyncratic psychic mechanism was a central subject of laboratory experiments and deliberations, a subject I address in Part II.

48. I. O. Lobasov, *Otdelitel'naia rabota zheludka sobaki* (Academy diss., St. Petersburg, 1896), 30–31, 32–33.

49. TsGIA SPb 2282.56.1: 98.

50. P. P. Khizhin, *Otdelitel'naia rabota zheludka sobaki* (Academy diss., St. Petersburg, 1894), 153.

51. G. Ushakov, "Laboratoriia Pavlova v Institute eksperimental'noi meditsiny," in Kreps, *Pavlov v vospominaniiakh,* 248.

52. I. P. Pavlov, *Lektsii po fiziologii* [1911–13], in *PSS*, 5: 26.

53. Pavlov, *Lektsii*, 11–12; *Lectures* (trans.), ix.

54. A. A. Val'ter, *Otdelitel'naia rabota podzheludochnoi zhelezy* (Academy diss., St. Petersburg, 1897), 35.

55. A. S. Sanotskii, *Vozbuditeli otdeleniia zheludochnogo soka* (Academy diss., St. Petersburg, 1892), 19, 16, 11, 51, 39. I have discovered only two exceptions to this pattern of the use of *I* and *we*: both Popel'skii and Tolochinov refer in their work to "my" decisions and conclusions. Each subsequently clashed with the chief. See L. B. Popel'skii, *O sekretorno-zaderzhivaiushchikh nervakh podzheludochoi zhelezy* (Academy diss., St. Petersburg, 1896); and I. Tolotschinoff [Tolochinov], "Contribution à l'étude de la physiologie et de la psychologie des glandes salivaires," in *Comptes Rendus du Congres des Naturalistes et Médecins du Nords* (Helsingfors [Helsinki], 1902), 42–46.

56. Babkin, *Pavlov* (MS), chap. 7, p. 9.

57. I. S. Tsitovich, "Kak ia uchilsia i rabotal u Pavlova," in Kreps, *Pavlov v vospominaniiakh*, 260; and Boris Babkin, *Pavlov: A Biography* (Chicago: University of Chicago Press, 1949), 116–17. Babkin adds, "Pavlov was not greatly interested in the general education in physiology even of his most earnest pupils. Once, at the very beginning of my work in his laboratory, I asked his advice on how best to learn physiology. He quickly replied: 'Read the *Ergebnisse* [*der Physiologie*] and so approach the subject gradually' and at once turned the conversation to laboratory matters." See Babkin, *Pavlov*, 117.

58. The research on the psychic secretion of the salivary gland, which eventually shifted laboratory investigations to the study of conditional reflexes, is the subject of Chapter 7. Pavlov's response to the discovery of secretin is discussed briefly in that chapter.

59. For the work to improve the dog-technology, see V. N. Vasil'ev, *O vliianii raznogo roda edy na deiatel'nost' podzheludochnoi zhelezy* (Academy diss., St. Petersburg, 1893); and Iu. M. Iablonskii, *Spetsificheskoe zabolevanie* (Academy diss., St. Petersburg, 1894). For the work on exciters of pancreatic secretion, see I. L. Dolinskii, *O vliianii kislot na otdelenie soka podzheludochnoi zhelezy* (Academy diss., St. Petersburg, 1894); I. O. Shirokikh, "Spetsificheskaia vozbudimost' slizistoi obolochki pishchevaritel'nago kanala," *Arkhiv Biologicheskikh Nauk* 3, no. 5 (1895); and N. I. Damaskin, "Deistvie zhira na otdelenie podzheludochnogo soka," *Trudy obshchestva russkikh vrachei za 1895–96* 63 (February 1896): 7–14. For elucidation of the secretory patterns of the pancreas, see Val'ter, *Otdelitel'naia rabota;* for verification of these patterns, see A. R. Krever, *K analizu otdelitel'noi raboty podzheludochnoi zhelezy* (Academy diss., St. Petersburg, 1899). Further elucidation of nervous control is provided in Popel'skii, *O sekretorno-zaderzhivaiushchikh;* and W. Sawitsch [V. V. Savich], "Die Wirkung des Wagus auf Pancreas," in *Comptes Rendus du Congres des Naturalistes et Médecins du Nords* (Helsingfors [Helsinki], 1902), 41–42.

60. I discuss the verification of Val'ter's results by Krever and Bukhshtab in Chapter 6.

61. P. Borissow [P. Borisov] and A. Walther [A. Val'ter], "Zur analyse der Saure-wirkung auf die Pancreassecretion," in *Comptes Rendus du Congres des Natural-istes et Médecins du Nord* (Helsingfors [Helsinki], 1902), 42; V. V. Savich, "Mekha-nizm otdeleniia podzheludochnago soka," *Trudy obshchestva russkikh vrachei za 1903–1904* 72 (November-December 1904): 99–103; Ia. A. Bukhshtab, "O rabote podzheludochnoi zhelezy posle pererezki vnutrennostnykh i bluzhdaiushchikh nervov," *Trudy obshchestva russkikh vrachei za 1903–1904* 71 (March-May 1904): 72–78.

62. One way of reading Gerald Geison's account of the competition between Jean-Joseph Touissant and Louis Pasteur is as a mismatch between a workshop and a factory. Touissant, in his workshop, was able to pursue only one line of investi-gation at a time, while Pasteur, in his factory, pursued several (including Touis-sant's). See Geison, *The Private Science of Louis Pasteur* (Princeton: Princeton Uni-versity Press, 1995), 145–76.

63. Ivan Pavlov to Vladimir Pavlov, May 23 [1912], in *Perepiska Pavlova*, ed. E. M. Kreps (Leningrad: Nauka, 1970), 427.

64. Babkin, *Pavlov* (MS), chap. 7, p. 11.

65. See, for example, N. D. Strazhesko, "Vospominaniia o vremeni, proveden-nom v laboratorii Ivana Petrovicha Pavlova," in Kreps, *Pavlov v vospominaniiakh*, 225.

66. Babkin, *Pavlov* (MS), chap. 11, pp. 19–20.

67. I. S. Tsitovich, "Kak ia uchilsia," 255.

68. Ibid.

69. Babkin, *Pavlov* (MS), chap. 13, p. 4.

70. Ibid., chap. 7, p. 11.

71. Ibid., chap. 14, p. 1.

72. A. F. Samoilov, "Obshchaia kharakteristika issledovatel'skogo oblika I. P. Pavlova," in Kreps, *Pavlov v vospominaniiakh*, 203–4.

73. See, for example, V. V. Savich, "Ivan Petrovich Pavlov: biograficheskii ocherk," in *Sbornik posviashchennyi 75-letiiu akademika I. P. Pavlova* (Leningrad, 1924), 24.

74. Babkin, *Pavlov*, 112.

75. L. A. Orbeli, "Pamiati Ivana Petrovicha Pavlova," in Kreps, *Pavlov v vospom-inaniiakh*, 163–64.

76. V. P. Kashkadamov, "Iz vospominanii o rabote v Institute eksperimental'noi meditsiny (1894–1897 gg.)," in Kreps, *Pavlov v vospominaniiakh*, 109.

77. Orbeli, "Pamiati," 164.

78. Tsitovich, "Kak ia uchilsia," 256.

79. Orbeli, "Pamiati," 171. For the final product of this research, see V. N. Boldyrev, *Periodicheskaia rabota pishchevaritel'nago apparata pri pustom zheludke* (Academy diss., St. Petersburg, 1904). This episode also demonstrates how the ac-

knowledged importance of the psyche could be used to explain away discordant results.

80. On Pavlov's temper, see, for example, Tsitovich's recollection (in "Kak ia uchilsia," 263) that, when dissatisfied, Pavlov frequently screamed at coworkers and at himself: "Those surrounding him at such times froze in their place, since at such a moment it was easy to fall victim to his hot hand."

81. My thanks to Gerald Geison for suggesting the term *literary products,* which he uses in his unpublished paper "Organization, Products, and Marketing in Pasteur's Scientific Enterprise" (1996).

82. W. N. Boldyreff [V. N. Boldyrev], "I. P. Pavlov as a Scientist," *Bulletin of the Battle Creek Sanitarium* 24 (1929): 224.

83. Tsitovich, "Kak ia uchilsia," 263.

84. Babkin, *Pavlov* (MS), chap. 11, p. 21.

85. Tsitovich, "Kak ia uchilsia," 263.

86. We have already encountered accounts of Pavlov's memory for praktikanty's data, in the recollections of Babkin (*Pavlov,* 112) and Orbeli ("Pamiati," 163–64).

87. Geison, *Private Science,* 237.

88. Babkin, *Pavlov* (MS), chap. 11, p. 21. For an especially dramatic change in direction in a dissertation, see Krever's *K analizu otdelitel'noi raboty.* See also Khizhin, *Otdelitel'naia rabota,* 104 (for a tentative suggestion) and 117 (where, in a summary, the suggestion becomes a "quite definite conclusion"); similarly, compare Lobasov, *Otdelitel'naia rabota,* 89 and 98.

89. Sanotskii, *Vozbuditeli otdeleniia,* 9.

90. Val'ter, *Otdelitel'naia rabota,* 23.

91. Ia. A. Bukhshtab, *Rabota podzheludochnoi zhelezy posle pererezki bluzhdaiushchikh i vnutrennostnykh nervov* (Academy diss., St. Petersburg, 1904), 45–46.

92. N. P. Kazanskii, *Materialy k eksperimental'noi patologii i eksperimental'noi terapii zheludochnykh zhelez sobaki* (Academy diss., St. Petersburg, 1901), 22.

93. Vasil'ev, *O vliianii raznogo roda edy,* 23; Krever, *K analizu,* 20 (I provide a more detailed discussion of this in Chapter 6); Zavriev, *Fiziologiia i patologiia,* 92; and Kazanskii, *Materialy,* 27.

94. Kazanskii, *Materialy,* 23–24.

Chapter 4. The Remarkable Druzhok

1. E. A. Ganike, "Vospominaniia ob Ivane Petroviche Pavlove," in SPF ARAN 259.4.82; republished as "Ob odnoi mechte Ivana Petrovicha," in *I. P. Pavlov v vospominaniiakh sovremennikov,* ed. E. M. Kreps (Leningrad: Nauka, 1967), 76–77.

2. I. P. Pavlov, "Innervatsia podzheludochnoi zhelezy" [1888], in *PSS,* 2, pt. 1: 96–132.

3. As with his work on the pancreatic gland, Pavlov first published a short

"preliminary communication" in both Russian and German in 1889. The longer report appeared in Russian in 1890: I. P. Pavlov and E. O. Shumova-Simanovskaia, "Innervatsia zheludochnykh zhelez u sobaki," republished in *PSS*, 2, pt. 1: 175–96. It was also published, with some interesting changes, in German in 1895; the editors of *PSS* compare the Russian and German variants (pp. 196–99). The physiological argumentation in this article was probably largely or even exclusively Pavlov's: as the article itself made clear, the central line of analysis followed directly from his long-standing investigatory interests and inclinations. Judging by her background, training, and subsequent research, Shumova-Simanovskaia's contribution was probably largely limited to the chemical analysis of the gastric secretions obtained during experiments.

For Pavlov's comments to his wife about his difficulties in establishing the existence of vagal control over the gastric glands, see Chapter 2.

4. R. Heidenhain, *Handbuch der Physiologie der Absonderung und Aufsaugung*, vol. 5, pt. 1, of L. Hermann's, *Handbuch der Physiologie* (Leipzig: F. C. W. Vogel, 1883). In his article with Shumova-Simanovskaia ("Innervatsia zheludochnykh zhelez u sobaki"), Pavlov cited the Russian edition of this work: Heidenhain, *Fiziologiia*, 150.

5. For Heidenhain's review of these experiments, see his *Fiziologiia*, 142–50.

6. For example, in the 1891 edition of his famous physiology text, Michael Foster, following Heidenhain, regarded the possibility of central nervous involvement in gastric secretion with even greater skepticism than he had in earlier editions. For Foster, there were "no facts which afford satisfactory evidence" of a role for the central nervous system in gastric secretion. The "secretion of quite normal gastric juice" from the Heidenhain sac (in which the vagal nerves were completely severed) demonstrated the "subordinate value of any connection between the gastric membrane and the central nervous system," and "all attempts to provoke or modify gastric secretion by the stimulation of the nerves going to the stomach have hitherto failed." Michael Foster, *A Text-Book of Physiology*, 4th Amer. ed. (Philadelphia: Lea Brothers and Co., 1891), 338.

7. Heidenhain, *Fiziologiia*, 143–44; Heidenhain discusses the results obtained by Bidder and Schmidt, Richet, Schiff, and Braun.

8. Ibid., 144

9. Pavlov and Shumova-Simanovskaia, "Innervatsia zheludochnykh zhelez," 177.

10. In their article of 1890, Pavlov and Shumova-Simanovskaia introduced the term *sham-feeding* (*mnimoe kormlenie*) for this procedure. In teasing (*poddraznivanie*) experiments, on the other hand, the dog was shown the food but was not permitted to eat it.

11. Ibid., 178. After a complete vagotomy was performed, however, the dogs were condemned to severe physiological problems and certain death.

12. Ibid., 181, 178.

13. This admission was deleted from the German edition of the article, as were two previous sentences containing some theoretical ruminations.

14. Ibid., 178, 180.

15. N. Ia. Ketcher, "Refleks s polosti rta na zheludochnoe otdelenie," *Trudy obshchestva russkikh vrachei* 57 (1890–1891): 24–30. I have been unable to secure a copy of his doctoral dissertation. Ketcher's article contains the first-person-singular language noticeably absent in the later products of praktikanty in the Institute's Physiology Division. For example, "it seems to me that" (p. 24) and "my opinion" (pp. 26–27).

16. Ibid., 25–26.

17. Ibid., 25. Like Pavlov and Shumova-Simanovskaia, Ketcher worried about the long "latency" period of five to seven minutes between eating and gastric secretion. He proposed that this represented a "summation of separate irritations"—that is, secretion began only when a certain threshold had been crossed. Also like Pavlov and Shumova-Simanovskaia, Ketcher attributed to this latency period a certain "sense," suggesting that its purpose was to give the ptyalin in salivary juice time to act on starchy foods.

18. A. S. Sanotskii, *Vozbuditeli otdeleniia zheludochnago soka* (Academy diss., St. Petersburg, 1892), 10. Iurgens confirmed the earlier findings, concluding that the reflex from the roof of the mouth ceased entirely when the vagus was severed. See N. P. Iurgens, *O sostoianii pishchevaritel'nago kanala pri khronicheskom paraliche bluzhdaiushchikh nervov* (Academy diss., St. Petersburg, 1892).

19. Sanotskii, *Vozbuditeli*, 9.

20. The feeding procedures were the same as developed by Ketcher, with one difference: dogs were given a portion of meat to eat normally, although (owing to the esophagotomy) the meat did not reach the stomach and so had no nutritive value. The purpose of this, I surmise, was to keep all the mechanisms involved in the act of eating in good shape.

21. Ibid., 9. As we shall see below, this stricture left much room for interpretation.

22. Ibid., 19–20. Sanotskii also noticed that when dogs entered the laboratory, they often began to secrete gastric juice immediately. This, Sanotskii (and Pavlov) concluded, was also a "psychic secretion" that resulted from the excitation (and expectation of food) elicited by the experimental setting itself. Sanotskii discovered that the average acidity and proteolytic power of this initial secretion was virtually identical to that of the psychic secretion elicited by teasing experiments. As Sanotskii also discovered, this high level of proteolytic power distinguished psychic secretion from the gastric secretions elicited in the second stage of digestion, when food excited the nerves of the mucous membrane of the stomach. Thus he (and Pavlov) concluded that this initial secretion was the result of psychic excitation and named it *volitional secretion (proizvol'noe otdelenie)*. Ibid., 20–21. As we saw in Chapter 3, in 1904 V. N. Boldyrev finally convinced the chief that volitional

secretion was sometimes the product of periodical gastric activity that occurred without an external stimulus.

The method used in Pavlov's laboratory for measuring proteolytic power is described below (n. 51).

23. Ibid., 19–20. Sanotskii noted that in the one of twenty trials in which teasing had failed to elicit a psychic secretion, the dog had eaten a mere thirteen hours before.

24. Ibid., 26.

25. *Positive* results with experiments of this type would later be important in research on conditional reflexes.

26. For Sanotskii's account of these experiments, see ibid., 34–39. Of the three dogs that were force-fed balls of pitch and mustard-soaked meat, one, in fact, did manifest gastric secretion. Sanotskii dismissed this result, however, with the observation that the amount of secretion was "not so great as after sham-feeding." Furthermore, the dog's personality and physiological state made it an unreliable experimental animal: "This last dog, first, was always distinguished by a rather great impressionability toward everything that was in any way linked with the question of food; and, on the other hand, these experiments upon it coincided with the moment when the animal began to manifest a tendency toward a progressive loss of weight. At the time of the experiments, the weight loss was so insignificant as to go unnoticed; only when it began to progress, moreover, when the extraordinary irritability of the gastric glands was striking (a type of hypersecretion at any slight pretext) did it become probable that during the time of these experiments, too, the animal was not in a normal state. Two months after these experiments the dog perished." Ibid., 39.

27. Sanotskii fed the esophagotomized dogs through a fistula, a procedure that was later seen as unreliable because of the possibility that this very process might arouse a psychic secretion.

28. For Sanotskii's experiments and reasoning, see ibid., 45–85. His approach was to discredit by experimentation any other possible mechanisms (such as the mechanical stimulation of the stomach wall, the effect of saliva, Schiff's theory that food products in the blood had a "peptogenic" effect on inactive "propepsin").

29. Ibid., 84–85. The word *svoeobraznyi* appears very frequently in laboratory descriptions of the psyche as a secretory mechanism. It can mean either "distinctive" or "idiosyncratic." Because my translation varies with the context, I include the original Russian word to alert the reader to my own interpretive decisions.

30. Ibid., 43.

31. I discuss this investigation and its clinical aftermath in Chapter 8.

32. My thanks to Professor Graham Dockray of the University of Liverpool for showing me R. A. Gregory's film about physiological surgery, for discussing with me the nature of this operation, and for impressing upon me the difficulties involved in performing it for the first time.

33. P. P. Khizhin, *Otdelitel'naia rabota zheludka sobaki* (Academy diss., St. Petersburg, 1894), 14.

34. Ibid., 24–26. For Pavlov's speech to the Society of Russian Physicians, see "K khirurgicheskoi metodike issledovaniia sekretornykh iavlenii zheludka" [March 1894], in *PSS,* 2, pt. 1: 276–81.

35. Ivan Pavlov to Serafima Pavlova, June 5 [1894], and E. A. Ganike, "Vospominaniia ob Ivane Petrovichem Pavlove," both in SPF ARAN 259.4.82; Boris Babkin, *Pavlov: A Biography* (Chicago: University of Chicago Press, 1949), 99. Some thirty years later, Pavlov spoke warmly about Khizhin's courage in the face of these repeated failures, which "threatened him with complete failure in his career." See I. N. Zhuravlev, "Moi vpechatleniia ob I. P. Pavlove I ego laboraotorii," in Kreps, *Pavlov v vospominaniiakh,* 91.

36. In referring to the experimental dogs—regardless of their sex—laboratory publications usually employed the feminine pronoun, since the word *dog* (*sobaka*) is feminine in Russian.

37. Khizhin, *Otdelitel'naia rabota,* 26–27.

38. Ibid., 12, 48–49. Pavlov also repeatedly made clear that the operation was his innovation, for example in his "K khirurgicheskoi metodike," 279

39. For Pavlov's remark that he had operated on Gordon, see his first public comments about the isolated stomach in his "K khirurgicheskoi metodike," 279. The closest he came to a similar pronouncement about Druzhok was his reference to the changes that "we (I and doctor Khizhin)" had introduced to the Heidenhain sac, in *Lektsii,* 32; *Lectures* (trans.), 11. (Unless otherwise noted, all translations of *Lektsii* are my own.)

40. Ganike, "Ob odnoi mechte"; A. F. Samoilov, "Obshchaia kharakteristika issledovatel'skogo oblika I. P. Pavlova," in Kreps, *Pavlov v vospominaniiakh.* Babkin (*Pavlov,* 97) strongly implies that Pavlov operated on Druzhok, but Babkin was not present in the laboratory at the time.

41. See Khizhin's *Otdelitel'naia rabota,* 12; and "Otdelitel'naia rabota zheludka u sobak," *Trudy obshchestva russkikh vrachei 1894–1895 gg* 61 (September 1894): 19. In the latter report, Khizhin credited Pavlov by name with the idea for the isolated sac (p. 19) and for a surgical innovation to overcome one obstacle in the operation (p. 21). But regarding the actual operation on Druzhok, he said only, "we succeeded in achieving a complete and solid success" and "we finally succeeded in acquiring a dog" with a fully isolated stomach (p. 20). Another interesting silence concerns Druzhok's "birthplace," as discussed later.

42. Khizhin frequently noticed that Druzhok's gastric glands were active even before the dog was excited with food during experiments; this was attributed to volitional secretion.

43. Khizhin, *Otdelitel'naia rabota,* 31–32.

44. Ibid., 30–31.

45. Ibid., 24.

46. Pavlov's calendar book for this period is filled with references to various administrative tasks: meetings with Prince Ol'denburgskii and other Institute personages, purchasing equipment from an instrument factory, and so forth. Very few notes concern scientific subjects and none mention Khizhin and Druzhok. This notebook is preserved in SPF ARAN 259.1.59/4.

47. Khizhin, *Otdelitel'naia rabota,* 33.

48. Khizhin conceded that the secretion of the Pavlov sac was quantitatively small, since the sac only encompassed about one-fifth of the surface of the normal stomach, but he insisted that this did not influence the essential correspondence between the sac's secretions and those of the intact stomach.

49. Ivan Pavlov to Serafima Pavlova, June 3 and June 5 [1894], copy in SPF ARAN 259.7.1300/2.

50. Khizhin's logic was described sympathetically by Lobasov, who inherited Druzhok from him and conducted experiments refuting Khizhin's results. See I. O. Lobasov, *Otdelitel'naia rabota zheludka sobaki* (Academy diss., St. Petersburg, 1896), 68–69. In his *Text-Book of Physiology,* Foster related the well-known fact that "the essential property of gastric juice is the power of dissolving proteid [protein] matters and of converting them into a substance called peptone" (p. 309).

51. The Mett method for determining proteolytic power was based on the rate of proteolysis of coagulated egg white. Fluid egg white was coagulated in fine glass tubes, the tubes placed in the fluid to be investigated, then measurements taken at desired time intervals to determine how much of the egg white had been consumed, in millimeters. This provided a quantitative measure of proteolytic power. The laboratory view changed over time on whether, for purposes of comparison, this number (in millimeters) or its cube (cubic millimeters) best expressed the proteolytic power of a secretion. (S. G. Mett had earlier worked with Pavlov in Botkin's laboratory.)

52. Khizhin, *Otdelitel'naia rabota,* 132. It is interesting that neither Khizhin nor Pavlov was unsettled by the fact that in these two experiments, different quantities of peptone produced the same secretory result (the solution for June 1 contained 10 grams of peptone; that for June 3, 20 grams). This is one of many examples of apparent contradictions within the same text, since in other parts of the dissertation Khizhin argues for a fairly precise relationship between the quantity of exciting substances (specific foods) and the quantity of gastric secretion.

53. To avoid experimental error introduced by impurities in the peptone preparation, Lobasov used "pure peptone" supplied by Nencki's Chemical Division at the Institute.

54. Ibid., 120. As we have seen, Sanotskii used such reasoning in 1892 to reinterpret Ketcher's experiments with force-feeding.

55. Ibid., 96.

56. Lobasov, *Otdelitel'naia rabota,* 71–74.

57. In *Lektsii,* Pavlov wrote about their "erroneous" conclusion that peptone ex-

cited the secretory apparatus of the stomach. See *Lektsii,* 131; *Lectures* (trans.), 96–97. As so often occurs in the history of science, this epilogue to the peptone experiments is a prelude to another epilogue: current physiology textbooks consistently include peptone as an exciter of the gastric glands.

58. Pavlov, *Lektsii,* 130; *Lectures* (trans.), 96. Though the word *protein* was coming into common usage by the 1890s, Pavlov always used *proteid,* as did Thompson in the English translation of *Lectures* in 1902.

59. Sanotskii, *Vozbuditeli,* 85, 40. The language remained much the same in the more rounded definition offered in Lobasov's dissertation of 1896: psychic secretion was not a simple reflex but rather the "result of a psychic moment—of desire, the expectation of food, and the subsequent pleasure of eating . . . The secretion of juice during sham-feeding or—what is the same—during the actual act of eating, is a result of a distinctive (*svoeobraznyi*) psychic process similar to that by which nature has . . . facilitated the manifestation of all the main inclinations inherent to animals. This process consists of an instinctive desire for food, of the aspiration (*stremlenie*) to satisfy an appetite for it. The significance of the act of eating itself consists in this: eating enlivens representations of food and appetite (the opinion of Professor Pavlov)." See Lobasov, *Otdelitel'naia rabota,* 32.

60. Sanotskii, *Vozbuditeli,* 22–23.

61. Ibid., 20, 39.

62. Khizhin, *Otdelitel'naia rabota,* 43.

63. Ibid., 40–41.

64. Pavlov, "O vzaimnom otnoshenii fiziologii i meditsiny v voprosakh pishchevareniia" [1894], in *PSS,* 2, pt. 1: 257.

65. Khizhin, *Otdelitel'naia rabota,* 96. Note that Khizhin had earlier decided against precisely this same interpretive option in his analysis of the results with peptone: he decided that drops of peptone landing on the dog's tongue—leading Druzhok to lick his lips—did *not* reflect excitation of the dog's psyche. The secretory results of this experiment were instead attributed to the excitation by peptone of the nerves of the mucous membrane of the stomach.

66. Judging from the information in Khizhin's and Lobasov's theses, they usually required three to five minutes to feed Druzhok through a cannula.

67. If this explanation were accepted, it would undermine the assumption that experiments in which a dog was fed through a cannula mirrored the results of nervous-chemical secretion during normal feeding. It would also contradict the laboratory view that all secretion was "purposive." The centrality of this view is discussed in Chapter 5.

68. Khizhin, *Otdelitel'naia rabota,* 98.

69. Ibid., 120, 148. This disclaimer appears in the part of the thesis immediately following Khizhin's discussion of experimental results. Such tentativeness is entirely missing from the conclusion to the thesis, however, where the specific excitability of the nerves of the mucous membrane of the stomach is accepted as fact.

Indeed, the distinctive response of these nerves to different foods had already become the explanatory basis for the characteristic secretory curves, as we shall see in Chapter 5.

70. Lobasov, *Otdelitel'naia rabota,* 23.

71. Ibid., 33. Notice Lobasov's use of the terms *secretory work* and *typicality.* As we shall see in Chapter 5, these reflected the emergence of the guiding metaphor of digestive system as complex chemical factory. Pavlov had first presented this metaphor publicly in a discussion of Khizhin's research in December 1894.

72. Lobasov, *Otdelitel'naia rabota,* 15. Druzhok's digestive tract was, Lobasov conceded, sometimes harmed by "clumsy or constant experiments with certain irritators (spirits, Liebig extract, fat)."

73. Ibid., 25.

74. Ibid., 25–26. Professor Charles Yeo of The Johns Hopkins University tells me that such shrinkage is common in isolated sacs. My thanks to Dr. Yeo for discussing with me this and other aspects of digestive physiology.

75. Ibid., 27.

76. Or was it? As we shall see below, Druzhok's next praktikant, Volkovich, in his attempt to explain discordant data, suggested that even food inserted stealthily into the sleeping dog's stomach might produce a psychic secretion. For Lobasov's description of this procedure and his confidence in it, see ibid., 46.

77. Ibid., 129. Lobasov also relied on previous experiments conducted by another praktikant, Ushakov, which indicated that "sham-feeding with milk elicits either no secretion or an entirely insignificant one."

78. Ibid., 130. Lobasov reported that the mean proteolytic power (by the Mett method) was 3.25 millimeters for experiments with normal feeding compared with 2.89 millimeters when milk was placed directly in the stomach. The mean for feeding in the first hour—when psychic secretion was, according to laboratory doctrine, most pronounced—was 4.21 millimeters (compared with 3.98 for direct placement in the stomach).

79. Ibid.

80. A. N. Volkovich, *Fiziologiia i patologiia zheludochnykh zhelez* (Academy diss., St. Petersburg, 1898), 38.

81. Ibid., 41–42. Pavlov himself, in his remarks during a discussion of Volkovich's report to the Society of Russian Physicians in March 1898, noted simply that "as a consequence of prolonged experimentation upon him, Druzhok developed atrophy of the stomach membrane." See *PSS,* 6: 117.

82. Volkovich, *Fiziologiia i patologiia,* 69.

83. Ibid., 57–58, 68.

84. I discuss Pavlov's attitude about the relationship between laboratory and clinic in Chapter 9.

85. I have identified the following dog-technologies with an isolated sac (with their date of creation): Riabchik (December 1897), for use by Soborov; Pestryi (Oc-

tober 1899) and Laska (February 1900), for use by Kazanskii; Volchok (n.d.) and Katashka (n.d.), for use by Virshubskii; and Seryi (n.d.) and Sulema (n.d.), for use by Zavriev. The isolated-sac operation became routinized in these years, and one such dog-technology was even made available for Pavlov's lectures to medical students.

86. See I. K. Soborov, *Izolirovannyi zheludok pri patologicheskom sostoianii pishchevaritel'nogo kanala* (Academy diss., St. Petersburg, 1899); Ia. Kh. Zavriev, *Materialy k fiziologii i patologii zheludochnykh zhelez* (Academy diss., St. Petersburg, 1900); I. P. Zhegalov, *Otdelitel'naia rabota zheludka pri pereviazke protokov podzheludochnoi zhelezy i o belkovom fermente v zhelchi* (Academy diss., St. Petersburg, 1900); S. S. Zimnitskii, *Otdelitel'naia rabota zheludochnykh zhelez pri zaderzhke zhelchi v organizme* (Academy diss., St. Petersburg, 1901); and N. P. Kazanskii, *Materialy k eksperimental'noi patologii i eksperimental'noi terapii zheludochnykh zhelez sobaki* (Academy diss., St. Petersburg, 1901).

87. Pavlov's speech to the Society, "Sovremennoe ob"edinenie v eksperimente glavneishikh storon meditsiny na primere pishchevareniia" [1899], is in *PSS*, 2, pt. 2: 247–84. I discuss this speech in Chapter 9. The isolated sac proved to be a disappointingly unreliable tool for investigation of these questions. When the large stomach was damaged—whether by ice, scalding water, or alcohol—the secretions from the small sac did not mirror but rather compensated for secretions in the large stomach. Pavlov referred to this problem in passing in 1899 but seemed undeterred by it. See Pavlov, "Sovremennoe ob"edinenie," 268–69. One year later, however, Zavriev (in *Materialy k fiziologii*) commented that this phenomenon invalidated the earlier work by Volkovich and Soborov—work that Pavlov had lauded in his earlier speech. By my reading of laboratory dynamics, Zavriev's verdict no doubt reflected the chief's as well.

88. I. P. Pavlov, "Laboratornye nabliudeniia nad patologicheskimi refleksami s briushnoi polosti" [1898], in *PSS*, 1: 553–54.

89. I. P. Pavlov, "Vystupleniia v preniiakh po dokladu A. P. Sokolova" [1902], in *PSS*, 6: 197.

90. Pavlov was targeted by Russia's antivivisectionists in the 1890s and his laboratory sustained at least one unexpected inspection by Baroness Meiendorf, the well-connected head of the Russian Society for the Protection of Animals. In 1903 Pavlov served on the Military-Medical Academy's committee charged with replying to Meiendorf's insistence that animal experimentation be strictly limited and monitored by her Society. The committee's conclusions and Pavlov's separate statement are republished as "Zakliuchenie komissii o vivisektsii i osoboe mnenie I. P. Pavlova," in *PSS*, 6: 214–26. I address this interesting chapter in Pavlov's life in my forthcoming biography.

91. Here, however, it is difficult to distinguish rhetoric from reality. Pavlov always emphasized the necessity of modern, expansive facilities for the success of

physiological surgery, yet he never used Druzhok as an example. Available sources do not make clear whether Druzhok's isolated sac was created in the new surgical complex or in the single operating room available in the Physiology Division's original building. Pavlov was still consulting with the architect of the new laboratory wing in April 1894, when this operation was conducted, but perhaps the surgical complex was already operational. Even if it was not, Pavlov considered the single operating room in his original laboratory superior to that in the Botkin laboratory, especially after Massen devised procedures for improving it hygienically. Pavlov and Shumova-Simanovskaia had succeeded in esophagotomizing dogs in Botkin's laboratory, but these dogs became "progressively exhausted" and soon died. Sanotskii attributed his greater success in preserving the health of esophagotomized dogs to the hygienic conditions in the operating room of Pavlov's original Physiology Division building. Furthermore, he attributed his ability to consistently elicit a "psychic secretion" from these dogs to their more robust health.

92. Identification of the two important exciters, psychic and nervous-chemical, that Druzhok was created to unite also owed much to the research by P. N. Konovalov and N. P. Iurgens in 1891–93. Iurgens conducted acute experiments on the nervous control of gastric secretion; Konovalov followed up Sanotskii's experiments on the dynamics of appetite in esophagotomized dogs. I discuss Konovalov's research, and the clinical product it helped generate, in Chapter 8.

Chapter 5. From Dog to Digestive Factory

1. I. P. Pavlov, "O vzaimnom otnoshenii fiziologii i meditsiny v voprosakh pishchevareniia" [1894], in *PSS*, 2, pt. 1: 245–74. Pavlov's speech was summarized very favorably in the newspaper *Novoe Vremia*, December 30 [January 11], 1894 [1895], no. 6766: 4.

2. George Johnson, *Fire in the Mind: Science, Faith, and the Search for Order* (New York: Vintage Books, 1995), 4–5.

3. P. P. Khizhin, *Otdelitel'naia rabota zheludka sobaki* (Academy diss., St. Petersburg, 1894). The word *work* appears on pp. 12, 119, and 149–50.

4. Pavlov, *Lektsii*. Typical pre-1894 titles for laboratory publications were A. S. Sanotskii's *Exciters of the Secretion of Gastric Juice* (*Vozbuditeli otdeleniia zheludochnogo soka*) (Academy diss., St. Petersburg, 1892); and I. P. Pavlov's "On the Surgical Method of Investigation of the Secretory Phenomena of the Stomach" (K khirurgicheskoi metodike issledovaniia sekretornykh iavlenii zheludka) [March 1894], in *PSS*, 2, pt. 1: 275–76. After the completion of Khizhin's thesis in fall 1894, the word *work* became standard. Examples are I. O. Lobasov, *The Secretory Work of the Stomach of the Dog* (*Otdelitel'naia rabota zheludka sobaki*) (Academy diss., St. Petersburg, 1896); I. P. Pavlov, "On the Secretory Work of the Stomach during Starvation" (Ob otdelitel'noi rabote zheludka pri golodanii) [1897], in *PSS*, 2, pt.

2: 226–30; and A. M. Virshubskii, *The Work of the Gastric Glands with Various Sorts of Fatty Food* (*Rabota zheludochnykh zhelez pri raznykh sortakh zhirnoi pishchi*) (Academy diss., St. Petersburg, 1900).

5. Pavlov, "O vzaimnom otnoshenii fiziologii i meditsiny," 250.

6. The implications of the factory metaphor were in this respect quite different from those of the "motor" metaphor adopted, for example, by Helmholtz and Marey. See Anson Rabinbach, *The Human Motor: Energy, Fatigue, and the Origins of Modernity* (Berkeley: University of California Press, 1992).

7. In an unpublished section of his manuscript, Babkin explains Pavlov's attitude toward statistics as follows: "When he observed some experiment that deviated from the normal course, he was not usually satisfied in obtaining only a theoretical explanation of this deviation, but planned a new experimental attack. From this stems his reluctance to use the statistical method in physiology, which meekly accepts variations in results and is satisfied with a middle course, thus making further experimental development of a question unnecessary." See Babkin, *Pavlov* (MS), chap. 12, p. 7. This reasoning mirrored Bernard's.

8. Pavlov, "O vzaimnom otnoshenii fiziologii i meditsiny," 261.

9. A. N. Volkovich, *Fiziologiia i patologiia zheludochnykh zhelez* (Academy diss., St. Petersburg, 1898), 10. Even when Volkovich purportedly summarized Khizhin's conclusions, he did so in language more appropriate to his own: "Doctor Khizhin demonstrated that 'to each type of food—meat, bread, milk—there is always its own, completely determined work of the gastric glands with relation to the qualities of the juice, its quantity, the course of secretion, and the duration of the entire secretory process'" (ibid., 21; see also p. 10). As we have seen, however, Khizhin did not argue that each type of food had its own course of secretion. He concluded, precisely to the contrary, that the curve for the amount of secretion for different foods was essentially the same (differing, however, in its proteolytic power).

10. For a survey of Russian attitudes toward the factory, see M. I. Tugan-Baranovsky, *The Russian Factory in the Nineteenth Century* [1907], trans. Arthur Levin and Clara Levin (Homewood, Ill.: Richard D. Irwin, 1970), 413–48. For a good example of anti-factory attitudes among conservative commentators, see the article by "Sigma," "Na staleliteinom zavode," *Novoe Vremia* July 3 [July 15], 1894, no. 6588: 2. Aside from the common objections I mention in the text, Sigma observes that the "colossal successes" of contemporary technology, as embodied in the factory, inevitably gave birth to "a class of people who tend to exaggerate the power of the human mind." Such people were seduced by the notion that man could be the mechanic not only of the factory but also of society, morals, and habits. Sigma noted also that the population of the factory was divided into "geniuses and automatons (*avtomaty*)"—the former "free and wealthy," the latter "dependent and poor." He comforted himself with the observation that, notwithstanding the aspirations of those infatuated with the factory, "man is not a machine and society is not a factory."

11. Pavlov met Mendeleev in the 1870s through his brother, Dmitrii Pavlov, who worked in Mendeleev's chemistry laboratory at St. Petersburg University, and took Mendeleev's course in chemistry.

12. D. Mendeleev, "Zavody," in *Entsiklopedicheskii slovar' Brokgauza i Efrona*, vol. 12 (St. Petersburg, 1894).

13. Ibid., 100–104.

14. Ibid., 101–2

15. Pavlov, "O vzaimnom otnoshenii fiziologii i meditsiny," 252.

16. See Raphael Samuel, "Workshop of the World: Steam Power and Hand Technology in Mid-Victorian Britain," *History Workshop: A Journal of Socialist Historians* 3 (spring 1977): 6–72.

17. Khizhin, *Otdelitel'naia rabota*, 118, 119.

18. Vladimir Dal', *Tolkovyi slovar' zhivogo velikorusskogo iazyka* [1882] (Moscow: Russkii Iazyk, 1990), 4: 618. Pavlov first used the word *shablonno* with reference to the digestive system in his speech of December 1894. See Pavlov, "O vzaimnom otnoshenii fiziologii i meditsiny," 254. The noun *shablon* was later employed to refer to the mass, clichéd art forms that we refer to as *kitsch*.

19. Khizhin, *Otdelitel'naia rabota*. Khizhin describes the secretions of the digestive glands as *pravil'nye* ("regular" and "correct") or reflecting *pravil'nost'* ("regularity" and "correctness") (pp. 56, 58, 59, 60, 116, and 117 [twice]); in three of these cases, the secretions express "remarkable *pravil'nost'*." Secretory activity is *zakonno* (lawful) and expresses *zakonnost'* (lawfulness) (pp. 75, 76).

20. Ibid., appended "propositions."

21. *Trudy obshchestva russkikh vrachei za 1893–94* 61 (September 1894): 39.

22. Ibid., 43.

23. In his thesis, *Otdelitel'naia rabota*, Khizhin clearly states that secretory activity is purposive. The glands secrete "the amount of juice needed for digestion of the given portion of food" (p. 59). He also notes that "nature, with special reason" provides for infants to be fed milk (p. 150): digestion of milk requires no psychic secretion, and Khizhin assumes that an infant is incapable of psychic responses.

24. By using the term *construction* I do not mean to imply that these curves were conjured from whole cloth. Rather, I want to draw attention to the way in which the relationship between experimental results and interpretive claims is similar to that between the materials used in a building and the shape of the final edifice. In both cases, the primary material limits and to some extent shapes—but does not wholly determine—the final product. In both cases, the vision of the directing mind (whether scientist or architect) and a host of other factors (obstacles that emerge during the construction process and decisions about how to overcome them, available tools and technologies, and so forth) play an important role.

25. The data charts are in Khizhin, *Otdelitel'naia rabota,* 53 (mixed food), 71 (meat), 88 (bread), and 93 (milk). Druzhok was fed mixed food in various quan-

tities: (1) 200 cubic centimeters of milk, 25 grams of meat, and 25 grams of bread (eleven experiments); (2) 300 cubic centimeters of milk, 50 grams of meat, and 50 grams of bread (six experiments); and (3) 600 cubic centimeters of milk, 100 grams of meat, and 100 grams of bread (twenty-six experiments).

26. Halbert L. Dunn, "Application of Statistical Methods in Physiology," *Physiological Reviews* 9 (1929): 276. My thanks to Harry Marks for bringing this source to my attention. On the early history of statistical reasoning and physiology, see William Coleman, "Experimental Physiology and Statistical Inference: The Therapeutic Trial in Nineteenth-Century Germany," in *The Probabilistic Revolution,* ed. Lorenz Krüger, Gerd Gigerenzer, and Mary S. Morgan (Cambridge, Mass.: MIT Press, 1987), 2: 201–26. Coleman concludes that "the serious use of statistical methods in experimental physiology" began only after 1900 (p. 201).

27. For a thought-provoking discussion of some general issues raised by graphs and curves, see Roger Krohn, "Why Are Graphs so Central in Science?" *Biology and Philosophy* 6, no. 2 (April 1991): 181–203.

28. Volkovich, *Fiziologiia i patologiia,* 25.

29. For Pavlov in 1897, curves *b* and *g, c* and *e,* and *d* and *f* were "stereotypical" pairs, establishing the identity of gastric secretory patterns in two different dogs: *b* and *g* are meat curves, *c* and *e* are bread curves, *d* and *f* are milk curves. Curve *a* represents the results of a sham-feeding trial.

30. Khizhin, *Otdelitel'naia rabota,* 107. So, 200 grams of mixed food generated 62.7 cubic centimeters of gastric juice, 200 grams of meat generated 56.9, 200 grams of bread generated 33.6, and 200 cubic centimeters of milk generated 37.0. Khizhin's inclusion of mixed food (mixtures of meat, bread, and milk in various quantities) in this list seems strange: after all, if each distinct food elicited a specific secretory response, one would expect the response to mixed food to be simply a combination of those specific responses. The use of these data attests to the fact that Khizhin (and Pavlov) conducted these experiments on mixed food in April and May, before they arrived at the hypothesis that the glands responded distinctively to different foods. These results for mixed food provide a good example of the numerous "roads not taken" in the work of Pavlov and his coworkers—the questions *not* posed to the data. Why, one might ask, should 200 grams of mixed food elicit a greater secretory response than 200 grams of any food from which the mixture is composed?

31. The paucity of concrete data in Khizhin's thesis makes any comparison of the results of individual trials impossible here. I make such comparisons wherever possible.

32. By 1896, these "opposite poles" were defined as follows: bread elicits a strong psychic secretion and is weak in chemical exciters; milk elicits a weak psychic secretion and is rich in chemical exciters.

33. Khizhin, *Otdelitel'naia rabota,* 108.

34. Ibid., 80. As we shall see in Chapter 6, Pavlov explicitly stated in *Lectures* that

if a result could be obtained "in two out of five" experiments, "or about that," it must be considered an expression of an underlying law.

35. This was a problem only because Pavlov's scientific vision raised this difference between dogs to the status of an important discrepancy.

36. Khizhin, *Otdelitel'naia rabota,* 51.

37. Dividing Gordon's secretory response into thirds, Khizhin noted the "remarkable regularity" of the results. In the first one-third of the digestive act, Gordon's digestive glands produced from 55.8 to 76.5 percent of the total juice; in the second one-third, 17.4 to 35.2 percent; and in the final one-third, 3.3 to 14.6 percent. Khizhin translated this into a general pattern: about two-thirds of the juice was secreted in the first digestive period, about one-quarter during the second, and about one-twentieth in the third. Ibid., 57.

38. Ibid., 59–60.

39. That these results were indeed noted carefully in the protocols is apparent not only from the memoir literature discussed in Chapter 3 but also from the citation of these results years afterward in laboratory publications.

40. The "minimum" results reported for each hour do not necessarily come from the same experiment, nor do the "maximum" results. Rather, they are the lowest and highest values obtained in the entire experimental series. We can use them here to gain some appreciation of the range of results for any single hour but not to reproduce the results of individual experiments.

41. Khizhin, *Otdelitel'naia rabota,* 71. Khizhin rarely provided sufficient data to judge the range of results, but when he did, the range is almost always large. Compare, for example, the data for proteolytic power in five experiments with 200 grams of bread, in Khizhin, *Otdelitel'naia rabota,* 86–88 (data in millimeters):

HOUR	AVERAGE	LOWEST	HIGHEST	RANGE
1	6.10	5.23	7.56	2.33
2	7.97	6.50	8.78	2.28
3	7.51	6.37	8.81	2.44
4	6.19	3.56	8.36	4.80
5	5.29	2.50	8.12	5.62
6	5.72	4.80	7.00	2.20
7	5.48	4.79	6.62	1.83
8	5.50	4.62	6.00	1.38

42. Khizhin introduces data dividing the secretory response into the first, second, and third portions of the digestive act. By using this larger time period for plotting his data—as opposed, say, to the usual hourly periods—he "smoothed out" hourly differences and maximized similarities between the results with meat and mixed food.

43. Ibid., 83. It is possible, of course, that Khizhin attempted experiments with larger quantities of bread but was so dissatisfied with the results that he did not report them in his thesis.

44. For example, the data Khizhin provides for the amount of secretion elicited by the ingestion of bread indicate that the results frequently varied by about 33 percent from experiment to experiment. He also noted that the secreted fluids elicited by these experiments were very thick in the first hour, but "this phenomenon was manifested each time with different intensity." See Khizhin, *Otdelitel'naia rabota,* 84–85. The relatively few cases in which Khizhin provides sufficient data to make a judgment suggest that a wide range of results was typical. For another example, he admits to a great range in results and explains this as an outcome of uncontrolled variables (p. 64), and he presents data demonstrating that behind arithmetical averages of 2.67, 2.34, and 3.36 millimeters for proteolytic power during the first three hours of experiments with mixed food lay results for individual experiments ranging from about 1.0 to 6.0 millimeters in any single hour (p. 65).

45. Ibid., 89. Khizhin conceded that secretion after a meal of bread continued longer than after a meal of meat, but treated this as insignificant.

46. Lobasov, *Otdelitel'naia rabota zheludka sobaki,* 62, 62, 89, 101 (twice), 109, 129, 133, 139 (three times), 146, 154, 157, 158, 160, and the title. Interestingly, the word *shablon* is used only once, pejoratively, to describe scientific investigations that are insufficiently flexible and adapted to the task at hand. This is precisely the same meaning of that word in Khizhin's and Pavlov's usage—that is, they insist that digestive secretion does not proceed *shablonno.*

For *prisposoblennost'* ("adaptedness" or "adaptiveness") or *prisposoblenie* (adaptation), see ibid., 62, 63, 89, 101 (twice), 129, 133, 139 (three times), 146, 154, 157, 158, 160, and the title page; for *slozhno* (complex), 21, 23 (twice), 101, 139, and 158; for *tonkost'* (subtlety), 101; for the "end goal" (*konechnaia tsel'*), 1; for *bread juice* and so forth, 102 and 110. Lobasov once refers to experiments that are conducted *pravil'no,* that is, correctly (p. 64).

47. Khizhin, *Otdelitel'naia rabota,* 116–17.

48. Lobasov, *Otdelitel'naia rabota,* 21–22. Unlike Khizhin, Lobasov does not analyze the secretory response to "mixed food," which by this time had ceased to be a meaningful category.

49. Khizhin, *Otdelitel'naia rabota,* 76.

50. For Lobasov's analysis of the milk curve, see his *Otdelitel'naia rabota,* 129–30.

51. Ibid., 34. This counting procedure, of course, did not eliminate the differences that had led Khizhin to distinguish between "psychic" and "usual" secretions. It did, however, conceal them. Had secretory curves been plotted for fifteen-minute rather than one-hour intervals, this change in counting procedure would have accomplished little.

52. Ibid., 41–44.

53. Ibid., 41.

54. Volkovich cites Lobasov's analysis of the secretory response to nonfat milk in *Fiziologiia i patologiia*, 12, 14.

55. See Lobasov, *Otdelitel'naia rabota*, 19, 21, 40.

56. Among the experimental arguments for this ability of starch to heighten proteolytic power was a comparison of the secretory response to meat and to meat mixed with starch. According to Lobasov, the mixture elicited secretion with a significantly higher proteolytic power—substantially higher than that elicited by a meal of meat and resembling that elicited by a meal of bread. The mixture of meat and starch, Lobasov noted, played "the role of synthesized bread." Lobasov further noted that, in four experiments, Druzhok responded to the ingestion of a starchy mass (composed of arrowroot) with a secretion that much resembled the bread curve for both the amount of secretion and proteolytic power. Ibid., 103–4.

57. Pavlov's longtime coworker and disciple Boris Babkin later noted that "we know now that this effect [gastric secretion] is probably due in part to the presence of histamine in the meat extract." As for Lobasov and Pavlov's explanation: "No experimental proof of this rather doubtful theory was given." See Babkin, *Pavlov,* 226–27. Analogous results in experiments on pancreatic secretion in 1902 would lead Bayliss and Starling to announce their discovery of a humoral mechanism, as I discuss in Chapter 9.

58. Lobasov, *Otdelitel'naia rabota,* 100, 102.

59. For each of these examples one could analyze the data from supporting experimental trials to demonstrate a certain selectivity at work. Experimental trials measuring the amount and proteolytic power of psychic secretion, the normal digestion of various foods, and the secretory response to various foods placed in the stomach generated the same range of data manifested in the trials analyzed elsewhere in this book. For Lobasov's analysis of the food curves, see ibid., esp. 101–33.

60. Lobasov gives the average results for proteolytic power of gastric juice secreted during the entire digestive act as 3.53 millimeters for 400 grams of meat, 3.76 for 200 grams of meat, and 4.46 for 100 grams of meat. The inverse relationship between the amount of meat and proteolytic power presumably resulted from the fact that a larger piece of meat contained more extractive substances than a smaller one and therefore elicited more chemical juice. Because chemical juice contained less ferment than appetite juice, this heightened secretion brought *down* the proteolytic power of the gastric juice as a whole.

61. Ibid., 145, 157. This conclusion involved an important interpretive judgment. What, precisely, was a "small fluctuation" as opposed to the "large fluctuations" that ostensibly distinguished the secretory responses to various foods and defined the secretory curves themselves? As we shall see in Chapter 9, some future investigators would point to these fluctuations as evidence for the existence of humoral mechanisms.

62. A. A. Val'ter, *Otdelitel'naia rabota podzheludochnoi zhelezy* (Academy diss., St. Petersburg, 1897), 38.

63. V. N. Vasil'ev, *O vliianii raznago roda edy na deiatel'nost' podzheludochnoi zhelezy* (Academy diss., St. Petersburg, 1893).

64. Val'ter, *Otdelitel'naia rabota podzheludochnoi zhelezy,* 177.

65. See ibid., 66–67, 110.

66. Ibid., 105. The tables that Val'ter appends to his thesis (ibid., 180–82) show the variations to be even greater: for milk, 37.25 to 72.5 cubic centimeters; for bread, 120.0 to 215.25; for meat, 81.25 to 199.5.

67. In 1897 Pavlov published his report on the influence of starvation on digestive secretion. He reported that, after four or five days without food or water, a dog ceased to respond to sham-feeding with a gastric secretion. If, however, water was then poured into the animal's stomach, within two or three days sham-feeding again elicited a normal secretory effect. This, Pavlov claimed, demonstrated the dependence of secretion on the water content in an animal's body. See Pavlov, "On the Secretory Work of the Stomach" (Ob otdelitel'noi rabote zheludka).

68. Val'ter, *Otdelitel'naia rabota podzheludochnoi zhelezy,* 109.

69. Ibid., 110.

70. Volkovich, *Fiziologiia i patologiia,* 15–16. In his thesis, Volkovich acknowledged begrudgingly that critics were justified in their skepticism that experiments on a single dog had established "the general physiological laws of digestion." Yet neither the chief (who probably wrote this uncharacteristically argumentative section of the thesis) nor his praktikant seem to have doubted seriously that the results with a second dog would prove "stereotypical." "Tens of experiments repeated by various investigators" on Druzhok had, after all, yielded "results that corresponded one with the other, often even in their smallest details." Despite this "weighty refutation" of the critics, it was "extraordinarily important to repeat these experiments on another dog," to "prove clearly yet again that the very same results acquired with Druzhok can be acquired on any dog, and so once again to confirm the consistency and lawfulness of earlier conclusions." Ibid.

71. Ibid., 17.

72. The experiments on milk, which addressed the more complex issue of the effect of fat on gastric secretion, followed in December 1897, by which time *Lectures* had already been published.

73. Jurgen Renn used the expression "watch the representation take over" in the presentation of his joint work with Tilman Sauer, "The Private and Public Knowledge of Albert Einstein" (paper presented at the conference "Reworking the Bench: Research Notebooks in the History of Science," Max-Planck-Institut fur Wissenschaftsgeschichte, Berlin, November 11–14, 1998). For the use of this concept, see Hans-Jorg Rheinberger, *Toward a History of Epistemic Things: Synthesizing Proteins in the Test Tube* (Stanford: Stanford University Press, 1997). I have taken the expression "thick things" from Ken Alder's creative adaptation of Clifford Geertz's

concept of "thick descriptions," in Alder's "Making Things the Same: Representation, Tolerance and the End of the *Ancien Régime* in France," *Social Studies of Science* 28, no. 4 (August 1998): 502–5. By "thick things," Alder writes, "I mean to invoke two aspects of material artifacts. First, the difficulty of consistently shaping the material world into a working artifact, or what one early modern technologist called the 'resistance and obstinacy of matter.' And second, the related challenge of assimilating ordinary artifacts to any idealized representation in such a way that their qualities can be captured in their entirety" (p. 503). My thanks to Harry Marks for bringing this article to my attention.

74. R. Heidenhain, *Handbuch der Physiologie der Absonderung und Aufsaugung*, vol. 5, pt. 1, of L. Hermann's, *Handbuch der Physiologie* (Leipzig: F. C. W. Vogel, 1883), 145, 157, 183; and Michael Foster, *A Text-Book of Physiology*, 4th Amer. ed. (Philadelphia: Lea Brothers and Co., 1891), 369 (where Foster uses a curve for pancreatic secretion that was first presented by Bernstein in 1869).

75. Heidenhain found a somewhat higher level of secretion than had Bernstein and attributed this to their use of different types of fistulas. In his experiments, the period of maximally rapid secretion lasted longer than it had in Bernstein's, a minor discrepancy that Heidenhain simply noted. Heidenhain's curves differ in some interesting respects from Pavlov's. For Heidenhain, secretion began immediately after the dog took the food, accelerated to its peak sometime during the first three hours ("sometimes earlier, sometimes later"), declined through the fifth or seventh hours, and then rose again, to a second and lower peak, from the ninth to the twelfth hour. Secretion, he noted, could continue for as long as seventeen hours after ingestion. Pavlov attributed the initial peak in gastric secretion to the action of the psyche, which was presumably absent in the vagotomized Heidenhain pouch. Pavlov never, to my knowledge, dealt with this contradiction.

76. The difference in vagal innervation does not affect my point here.

Chapter 6. The Physiology of Purposiveness

Here, as in earlier chapters, in referring to Pavlov's *Lectures,* I provide references for both the original Russian publication (*Lektsii*) and the authorized English translation (*Lectures* [trans.]). Unless otherwise noted, all translations from *Lektsii* in the text and notes are my own. Thompson's English edition is sometimes inaccurately translated and consistently tones down Pavlov's references to "purposiveness."

Where page numbers for *Lectures* are included in the text, I give first the page in *Lektsii* followed by the page in *Lectures* (trans.); for example, pp. 57; 36.

1. Pavlov's original lectures were delivered at the request of Sergei Luk'ianov, director of the Institute. Pavlov replaced I. R. Tarkhanov as the Military-Medical Academy's professor of physiology in 1895 and lectured there on digestive physiology for decades.

2. For a discussion of Pavlov's interactions with physicians, see Chapter 9.

3. Pavlov, *Lektsii*, 12; *Lectures* (trans.), second page of preface. In *Lectures*, Pavlov frequently warned that these experiments might not work properly during public presentations, since the novel setting affected the psyche of the experimental animal. He offered suggestions for overcoming this problem and, in later years, used particular dogs, well-adapted to this purpose, in such lectures.

4. On "witnessing," see Steven Shapin and Simon Schaffer, *Leviathan and the Air Pump: Hobbes, Boyle, and the Experimental Life* (Princeton: Princeton University Press, 1985). Pavlov did not, of course, demonstrate all his experiments before these audiences. For example, the isolated-sac operation did not become routinized until after 1897 and, in any case, a demonstration of the various characteristic secretory curves would have taken much too long for a class session.

5. Pavlov, *Lektsii*, 20; *Lectures* (trans.), 2.

6. See M. I. Tugan-Baranovsky, *The Russian Factory in the Nineteenth Century* [1907], trans. Arthur Levin and Clara Levin (Homewood, Ill.: Richard D. Irwin, 1970),171–214, 363–412.

7. In Thompson's English edition, both the distinction between *factory* and *manufactory* and the reference to *kustarnyi lad* disappear. Pavlov's use of the factory metaphor is also muffled by the translation of Pavlov's introductory phrase, *pishchevaritel'nyi kanal est', ochevidno, khimicheskii zavod*, meaning "the digestive canal is, obviously, a chemical factory," as "the digestive canal may be compared to a chemical factory." *Lectures* (trans.), 2.

8. Pavlov, *Lektsii*, 57; *Lectures* (trans.), 36.

9. For some of the many other references to purposiveness, see *Lektsii*, 52, 66, 172; *Lectures* (trans.), 23, 63–64, 129.

10. Pavlov, *Lektsii*, 66. In *Lectures* (trans.), 45, the Russian word for "mind" (*um*) is translated as "instinct."

11. Another issue was the "hidden period" of five to ten minutes between the initial excitation of the gastric glands and the first drops of gastric juice. One must, Pavlov wrote, "recognize in this some special purpose." He speculated that the purpose might be to allow the salivary ferment time to work "unimpeded," but conceded that "such an explanation cannot be regarded as very convincing, as the issue concerns a fact that has still not been subjected to systematic scientific analysis." *Lektsii*, 59; *Lectures* (trans.), 38.

12. In Thompson's English translation of *Lectures*, the word *tselesoobraznost'*, meaning "purposiveness," is translated as "lawfulness." Pavlov offers two "striking" examples of the purposiveness of specific features of the secretory curves. First, he claims that the relative amounts of pepsin elicited by equivalent amounts of proteins in milk, bread, and meat correspond to "the results of the physiological-chemical investigation of the digestibility of all these [various] proteids." Second, he suggests that the vegetable protein in bread, which requires much ferment for its digestion, receives this ferment in the form of a relatively small quantity of

high-ferment juice in order to avoid two problems that might otherwise result from an excess of hydrochloric acid: damage to the organism and a juice less efficient in digesting starch. *Lektsii,* 58–59; *Lectures* (trans.), 37–38.

13. As indicated by the quotation at the beginning of this section, Pavlov had earlier referred to this as an a priori truth.

14. Claude Bernard, *An Introduction to the Study of Experimental Medicine* [1865], trans. Henry Copley Greene (New York: Dover Publications, 1957), 135. My thanks to William Rothstein for bringing this passage on averages to my attention.

15. P. P. Khizhin, *Otdelitel'naia rabota zheludka sobaki* (Academy diss., St. Petersburg, 1894), 71–72.

16. I use here the two experiments reported in *Lectures* plus the "absolute minimum" and "absolute maximum" results reported in Khizhin's thesis. I have calculated the total secretion in the fifth experiment by using, in addition to these four results, the "average" result reported by Khizhin for the five experiments taken together. See Khizhin, *Otdelitel'naia rabota,* 71.

17. I have calculated the results for the first two hours of the missing three experiments (i.e., those not in Figure 19) as follows. Where Khizhin's "maximum" and "minimum" results do not match results in the two experiments for which we have precise information, these must represent results from two other experiments. The results of the fifth experiment can then be computed using Khizhin's "average" result for the series of five experiments as a whole. This information is in Khizhin's thesis (ibid., 71). The two meat curves that, in Khizhin's results, peaked in the second hour may have been among those that, with Lobasov's new computation of the first hour (see Chapter 5), were made to peak in the first hour.

18. Pavlov, *Lektsii,* 42; *Lectures* (trans.), 22.

19. Val'ter's table also includes data on the duration of pancreatic secretion, which ranged from 3.25 to 6.5 hours. This also favored the choice of trials 1 and 5 (in both, secretion continued for 4.75 hours) and militated against the possible alternatives of trials 19 (3.75 hours), 22 (4 hours), 23 (4.25 hours), or 32 (3.75 hours). Consideration of this further factor also clarifies Pavlov's reasons for not excluding trial 1 and pairing two other trials. Val'ter's thesis was completed when *Lectures* was already in press, and it first appeared in the Military-Medical Academy's doctoral dissertation series in 1897. It was republished for an international audience in 1899, in a French translation that appeared in the Institute's journal. In this version the comprehensive data charts included in his doctoral thesis disappeared, replaced by data for selected experiments and by curves (in the appendix) for his six "best" trials. For *Lectures,* Pavlov had chosen the two "best" of these. See A. A. Val'ter, "Excitabilité sécrétoire spécifique de la muqueuse du canal digestif. Cinquième mémoire. Sécrétion pancréatique," *Archives des sciences biologiques* 7 (1899): 1–86.

20. Pavlov, *Lektsii,* 42; *Lectures* (trans.), 22. The original Russian phrase is *v dvukh opytakh iz piati ili okolo togo,* which translates as "two experiments out of

five, or about that." In the German edition of *Lectures,* translated by Val'ter (J. P. Pawlow, *Die Arbeit der Verdauungsdrüsen,* trans. A. Walther [Wiesbaden: J. F. Bergmann, 1898], 28), the phrase became "two experiments out of about five" ("in zwei Versuchen von etwa fünf "). In Thompson's English edition (p. 22) it became simply "two experiments out of five."

21. Pavlov, *Lektsii,* 43; *Lectures* (trans.), 22–23.

22. It is interesting, however, that Pavlov decided to take these results from Lobasov's unpublished experimental protocols, rather than using the two trials with this same quantity of meat that Khizhin reported in his thesis. The results of Khizhin's two experiments did not resemble each other as closely as did the trials Pavlov chose from Lobasov's protocols. See Khizhin, *Otdelitel'naia rabota,* 41, 48.

23. Pavlov, *Lektsii,* 49; *Lectures* (trans.), 29.

24. In only three trials was Val'ter able to measure all three of the pancreatic ferments. He reported the complete hourly results for only the two later trials. He did, however, provide the average values for the three ferments in all three experiments, which permits us to compute the results for the missing trial and to understand Pavlov's choice of the latter two (units for each type of ferment content as in my Figure 21):

TRIAL	FAT-SPLITTING	AMYLOLYTIC	PROTEOLYTIC
20	4.0	7.5	7.7
27	4.88	5.88	9.5
28	4.75	6.12	9.3

These results are reported in A. A. Val'ter, *Otdelitel'naia rabota podzheludochnoi zhelezy* (Academy diss., St. Petersburg, 1897), 180.

25. Pavlov, *Lektsii,* 55; *Lectures* (trans.), 34.

26. A. N. Volkovich, *Fiziologiia i patologiia zheludochnykh zhelez* (Academy diss., St. Petersburg, 1898), 26.

27. Ibid., 24. In his thesis Volkovich acknowledged that laboratory conclusions had been criticized for their dependence on results with a single dog and argued that his trials with Sultan should lay such doubts to rest: "Numbers will speak more eloquently than words" (p. 18). These numbers demonstrated conclusively that "gastric work is repeated *pravil'no* (regularly) to the greatest degree, and all the characteristic particularities of the secretory curve—both qualitative and quantitative—acquired on one animal are reproduced by the other with striking accuracy" (p. 24).

28. For example, Volkovich offered two possible explanations for Sultan's consistently producing more secretion than did Druzhok. First, this could be the result of the dogs' differing "individuality"—that is, of "a greater or lesser receptivity of nerve endings to irritation." This was unlikely, however, since such a

difference should also manifest itself with respect to the trophic nerves that controlled proteolytic power. In other words, if Sultan was more sensitive to irritation than was Druzhok, this should have resulted in both a greater amount of secretion and juice of a higher proteolytic power. The second, more likely explanation was that Sultan's isolated sac was larger than Druzhok's. The experimenters, of course, had no way of judging directly the relative size of the dogs' isolated sacs. Volkovich argued, however, that the differences in amount of secretion remained consistent from food to food: "for example, with Druzhok, the average amount of juice poured upon meat in the first hour is $1\frac{1}{2}$ times that poured on bread, and 4 times greater than that poured on milk; these very same relations are preserved entirely with Sultan." As is clear from the curves in my Figure 24, had he compared the results for other hours (or any hour in the results with milk), this argument would have been more difficult to sustain.

Similarly, Volkovich chose not to compare the overall average data for proteolytic power in Druzhok's and Sultan's secretory responses to different foods, a comparison that would have complicated the case for stereotypicity. He was confronted with the following data for proteolytic power, in millimeters: Druzhok— bread, 6.51; milk, 3.25; meat, 2.98; Sultan—bread, 5.66; milk, 3.5; meat, 3.78. Volkovich might have argued that the absolute differences here were small (although they exceeded the laboratory's assumption of a maximum 10 percent margin of error). The results with Sultan, however, violated one of the primary features of the characteristic curves for proteolytic power: in contrast to trials with Druzhok, proteolytic power for meat was higher than for milk. For Volkovich's discussion, see his *Fiziologiia i patologiia,* 26–27. Pavlov did not address these issues in *Lectures.*

29. A. R. Krever, *K analizu otdelitel'noi raboty podzheludochnoi zhelezy* (Academy diss., St. Petersburg, 1899), 20–21.

30. Ibid., 23–24.

31. Ibid., 30.

32. Ia. A. Bukhshtab, *Rabota podzheludochnoi zhelezy posle pererezki vluzhdaiuzhikh i vnutrennostykh nervov* (Academy diss., St. Petersburg, 1904), 57–60.

33. Ibid., 68.

34. In the 1890s, a very small percentage of Russian physicians were women (about 550 of 16,000), but a disproportionate number of them may well have resided in St. Petersburg, home of the Women's Medical Institute. See Nancy Mandelker Frieden, *Russian Physicians in an Era of Reform and Revolution, 1856–1905* (Princeton: Princeton University Press, 1981), 323. Pavlov's audience may also have included female *fel'dshery* (medical paraprofessionals). Finally, the Higher Women's Courses, a university for women, thrived in St. Petersburg, and its students may have attended Pavlov's lectures out of sheer interest. In the German and English editions of *Lectures,* Pavlov's "Ladies and Gentlemen" becomes simply "Gentlemen."

35. For Pavlov's interactions with members of the Society, see Chapter 9.

36. I discuss Pavlov's previous publication record and the translation history of *Lectures* in Chapter 9.

37. Pavlov, *Lektsii,* 172; *Lectures* (trans.), 129. Thompson's English translation renders "artistic mechanism" (*khudozhestvennyi mekhanizm*) as "skilled mechanism" and replaces Pavlov's reference to *purposiveness* with the more scientifically acceptable notion of *adaptation:* "Instead of a crude indefinite scheme, we see now the outlines of a skilled mechanism which, as with everything in nature, proves itself to be adapted with the utmost delicacy and in the most suitable manner to the work which it has to perform."

38. Pavlov writes explicitly that "Blondlot's book and Heidenhain's article comprise almost all physiology's important achievements for more than 50 years regarding the conditions and mechanism of the secretory work of the stomach during digestion." See Pavlov, *Lektsii,* 151; *Lectures* (trans.), 111. He is referring to Nicolas Blondlot's *Traité analytique de la digestion* (1843); and Heidenhain's "Ueber die Absonderung der Fundusdrusen des Magens" (1879), later republished in L. Hermann's *Handbuch der Physiologie* (Leipzig: F. C. W. Vogel, 1883).

39. Pavlov, *Lektsii,* 11; *Lectures* (trans.), ix.

40. John Harley Warner, "Ideals of Science and Their Discontents in Late Nineteenth-Century American Medicine," *Isis* 82 (1991), 454–78.

41. Pavlov, *Lektsii,* 176; *Lectures* (trans.), 132.

42. For example, Pavlov noted that the laboratory had failed to provide any experimental justification for the common therapeutic use of neutral and alkaline salts of sodium, yet he did not advocate that physicians cease to use them. Rather, he concluded that the action of these salts was still poorly understood. See *Lektsii,* 191–92; *Lectures* (trans.), 145. The one issue on which Pavlov uncompromisingly denied traditional clinical views was on mechanical exciters of the stomach—a view that he considered emblematic of an outmoded, mechanistic concept of the digestive system. See Chapter 9 for Pavlov's discussion of this issue with one clinician.

43. Pavlov added, "More established nations, for example the English, have made a sort of cult of the act of eating." The correct attitude toward eating lay between these two extremes: "Don't get carried away, but pay the attention necessary."

Chapter 7. From the Machine to the Ghost Within

1. For an exception, and a substantially different interpretation than the one I offer here, see David Joravsky, *Russian Psychology: A Critical History* (Oxford: Basil Blackwell, 1989), 134–48.

2. See S. A. Ostrogorskii, *Temnyi punkt v innervatsii sliunnykh zhelez* (Academy diss., St. Petersburg, 1894).

3. Glinskii had begun his doctoral thesis (on the physiology of the intestines) under Pavlov's supervision in the Botkin laboratory and completed it in the Institute's Physiology Division in 1891. According to the biographical information on Glinskii in D. G. Kvasov and A. K. Fedorova-Grot, *Fiziologicheskaia shkola I. P. Pavlova* (Leningrad: Nauka, 1967), 86–87, his subsequent work on the salivary glands "was produced by Glinskii independently" and not at Pavlov's instructions. Glinskii published neither a description of the fistula he developed nor any account of his experimental work with it. Pavlov reported on these to the Society of Russian Physicians in 1895, and again in *Lectures*. In each report, the chief made clear that the operation belonged not to him and not to "the laboratory" but to Glinskii alone. See Ivan Pavlov's "Ob opytakh doktora Glinskogo nad rabotoi sliunnykh zhelez," in *PSS*, 2, pt. 1: 282–83; *Lektsii*, 95–97; *Lectures* (trans.), 68–69. The comments on purposiveness and the dryness of food are from Pavlov, "Ob opytakh," 282.

4. Pavlov, *Lektsii*, 94; *Lectures* (trans.), 66–67.

5. Pavlov, *Lektsii*, 94; *Lectures* (trans), 66.

6. S. G. Vul'fson, *Rabota sliunnykh zhelez* (Academy diss., St. Petersburg, 1898), 17.

7. The differences among these two reports and the final thesis testify to Vul'fson's and Pavlov's growing appreciation of the determinant role of the psyche, but for our purposes we can consider all three together. Vul'fson's first report to the Society of Russian Physicians, in October 1897, was published as "O psikhicheskom vliianii v rabote sliunnykh zhelez," *Trudy obshchestva russkikh vrachei* 65 (October 1897): 110–13; his report of March 1898 appeared as "Rabota sliunnykh zhelez," *Trudy obshchestva russkikh vrachei* 65 (March-May 1898): 451–59.

8. Vul'fson, *Rabota sliunnykh zhelez*, 27–28.

9. Vul'fson, "Rabota sliunnykh zhelez," 453–55.

10. Vul'fson, *Rabota sliunnykh zhelez*, 55.

11. Vul'fson, "O psikhicheskom vliianii," 113.

12. Vul'fson, *Rabota sliunnykh zhelez*, 53. This formulation was stronger than that of his report in October 1897, when he referred to salivary reactions as "psycho-physiological phenomena." See Vul'fson, "O psikhicheskom vliianii," 112.

13. Vul'fson, *Rabota sliunnykh zhelez*, 53–54.

14. Ibid., 43, 56; Vul'fson, "Rabota sliunnykh zhelez," 457.

15. Vul'fson, *Rabota sliunnykh zhelez*, 54.

16. Vul'fson, "Rabota sliunnykh zhelez," 458.

17. Ibid., 456. In his earlier report of October 1897 he had been less emphatic, arguing that the link between this "physiological process and the psychic world" would demand increasing attention from scientists. Vul'fson, "O psikhicheskom vliianii," 113.

18. Pavlov, remarks on Vul'fson's "Rabota sliunnykh zhelez," 458–59.

19. The scanty information available on Snarskii's life comes from several

sources: the curriculum vitae in his thesis, *Analiz normal'nykh uslovii raboty sli-unnykh zhelez u sobaki* (Academy diss., St. Petersburg, 1901); Kvasov and Fedorova-Grot's short account in *Fiziologicheskaia shkola I. P. Pavlova,* 226–27; and a few lines in V. N. Andreeva and E. A. Kosmachevskaia, "Nauchnye sviazi shkol I. P. Pavlova i V. M. Bekhtereva," *Zhurnal vysshei nervnoi deiatel'nosti* 42, no. 5 (1992): 1039–45. Snarskii emigrated from revolutionary Russia in 1923 and his subsequent fate is unknown.

20. Pavlov's assistant Nikolai Kharitonov was also treated for alcoholism at the Charity Home. Babkin, *Pavlov* (MS), chap. 10, pp. 1–2.

21. That Snarskii was at the time working in Bekhterev's clinic makes it even more likely that Pavlov actively recruited him. Otherwise, it would have made more sense for Snarskii to continue his work and complete a doctoral dissertation with Bekhterev.

22. As Pavlov later recalled, "The scientific spirit of the [Charity Home] made itself known, not only in the scientific investigations of its Director, but also in the scientific activity of the other physicians in the institution, many of whom performed experimental work in my laboratory and others at the Institute of Experimental Medicine (A. T. Snarskii, I. F. Tolochinov, A. I. Iushchenko)." See I. P. Pavlov, untitled declaration, November 17, 1924, in SPF ARAN 259.7.140.

I have also used here the short biographical note on Timofeev by N. G. Ozeretskovskaia, which accompanies Pavlov's declaration in the same archive. In a scientific report of 1898 appearing in the house publication of Bekhterev's clinic, Iushchenko noted that he had performed his experimental work in Pavlov's laboratory. See A. I. Iushchenko, "O reflektornykh tsentrakh v uzlakh simpaticheskoi nervnoi sistemy, ob otnoshenii nizhniago bryzheechnago uzla k innervatsii mochevogo puzyria i ob avtomaticheskikh dvizheniiakh posledniago," in *Otchety nauchnogo sobraniia vrachei Sankt-Peterburgskoi kliniki dushevnykh i nervnykh boleznei za 1895–96 i 1896–1897* (St. Petersburg, 1898), 95–97.

Pavlov later related to a coworker the following: "I had an acquaintance in the Academy, the psychiatrist T[imofeev]. Every Sunday I spent at his place in Udel'-naia, informing him of the entire course of our work. He died and did not understand. Dualism is firmly entrenched in people." See D. A. Biriukov, "I. P. Pavlov (po zapisam v dnevnike)," in *I. P. Pavlov v vospominaniiakh sovremennikov,* ed. E. M. Kreps (Leningrad: Nauka, 1967), 60.

23. Snarskii, *Analiz,* 4.

24. Ivan Sechenov's article "Reflexes of the Brain" (1863) was republished as a book of the same name: *Refleksky golovnogo mozga* (St. Petersburg, 1866). See my brief discussion of this work in Chapter 2.

25. Snarskii, *Analiz,* 4.

26. Even in humans, Snarskii argued, salivation was "not a volitional act." Thinking of food only led a person to salivate if hungry (and not, for example, af-

ter a filling meal). So, "obviously, the purely physiological state of hunger or satiation plays the decisive role in these cases." Ibid., 6.

27. Snarskii supported this conclusion by using the experiments of a previous
praktikant, V. V. Nagorskii, who had not completed his doctoral dissertation. In
his own experiments excising various nerves, Snarskii determined that the salivary response was dependent on the glossopharyngeal, lingual, and trigeminal
nerves.

28. Snarskii, *Analiz,* 52–54.

29. Account of Snarskii's defense of his dissertation, in *Izvestiia Imperatorskoi
Voenno-Meditsinskoi Akademii* 5, no. 1 (September 1902): 69.

30. Snarskii, *Analiz,* 54.

31. Account of Snarskii's defense, 70.

32. I. P. Pavlov, remarks to coworkers, transcribed by P. S. Kupalov, December
27, 1926, in SPF ARAN 259.1.203.

33. See Daniel P. Todes, "From Radicalism to Scientific Convention: Biological
Psychology in Russia from Sechenov to Pavlov" (doctoral diss., University of Pennsylvania, 1981); Daniel P. Todes, "Biological Psychology and the Tsarist Censor: The
Dilemma of Scientific Development," *Bulletin of the History of Medicine* 58 (1984):
529–44; Joravsky, *Russian Psychology.*

34. On the "men of the eighties," see Kendall Bailes, *Science and Russian Culture
in an Age of Revolution: Vernadsky and His Scientific School, 1863–1945* (Bloomington: Indiana University Press, 1989). On the depolarization in the 1880s of formerly
sharp ideological disputes concerning the biology of mind, see Todes, "From Radicalism to Scientific Convention," 331–88.

35. This is not to say that these medical disciplines became thoroughly apolitical, but they now related to other issues and in other ways than in the 1860s. See,
for example, Laura Engelstein, *The Keys to Happiness: Sex and the Search for
Modernity in Fin-de-Siècle Russia* (Ithaca, N.Y.: Cornell University Press, 1992).

36. L. A. Orbeli, "Pamiati Ivana Petrovicha Pavlova," in Kreps, *I. P. Pavlov v
vospominaniiakh,* 172.

37. A. A. Trzhetseskii, "Achylia gastrica," *Russkii arkhiv patologii, klinicheskoi
meditsiny i bakteriologii* 9 (1900): 185.

38. P. Ia. Borisov, "Znachenie razdrazheniia vkusovykh nervov dlia pishchevareniia," *Russkii Vrach,* 1903, no. 3: 871. For another example, see L. B. Popel'skii, "Ob
osnovnykh svoistvakh podzheludochnago soka," *Russkii Vrach,* 1903, no. 16: 8. In
this article the former coworker argued that "the gastric juice that appears during
the act of eating should be seen as a reflex elicited by the impressions received from
our sensory organs. Therefore, a more suitable name [than psychic juice] will be
'initial juice,' as opposed to the 'secondary' [juice] that appears as a reflex from the
mucous membrane of the stomach and intestines."

39. A persistent but still unsubstantiated report attributes Bekhterev's death to

Stalin's hand, after Bekhterev diagnosed the Soviet leader as a paranoid schizophrenic.

40. This orientation is clearly reflected in Bekhterev's definition of one of his specialties, neurology, as including "not only anatomy, embryology and physiology of the nervous system, but also experimental psychology or so-called psychophysics, which serves as a link between physiology of the nervous system and empirical psychology. Neurology stands in an equally close relationship to those areas of medicine, such as neuropathology, psychiatry and forensic psychopathology, which have as their goal, so to speak, the practical application of neurological knowledge at the patient's bedside." This passage from Bekhterev's letter to the administration of Kazan University in 1891 is quoted in M. K. Korbut, *Kazanskii Gosudarstvennyi Universitet* (Kazan, 1930), 2: 162.

41. V. N. Andreeva, "Ivan Petrovich Pavlov kak opponent i retsenzent," *Trudy Instituta Istorii Estestvoznaniia i Tekhniki* 41, no. 10 (1961): 294–323; V. N. Andreeva and E. A. Kosmachevskaia, "Nauchnye sviazi skhkol I. P. Pavlova i V. M. Bekhtereva," *Zhurnal vysshei nervnoi deiatel'nosti* 42, no. 5 (1992): 1039–45.

42. A. V. Gerver, "O vliianii golovnogo mozga na otdelenie zheludochnago soka," *Trudy obshchestva russkikh vrachei* [November and December 1899] 66 (1900): 142–43, 153–56, 158–62, 165–68. His comment on "nothing other than a reflex" is on p. 142.

43. Discussion of Gerver's "O vliianii," reprinted in *PSS*, 6: 150. As Pavlov pondered the "psychology of the salivary gland," he was clearly taken, at least for about a year, by this distinction between desire and thought. It appears in more elaborated form in one praktikant's doctoral thesis of 1900: "Truly, one must differentiate the *desire* to eat from the *thought* about food. Various food preparations, such as meat, bread, milk, and so forth, can to an identical degree satisfy the desire to eat, but do not answer identically to the conscious thought about one or another food; only in the latter case [does the dog] acquire the enjoyment of food; in the former, there is only satiation, or a slaking of hunger." For example, the coworker continued, during a famine people eat wood, skin, grass, and so forth, but this satisfies "only the torturous sense of hunger; that is, the sense of emptiness in the stomach—but [it does not engage the] appetite, which is the thought about tasty and welcome food." See A. M. Virshubskii, *Rabota zheludochnykh zhelez pri raznykh sortakh zhirnoi pishchi* (Academy diss., St. Petersburg, 1900), 6; emphasis in original.

44. Discussion of Gerver's "O vliianii," 150–51.

45. *Russkii Vrach*, 1902, no. 23: 884. This report is from an article published by Bekhterev in *Archiv für Anatomie und Physiologie* in that same year. Other articles by Bekhterev and his collaborators on issues relevant to the Pavlov laboratory include Ia. P. Gorshkov, "O lokalizatsii tsentrov vkusa v mozgovoi kore" (On the Localization of Taste Centers in the Brain Cortex), *Obozrenie psikhiatrii, nevrologii, i eksperimental'noi psikhologii*, 1900, no. 10: 737–42; V. M. Bekhterev, "O korkovom

zritel'nom tsentre" (On the Cortical Visual Center), *Obozrenie psikhiatrii, nevrologii, i eksperimental'noi psikhologii,* 1901, no. 7: 575–79; V. M. Bekhterev, "O refleksakh v oblasti litsa i golovy" (On Reflexes of the Face and Head), *Obozrenie psikhiatrii, nevrologii i eksperimental'noi psikhologii,* 1901, no. 9: 552–55; and, more generally, Bekhterev's *Psikhika i Zhizn'* (The Psyche and Life) (St. Petersburg, 1902) and the first volume of his *Osnovy ucheniia o funktsiiakh mozga* (Fundamentals of the Study of Brain Functions) (St. Petersburg, 1903).

46. Among the propositions that Tolochinov defended in his thesis was one on the future of psychology: "The psychiatric method, together with experiment and self-observation, should play a prominent role in psychology." I. F. Tolochinov, *O patologo-anatomicheskikh izmeneniiakh iader cherepnykh nervov i otnosiashchikh-siai k nim nervnykh volokon mozgovogo stvola pri narastaiushchem paralichnom slaboumii* (Academy diss., St. Petersburg, 1900), 209.

47. I. P. Pavlov, remarks to coworkers, December 27, 1926, in SPF ARAN 259.1.203.

48. I. Tolotschinoff [Tolochinov], "Contribution à l'étude de la physiologie et de la psychologie des glandes salivaires," in *Forhandlingar vid Nordiska Natur-forskare och Lakermotet,* July 1902, p. 43. Similarly, he writes, "It is impossible for me in my report to expose in detail all of my work" (p. 44). This first-person-singular language is, of course, even more insistent in the report and reminiscences about his work on conditional reflexes published in 1912–14, by which time he and Pavlov had broken relations and were quarreling about priority.

49. The priority dispute that later erupted concerned not so much who did what as who should get credit for what. Pavlov's comment about Tolochinov's priority is from remarks to coworkers, December 27, 1926.

50. I. F. Tolochinov, "Pervonachal'naia razrabotka sposoba uslovnykh refleksov i obosnovanie termina 'uslovnyi refleks,'" *Russkii Vrach,* 1912, no. 31: 1277.

51. Boris Babkin, *Pavlov* (MS), chap. 16, pp. 14–15. Like Tolochinov later, Babkin saw the discovery of "extinction" as Tolochinov's key accomplishment: "Nevertheless, these experiments established a series of fundamentally important facts concerning the extinction of conditioned reflexes, their subsequent restoration, and so on, and opened the way for a great many other studies requiring an increasingly difficult and more exact technique." Ibid.

52. The term *representation* appeared in the laboratory's literary products throughout the 1890s, but the term *association* originated, as we have seen, with Snarskii. Information about Tolochinov's experiments comes from his own articles, one contemporaneous ("Contribution à l'étude de la physiologie") and four published a decade later: "Pervonachal'naia razrabotka sposoba uslovnykh refleksov i obosnovanie termina 'uslovnyi refleks,'" *Russkii Vrach,* 1912, no. 31: 1277–82; "K voprosu ob osnovnykh proiavleniiakh uslovnykh i bezuslovnykh refleksov v pervonachal'noi razrabotke ikh metoda. Ne prigodnost' terminologii tormazheniia i rastormazhivaniia. Teoriia pobochnykh reflektornykh dug," *Russkii*

Vrach, 1913, no. 1: 20–24; "K voprosu ob osnovnykh proiavleniakh uslovnykh i bezuslovnykh refleksov v pervonachal'noi razrabotke ikh metoda," *Russkii Vrach,* 1913, no. 2: 54–57; "K voprosu ob osnovnykh proiavleniiakh uslovnykh i bezuslovnykh refleksov v pervonachal'noi razrabotke ikh metoda. Neprigodnost' terminologii tormozheniia i rastormazhivaniia. Teoriia pobochnykh reflektornykh dug," *Russkii Vrach,* 1913, no. 10: 333–35; and "Pervonachal'noe primenenie metoda uslovnykh refleksov k izsledovaniiu tsentrov kory bol'shogo mozga u sobak," *Nevrologicheskii vestnik,* 19, no. 2 (1912): 410–45.

The latter five articles contain protocols of his experiments, rearranged according to a series of arguments that emerged only some years after they were conducted. Tolochinov's narrative in these later articles was also shaped by his attempt to claim priority for the initial research on conditional reflexes and appeared at a time when the Pavlov and Bekhterev laboratories were locked in a bitter polemic. As a source of information about the actual process of research, then, they must be used carefully and skeptically. In my narrative I use with greatest confidence those aspects of the articles that either are irrelevant to or contradict the purposes for which Tolochinov wrote them. The comment in the text is from Tolochinov, "Pervonachal'naia razrabotka, 1281, 1278.

53. Tolochinov, "K voprosu" (1913, no. 2), 57.

54. In his articles, Tolochinov does not present these protocols in chronological order; rather, he arranges them to make his own scientific arguments. I have rearranged them in chronological order in an attempt to provide the framework for a rough (and no doubt incomplete) account of experimental developments.

55. Tolochinov, "K voprosu," (1913, no. 10), 333.

56. Tolochinov, "Pervonachal'naia razrabotka," 1277.

57. See, for example, V. M. Bekhterev's "O fenomene kolennoi chashki, kak raspoznavatel'nom priznake nervnykh boleznei, i o drugikh srodnykh iavleniiakh" (On the Phenomenon of the Knee Cap as a Diagnostic Sign of Nervous Illnesses), *Obozrenie psikhiatrii, nevrologii, i eksperimental'noi psikhologii,* 1896, no. 3: 171–76; "O refleksakh v oblasti litsa i golovy" (On Reflexes of the Face and Head), *Obozrenie psikhiatrii, nevrologii, i eksperimental'noi psikhologii,* 1901, no. 9: 552–54; "O nekotorykh reflektornykh iavleniiakh" (On Several Reflexive Phenomena), *Otchety nauchnogo sobraniia vrachei S. Peterburgskoi kliniki dushevnykh i nervnykh boleznei za 1900–1901* (St. Petersburg, 1901), 44–50; and "O glaznom reflekse" (On the Eye Reflex), *Obozrenie psikhiatrii, nevrologii, i eksperimental'noi psikhologii,* 1901, no. 11: 801–4.

58. This trial was conducted on Ryzhaia on February 18, 1902; see Tolochinov, "K voprosu" (1913, no. 2), 56.

59. One example is a trial with Ryzhaia on March 20, 1902. Here Tolochinov was making the point that any strong stimulus would cause even an unrelated conditional response to become stronger.

TIME	EXCITER	SECRETION (CC)
5:05	Toast waved before dog	0.05
5:08	Fingers coated with meat powder waved before dog	0.2
5:10	Fingers coated with meat powder waved before dog	0.05

[Tolochinov assumes that both conditional reflexes are declining toward extinction. He now exposes the dog to the "strong scent" of mustard oil.]

5:13	Open bottle of mustard oil placed several centimeters from dog's nose	0.6
5:17	Toast waved before dog	0.2
5:20	Fingers coated with meat powder waved before dog	0.2
5:23	Open bottle of mustard oil placed several centimeters from dog's nose	0.6
5:28	Toast waved before dog	0.4
5:30	Fingers coated with meat powder waved before dog	0.2

Note that the secretory response to fingers coated with meat powder did not actually rise after the second use of mustard oil at 5:23, but remained, rather, at 0.2 cubic centimeters. For Tolochinov, however, this was a positive result: he assumed the secretory response to fingers with meat powder had a tendency to decline toward extinction, so, had the dog not been exposed to mustard oil at 5:23, this secretory response would have been *less* than 0.2 cubic centimeters. Thus the 0.2 cubic centimeters of secretion at 5:30 reflected the fact that the strong stimulus of mustard oil had compensated for the decline of the conditioned reflex to fingers coated with meat powder. This trial is reported in Tolochinov, "K. voprosu" (1913, no. 2), 56.

60. In retrospect, it is fitting that the abstract of Tolochinov's "Contribution à l'étude de la physiologie" was published in the Helsingfors conference proceedings just below the abstract of a paper by two other coworkers, Borisov and Val'ter, who sought to defend Pavlovian digestive physiology in the new climate already created by Bayliss and Starling's discovery of secretin. Though not evident to the participants, these two papers represented the future and the past of Pavlov's labora-

tory enterprise. See P. Borissow and A. Walther, "Zur Analyse der Saurewirkung auf die Pankreassekretion," in *Forhandlingar vid Nordiska Naturforskare och Lakermotet*, July 1902, 42. (Note that Helsingfors [now Helsinki] was at this time within the Russian empire. Finland became an independent republic in 1919, with Helsinki as its capital.)

61. Tolochinov, "Contribution à l'étude de la physiologie," 43–44. Tolochinov's use of *we* here, as opposed to the first-person-singular language in other parts of his report, seems to be a deliberate collegial reference to Pavlov's conceptual and terminological contribution.

62. Ibid., 44–45. Tolochinov and Pavlov judged the dog's desire for food by its motor reactions, and these did not always correspond to the secretory response to the food substance.

63. Recall that the frontal lobes of one dog, Pudel', were extirpated at the very beginning of Tolochinov's experiments, in December 1901—that is, before Tolochinov had established the "normal" pattern of conditional reflexes in general, let alone in that dog. After the Helsingfors conference, experiments on Pudel' continued; a new dog, Milordka, was subjected to the standard experiments and then, in December 1902, its purported salivary center was ablated; and Voron's frontal lobes were removed in an operation in February 1903. Pavlov did not mention the results of these experiments in any of his written work. Tolochinov published the protocols of these experiments in 1912, claiming they demonstrated that extirpation of the purported salivary center made "no essential difference" in the development and manifestation of conditional reflexes. With these experiments, he claimed, he was the first person to use the method of conditional reflexes to investigate "the physiological significance of several divisions of the brain cortex and to destroy the scientific illusion of the cortical salivary center of [Bekhterev and his laboratory] and the synthetic center of Flechsig." These experiments are described in Tolochinov, "Pervonachal'noe primenenie metoda."

64. I. S. Tsitovich, "Vospominaniia ob akademike Ivane Petroviche Pavlove," in SPF ARAN 259.7.77: 14.

65. W. M. Bayliss and E. H. Starling's discovery was reported in "The Mechanism of Pancreatic Secretion," *Journal of Physiology* 28 (1902): 325–53. The quotation in the text is from p. 353. I discuss the humoralist critique of Pavlov's work in Chapter 9.

66. Sergei Gruzdev, in *Russkii Vrach*, 1902, no. 14: 546–47.

67. V. V. Savich, "Ivan Petrovich Pavlov: biograficheskii ocherk," in *Sbornik posviashchennyi 75-letiiu akademika I. P. Pavlova* (Leningrad, 1924), 11–12.

68. See Borissow and Walther, "Zur Analyse der Saurewirkung." The relative importance of nervous and humoral mechanisms in both pancreatic and gastric secretions continued to be a source of contention for decades. In this discussion—which deserves a systematic historical study—complex physiological issues were not settled simply by experimental results. According to Babkin, the discovery of

secretin "shook the very foundation of [Pavlov's] teaching of the exclusive nervous regulation of the secretory activity of the digestive glands, a concept which seemed to be established so solidly and supported by so many experimentally proved facts." Pavlov "did not give up at once the idea of the exclusive nervous regulation of pancreatic secretion but rather the reverse," changing his position after reading (in the fall or winter of 1902–3) Bayliss and Starling's complete account of their experiments. Babkin adds that the enthusiasm for humoralism became so great that "the function of the pancreatic secretory nerves was neglected completely" and "it required several years of work on the part of my colleagues and myself" to correct this error. See Babkin, *Pavlov,* 228; also 229–30.

69. Boris Babkin, for example, devoted the rest of his career—in Russia, the United States, and Canada—to studies of the digestive system. Even after Pavlov turned to work on conditional reflexes, research on digestion continued in his smaller laboratory at the Military-Medical Academy, although this proceeded, by all accounts, without his active engagement.

70. Pavlov had little taste for conducting such a review even in 1923 for the second edition of *Lectures.* The publisher had agreed to his wish that this publication be "a stereotypical reproduction of the 1897 edition." Rather than revise the body of his text, Pavlov merely added a preface. Here he informed the reader that the years since 1897 had introduced "several changes and corrections, both to the factual (to a lesser degree) and theoretical (of course, much more) part of *Lectures.*" He then explained, in two paragraphs, that psychic secretion was now considered a conditional reflex and that Bayliss and Starling had broadened scientific understanding of the mechanisms of the digestive glands to include, "together with the indubitable nervous mechanism," a humoral mechanism as well. Pavlov, *PSS,* 2, pt. 2: 13–14.

71. On the eve of his completion of *Lectures,* Pavlov wrote a letter to Tsion that both hinted at some weariness with digestive physiology and made clear that any new line of investigation would need to make good use of the capital already accumulated in his physiology factory: "I still cannot part with digestion and therefore mainly spend my time and labor on it. In the laboratory of the Institute of Experimental Medicine I have at my disposal the means and setting for operating upon and keeping animals—such as barely exist anywhere else—and therefore consider myself obligated to drag from my animals whatever I can. There are now at my disposal 20–30 dogs . . . that have been operated upon in various ways in different parts of their digestive canal." I. P. Pavlov to I. F. Tsion, September 13 [1897], in SPF ARAN 259.7.167; reprinted in E. M. Kreps, ed., *Perepiska Pavlova* (Leningrad: Nauka, 1970), 56–57.

72. Savich, "Ivan Petrovich Pavlov," 18.

73. A. F. Samoilov, "Obshchaia kharakteristika issledovatel'skogo oblika I. P. Pavlova," in Kreps, *Pavlov v vospominaniiakh,* 214.

74. N. D. Strazhesko, "Vospominaniia o vremeni provedennom v laboratorii Ivana Petrovicha Pavlova," in Kreps, *Pavlov v vospominaniiakh,* 226.

75. I. P. Pavlov, *Zapisi lektsii po fiziologii pishchevareniia prochitannykh Ivanom Petrovichem Pavlovym v 1902–1903 gg.* (Notes of Lectures on the Physiology of Digestion Delivered by Ivan Petrovich Pavlov in 1902–1903), in SPF ARAN 259.1.78: 18–18ob. The lecture notes are dated the first week of September 1902. Pavlov's annual report mentions that he delivered lectures on the physiology of digestion to teachers in Pavlovsk and to sanitary physicians in St. Petersburg's Duma building. TsGIA SPb 2282.1.201: 15.

76. It is quite possible that Pavlov was initially reluctant to assign any but the most promising praktikanty to the task of nurturing his "vulnerable child."

77. I. P. Pavlov, "O psikhicheskoi sekretsii sliunnykh zhelez" [1904], in *PSS*, 3, pt. 1: 45. My translation differs little from the standard English translation in Ivan Pavlov, *Lectures on Conditioned Reflexes: Twenty-five Years of Objective Study of the Higher Nervous Activity (Behavior) of Animals*, trans. W. Horsley Gantt (New York: International Publishers, 1928), 65.

78. Tolochinov, "Pervonachal'naia razrabotka," 1281.

79. Horsley Gantt, who in 1928 produced an English translation of the first collection of Pavlov's work in this field, claimed he was just following an already-established convention: "Condition*al* (*ooslovny*) [*uslovnyi*] and not condition*ed* is Prof. Pavlov's term, but as *conditioned reflex* has become fixed in English usage instead of conditional reflex, we adhere to the term conditioned." See translator's footnote, in Pavlov, *Lectures on Conditioned Reflexes*, 79; emphasis in original. Gantt was referring here to the English translation of Pavlov's second major volume on this subject by G. V. Anrep (a native Russian speaker), published the year before Gantt's translation of the earlier volume appeared: I. P. Pavlov, *Conditioned Reflexes: An Investigation of the Physiological Activity of the Cerebral Cortex*, trans. G. V. Anrep (New York: Dover Publications, 1927). Gantt also remarks in his translator's note that "in French and German translation, Prof. Pavlov's original term (conditional) has been preserved." See Pavlov, Lectures on conditional reflexes, 79.

80. Tolochinov, "Contribution à l'étude de la physiologie," 44, 45.

81. Tolochinov, "Pervonachal'naia razrabotka," 1281; emphasis in original.

82. Some of Tolochinov's experimental trials seem designed to test the possibility of a sort of "inverse square law" for his "reflex at a distance." For example, on December 8, 1901, he tested the secretory effect of mustard oil held at various, carefully measured distances from the dog's nose. The gravity analogy also played a role in Pavlov's ideas about conditional reflexes. He wrote, for example, about the "attraction" of a weaker nervous impulse to a stronger nervous impulse.

83. Orbeli, "Pamiati Ivana Petrovicha Pavlova," 172.

84. For example, in his introduction to the Russian translation of Robert Tigerstedt's textbook of physiology, Pavlov wrote, "The physiologist, like every specialist, forms the habit of certain means of expression, of presentation, which are not

always successful, and are sometimes tentative (*uslovnyi*) and, finally, outdated." This was written in October 1900 and published in 1901; reprinted in *PSS*, 6: 164. The only relevant contemporary usage of the word *bezuslovnyi* (antonym of *uslovnyi*) I have been able to find is in Lobasov's doctoral thesis, where it appears as a synonym for "indubitable": "But if we are convinced of the indubitable (*bezuslovnyi*) importance of the psychic moment as an exciter of initial secretion, we should at the same time be convinced also that the psychic moment is not the only condition that determines the secretory work of the gland." See I. O. Lobasov, *Otdelitel'naia rabota zheludka sobaki* (Academy diss., St. Petersburg, 1896), 63. In my fruitful discussion with Anna Krylova, a historian of modern Russia and a native Russian speaker, on the meaning of *uslovnyi*, she suggested that the only context in which this word has acquired the meaning "condition*ed*" is that created by Pavlov. In other words, Dr. Krylova thinks *uslovnyi* always has the meaning "condition*al*" except in scientific works and common phrases created in the wake of Pavlov's research.

85. At this point, of course, an important shift has occurred in Pavlov's interpretation of, for example, the salivation elicited by a dog seeing meat. Formerly, he had considered this a psychic secretion; here he is using it as an example of a simple physiological reflex. As discussed later in the text, the concept of *conditional reflexes* required a revision of long-standing interpretations related to what were becoming unconditional reflexes. See I. P. Pavlov, "Eksperimental'naia psikhologiia i psikhopatologiia na zhivotnykh" [1903], in *PSS*, 3, pt. 1: 28. According to historian of physics Robert Kargon, Pavlov's analysis of "action at a distance" was on a solid footing from the perspective of contemporary physics. My thanks to him for discussing this with me.

86. Pavlov, "Eksperimental'naia psikhologiia," 29. Even after elaborating this distinction between conditional and unconditional reflexes, Pavlov continued for some time to pay implicit homage to Tolochinov's insight by references to "reflexes at a distance." In 1904 he even equated the two, referring to "the conditional reflex (the reflex at a distance)." See I. P. Pavlov, "O psikhicheskoi sekretsii sliunnykh zhelez (slozhno-nervnye iavleniia v rabote sliunnykh zhelez)," in *PSS*, 3, pt. 1: 43. Pavlov's distinction between the conditional and unconditional reflex left unresolved one important question: "What is the unconditional irritant in food substances? The factual material collected so far [is] insufficient for resolution of this question." See Pavlov, "O psikhicheskoi sekretsii," 54. In other words, what qualities of food have a "business relation" to the work of the salivary glands? In his Nobel Prize speech later that same year, Pavlov argued that these unconditional irritants were "firmness, dryness, a certain chemical composition, and so forth." Color and form, on the other hand, were only "signals" for these unconditional qualities. See I. P. Pavlov, "Pervye tverdye shagi na puti novogo issledovaniia" [1904], in *PSS*, 3, pt. 1: 61–62.

87. As Pavlov put it in 1904, "our old physiological reflex is constant, uncon-ditional, while the new reflex fluctuates (*kolebletsia*) all the time, and so is *condi-tional.*" Pavlov, "Nobelevskaia rech', proiznesennaia 12 dekabria 190 g. v Stok-gol'me," in *PSS*, 2, pt. 2: 364.

88. Pavlov, "Eksperimental'naia psikhologiia," 29–30.

89. Ibid., 30.

90. Ibid.

91. Pavlov, "O psikhicheskoi sekretsii," 42, 52.

92. Pavlov, "Nobelevskaia rech'," 62.

93. B. P. Babkin, *Opyt sistematicheskago izucheniia slozhno-nervnykh (psikhi-cheskikh) iavlenii u sobaki* (Academy diss., St. Petersburg, 1904), 3.

94. Here I am using *trust* as in Theodore Porter's *Trust in Numbers: The Pur-suit of Objectivity in Science and Public Life* (Princeton: Princeton University Press, 1995).

95. Vul'fson, "Rabota sliunnykh zhelez," 458.

96. For Geiman's procedures and their effect on salivation, see his thesis, *O vli-ianii razlichnago roda razdrazhenii polosti rta na rabotu sliunnykh zhelez* (Academy diss., St. Petersburg, 1904), 7–9 (his emphasis on the importance of experiment 39 is on p. 9). He provides the date for experiment 38 as March 7, 1903 (p. 122). Al-though Pavlov defended Geiman's conclusions vigorously, he did hint that exper-imental confirmation was not yet conclusive: "If these facts are confirmed by fur-ther investigations, everything will be reduced to simple reflexive relations." See *PSS*, 6: 202. Geiman's thesis indicates that his topic was chosen in part because of the research by Henri and Malloizel in the physiological laboratory at the Sor-bonne. These French investigators had disputed some of Vul'fson's conclusions (most importantly, his conclusion that the salivary response depended on the dry-ness of the food). As Geiman noted, the salivary glands had been much more widely investigated than the gastric and pancreatic glands because they were much more easily accessible to researchers. See Pavlov, *O vliianii,* 1–3. It is also interest-ing that Geiman left unmentioned in his thesis Vul'fson's earlier conclusions about the central role of the psyche, casting his contribution, instead, as an explanation of facts first noted by this previous praktikant.

97. Discussion of Geiman's report before the Society of Russian Physicians, February 20, 1903, reprinted in *PSS*, 6: 201–2.

98. Ibid., 201–5.

99. For the 1906 version, see Pavlov, "Estestvenno-nauchnoe izuchenie tak nazyvaemoi dushevnoi deiatel'nosti vysshikh zhivotnykh" [1906], in *PSS*, 3, pt. 1: 64–65.

100. Pavlov, preface to *Dvadtsatiletnii opyt ob"ektivnogo izucheniia vysshei nervnoi deiatel'nosti zhivotnykh* [1923], in *PSS*, 3, pt. 1: 13–14.

101. Pavlov, *Dvadtsatiletnii opyt,* 13–15. Pavlov told the same tale, with added de-tail, in his remarks to coworkers, December 27, 1926.

102. The language of Snarskii's thesis suggests that praktikant and chief enjoyed good working relations. Although he occasionally indulges in first-person-singular language, Snarskii concludes with the standard acknowledgment of Pavlov's participation and intellectual guidance: "I extend my especially respectful and profound gratitude to the most esteemed professor Ivan Petrovich Pavlov, who not only gave me the theme for elaboration, but also always, with fervent interest, followed each step of the work and illuminated the path for me with new ideas and instructions." Snarskii, *Analiz*, 56.

103. In 1904, four coworkers in the Pavlov laboratory reviewed the development of the laboratory's lines of investigation. Ganike summarized Snarskii's thesis respectfully and at length, noting his conclusion that "the psyche is not needed for the regulation of salivation, although there can be no doubt that the reflexive apparatus of salivation by the path of associations is tightly linked with the conscious centers of the higher sense organs." V. N. Boldyrev, V. I. Vartanov, E. A. Ganike, and A. P. Sokolov, "Referaty trudov professora I. P. Pavlova i ego uchenikov," *Arkhiv Biologicheskikh Nauk* 11, suppl. (1904): 9.

104. In his English translation of Pavlov's tale, Gantt added a supportive footnote: "In the light of our present exact knowledge of nervous processes, Snarsky's account seems almost humorous." Gantt then translated two sentences from Snarskii's thesis that clearly reflected the praktikant's "subjective" approach: "Under the influence of inedible substances one can see that the quantity of saliva does not answer to the degree of pleasantness of the substances . . . After [ingesting] sand the dog licks vigorously and smacks his lips, and it is clear that the grimace of disgust is not so prominent as the desire to cleanse the mouth." See Pavlov, *Lectures on Conditioned Reflexes*, 38. As we have seen, had Gantt so desired he could easily have identified similar passages in the work of Vul'fson and Pavlov at the time.

105. Pavlov did, perhaps, understate Tolochinov's theoretical contribution, although this is difficult to judge given the nature of extant materials and the limited length and purpose of Pavlov's historical comments. In 1904 Pavlov noted that "the first investigations were carried out in our laboratory by doctor Tolochinov. His experiments demonstrated convincingly, it seems to me, that our subject can actually be studied with complete success in the indicated direction." Pavlov also cited Tolochinov's report of 1902. See Pavlov, *Dvadtsatiletnii opyt*, 42. In the 1906 narrative at the T. H. Huxley celebration, Tolochinov drops out of the short historical account, which concentrates on Snarskii's subjectivism. See Pavlov, "Estestvenno-nauchnoe izuchenie," 64–65. In the preface to his 1923 volume on conditional reflexes, Pavlov wrote that, having decided after his clash with Snarskii to approach psychic secretion objectively, "I set about the implementation of this decision with a new coworker, doctor I. F. Tolochinov." See Pavlov, *Dvadtsatiletnii opyt*, 14. Pavlov's earliest reports on this research, in 1903–4, contain a number of references to "the experiments of Dr. Tolochinov." In his private comments to his

coworkers, Pavlov combined an appreciation of Tolochinov's seminal contribution with an emphasis on his own heroic role (remarks to coworkers, December 27, 1926):

> Of course, after this episode [with Snarskii] I tried to find a person with whom one could go further. Such a person was found in the person of doctor Ivan Filippovich Tolochinov. Since this was, so to speak, a very torturous period for me, Ivan Filippovich became very close to my heart, became a person very close to me.
>
> But then there interceded a wretched Russian characteristic. He did not need a doctoral degree, he already was a doctor and worked with me purely from scientific interest. He completed the work, made a report in Helsingfors. I asked him to write up everything. He did not write it, whether from laziness, or because he did not completely believe in the path along which we had embarked, perhaps that was it. For 10 years he didn't touch these experiments [protocols]. When they had already grown up and begun to produce a stir, when there were also reports that were listened to with interest, then he got down to it.
>
> Well, of course, he was the first to put his hand to it; priority, one could say, belonged to him, but nothing remained of his work! I had prodded him much at the time: "Write, Ivan Filippovich, write." He did not write. But 10 years later he suddenly took to it and published in "Russkii Vrach" a presentation of his experiments with his thoughts. Published them, and these were pathetic reminiscences about things that never occurred. It was such a strange bit of writing, and it was so hard for me. I had to, not without forcing myself, write a letter against Ivan Filippovich and say that these memoirs mixed fact with fiction, and that I have not the least responsibility for them.
>
> This was one of the hard episodes. Then the thing developed on its own. My closest coworker was B. P. Babkin, who had done much work and defended his dissertation. And then the thing broke out onto the open road when V. N. Boldyrev and I discovered artificial conditional reflexes.

106. Babkin, *Pavlov* (MS), chap. 14, pp. 284–85. Babkin also recalled that the research on conditional reflexes had little appeal to one of Pavlov's key audiences: "The reading of reports of conditioned reflexes by Pavlov and his colleagues at the meetings of the Society of Russian Physicians was regarded by most of the members as an unwelcome necessity. Their attitude was that it was their duty out of respect for Pavlov to listen to these quasi-scientific papers, which were basically of little worth." Ibid.

107. These phrases come from Pavlov's annual reports on the activities of his Physiology Division, submitted to Prince Ol'denburgskii. TsGIA SPb 2282.1.201 [1903], 1.222 [1904], 1.239 [1905], 1.252 [1906], and 1.263 [1907].

Part III Laboratory Products

1. I. P. Pavlov, "Fiziologicheskaia khirurgiia pishchevaritel'nogo kanala" [1902], in *PSS*, 6: 286.

2. As Bruno Latour and Steve Woolgar pointed out in their *Laboratory Life: The Social Construction of Scientific Facts* (Beverly Hills: Sage, 1979), the desire to continue and expand their research is commonly a central motivation for scientists.

Chapter 8. Gastric Juice for Sale

1. True to the Mendeleevan distinction, Pavlov used the phrase *malaia fabrika zheludochnago soka,* "small gastric juice manufactory," but the more euphonious translation "small gastric juice factory" serves our purposes here. See Chapter 5.

2. In 1901, when a rumor circulated that Pavlov had won the Nobel Prize, Tsion praised him for, among other things, being true to the vision of a physiology-based scientific medicine: "I especially value in your scientific activity the fact that, standing at the head of a marvelously equipped laboratory in the Institute of Experimental Medicine, you have not allowed yourself to become attracted to the slippery path and shaky soil of now so fashionable bacteriology. You have remained true to the conviction that only experimental physiology can provide the soil for scientific medicine." SPF ARAN 259.2.1277: 1; reprinted in E. M. Kreps, ed., *Perepiska Pavlova* (Leningrad: Nauka, 1970), 61.

3. Ivan Pavlov and E. O. Simanovskaia, "Innervatsiia zheludochnykh zhelez u sobaki" [1890], in *PSS*, 2, pt. 1: 175–99; text quotation from p. 180.

4. For a detailed analysis of this history, see Frederic Lawrence Holmes, *Claude Bernard and Animal Chemistry: The Emergence of a Scientist* (Cambridge, Mass.: Harvard University Press, 1974), 141–214; esp. 160–78; and Horace W. Davenport, *A History of Gastric Secretion and Digestion: Experimental Studies to 1975* (New York: Oxford University Press, 1992), 78–84.

5. E. O. Shumova-Simanovskaia, "O zheludochnom soke i pepsine u sobak," *Arkhiv Biologicheskikh Nauk* 2 (1893), no. 3. The chemical investigation of gastric juice was also taken up by other investigators at the Institute of Experimental Medicine, including M. V. Nencki (head of the Chemistry Division) and E. S. London.

6. S. M. Luk'ianov, *Osnovaniia obshchei patologii i pishchevareniia* (St. Petersburg, 1897).

7. Frank Woodbury, *On Disordered Digestion and Dyspepsia* (Detroit: George S. Davis, 1889), 16–22. Woodbury was a Fellow of the College of Physicians of Philadelphia. Dyspepsia still awaits its first systematic historical account. For indications of the richness of this subject, see Gert H. Brieger, "Dyspepsia: The American Disease? Needs and Opportunities for Research," in *Healing and History: Essays for George Rosen,* ed. Charles E. Rosenberg (New York: Science History Publications, 1979), 79–90.

8. Luk'ianov, *Osnovaniia obshchei patologii*, 107.

9. See "Rapport sur la pepsine fait a la Société de pharmacie de Paris," *Journal de pharmacie et de chimie* 1, ser. 4 (1865): 81–126. The comments about "German pepsin" are on pp. 82–83.

10. For one of the periodic attempts to evaluate these different preparations, see M. A. Petit, "Sur la préparation de la pepsin," *Journal de pharmacologie et de chimie* 12 (1880): 85–94.

11. P. N. Konovalov, *Prodazhnye pepsiny v sravnenii s normal'nym zheludochnym sokom* (Academy diss., St. Petersburg, 1893), 47.

12. P. N. Konovalov's recollections, written sometime before mid-1933, are in SPF ARAN 895.4.397.

13. Ibid., 4.397: 19–20

14. See Konovalov's *Prodazhnye pepsiny* and his report to the Society of Russian Physicians, "Razlichnye sorta prodazhnago pepsina v sravnenii s natural'nym zheludochnym sokom," *Trudy obshchestva russkikh vrachei* 59 (April-May 1893): 30–37.

15. Konovalov, *Prodazhnye pepsiny*, 5.

16. Until Boldyrev's research of 1903–4, the "volitional secretion" that Konovalov waited out was attributed to the appetite aroused by the dog's expectation of eating.

17. Some dogs who subsequently worked for many years in the "small gastric juice factory" did, however, acquire names.

18. The bactericidal function of gastric juice had been commented on earlier. One author even concluded that this was the main role of gastric juice. See B. I. Kiianovskii, "K voprosu o protivomikrobnom svoistve zheludochnago soka," *Vrach*, 1890, no. 38: 864–65, and no. 40: 915–17.

19. Doctoral dissertations in medicine contained an appendix of "propositions" that the author was prepared to defend orally. Konovalov's first proposition was "the right to the uncontrolled preparation and sale of factory medicaments of animal and plant origin often leads to abuse and can damage public health." Konovalov, *Prodazhnye pepsiny*, unnumbered page at end of text.

20. To my knowledge, this report from a children's hospital never appeared.

21. Konovalov, *Prodazhnye pepsiny*, 64.

22. A. A. Troianov, "O gastroenterostomii," *Trudy obshchestva russkikh vrachei* 60 (November 1893): 28.

23. Frémont's innovation was announced in *Gazette médicale de Paris*, 1895, no. 20: 235. The fistula developed by Ludwig Thiry was used to collect intestinal secretions.

24. Pavlov, "Istoricheskaia zametka ob otdelitel'noi rabote zheludka" [1896], in *PSS*, 2, pt. 1: 320–22; originally published in *Arkhiv Biologicheskikh Nauk* 4 (1896). Pavlov here emphasizes his own priority rather than the priority of "we, the laboratory." He uses the Russian phrase *ustanovlenyi mnoiu vmeste s moimi sotrud-*

nikami (literally, "established by *me* together with my coworkers") rather than the quite common phrase *ustanovlenyi nami vmeste s moimi sotrudnikami* (literally, "established by *us* together with my coworkers"). The same is true of his reference to the isolated stomach "developed by me together with doctor Khizhin."

25. Pavlov, "Istoricheskaia zametka," 322.

26. Pavlov, *Lektsii*, 30; *Lectures* (trans.), 10.

27. For the same reason, Maurice Hepp later named the gastric juice drawn from pigs *dyspeptine*. See his "Opotherapie gastrique par la Dyspeptine (suc gastrique naturel de porc)," *Le Progrés Médical* 18, ser. 3 (July–December 1903): 44.

28. A. M. Virshubskii, "Staroe i novoe v oblasti sekretornoi funktsii zheludka," *Gazeta Meditsina* 23–24 (1898): 4–14; and 25–26 (1898): 4–8.

29. Virshubskii, "Staroe i novoe," 25–26: 8.

30. Ibid., 5–6. For brevity, I have collapsed Virshubskii's seven-point outline into a three-point outline here.

31. Ibid., 6. I am not certain which two published references to the efficacy of gastric juice Virshubskii had in mind here. To my knowledge, before publication of his article there were four publications relevant to this topic: Konovalov, *Prodazhnye pepsiny* [1883]; Troianov, "O gastroenterostomii" [1893]; Pavlov, *Lectures* [1897]; and Frémont, in *Gazette médicale* [1895].

32. Virshubskii, "Staroe i novoe," 25–26: 6–7.

33. Here he referred to the investigations of E. S. London, who worked in the division of General Pathology at the Institute of Experimental Medicine.

34. Virshubskii, "Staroe i novoe," 25–26: 8. Virshubskii's invocation of the "natural" character of gastric juice implied that it was somehow preadapted for optimal digestion. Actually, as Virshubskii mentioned in passing, Konovalov had discovered that the normal acidity of gastric juice was 0.5 percent, not the 0.2 percent considered optimal for digestion. This not only posed a potential problem for Pavlov's doctrine of purposiveness but also implied that, like solutions of pepsins and hydrochloric acid, natural gastric juice might be too acidic to enhance digestion.

35. Ibid., 8.

36. Ibid. That same year, Andrei Volkovich included the following proposition in his doctoral dissertation: "Treatment with the gastric juice of dogs for dyspepsia in young children, and for the consequences of alcohol abuse in adults, has entirely satisfactory results." A. N. Volkovich, *Fiziologiia i patologiia zheludochnykh zhelez* (Academy diss., St. Petersburg, 1898), unnumbered page at end of text. He gave no source for these findings.

37. A. A. Finkel'shtein, "Lechenie estestvennym zheludochnym sokom," *Vrach*, 1900, no. 32: 963; Pavlov, *Lektsii*, 30; *Lectures* (trans.), 10; Konovalov, *Prodazhnye pepsiny*, 63; Virshubskii, "Staroe i novoe," 7–8; V. N. Boldyrev "Natural'nyi zheludochnyi sok, kak lechebnoe sredstvo, i sposob ego dobyvaniia," *Russkii Vrach*, 1907, no. 5: 154, 157.

38. TsGIA SPb 2282.1.102: 56.

39. Ibid., 1.113: 63. This draft of his report is dated May 19, 1899. Ol'denburgskii was here reproducing the phrase "heightened demand" from Pavlov's annual report. See ibid., 1.125: 270b.

40. For example, see the discussion of the Institute Council on April 26, 1900, about the use of an unproven medicine on people in TsGIA SPb 2282.1.145: 16.

41. The protocols of these discussions of the Institute Council for April 24, May 28, October 5, and October 27, 1898, and February 16 and April 13, 1899, are in ibid. 1.112 and 1.125.

42. For the decision on advertising, see the protocols of the November 10, 1898, meeting of the Institute Council in ibid., 1.112: 73. For the decision of the Administrative Council, see N. M. Gureeva, N. A. Chebysheva, eds., *Letopis' zhizni i deiatel'nosti akademika I. P. Pavlova*, commentary by V. L. Merkulov (Leningrad: Nauka, 1969), 76.

43. TsGIA SPb 2282.1.162: 640b-65. For Luk'ianov's earlier suggestion that Pavlov write instructions, see ibid., 1.162: 11.

44. Henri Huchard, in *Bulletin général de thérapeutique,* 1900, p. 543; cited in Boldyrev, "Natural'nyi zheludochnyi sok," 152.

45. Among the articles reporting the success of gasterine were P. E. Launois, "Hypopsie," *Bulletins et Mémoires de la Société Médicale des Hôpitaux,* January 12, 1900, pp. 44–49; M. Bardet, "Gasterine de Dr. Frémont," *Le Bulletin Médical,* 1900, no. 32; M. P. Le Gendre, "L'opotherapie gastrique au moyen du suc gastrique de chien a estomac isolé (gasterine de Frémont)," *Le Bulletin Médical,* 1900, no. 8: 89–90; M. P. Le Gendre, "L'opotherapie gastrique au moyen du suc gastrique de chien a estomac isolé (gasterine de Frémont)," *Bulletins et Mémoires de la Société Médicale des Hôpitaux,* January 26, 1900, pp. 61–70; M. P. Le Gendre, "De l'opotherapie gastrique par la gasterine dans divers cas d'insuffisance digestive ou d'infection gastro-intestinale avec retentissement sur le foie," *Bulletins et Mémoires de la Société Médicale des Hôpitaux de Paris,* January 11, 1901, pp. 685–89; Albert Mathieu, "Note sur l'emploi therapeutique de la gasterine," *Bulletins et Mémoires de la Société Médicale des Hôpitaux de Paris,* June 28, 1901, pp. 730–40.

46. *Vrach,* 1900, no. 6: 179; emphasis in original. The *Russkii arkhiv patologii* (Russian Archive of Pathology) (2 [1901], no. 4: 678) also reported on French developments, commenting that Frémont "praises highly the gastric juice of the dog as a remedy for poor secretory activity in the stomach."

47. A. A. Finkel'shtein, "Lechenie estestvennym zheludochnym sokom," *Vrach,* 1900, no. 32: 963–65.

48. Pavlov apparently first used the phrase *factory dogs* in print in 1905, by which time it was apparently well-established in his laboratory's lexicon. See his "Laboratornye nabliudeniia nad razmiagcheniem kostei u sobaki" [1905], in *PSS,* 6: 238.

49. I. S. Tsitovich, "Natural'nyi zheludochnyi sok i rezul'taty ego primeneniia u bugorchatkovykh bol'nykh," *Russkii Vrach,* 1907, no. 28: 962.

50. The composition of the dogs' meals was described by Boldyrev in his "Natural'nyi zheludochnyi sok," 157. I. K. Konarzhevskii visited the laboratory in July 1901 and later described the care and feeding of the factory dogs in detail. At that time, apparently, each dog subsisted on two daily meals of about a pound of meat, a pound and a half of oatmeal, and a quarter pound of bread. See I. K. Konarzhevskii, "Natural'nyi zheludochnyi sok sobak, Succus naturalis ventriculi canis, kak terapevticheskoe sredstvo pri lechenii khronicheskikh stradanii zheludka do raka posledniago vkliuchitel'no," *Russkii Meditsinskii Vestnik* 4 (1902), no. 11: 5. The hungry coworker was A. I. Smirnov, then a student at St. Petersburg University, earning an extra 40 rubles a month from his work in the gastric juice operation. On being informed by another coworker of Smirnov's depredations, Pavlov insisted on giving him money and "always showed concern over my life circumstances." See A. I. Smirnov's autobiographical statement, read to Iu. A. Vinogradov by Smirnov's spouse on March 16, 1969 (taped interview), in SPF ARAN.

51. See Boldyrev, "Natural'nyi zheludochnyi sok," 154; Nikolai Riazantsev, "O zheludochnom soke koshki," *Arkhiv Biologicheskikh Nauk* 3 (1894). On bull juice, see Riazantsev's short report in *Vrach*, 1898, no. 39: 1145. Riazantsev worked in a veterinary institute in Khar'kov, where he studied gastric secretion in various domesticated animals. He also popularized gastric juice treatment in Khar'kov; see, for example, his "The Acquisition of Gastric Juice by Prof. Pavlov's Method and Its Characteristics in Comparison with Commercial Pepsins," *Trudy Khar'kovskogo meditsinskogo obshchestva* 2 (1905): 27–30.

52. Babkin, *Pavlov*, 131. Smirnov describes at length the operation of the gastric juice factory; see his taped autobiographical statement, in SPF ARAN. See also Babkin, *Pavlov* (MS), chap. 14, pp. 3–4.

53. Pavlov, "Lektsii po fiziologii," in *PSS*, 5: 105–6.

54. Ibid., 107.

55. Boldyrev, "Natural'nyi zheludochnyi sok," 156.

56. Smirnov, taped autobiographical statement, in SPF ARAN. Babkin (*Pavlov*, 131) also mentions that the gastric juice factory was Pavlov's first stop in the laboratory.

57. TsGIA SPb 2282.1.190: 1.

58. For example, in 1901 the laboratory had 11 rubles and 79 kopecks left for the month of December. TsGIA SPb 2282.1.171.

59. By 1906 the gastric juice factory was making enough money for Pavlov to pay Ganike 500 rubles a year from the proceeds, in addition to Ganike's salary as Pavlov's laboratory assistant. See TsGIA SPb 2282.1.251: 13. In 1909, by which time gastric juice was bringing in about 4,000 rubles yearly, Pavlov raised Ganike's bonus to 1,000 rubles a year and allowed Ganike to hire an assistant, A. I. Smirnov. See ibid., 1.281. Between his management of the gastric juice operation and his performance of various technical tasks, Ganike expanded his operation gradually until it became an independent Physico-Physiological Division in 1916. This was es-

sentially a workshop attached to Pavlov's factory. Juice proceeds also allowed Pavlov to hire a mechanic in 1910 to assist in developing and maintaining technologies for research on conditional reflexes. In 1910 Pavlov reported to the Institute Council that he was paying I. G. Gerasimov 1,500 rubles a year from the revenues earned by the sale of gastric juice. See ibid., 306.1: 4.

60. Ivan Pavlov to Prince A. P. Ol'denburgskii, January 14, 1903 (typed copy), in SPF ARAN 259.4.234.

61. TsGIA SPb 2282.1.183: 3–40b.

62. Babkin, *Pavlov*, 69–70.

63. TsGIA SPb 2282.1.162: 71.

64. Troianov (in "O gastroenterostomii") reported favorable results using two tablespoons a day, and Boldyrev (in "Natural'nyi zheludochnyi sok") wrote (no doubt from coworkers' experiences in the laboratory) that "often" at least 200 grams a day was needed to produce any benefit. A Russian physician, I. K. Konarzhevskii, used gastric juice to treat twenty patients with various complaints in summer 1901. See Konarzhevskii, "Natural'nyi zheludochnyi sok sobak." Although he did not publish his results until 1902, he may well have apprised the laboratory of them earlier. He reported prescribing various dosages but usually about 130 grams a day, taken in equal portions with three meals.

65. A. Sokolov, quoted in Konarzhevskii, "Natural'nyi zheludochnyi sok," 8.

66. *Novosti i Birzhevaia gazeta*, January 3, 1904, p. 3.

67. Babkin, *Pavlov*, 70.

68. Boldyrev, "Natural'nyi zheludochnyi sok," 156. He also proposed (p. 152) that, due to the lack of advertising, the public was largely ignorant of the existence of this remedy.

69. TsGIA SPb 2282.1.200: 6 and 1.374: 230b. Clearly, five dogs would not have sufficed to produce fifteen thousand flagons a year, but I have been unable to discover material on this expansion of the gastric juice factory.

70. For example, after Pavlov won the Nobel Prize in 1904 he refused to invest it according to the advice of a well-placed friend—or even to allow his wife to do so with her share of the Prize money—because he considered the market an unsuitable place for money obtained for scientific achievements.

71. Boldyrev, "Natural'nyi zheludochnyi sok," 156. See also Tsitovich, "Natural'nyi zheludochnyi sok."

72. For Pavlov's comments in *Lectures* on the relationship between physiologist and physician, see Chapter 6. For his comments on this subject in discussions at the Society of Russian Physicians, see Chapter 9. The one widespread clinical belief that Pavlov criticized uncompromisingly was the view that mechanical stimulation of the stomach elicited gastric secretion. He viewed this belief as incompatible with his own view of a purposive and precise digestive factory. Furthermore, if this belief were true, it would undermine confidence in the Pavlov isolated

sac as a mirror of normal digestive processes (since food did not come into contact with the isolated sac itself).

73. Pavlov, "O vzaimnom otnoshenii fiziologii i meditsiny v voprosakh pishchevareniia. Chast' I" [1894], in *PSS,* 2, pt. 1: 254–56, 261–62, 266–67.

74. Pavlov, *Lektsii,* 179–80; *Lectures* (trans.), 134–39.

Chapter 9. Hail to the Chief

1. Nancy Mandelker Frieden, *Russian Physicians in an Era of Reform and Revolution, 1856–1905* (Princeton: Princeton University Press, 1981), 113. The Society originated in 1865 as the Society of Naturalists and Physicians, in 1885 was transformed into the Moscow-Petersburg Medical Society, and in 1887 became the national Society of Russian Physicians in Memory of N. I. Pirogov.

2. Bekhterev's coworkers delivered their reports primarily to the Scientific Conferences of Physicians of the St. Petersburg Clinic for Mental and Nervous Illnesses (over which Bekhterev presided) and published most frequently in Bekhterev's journal, *Obozrenie psikhiatrii, nevrologii, i eksperimental'noi psikhologii* (Review of Psychiatry, Neurology, and Experimental Psychology). We saw in Chapter 7 the beginnings of the conflict that ensued when Bekhterev invaded Pavlov's turf at the Society of Russian Physicians in 1899 and 1903. That conflict escalated into a social sensation in 1908.

3. L. A. Orbeli, *Vospominaniia* (Moscow: Nauka, 1966), 35–36.

4. I. P. Pavlov, "Sovremennoe ob"edinenie v eksperimente glavneishikh storon meditsiny na primere pishchevareniia" [1899], in *PSS,* 2, pt. 2: 270.

5. Pavlov introduced his remarks after one praktikant's report of March 3, 1894, by stating that "I want to explain the reasons why this physiological report is being given to our Society, which consists primarily of representatives of clinical medicine." I. P. Pavlov, comments on S. A. Ostrogorskii's report to the Society of Russian Physicians, in *PSS,* 6: 38–39.

6. For the constant clinical themes in Pavlov's comments, see also the recorded discussions of reports presented at meetings of the Society of Russian Physicians by the following praktikanty, all in *PSS,* 6 (pages in parentheses): I. L. Dolinskii, February 1894 (35–38); P. P. Khizhin, September 1894 (40–45); I. O. Lobasov, May 1895 (69–70); L. B. Popel'skii, October 1896 (77–79); A. N. Volkovich, March 1898 (116–19); A. S. Serdiukov, September 1898 (122–27); P. E. Kachkovskii, January 1899 (137–40); I. K. Soborov, April 1899 (140–43); Ia. Zavriev, November 1899 (145–47); A. M. Virshubskii, April 1900 (155–57); I. P. Zhegalov, December 1900 (176–78); A. P. Sokolov, March 1901 (184–86); S. I. Lintvarev, April 1901 (187); and N. Kazanskii, May 1901 (187–89).

7. I. P. Pavlov, comments in discussion of P. P. Khizhin's report to the Society of Russian Physicians, September 29, 1894, in *PSS,* 6: 43.

8. Ibid., 43.

9. A. A. Troianov, comments in discussion of P. P. Khizhin's report to Society of Russian Physicians, September 29, 1894, in *PSS*, 6: 45; Pavlov, comments on Khizhin's report, 45.

10. I. P. Pavlov, comments in discussion of I. O. Lobasov's report to the Society of Russian Physicians, May 13, 1895, in *PSS*, 6: 70.

11. I. P. Pavlov, comments in discussion of A. N. Volkovich's report to the Society of Russian Physicians, March 5, 1898, in *PSS*, 6: 117–18.

12. Exchange reported in A. A. Val'ter, "Rabota podzheludochnoi zhelezy pri miase, khlebe, moloke i pri vlivanii kisloty," *Trudy obshchestva russkikh vrachei*, September 1896, pp. 38, 39. (This exchange is omitted in the edited version published in *PSS*, 6: 75–77.)

13. Ibid., 41.

14. I. P. Pavlov, comments in discussion of A. S. Serdiukov's report to the Society of Russian Physicians, September 24, 1898, in *PSS*, 6: 126.

15. I. P. Pavlov, "O vzaimnom otnoshenii fiziologii i meditsiny v voprosakh pishchevareniia. Chast' II" [1895], in *PSS*, 2, pt. 1: 272.

16. I. P. Pavlov, comments in discussion of I. K. Soborov's report to the Society of Russian Physicians, April 29, 1899, in *PSS*, 6: 141.

17. I. P. Pavlov, comments in discussion of S. I. Lintvarev's report to the Society of Russian Physicians, April 26, 1901, in *PSS*, 6: 187.

18. I. P. Pavlov, comments in discussion of A. I. Bulavintsov's report "Psychic Gastric Juice in People" to the Society of Russian Physicians, February 6, 1903, in *PSS*, 6: 200–201. Pavlov referred here ("clinicians have already attempted") to the experiments of Schüle, who claimed to have demonstrated that the psyche played no role in gastric secretion in humans.

19. I. P. Pavlov, "Rech', proiznesennaia po povodu izbraniia tovarishchem predsedatelia obshchestva russkikh vrachei v S. Peterburge" [October 7, 1893], in *PSS*, 6: 28. This speech was originally published in *Trudy obshchestva russkikh vrachei* 60 (October 1893): 1–2. The quotation in the text provides a good example of Pavlov's use of the word *purposiveness* as a positive description of human endeavors. As we have seen, he routinely invoked this same word in his description of the digestive system.

20. Pavlov, "O vzaimnom otnoshenii fiziologii i meditsiny v voprosakh pishchevareniia. Chast' I" [1894], in *PSS*, 2, pt. 1: 246.

21. See Pavlov's earlier comments to the Society on Tarkhanov's report, in *PSS*, 6: 47–48.

22. Pavlov, "O vzaimnom otnoshenii fiziologii," 263. This was the conclusion to the first part of Pavlov's report in December 1894.

23. This discussion took place after the second part of Pavlov's report on January 12, 1895, and was first published in *Trudy obshchestva russkikh vrachei v S.-Pe-*

terburge 61 (January 1895): 175–79; republished in *PSS*, 6: 49–52. Sirotinin's comment is in *PSS*, 6: 51.

24. Pavlov, "Rech' po povodu izbraniia," 28–29.

25. Ibid., 29.

26. Pavlov, "O vzaimnom otnoshenii fiziologii," 245–46. Note that Pavlov has subtly changed his characterization of the two groups (physicians and physiologists). In 1893 he had spoken of the relationship between practicing physicians and "theoretical physicians dedicating themselves to the laboratory." Now, in 1894, he characterizes the latter as "physiology" or "theoretical-laboratory knowledge." For the physician as "mechanic of the human body," see ibid., 247. For the use of this metaphor in U.S. discourse on the relationship between laboratory and clinic, see John Harley Warner, "Ideals of Science and Their Discontents in Late Nineteenth-Century American Medicine," *Isis* 82 (1991): 458.

27. Pavlov, "O vzaimnom otnoshenii fiziologii," 246, 248.

28. Pavlov, "Sovremennoe ob"edinenie v eksperimente glavneishikh storon meditsiny na primere pishchevareniia" (speech to the Society of Russian Physicians, 1899), reprinted in *PSS*, 2, pt. 2: 247–82. This was originally published in *Trudy obshchestva russkikh vrachei v S.-Peterburge* 67 (November–December 1900): 197–242.

29. Pavlov, "Sovremennoe ob"edinenie," 275.

30. For Pavlov's first report to the Society on this subject, see "Patologo-terapevticheskii opyt nad zheludochnym otdeleniem sobaki," *Trudy obshchestva russkikh vrachei v S.-Peterburge* 64 (May 1897): 581–89; reprinted in *PSS*, 2, pt. 2: 219–25.

31. Pavlov, "Sovremennoe ob"edinenie," 275, 271.

32. I. P. Pavlov, "O vyzhivanii sobak s pererezannymi bluzhdaiushchimi nervami" [1896], *PSS*, 1: 536–49. For the horrified response of one antivivisectionist (and a good description of Pavlov's presentation to the Society), see "Nauchnoe 'naslazhdenie,'" *Vestnik Rossiiskago obshchestva pokrovitel'stva zhivotnykh* 7, no. 9–10 (September-October 1895): 249–65. For Pavlov's first report on this subject, based on experiments conducted with praktikant Kachkovskii, see "O smerti zhivotnykh vsledstvie pererezki bluzhdaiushchikh nervov," in *PSS*, 1: 530–35.

33. Pavlov, "Sovremennoe ob"edinenie," 271.

34. Ibid., 272.

35. Ibid., 284.

36. See, for example, V. F. Orlovskii, "Obzor rabot po pishchevaritel'nym organam za 1899 god," *Izvestiia Voenno-Meditsinskoi Akademii* 3 (September 1901): 49–60. For other annual reviews, see *Izvestiia Voenno-Meditsinskoi Akademii* 4 (January 1902): 41–53; 6 (January 1903): 40–58; 8 (February 1904): 178–89; and 8 (March 1904): 304–7.

37. See A. M. Virshubskii, "Obobshchenie noveishikh faktov iz fizologii pishchev-

areniia s tochki zreniia printsiopov: a) spetsificheskoi vozbudimosti i b) vzai-modeistviia funktsii organov," *Praktikcheskaia meditsina* 7 (1900): 129–38. Former praktikant I. O. Shirokikh became a professor at the Novoaleksandriiskii Institute of Agriculture and Forestry, and he turned his scientific expertise to agricultural topics in "Postanovka khoziaistvennykh opytov po vyiasneniiu naivygodneishego ispol'zovaniia dlia kormov polevykh otbrosov (solomy i miakiny) pri molochnom i miasnom skotovodstve," in *Dnevnik XI s"ezda estestvoispytatelei i vrachei* (St. Petersburg, 1901); *Dannye po issledovaniiu russkogo slivochnogo masla* (Warsaw, 1903); and "Znachenie idei I. P. Pavlova dlia razvitiia ucheniia o kormlenii sel'skokhozi-aistvennykh zhivotnykh," *Arkhiv Biologicheskikh Nauk* 11, suppl. (1904): 45–48.

38. On London's investigations, see Horace W. Davenport, *A History of Gastric Secretion and Digestion: Experimental Studies to 1975* (New York: Oxford University Press, 1992), 329–41.

39. Boris Babkin, *Pavlov: A Biography* (Chicago: University of Chicago Press, 1949), 83.

40. The praktikanty who became physiologists (with their dates in the laboratory) were A. F. Samoilov (1892–95), L. B. Popel'skii (1896–97), A. A. Val'ter (1896–1902), V. V. Savich (1900–1904, 1907, 1915), V. N. Boldyrev (1900–1911), B. P. Babkin (1902–4, 1912), L. A. Orbeli (1901–15), and I. S. Tsitovich (1901–3, 1911). Val'ter, Savich, Boldyrev, Orbeli, and Babkin developed long-term working relations with the laboratory that were atypical for praktikanty; the chief clearly perceived them as the beginnings of a "Pavlov school." Val'ter died in 1902. Boldyrev and Babkin emigrated after the Bolsheviks took power in 1917 and built successful careers as physiologists in the United States and Canada. With Pavlov's help, Savich became a professor of pharmacology (a discipline that Pavlov viewed as properly the province of physiology). Orbeli became a renowned physiologist and powerful scientific entrepreneur, inheriting his mentor's empire after Pavlov's death.

41. Serafima Pavlova, *Vospominaniia*, in SPF ARAN 259.1.169: 318.

42. Iu. Laudenbakh, review of *Lektsii o rabote glavnykh pishchevaritel'nykh zhelez*, in *Russkii arkhiv patologii, klinicheskoi meditsiny i bakteriologii* 6, no. 5 (1897): 599–600.

43. S. S. Salazkin, "Staroe i novoe v oblasti pishchevareniia," *Russkii arkhiv patologii, klinicheskoi meditsiny i bakteriologii* 6, no. 5 (1897): 589. Salazkin is not listed among Pavlov's praktikanty in *Fiziologicheskaia shkola*, but on the basis of an archival source that I have not seen, he is included in N. M. Gureeva and N. A. Chebysheva, eds., *Letopis' zhizni i deiatel'nosti akademika I. P. Pavlova*, commentary by V. L. Merkulov (Leningrad: Nauka, 1969), 192. Salazkin subsequently became chief of the Biochemistry Division of the Institute of Experimental Medicine and director of the Institute from 1927 to 1931.

44. I. R. Tarkhanov, "Fiziologiia," in *Rossiia: Entsiklopedicheskii Slovar'*, ed. F. A. Brokgauz and I. A. Efron [1898] (Leningrad: Lenizdat, 1991), 768. Tarkhanov added further faint praise and an expression of jealousy about Pavlov's institutional po-

sition: "The arena for Pavlov's scholarly activity is not only the Military-Medical Academy, but also the Institute of Experimental Medicine, in which he heads the wealthy physiological division. Thanks to the wealthy scientific setting of the physiological laboratory of this Institute and the lavish sums available to produce scientific investigations, the majority of Pavlov's works, and those of his students, are produced within the walls of this Institute, to which science is already indebted for many valuable and important investigations." Ibid.

45. S. I. Chir'ev, *Fiziologiia cheloveka* (Kiev, 1902), 303–34. Chir'ev cited Heidenhain on the qualities of gastric juice; Beaumont, Basov, Bernard, Heidenhain, and others on means of obtaining it; and Bernard, Virchow, and others on its role in digestion. Tarkhanov, in his *Populiarnye lektsii po fiziologii v sviazi s gigienoi* (Popular Lectures on Physiology in Connection with Hygiene) (St. Petersburg, 1906), made no mention of Pavlov's work. Only after Pavlov received the Nobel Prize did his contributions shape the presentation of digestive physiology in Russian textbooks. See, for example, V. I. Vartanov, *Kurs fiziologii* (St. Petersburg, 1906); V. F. Verigo, *Osnovy fiziologii cheloveka i vysshikh zhivotnykh,* vol. 2 (St. Petersburg, 1909); and V. Ia. Danilevskii, *Fiziologiia cheloveka,* vol. 1 (Moscow, 1913).

46. M. Iu. Gol'dshtein, "Estestvoznanie i tekhnika v xix veke. Ocherk.," *Niva,* suppl., 1901, pp. 340–91. For Gol'dshtein's review of nineteenth-century developments in physiology, see ibid., 371–73 (the Russian trio is mentioned on p. 372).

47. This excerpt from the committee's report is from Gureeva and Chebysheva, *Letopis',* 86; Merkulov (in his commentary) gives the archival source as SPF ARAN 1.1a.147.

48. J. S. Edkins, "Mechanism of Secretion of Gastric, Pancreatic, and Intestinal Juices," in *Text-Book of Physiology,* vol. 1, ed. E. A. Schäfer (Edinburgh: Young J. Pentland, 1898), 548.

49. Ibid., 541, 545–46. Edkins referred to works both by "Chischin" (whom he cited according to a report in a German scientific journal) and "Khigine" (whom he cited according to the French version of his doctoral thesis in *Archives des sciences biologiques*)—and apparently did not realize these were one and the same (Khizhin).

50. J. P. Pawlow, preface to *Die Arbeit Der Verdauungsdrusen,* trans. A. Walther (Wiesbaden: J. F. Bergmann, 1898), ix.

51. In his preface (ibid.) Pavlov acknowledged his good fortune that "an active member of our laboratory, Dr. Val'ter, agreed to take upon himself the labor of translating this book. He was in a position to convey, not only that which he understood, but also that which he himself had seen and experienced."

52. W[ilhelm] O[stwald], review of J. P. Pawlow, *Die Arbeit der Verdauungsdrusen,* in *Zeitschrift für Physikalische chemie* 26 (1898), 757. Ostwald was coeditor of this journal. Ten years later he addressed the nature of scientific creativity in his popular book *Grosse Manner* (Leipzig: Akademische Verlagsgesellschaft, 1909); in this work he identified distinctive "classical" and "romantic" scientific types. The

book caused quite a stir—four editions were published by 1910—and was quickly translated into various languages, including Russian. According to Babkin, "In Pavlov's laboratory everybody including Pavlov had read it, and it was hotly discussed," particularly with respect to the chief's style of work (Babkin, *Pavlov,* 144). Those discussions provide the basis for Babkin's chapter "Classicists and Romanticists in Science" (*Pavlov,* 144–51).

For other references to Pavlov's "pupils" and "the Pavlov school," see H. Munk's review in *Centralblatt für Physiologie,* vol. 26, *Literatur 1898* (Leipzig: Franz Deuticke, 1899), 552; J. Boas's review in *Archiv für Verdauungs-Krankheiten* 5 (1899): 300–301; Gofman's review in *Archiv für Verdauungs-Krankheiten* 4 (1898): 98–99; Penzoldt-Erlagen's review in *Munchener medicinische Wochenschrift* 46, no. 1 (1899): 16–17; anonymous review in *Lancet* 2 (December 6, 1902): 1552–53; Lafayette Mendel, "Professor Pawlow's Researches on the Physiology of Secretion," *Science* (new ser.) 14, no. 356 (1901): 647–49 (this article resulted from the false rumor that Pavlov had been awarded the Nobel Prize); and J. P. Langlois, "Iwan Petrowitsch Pawlow," *La Presse Médicale,* March 25, 1905. Theodor Rosenheim, who reviewed *Lectures* for *Litteratur-Belage No. 16 der Deutschen Medicinischen Wochenschrift,* July 13, 1899, was an exception: he referred to Pavlov's "coworkers."

53. Boas review, *Archiv für Verdauungs-Krankheiten;* Munk review, *Centralblatt für Physiologie;* anonymous review in *British Medical Journal,* January 17, 1903, p. 151.

54. For example, Rosenheim, a specialist on the treatment of digestive disorders, noted Pavlov's conclusion that "to each type of food corresponds a determined hourly course of secretion and a characteristic alternation of the quality of the juice." See Rosenheim review, *Litteratur-Belage.* The reference to "definite periodic laws" is from review, *Lancet.*

55. Boas review, *Archiv für Verdauungs-Krankheiten;* Munk review, *Centralblatt für Physiologie;* review, *Lancet.*

56. See, for example, Gofman's review in *Archiv für Verdauungs-Krankheiten,* 98–99; and review, *British Medical Journal,* 153. (I have used here the wording in the English edition of *Lectures,* as quoted in the *British Medical Journal.*)

57. Boas review, *Archiv für Verdauungs-Krankheiten;* Rosenheim review, *Litteratur-Belage*

58. Mendel, "Professor Pawlow's Researches," 648, 649 (Foster is quoted on p. 649).

59. William H. Howell, *An American Text-Book of Physiology,* 2d ed. rev. (Philadelphia: W. B. Saunders, 1900), 1: 237, 242.

60. William H. Howell, *An American Text-Book of Physiology* (Philadelphia: W. B. Saunders, 1896), 181; Howell, *American Text-Book of Physiology* (1900), 240–41. Similarly, "Dolinsky, working upon dogs under more favorable conditions" (1896, p. 177) becomes "Dolinsky, working upon dogs by Pawlow's methods" (1900, p. 236).

61. This comment is from Graham Lusk's introduction to the English edition, Robert Tigerstedt, *A Text-Book of Human Physiology* (New York: D. Appleton and Co., 1906), xi.

62. Robert Tigerstedt to I. P. Pavlov, April 16, 1898, in SPF ARAN 259.2.1017. Tigerstedt is referring to his *Lehrbuch der Physiologie des Menschen* (Leipzig: Hirzel, 1898).

63. Pavlov's preface to the Tigerstedt volume is reprinted in *PSS*, 6: 163–64; text quotation from p. 164; notes and additions to the Russian edition follow on pp. 165–71.

64. On this relationship, see V. L. Merkulov, "Materialy o druzhbe I. P. Pavlova s R. Tigershtedtom," *Fiziologicheskii zhurnal SSSR* 14, no. 9 (1959): 1162–65.

65. F. Riegel, "Ueber medicamentose Beeinflussung der Magensaftsecretion," *Zeitschrift für Klinische Medicin* 37 (1899): 381–402; and Walther Clemm, "Ueber die Beeinflussung der Magensaftabschediung durch Zucker," *Therapeutische Monatshafte* 15 (August 1901): 405–13. For Riegel's purposes, it was not important to reconcile the varying secretory curves obtained from different dogs. He noted simply that "the fluctuations in these data vary in different animals, differing according to the size of the animal, the size of its small stomach, the appetite and so forth. Under equivalent experimental conditions, however, the amount and course of gastric secretion is always approximately the same." The important point was not the precise nature of the normal curve for gastric secretion but, rather, the influence of atropine and pilocarpine on gastric secretion. For Riegel's curves, see his "Ueber medicamentose," 386.

66. Prof. Schüle, "In wie weit stimmen die Experimente von Pawlow am Hunde mit den Befunden am normalen menschlichen Magen uberein?" *Deutsches Archiv für Klinische Medicin* 71 (1901): 111–32. Schüle also concluded that, in humans, the nature of the food had little effect on the quality of gastric secretion. He did, however, affirm Pavlov's argument about the unimportance of mechanical irritants and attributed gastric secretion entirely to nervous-chemical mechanisms.

67. Hornborg, a Finnish investigator, supported Pavlov's view. He reported the case of a boy with a "stricture of the esophagus and a fistula in the stomach. Food when chewed and swallowed did not reach the stomach, but was regurgitated; it caused, nevertheless, an active psychical secretion in the empty stomach"; described in W. H. Howell, *A Text-Book of Physiology for Medical Students and Physicians* (Philadelphia: W. B. Saunders, 1909), 747. As we shall see in Chapter 10, both Schüle's and Hornborg's results played a part in the Nobel Committee's deliberations about Pavlov. Investigations by one Russian physician also confirmed the importance of appetite, to Pavlov's great delight. See the discussion of A. I. Bulavintsov's report to the Society of Russian Physicians, February 6, 1903, in *PSS*, 6: 200–201.

68. Quotation from Boas review in *Archiv für Verdauungs-Krankheiten*. On the German visitors to Pavlov's laboratory, see V. L. Merkulov, "Rabota nemetskikh i

abstro-vengerskikh uchenykh v fiziologicheskom otdele Instituta Eksperimental'noi Meditsiny pod rukovodstvom I. P. Pavlova 1902–1908 gg.," *Ezhegodnik Instituta Eksperimental'noi Meditsiny* 4 (1961): 511–20.

69. For Orth's letter of 1903, see SPF ARAN 259.9.1216; for Emerson's letter of
1903, see E. M. Kreps, ed., *Perepiska Pavlova* (Leningrad: Nauka, 1970), 210; for
Benedict's letters of 1907–30, see SPF ARAN 259.9.69. A Russian translation of
Benedict's letters (which were written in English) is in Kreps, *Perepiska*, 219–25.

70. This correspondence remains in Pavlov's personal papers in SPF ARAN: F.
Rollin, 254.9.1232; H. J. Hamburger, 254.9.182; Carl Lewin, 254.9.1188; Paul Mayer,
254.9.1375; an unidentified German, Lieff, 259.9.1374; R. Tigerstedt, 259.2.1117; and
Karl Mörner, 259.9.1206.

71. The relevant correspondence is in SPF ARAN: Walther Straub, 254.9.1286;
Waldemar Koch, 254.9.1116; Hermann Munk, 254.9.1208; Johann Orth, 254.9.1216;
G. Stewart, 254.9.1182; Ernest Stadler, 254.9.1284; F. A. Steeksma, 254.9.1285; and
Alois Velich, 254.9.1119. This list of scientist-visitors does not include more infor-
mal visitors, such as Wilhelm Roux, or those who visited shortly after 1904, such
as Dublin physiologist (and translator of *Lectures*) W. H. Thompson.

72. See David M. Matthews, *Protein Absorption: Development and Present State
of the Subject* (New York: Wiley-Liss, 1991). My thanks to Joseph Fruton for bringing this source to my attention.

73. See Merkulov, "Rabota nemetskikh." After Friedenthal returned to Berlin,
his chief, Hermann Munk, wrote a letter to Pavlov requesting permission to visit
the laboratory in order to study the creation of a dog with an isolated sac. It is not
clear whether this trip ever occurred. See SPF ARAN 254.9.1182.

74. F. Steeksma to I. P. Pavlov, January 8, 1905, from Merkulov's Russian translation in Kreps, *Perepiska*, 214–15.

75. Walther Gross to I. P. Pavlov, June 5, 1902, and April 4, 1905; from Merkulov's
Russian translation in Kreps, *Perepiska*, 204, 205.

76. See Bernt Lonnqvist, "Beitrage zur Kenntsniss der Magensaftabsonderung,"
Skandinavisches Archiv für Physiologie 1 (1906): 194–262. Both Lonnqvist and Gross
used a dog-technology that Pavlov designed and created for investigation of the
chemical excitation of the gastric glands. See Davenport, *History of Gastric Secretion and Digestion*, 192–93.

77. On Gross's research, see Davenport, *History of Gastric Secretion and Digestion*, 192–94; and Merkulov, "Rabota nemetskikh," 512–14.

78. The statement about the importance of Cohnheim's work is from the ded-
ication in Matthews, *Protein Absorption*. For Cohnheim's discovery of erepsin, see
ibid., 39–43. Otto Cohnheim was the son of the eminent pathologist Julius Cohnheim. On Cohnheim and erepsin, see ibid., 39–43, 373.

79. Otto Cohnheim to I. P. Pavlov, November 3, 1902; June 9, 1907; November
18, 1910; from the Russian translations in Kreps, *Perepiska*, 206–7; 208; 209. In 1910
Cohnheim informed Pavlov that another German scientist, a certain Professor

Weinbruk from Wiesbaden, was shortly to visit St. Petersburg and requested that Pavlov show him his laboratory. Either Cohnheim or Gross apparently interested some of Pavlov's coworkers in Paul Ehrlich's popular side-chain theory, but Pavlov himself was resolutely uninterested.

80. Emil Abderhalden to I. P. Pavlov, October 10, 1904; September 22, 1905; September 9, 1924; from L. O. Zeval'd's Russian translations in Kreps, *Perepiska*, 210–11; 211–12; 212.

81. The commemorative volume, which appeared in both Russian and French, was published as a supplement to the 1904 edition of *Arkhiv Biologicheskikh Nauk* (Archive of Biological Sciences). It was compiled and in press before Pavlov was awarded the Nobel Prize in late October 1904 but was not published until December of that year. The photographs in Figures 6 through 10 and 12 in Chapter 3 are from this album.

82. J. P. Langlois, "Iwan Petrowitsch Pawlow," *La Presse Médicale*, March 25, 1905, pp. 186–87. This article appeared after Pavlov received the Nobel Prize but makes no mention of it. Langlois refers, rather, to the celebration of Pavlov's jubilee at the Society of Russian Physicians on December 16 (two weeks after the Nobel ceremonies). My impression is that, for Langlois, Pavlov's position at the "pinnacle" predated this award.

83. Despite the entreaties of the compilers of the volume, only six of twenty-one scientific contributions mentioned Pavlov's work.

84. On the historical development of the biochemistry of ferments and enzymes, see Joseph S. Fruton, *Molecules and Life: Historical Essays on the Interplay of Chemistry and Biology* (New York: Wiley-Interscience, 1972).

85. Before 1899 the laboratory's only doctoral thesis on the intestines was D. L. Glinskii's *K fiziologii kishek* (Academy diss., St. Petersburg, 1891).

86. N. P. Shepoval'nikov, *Fiziologiia kishechnago soka* (Academy diss., St. Petersburg, 1899); Pavlov, "Sovremennoe ob"edinenie," 257–58.

87. Pavlov, "Sovremennoe ob"edinenie," 258. It had been demonstrated years earlier that bile contained another ferment that acted on the fat-splitting ferment in pancreatic juice.

88. Pavlov, comments on discussion of I. I. Lintvarev's report to the Society of Russian Physicians, March 8, 1901, in *PSS*, 6: 180.

89. V. N. Boldyrev, V. I. Vartanov, E. A. Ganike, and A. P. Sokolov, "Referaty trudov professora I. P. Pavlova i ego uchenikov," *Arkhiv Biologicheskikh Nauk* 11, suppl. (1904): 77.

90. Babkin describes the "final form" of the doctrine that pancreatic secretion adapted over time to various diets: "On a meat diet, trypsin and lipase were in an active state and did not require enterokinase and bile, respectively, for their activation. On a prolonged bread-and-milk diet these two pancreatic enzymes gradually acquired a zymogenic form: that is, they were in an inactive state, and without the help of their activators—enterokinase and bile—they produced only a

weak effect on their corresponding substrates. Amylase under all circumstances was secreted in an active form." See Babkin, *Pavlov,* 243. See also I. I. Lintvarev, *Vliianie razlichnykh fiziologicheskikh uslovii na sostoianie i kolichestvo fermentov v soke podzheludochnoi zhelezy* (Academy diss., St. Petersburg, 1901).

91. See, for example, Edkins, "Mechanism of Secretion," 551; and Howell, *Text-Book of Physiology* (1896), 176.

92. This was Babkin's assessment in *Pavlov,* 243.

93. C. Delezenne and A. Frouin, "La sécrétion physiologique du pancréas ne posséde pas d'action digestive propre vis-à-vis de l'albumine," *Comptes rendus hebdomadaires des séances et mémoires de la Société de Biologie* 54 (1902): 693.

94. C. Delezenne and A. Frouin, "Nouvelles observations sur la sécrétion physiologique du pancréas. Le suc pancréatique des bovides," *Comptes rendus hebdomadaires des séances et mémoires de la Société de Biologie* 55 (1903): 455.

95. B. P. Babkin, "K voprosu ob otdelitel'noi deiatel'nosti podzheludochnoi zhelezy," *Izvestiia Voenno-Meditsinskoi Akademii* (1904), as summarized in Boldyrev et al., *Referaty trudov,* 77.

96. William Howell, one of Pavlov's greatest admirers, captured this tendency in the preface to the third edition of his textbook of physiology (1909): "During recent years chemical work in the fields of digestion and nutrition has been very full, and as a result theories hitherto generally accepted have been subjected to criticism and alteration, particularly as the important advances in theoretical chemistry and physics have greatly modified the attitude and point of view of the investigators in physiology. Some former views have been unsettled and much information has been collected which at present it is difficult to formulate and apply to the explanation of the normal processes of the animal body. It would seem that in some of the fundamental problems of metabolism physiological investigation has pushed its experimental results to a point at which, for further progress, a deeper knowledge of the chemistry of the body is especially needed. Certainly the amount of work of a chemical character that bears directly or indirectly on the problems of physiology has shown a remarkable increase within the last decade." William H. Howell, *A Text-Book of Physiology,* 3d ed., thoroughly revised (Philadelphia: W. B. Saunders, 1909), 4.

97. Babkin, *Pavlov* (MS), chap. 10, p. 4. Babkin dated this scene at "about 1900." His first year in the Pavlov laboratory was 1902; Popel'skii's articles criticizing Pavlov appeared in 1901–3. This portion of Babkin's text was excised by his editor and doesn't appear in the published biography.

98. Babkin, *Pavlov* (MS), chap. 10, pp. 3, 4–5.

99. L. V. Popel'skii, *O sekretorno-zaderzhivaiushchikh nervakh podzheludochnoi zhelezy* (Academy diss., St. Petersburg, 1896), 118.

100. Popel'skii, *O sekretorno-zaderzhivaiushchikh,* unnumbered page following p. 118.

101. See Popel'skii's second proposition, ibid.; and Pavlov, comments on Po-

pel'skii's report to the Society of Russian Physicians, in *PSS*, 6: 77–79. Pavlov explained that the salivary glands lacked this "antagonistic" apparatus because "when the salivary glands are working physiologically, it rarely occurs that they must inhibit their activity; in the abdominal glands [i.e., the gastric and pancreatic glands], on the other hand, where the task concerns subtle work, inhibition is frequently put into motion, depending upon the sort of food; therefore, here [in both the gastric and pancreatic glands] both systems are developed identically." See Pavlov, comments on Popel'skii's report, 79.

102. L. B. Popel'skii, "Reflektornyi tsentr podzheludochnoi zhelezy," *Trudy obshchestva russkikh vrachei* 67 (November 1900): 616–29; and L. Popielski, "Ueber das peripherische reflectorische Nervencentrum des Pankreas," *Pflügers Archive* 86 (1901): 215–46.

103. Lev Popel'skii, *Istoricheskii ocherk kafedry fiziologii v Imperatorskoi Voenno-Meditsinskoi Akademii za 100 let (1798–1898)* (St. Petersburg: Ministry of Internal Affairs, 1899).

104. On Sechenov, Tsion, Tarkhanov, see ibid., 51, 90, 120; on Pavlov, see ibid., 155. Popel'skii was more enthusiastic about Pavlov's lectures, which "are distinguished by simplicity and clarity," developed the theme that students, as future physicians, must understand the human organism as a machine, and were so thoroughly grounded in experimental demonstrations that the course could be termed "Physiology in experiments." See ibid., 142–43, 155, 157.

105. Babkin, *Pavlov* (MS), chap. 10, p. 4. Orbeli recalled that Pavlov supported Popel'skii's candidacy, which failed because of discrimination against Poles. See Orbeli, *Vospominaniia*, 46–47.

106. See L. V. Popel'skii's "O tselesoobraznosti v rabote pishchevaritel'nykh zhelez (Predvaritel'noe soobshchenie)," *Vrach*, 1901, no. 50: 1–5; "O perifericheskom otrazhennom tsentre zheludochnykh zhelez," *Vrach*, 1901, no. 51: 1–4; "Prichiny raznoobraziia svoistv podzheludocnago soka v otnoshenii belkago brodila," *Russkii Vrach*, 1902, no. 18: 679–84; "O tselesoobraznosti v rabote podzheludochnoi i zheludochnykh zhelez," *Russkii Vrach*, 1902, no. 35: 1242–48; "Ob osnovnykh svoistvakh podzheludochnago soka," *Russkii Vrach*, 1903, no. 16: 1–11; and "Vkus i potrebnosti organizma," *Russkii Vrach*, 1903, no. 49: 1737–40. The first, preliminary communication listed above was republished in German as L. Popielski, "Ueber die Zweckmassigkeit in der Arbeit der Verdauungsdrusen," *Deutsche Medicinische Wochenschrift* 28 (1902): 864–65. Popel'skii related the results of some of his own experiments, and developed his critique of Val'ter's thesis, in L. Popielski, "Ueber die Grundeigenschaften des Pankreassaftes," *Centralblatt für Physiologie* 17, no. 3 (1903): 65–70. According to Orbeli, the physiological laboratory in which Popel'skii worked had been created at Pavlov's suggestion. See Orbeli, *Vospominaniia*, 47.

107. Popel'skii, "O tselesoobraznosti v rabote" (1901), 5.

108. Popel'skii, "O tselesoobraznosti v rabote" (1902), 1242.

109. Popel'skii, "Prichiny raznoobraziia," 682; emphasis in original.

110. Popel'skii presented his experimental data in ibid., 679–80, 683. He rejected the Mett method favored by the Pavlov laboratory as insufficiently sensitive; instead he used a fibrin preparation to measure proteolytic power.

111. "The quantity and quality of pancreatic juice is determined by the amount and strength of the irritant. If one keeps in mind that the irritant can be either *weak* or *strong,* and that the amount can be either *small* or *large,* [one sees that] there is an entire series of pancreatic juices with the most varied qualities, and that each sort of food will have its own sort of juice." Popel'skii, "Prichiny raznoobraziia," 683; emphasis in original.

112. Ibid.

113. Ibid., 684; emphasis in original.

114. Popel'skii's "O tselesoobraznosti v rabote" (1901); "Prichiny raznoobraziia"; "O tselesoobraznosti v rabote" (1902); "Ob osnovnykh"; and "Vkus i potrebnosti organizma."

115. Popielski's "Ueber die Zweckmassigkeit"; and "Ueber die Grundeigenschaften."

116. For a short sketch of Popel'skii's career, see A. Laskiewicz, "Leon Popielski, Founder of the Polish School of the Physico-Pharmacological Research," *Bulletin of Polish Medical Science and History* 110 (1967): 147–48.

117. The phrase *humoralist tide* is from Arthur Silverstein, "Cellular versus Humoral Immunity: Determinants and Consequences of an Epic Nineteenth-Century Battle," in his *A History of Immunology* (San Diego: Academic Press, 1989), 38–58.

118. C. J. Martin, "Obituary: Ernest Henry Starling," *British Medical Journal* 1 (1927): 900–905; quoted in R. A. Gregory, "The Gastrointestinal Hormones: An Historical Review," in *The Pursuit of Nature: Informal Essays on the History of Physiology,* ed. A. L. Hodgkin et al. (Cambridge: Cambridge University Press, 1977), 107.

119. W. M. Bayliss and E. H. Starling, "The Mechanism of Pancreatic Secretion," *Journal of Physiology* 28 (1902): 325–53.

120. For the initial report of Bainbridge's research, see F. A. Bainbridge, "On the Adaptation of the Pancreas," *British Medical Journal* 1 (1904): 778–81. Edkins reported to the Royal Society on his results in May 1905 and published them the following year, in J. S. Edkins, "The Chemical Mechanism of Gastric Secretion," *Journal of Physiology* 34 (1906): 133–44. My thanks to Tilli Tansey for sharing with me her knowledge of these two scientists' institutional circumstances.

121. I am using here the abstract of Bayliss and Starling's Croonian Lecture of 1904, "The Chemical Regulation of the Secretory Process," *Nature* 70 (May-October 1904): 65–68; and the text of Starling's "The Chemical Correlation of the Functions of the Body," published in 1905 in four issues of *Lancet:* August 5, pp. 339–41; August 12, pp. 423–25; August 19, pp. 501–3; and August 26, pp. 579–83. In their joint lecture of 1904, the British physiologists coined the word *hormone,* from the

Greek word meaning "arouse to activity." "These chemical messengers," they explained, "have to be carried from the organ where they are produced to the organ which they affect by means of the blood stream, and the continually recurring physiological needs of the organism must determine their repeated production and circulation throughout the body." Quoted in Gregory, "Gastrointestinal Hormones," 110.

122. To my knowledge, there is no systematic historical account of Bayliss and Starling's experimental practices.

123. Bayliss and Starling, "Mechanism of Pancreatic Secretion," 343, 353.

124. Starling, "Chemical Correlation," 424. In this passage he seems to be speaking for Bayliss as well. Starling did mention that "other workers, however, such as Fleig and Wertheimer, believed that both mechanisms are at work—namely, that through the mucous membrane of the intestine the pancreas can be excited to secrete by both nervous and chemical means." According to Babkin, writing later about the polarized discourse that followed, the Pavlov laboratory "was engaged in the rehabilitation of the role of the nervous system in the secretory processes of the digestive glands. This was a natural reaction against the militant attitude of the English physiologists, led by Bayliss and Starling, who absolutely denied all participation of the nervous system in some of the secretory processes, such as the activity of the pancreas or the second phase of gastric secretion. It is quite understandable that every indication that the nervous system might take part in the regulation of the secretory processes was quickly noted and eagerly explored by those who had actually observed so many examples of its action on the digestive glands." See B. P. Babkin, *Secretory Mechanism of the Digestive Glands* (New York: Paul B. Hoeber, 1944), 471. Elsewhere he writes that "it required several years of work on the part of my colleagues and myself and actual demonstration of the secretory effect of vagus nerve stimulation to persuade [Bayliss and Starling] that the pancreatic gland possessed secretory nerves." See Babkin, *Pavlov*, 230.

125. Starling, "Chemical Correlation," 501.

126. Bayliss and Starling, "Chemical Regulation," 66.

127. Ibid., 66–67.

128. Ibid., 67. Starling also cites Popel'skii's critique of Pavlov in "Chemical Correlation," 503. Bayliss and Starling did accept the view, also originating with the Pavlov laboratory, that the pancreatic gland adapted over time to a specific diet, and this they accounted for (following Bainbridge) by a humoral mechanism. See Bayliss and Starling, "Chemical Regulation," 67–68; and Starling, "Chemical Correlation," 503.

129. Starling, "Chemical Correlation," 502.

130. Bayliss and Starling, "Chemical Regulation," 67. In his address of 1905, Starling again noted that Val'ter's results "have been controverted by Popielski," but added that "if Walter's contentions are supported it will be interesting to determine whether the adaptation of the pancreatic activity to the nature of the food

is nervous in character as imagined by Pawlow or whether the mechanism in this case also is chemical." See Starling, "Chemical Correlation," 503.

131. Starling, "Chemical Correlation," 502.

132. The best sources are the nuanced "internalist" accounts provided by Babkin's historical references in his *Secretory Mechanism of the Digestive Glands;* by Gregory's "Gastrointestinal Hormones"; and by Davenport's *History of Gastric Secretion and Digestion,* 134–257.

133. William Maddock Bayliss, *Principles of General Physiology* (London: Longmans, Green, and Co., 1915), 370 (photograph), 371 (quotation).

134. Bayliss, *Principles,* 372.

Chapter 10. The Nobel Prize

My thanks to Johan Ledin for all translations of Nobel archival documents. For another analysis of the Nobel deliberations concerning Pavlov, also based on archival research, see George Windholz and James R. Kuppers, "Pavlov and the Nobel Prize Award," *Pavlovian Journal of Behavioral Science* 25 (1990): 155–62.

1. Contrary to my expectations, the extensive archival materials on the Nobel Committee's deliberations contain not the slightest hint that Pavlov's nationality—or, say, Russian-Swedish relations—played any role whatsoever in the fate of his candidacy.

2. *Protokoly zasedanii konferentsii Imperatorskoi Voenno-Meditsinskoi Akademii,* January 20, 1901, Rossiiskii Gosudarstvennyi Voenno-Istoricheskii Arkhiv 316.69. 38: 154

3. President Pachoutine [Pashutin] et al. of the Military-Medical Academy in St. Petersburg to the Nobel Committee, January 1901 (in French), Betänkande år 1901 angående J. P. Pawlow. P. M. Försändelser och Betänkanden, Nobel Committee Archives, Stockholm.

4. W. H. Howell to the Nobel Committee, January 2, 1901, Betänkande år 1901 angående J. P. Pawlow. P. M. Försändelser och Betänkanden.

5. See Goran Liljestrand, "The Prize in Physiology or Medicine," in *Nobel: The Man and His Prizes,* ed. H. Schück (Amsterdam: Elsevier Publishing, 1962), 152. Committee representatives also made a site visit to the facilities of one of Pavlov's competitors, Niels Finsen, in order to witness firsthand the results of his light therapy for lupus vulgaris.

6. Quoted in Liljestrand, "Prize," 136.

7. Liljestrand, "Prize," 139; emphasis added.

8. "Rapport afgifven till den Medicinska Nobelkomitens fran J. E. Johansson och Robert Tigerstedt, Juli 1901," p. 1, Betänkande år 1901 angående J. P. Pawlow. P. M. Försändelser och Betänkanden.

9. Ibid., 4. Experiments to demonstrate the varying characteristic secretory curves would have taken one day for each foodstuff tested. Johansson and Tiger-

stedt's comment about the amount of gastric acid does not include quantitative data and suggests that Pavlov did not measure the proteolytic power of the gastric secretions elicited in these experiments.

10. "Rapport fran Johansoon och Tigerstedt," 8.

11. Liljestrand, "Prize," 152.

12. "Rapport fran Johansoon och Tigerstedt," 8. Windholz and Kuppers first interpreted this comment as a counterargument on Pavlov's behalf in their "Pavlov and the Nobel Prize Award," 156.

13. Robert Tigerstedt, memo to the Nobel Committee, July 1901, p. 1, Betänkande år 1901 angående J. P. Pawlow. P. M. Försändelser och Betänkanden.

14. K. A. H. Mörner, memo to the Nobel Committee, July 30, 1901, Betänkande år 1901 angående J. P. Pawlow. P. M. Försändelser och Betänkanden.

15. As Windholz and Kuppers point out (in "Pavlov and the Nobel Prize Award," 157), Pavlov's record of self-authored publications was indeed unimpressive compared with that of other physiologists: "For instance, Tigerstedt, who was four years younger than Pavlov, had already written two specialized treatises on physiology and five textbooks . . . whereas Pavlov's only book consisted of read lectures." The same point could be made about many of Pavlov's other admirers, such as Langlois and Howell.

16. Robert Tigerstedt, appendix to memo to the Nobel Committee, September 8, 1901, "P. M. angående prof. J. P. Pawlowa arbeten," in P. M. Försändelser och Betänkanden.

17. I. P. Pavlov to I. F. Tsion, September 4 [1901], in SPF ARAN 259.7.167: 5; reprinted in E. M. Kreps, ed., *Perepiska Pavlova* (Leningrad: Nauka, 1970), 57. For the story of Pavlov's supposed prize, see *Izvestiia Voenno-Meditsinskoi Akademii*, September 1901, p. 90, which cites as its source the newspaper *Novoe Vremia*; and *Vrach*, 1901, no. 33: 1026. This rumor clearly enhanced interest in Pavlov's work, resulting, for example, in Lafayette Mendel's article "Professor Pawlow's researches on the physiology of secretion," *Science* (new ser.) 14, no. 356 (1901): 647–49.

18. Quoted in Liljestrand, "Prize," 185.

19. Ibid. Liljestrand also offers the following informed speculation. Like those who had nominated Behring, the faculty "evidently was of the opinion that the value of Behring's discovery had not been generally recognized until recently . . . It is also possible that at the very first distribution it was considered desirable to honour an achievement that was sufficiently well known not to create too much of a surprise. There may also have been some hesitation about dividing the prize between works on widely different subjects." Ibid.

20. Tigerstedt's letter to "Sehr gelehrter Herr Coleage," dated December 17, 1901, and his letter to "Lieber Freund," dated September 20, 1902, in SPF ARAN 259.2.1017 (in German); republished in Russian translation in Kreps, *Perepiska*, 194–95.

21. See Tigerstedt's letters to Pavlov of September 5, 1901; December 17, 1901; and

February 16, 1907; in SPF ARAN 259.2.1017; republished in Russian translation in Kreps, *Perepiska*, 193–94, 197–98.

22. Mörner's letter of thanks to Pavlov, dated October 26, 1902, is in SPF ARAN 259.9.1206.

23. Tigerstedt's nomination letter is dated October 26, 1901, and was sent from Stockholm. Tigerstedt was not on the faculty of the Karolinska Institute and so was not a member of the five-person Committee. Johansson, who was a member, probably informed Tigerstedt immediately of the Committee's decision. Pavlov was also nominated for the 1902 prize by L. Fredericq, W. Masius, and I. P. Nuel in Liège; V. Roth in Moscow; and C. von Voit in Munich.

24. See Tigerstedt's letters to Pavlov of March 26, 1902, and [month illegible] 14, 1902, in SPF ARAN 259.2.1017.

25. Boris Babkin, *Pavlov: A Biography* (Chicago: University of Chicago Press, 1949), 82.

26. Babkin addressed the influence of alkaline soaps on pancreatic secretion, and Sokolov presented new information on psychic secretion in the gastric glands.

27. K. A. H. Mörner, memo to the Nobel Committee, August 1902, p. 3, Betänkande år 1902 angående J. P. Pawlow. P. M. Försändelser och Betänkanden. Pavlov and S. V. Parashchuk's report "Edinstvo pepsina i khimozina" was soon published in both *Trudy obshchestva russkikh vrachei v S.-Peterburge* and *Bol'nichnaia gazeta Botkina*; reprinted in *PSS*, 2, pt. 2: 335–43. They responded to criticisms in another article published in the same two journals later that year. See Pavlov and Parashchuk, "Edinstvo pepsina i khimozina," in *PSS*, 2, pt. 2: 344–46. Pavlov seems to have employed here the same mode of reasoning he used in organ physiology: he perceived a basic regularity and reasoned from that regularity back to an underlying mechanism (in this case, he perceived that the proteolytic power associated with pepsin and the coagulating power associated with rennet seemed to vary in parallel, and so concluded that these were the same ferment).

28. O. Medin, memo to the Nobel Committee, July 1902, p. 2, Betänkande år 1902 angående J. P. Pawlow. P. M. Försändelser och Betänkanden.

29. Mörner memo, August 1902, p. 3.

30. K. A. H. Mörner, memo to the Nobel Committee, July 26, 1902, pp. 2–3, Betänkande år 1902 angående J. P. Pawlow. P. M. Försändelser och Betänkanden; Medin memo of July 1902, p. 1.

31. Mörner memo, August 1902, p. 5.

32. This account of the discussion is from the Committee's report to the Karolinska Institute's Faculty Collegium, September 25, 1902, in Protokoll m.m. Nobel Arenden, 1901–10.

33. I discuss these scientific developments in Chapter 9. Aside from Tigerstedt and Johansson, N. Uzinskii (Warsaw University), S. Leontiev (Kazan University), and V. Razumovskii (Kazan University) nominated Pavlov for the Nobel Prize in 1903.

34. K. A. H. Mörner, memo to the Nobel Committee, July 23, 1903, p. 14, Betänkande år 1903 angående J. P. Pawlow. P. M. Försändelser och Betänkanden.

35. As I noted in Chapter 9, scientists had been aware of the existence of zymogenic ferments since Heidenhain's research of the 1870s, but only with the vigorous growth of physiological chemistry in later decades did this fact command the great attention of researchers.

36. Mörner memo, July 23, 1903, p. 7. He was referring to L. Popielski, "Ueber die Zweckmassignkeit in der Arbeit der Verdauungsdrusen," *Deutsche Medicinische Wochenschrift* 28 (1902): 864–65, a German translation of the "preliminary report" first published in *Vrach* that same year. Popel'skii's longer and more detailed critique, published in *Russkii Vrach* later in 1902, was apparently never translated into a western language. I review Popel'skii's criticisms in Chapter 9.

37. Mörner memo, July 23, 1903, p. 7

38. On gastric and salivary secretion, see ibid., 8–9; for an account of Popel'skii's theory, see ibid., 8.

39. Ibid., 9, 14. Mörner did note (p. 9) that Popel'skii's statements should be viewed "with a certain caution" since he had earlier disputed the existence of enterokinase, only to subsequently concede error.

40. "P. M. ofver forslag till 1903 Ars prisutdelning inom prisgruppen fysiologi och medicin," 48, Nobel Committee Archive, Stockholm.

41. Mörner memo, July 23, 1903, p. 3.

42. "P. M. ofver forslag till 1903," pp. 48–49. Almquist supported Koch; Holmgren, Medin, Mörner, and Sundberg voted for Finsen.

43. Santesson also contributed an article to the 1904 volume of *Arkhiv Biologicheskikh Nauk* honoring Pavlov's twenty-fifth jubilee.

44. C. G. Santesson and J. E. Johansson, nominating letter to the Nobel Committee, January 23, 1904, Betänkande år 1904 angående J. P. Pawlow. P. M. Försändelser och Betänkanden

45. In a letter of September 20, 1902, Tigerstedt informed Pavlov that "it will perhaps interest you to know that a young physician here, Dr. Hornberg, has repeated your investigations on the psychic secretion of gastric juice [through experiments] on a patient with an (almost entirely) blocked esophagus—and has in all essentials confirmed your data. Since this investigation will serve as the theme of his doctoral dissertation, I ask you to send me the works on secretion of gastric juice that have been produced in your Institute and published only in Russian. With the help of my son-in-law, we will read them without difficulty." SPF ARAN 259.2.1017; republished in Russian translation in Kreps, *Perepiska*, 195.

46. Santesson and Johansson, nominating letter of January 23, 1904.

47. J. E. Johansson, report to the Nobel Committee, September 24, 1904, p. 1, Betänkande år 1904 angående J. P. Pawlow. P. M. Försändelser och Betänkanden.

48. As we have seen, the complete data were not, in fact, available in the dissertations. One could "fully follow" only a few selected experiments.

49. Johansson also mentioned Pavlov's latest research in which he used "salivary secretion as an indicator of so-called psychic phenomena" as an example of the continued fruitfulness of the Russian's methodological ingenuity. Pavlov had recently reported on this research to the international congress of physicians in Madrid (1903).

50. Nobel Committee report to the Karolinska Institute's Faculty Collegium, September 24, 1904, in Protokoll m.m. Nobel Arenden, 1901–10.

51. Karl Mörner to I. P. Pavlov, October 21, 1904, in SPF ARAN 259.2.575; republished in Russian translation in Kreps, *Perepiska*, 213–14. Babkin adds that the financial value of the Nobel Prize was equivalent to about 73,000 gold rubles or about 36,000 gold dollars—"a very considerable sum in those days." A friend from university days who had become a high-ranking state official offered Pavlov some potentially lucrative insider investment advice, but Pavlov instead deposited the money with the Nobels' St. Petersburg bank. The money gave the Pavlovs financial security and material comfort until the Bolsheviks confiscated it shortly after their seizure of power. See Babkin, *Pavlov*, 36.

52. Serafima Pavlova, *Vospominaniia*, in SPF ARAN 259.1.169: 318–19. Pavlov's use of the word *idol* carries an undercurrent of criticism of Serafima's religious faith. She also mentions that they first heard the good news from Tigerstedt. It seems unlikely that she would be mistaken about such a memorable moment, but Tigerstedt was not on the Nobel Committee and his letter of congratulations was sent weeks after Mörner's. Perhaps Tigerstedt informed the Pavlovs in some other way that left no archival trace. This possibility raises another, more intriguing question. To what extent was Tigerstedt privy to the Committee's deliberations in 1901–4, and what information might he have relayed to his "dear friend"? For example, might Pavlov have known how much Popel'skii's criticism had cost him in 1903?

53. Pavlova, *Vospominaniia*, 319–20.

54. K. A. H. Mörner's speech of December 10, 1904, was translated into Russian by Ganike and published in the first issue of *Russkii Vrach* for 1905. It is available in English at the Nobel e-museum: www.nobel.se/medicine/laureates/1904/press.html.

55. Pavlova, *Vospominaniia*, 319–20. As a state servant and faculty member at the Military-Medical Academy, Pavlov could indeed have worn various ornaments of rank, but he disliked doing so and wore them only with great reluctance at official state occasions.

56. Pavlov, "Nobelevskaia Rech'" [1904], in *PSS*, 6: 366.

57. Pavlova, *Vospominaniia*, 320.

58. The phrase *International Nobel Prize* is from the report on the fete at the Ol'denburgskiis' palace in *Novosti i Birzhevaia gazeta*, no. 344 (1904): 2.

Epilogue

1. Daniel P. Todes, "Pavlov and the Bolsheviks," *History and Philosophy of the Life Sciences* 17 (1995): 379–418.

2. On science and ideology in the Soviet Union, see David Joravsky, *Soviet Marxism and Natural Science* (New York: Columbia University Press, 1961); David Joravsky, *Russian Psychology: A Critical History* (Oxford: Basil Blackwell, 1989); Loren Graham, *Science, Philosophy, and Human Behavior in the Soviet Union* (New York: Columbia University Press, 1987); Mark Adams, "Eugenics in Russia, 1900–1940," in *The Wellborn Science: Eugenics in Germany, France, Brazil, and Russia,* ed. Mark Adams (New York: Oxford University Press, 1990), 153–216; and Nikolai Krementsov, *Stalinist Science* (Princeton: Princeton University Press, 1997).

3. See Todes, "Pavlov and the Bolsheviks."

4. See especially Pavlov's public speeches of 1918: "On the Mind in General," "On the Russian Mind," and "On the Foundations of Culture of Animals and Man," in SPF ARAN 259.1a.3, 1a.4, and 1a.5.

5. V. V. Savich, "Ivan Petrovich Pavlov: biograficheskii ocherk," in *Sbornik posviashchennyi 75-letiiu akademika I. P. Pavlova* (Leningrad, 1924), 18.

6. Pavlov's transcribed remarks are preserved in SPF ARAN 259.1.203. Pavlov was referring to *Lectures on the Work of the Large Hemispheres of the Brain* (Leningrad, 1927). His earlier book, *Twenty Years of Experience in the Objective Study of the Higher Nervous Activity of Animals* (Petrograd, 1923), was a compilation of his earlier reports and articles.

7. I. P. Pavlov, remarks at the University of Paris in 1925, in his *Neopublikovannye i maloizvestnye materialy I. P. Pavlova* (Leningrad: Nauka, 1975), 77.

SELECTED BIBLIOGRAPHY

Archives

Alan Mason Chesney Archive, The Johns Hopkins Medical Institutions, Baltimore
Dom-Muzei I. P. Pavlova, Ryazan
Gosudarstvennyi Arkhiv Rossiiskoi Federatsii, collection (*fond*) 109, Moscow
Kvartira-Muzei I. P. Pavlova, St. Petersburg
McGill University Archive, Montreal
Nobel Archives, Karolinska Institute, Stockholm
Rossiiskii Gosudarstvennyi Istoricheskii Arkhiv, collections (*fond*) 515, 565, 733, 777, 1289, St Petersburg
Rossiiskii Gosudarstvennyi Voenno-Istoricheskii Arkhiv, collection (*fond*) 316, Moscow
Sankt-Peterburgskii filial Arkhiva Rossiiskoi Akademii Nauk, collection (*fond*) 259 and collection of taped interviews, St. Petersburg
Tsentral'nyi Gosudarstvennyi Istoricheskii Arkhiv Sankt-Peterburga, collections (*fond*) 14, 565, 2232, 2282, St. Petersburg

Periodical Publications

Bol'nichnaia gazeta Botkina
Izvestiia Voenno-Meditsinskoi Akademii
Nevrologicheskii Vestnik
Novoe Vremia
Novosti i Birzhevaia gazeta
Obozrenie psikhiatrii, nevrologii, i eksperimental'noi psikhologii
Peterburgskaia zhizn'
Russkii arkhiv patologii, klinicheskoi meditsiny i bakteriologii
Russkii Vrach
Sankt Peterburgskie Vedomosti
Trudy obshchestva russkikh vrachei
Vestnik Rossiiskago obshchestva pokrovitel'stva zhivotnykh
Vrach

Other Primary Sources

Babkin, B. P. *Opyt sistematicheskago izucheniia slozhno-nervnykh (psikhicheskikh) iavlenii u sobaki.* Military-Medical Academy Doctoral Dissertation Series. St. Petersburg, 1904.

———. *Secretory Mechanism of the Digestive Glands.* New York: Paul B. Hoeber, 1944, 1950.

Bainbridge, F. A. "On the Adaptation of the Pancreas." *British Medical Journal* 1 (1904): 778–81.

———. "The Physiological Laboratory of the University of London." *Nature* 67 (12 March 1903): 441–42.

Bayliss, W. M. *Principles of General Physiology.* London: Longmans, Green, and Co., 1915.

Bayliss, W. M., and E. H. Starling. "The Chemical Regulation of the Secretory Process." *Nature* 70 (May-October 1904): 65–68.

———. "The Mechanism of Pancreatic Secretion." *Journal of Physiology* 27 (1902): 325–53.

Bekhterev, V. M. "O korkovom zritel'nom tsentre." *Obozrenie psikhiatrii, nevrologii, i eksperimental'noi psikhologii* 7 (1901): 575–79.

———. "O refleksakh v oblasti litsa i golovy." *Obozrenie psikhiatrii, nevrologii, i eksperimental'noi psikhologii* 9 (1901): 552–55.

———. *Osnovy ucheniia o funktsiiakh mozga.* St. Petersburg, 1903.

———. *Psikhika i Zhizn'.* St. Petersburg, 1902.

Bernar, Klod [Bernard, Claude]. *Fiziologiia serdtsa i otnoshenie ego k golovnomu mozgu.* Trans. N. Solov'eva. St. Petersburg: O. I. Bakst, 1867.

———. *Lektsii fiziologii i patologii nervnoi sistemy.* 2 vols. Trans. F. V. Ovsiannikov. St. Petersburg: N. Nekliudov, 1866–67.

———. *Vvedenie k izucheniiu opytnoi meditsiny.* Trans. N. Strakhov. St. Petersburg, 1866.

Bernard, Claude. *An Introduction to the Study of Experimental Medicine.* 1865. Trans. Henry Copley Greene. New York: Dover Publications, 1957.

———. "Sur la physiologie du coeur et ses rapports avec le cerveau." In *Leçons sur les propriétés des tissus vivants.* Paris: Baillière, 1866.

Boldyrev, V. N. "Natural'nyi zheludochnyi sok, kak lechebnoe sredstvo, i sposob ego dobyvaniia." *Russkii Vrach,* 1907, no. 5: 154, 157.

———. *Periodicheskaia rabota pishevaritel'nago apparata pri pustom zheludke.* Military-Medical Academy Doctoral Dissertation Series. St. Petersburg, 1904.

Boldyrev, V. N., V. I. Vartanov, E. A. Ganike, and A. P. Sokolov. "Referaty trudov professora I. P. Pavlova i ego uchenikov." *Arkhiv biologicheskikh nauk* 11, suppl. (1904).

Brokgauz, F. A., and I. A. Efron, eds. *Rossiia: Entsiklopedicheskii slovar'.* 1898. Reprint. Leningrad: Lenizdat, 1991.

Bukhshtab, Ia. A. *Rabota podzheludochnoi zhelezy posle pererezki bluzhdaiushchikh i vnutrennostnykh nervov.* Military-Medical Academy Doctoral Dissertation Series. St. Petersburg, 1904.

Clemm, Walther. "Ueber die Beeinflussung der Magensaftabschiedung durch Zucker." *Therapeutische Monatshafte* 15 (August 1901): 405–13.

Cyon, Elie de [Ilya Tsion]. *Methodik der Physiologischen Experimente und Vivisectionen.* Mit Atlas. 2 vols. Giessen: Ricker, 1876.

Damaskin, N. I. "Deistvie zhira na otdelenie podzheludochnogo soka." *Trudy obshchestva russkikh vrachei za 1895–96* 63 (February 1896): 7–14.

Delezenne, C., and A. Frouin. "La sécrétion physiologique du pancréas ne posséde pas d'action digestive propre vis-à-vis de l'albumine." *Comptes rendus hebdomadaires des séances et mémoires de la Société de Biologie* 54 (1902): 691–93.

———. "Nouvelles observations sur la sécrétion physiologique du pancréas. Le suc pancréatique des bovides." *Comptes rendus hebdomadaires des séances et mémoires de la Société de Biologie* 55 (1903): 455–58.

Dolinskii, I. L. *O vliianii kislot na otdelenie soka podzheludochnoi zhelezy.* Military-Medical Academy Doctoral Dissertation Series. St. Petersburg, 1894.

Dunn, Halbert L. "Application of Statistical Methods in Physiology." *Physiological Reviews* 9 (1929): 275–398.

Edkins, J. S. "The Chemical Mechanism of Gastric Secretion." *Journal of Physiology* 34 (1906): 133–44.

———. "Mechanism of Secretion of Gastric, Pancreatic, and Intestinal Juices." In *Text-Book of Physiology,* ed. E. A. Schäfer. Vol. 1, 531–58. Edinburgh: Young J. Pentland, 1898.

Finkel'shtein, A. A. "Lechenie estestvennym zheludochnym sokom." *Vrach,* 1900, no. 32: 963–65.

Foster, Michael. *A Text-Book of Physiology.* 4th Amer. ed. Philadelphia: Lea Brothers and Co., 1891.

Geidengain, Rudolf [Heidenhain, Rudolf]. *Fiziologiia otdelitel'nykh protsessov.* St. Petersburg, 1886.

Geiman, N. M. *O vliianii razlichnago roda razdrazhenii polosti rta na rabotu sliunnykh zhelez.* Military-Medical Academy Doctoral Dissertation Series. St. Petersburg, 1904.

Glinskii, D. L. *K fiziologii kishek.* Military-Medical Academy Doctoral Dissertation Series. St. Petersburg, 1891.

Gol'dshtein, M. Iu. "Estestvoznanie i tekhnika v xix veke. Ocherk." *Niva,* suppl. (1901): 340–91.

Gorshkov, Ia. P. "O lokalizatsii tsentrov vkusa v mozgovoi kore." *Obozrenie psikhiatrii, nevrologii, i eksperimental'noi psikhologii* 10 (1900): 737–42.

Heidenhain, R. *Handbuch der Physiologie der Absonderung und Aufsaugung.* Vol. 5, pt. 1 of *Handbuch Der Physiologie,* ed. L. Hermann. Leipzig: F. C. W. Vogel, 1883.

Howell, William H. *An American Text-Book of Physiology.* Philadelphia: W. B. Saunders, 1896, 1900, 1909.

Iablonskii, Iu. M. *Spetsificheskoe zabolevanie sobak, teriaiushchikh khronicheski sok podzheludochnoi zhelezy.* Military-Medical Academy Doctoral Dissertation Series. St. Petersburg, 1894.

Imperatorskii Institut Eksperimental'noi Meditsiny (1890–1910). St. Petersburg, 1911.

Iurgens, N. P. *O sostoianii pishchevaritel'nago kanala pri khronicheskom paraliche bluzhdaiushchikh nervov.* Military-Medical Academy Doctoral Dissertation Series. St. Petersburg, 1892.

Iushchenko, A. I. "O reflektornykh tsentrakh v uzlakh simpaticheskoi nervnoi sistemy, ob otnoshenii nizhniago bryzheechnago uzla k innervatsii mochevogo puzyria i ob avtomaticheskikh dvizheniiakh posledniago." In *Otchety nauchnogo sobraniia vrachei Sankt-Peterburgskoi kliniki dushevnykh i nervnykh boleznei za 1895–96 i 1896– 1897,* 95–97. St. Petersburg, 1898.

Kazanskii, N. P. *Materialy k eksperimental'noi patologii i eksperimental'noi terapii zheludochnykh zhelez sobaki.* Military-Medical Academy Doctoral Dissertation Series. St. Petersburg, 1901.

Khizhin, P. P. *Istoricheskii ocherk deiatel'nosti Ramonskoi lechebnitsy ee Imperatorskogo Vysochestva Printsessy E. M. Ol'denburgskoi so vremeni osnovaniia lechebnitsy.* Voronezh, 1893.

———. *Otdelitel'naia rabota zheludka sobaki.* Military-Medical Academy Doctoral Dissertation Series. St. Petersburg, 1894.

Konovalov, P. N. *Prodazhnye pepsiny v sravnenii s normal'nym zheludochnym sokom.* Military-Medical Academy Doctoral Dissertation Series. St. Petersburg, 1893.

Kreps, E. M., ed. *I. P. Pavlov v vospominaniiakh sovremennikov.* Leningrad: Nauka, 1967.

———. *Perepiska Pavlova.* Leningrad: Nauka, 1970.

Krever, A. R. *K analizu otdelitel'noi raboty podzheludochnoi zhelezy.* Military-Medical Academy Doctoral Dissertation Series. St. Petersburg, 1899.

Langlois, J. P. "Iwan Petrowitsch Pawlow." *La Presse Médicale,* 25 March 1905, pp. 185– 87.

Lewes, G. H. *The Physiology of Common Life.* Vol. 1. Edinburgh: Blackburn and Sons, 1859.

Lintvarev, I. I. *Vliianie razlichnykh fiziologicheskikh uslovii na sostoianie i kolichestvo fermentov v soke podzheludochnoi zhelezy.* Military-Medical Academy Doctoral Dissertation Series. St. Petersburg, 1901.

Lobasov, I. O. *Otdelitel'naia rabota zheludka sobaki.* Military-Medical Academy Doctoral Dissertation Series. St. Petersburg, 1896.

Lonnqvist, Bernt. "Beitrage zur Kenntsniss der Magensaftabsonderung." *Skandinavisches Archiv für Physiologie* 1 (1906): 194–262.

Luk'ianov, S. M. *Osnovaniia obshchei patologii i pishchevareniia.* St. Petersburg, 1897.

Mendel, Lafayette. "Professor Pawlow's researches on the physiology of secretion." *Science* 14, no. 356 (1901): 647–49.

Mendeleev, D. "Zavody." In *Entsiklopedicheskii slovar' Brokgauza i Efrona.* Vol. 12, 100– 104. St. Petersburg, 1894.

Mosolov, A. A. *Pri dvore poslednego imperatora: zapiski nachal'nika kantseliarii ministra dvora.* 1937. Reprint, St. Petersburg: Nauka, 1992.

Omelianskii, V. L., and L. A. Orbeli, eds. *Sbornik, posviashchennyi 75-letiiu akademika Ivana Petrovicha Pavlova.* Leningrad: Gosudarstvennoe Izdatel'stvo, 1924.

Ostrogorskii, S. A. *Temnyi punkt v innervatsii sliunnykh zhelez.* Military-Medical Academy Doctoral Dissertation Series. St. Petersburg, 1894.

Ostwald, Wilhelm. *Grosse Manner.* Leipzig: Akademische Verlagsgesellschaft, 1909.

Pavlov, I. P. *Conditioned Reflexes: An Investigation of the Physiological Activity of the Cerebral Cortex.* 1927. Trans. G. V. Anrep. New York: Dover Publications, 1960.

———. *Lectures on Conditioned Reflexes: Twenty-five Years of Objective Study of the*

Higher Nervous Activity (Behavior) of Animals. Trans. W. Horsley Gantt. New York: International Publishers, 1928.

————. *Neopublikovannye I maloizvestnye materialy I. P. Pavlova*. Leningrad: Nauka, 1975.

————. *Polnoe sobranie sochinenii*. 6 vols. Moscow: USSR Academy of Sciences, 1951–52.

————. *Programma chtenii dlia samoobrazovaniia*. 2d ed. St. Petersburg: Pedagogicheskii muzei voennykh uchebnykh zavedenii, 1897.

————. *The Work of the Digestive Glands*. Trans. W. H. Thompson. London: Charles Griffin and Co., 1902.

Pawlow, I. P. *Die Arbeit der Verdauungsdrüsen*. Trans. A. Walther. Wiesbaden: J. F. Bergmann, 1898.

Popel'skii, Lev. *Istoricheskii ocherk kafedry fiziologii v Imperatorskoi Voenno-Meditsinskoi Akademii za 100 let (1798–1898)*. St. Petersburg: Ministry of Internal Affairs, 1899.

————."Ob osnovnykh svoistvakh podzheludochnago soka." *Russkii Vrach*, 1903, no. 16: 1–11.

————. *O sekretorno-zaderzhivaiushchikh nervakh podzheludochoi zhelezy*. Military-Medical Academy Doctoral Dissertation Series. St. Petersburg, 1896.

————. "O tselesoobraznosti v rabote pishchevaritel'nykh zhelez (Predvaritel'noe soobshchenie)." *Vrach*, 1901, no. 50: 1–5.

————. "O tselesoobraznosti v rabote podzheludochnoi i zheludochnykh zhelez." *Russkii Vrach*, 1902, no. 35: 1242–48.

————. "Prichiny raznoobraziia svoistv podzheludochnago soka v otnoshenii belkovago brodila." *Russkii Vrach*, 1902, no. 18: 679–84.

Popielski, L. "Ueber die Grundeigenschaften des Pankreassaftes." *Centralblatt für Physiologie* 17, no. 3 (1903): 65–70.

————. "Ueber das peripherische reflectorische Nervencentrum des Pankreas." *Pflügers Archive* 86 (1901): 215–46.

————. "Ueber die Zweckmassigkeit in der Arbeit der Verdauungsdrusen." *Deutsche Medicinische Wochenschrift* 28 (1902): 864–65.

Pravila dlia storonnikh lits, zhelaiushchikh rabotat' v uchrezhdeniiakh Imperatorskago Instituta Eksperimental'noi Meditsiny. St. Petersburg, 1894.

Riegel, F. "Ueber medicamentose Beeinflussung der Magensaftsecretion." *Zeitschrift für Klinische Medicin* 37 (1899): 381–402.

Salazkin, S. S. "Staroe i novoe v oblasti pishchevareniia." *Russkii arkhiv patologii, klinicheskoi meditsiny i bakteriologii* 4, no. 5 (1897): 564–89.

Sanotskii, A. S. *Vozbuditeli otdeleniia zheludochnogo soka*. Military-Medical Academy Doctoral Dissertation Series. St. Petersburg, 1892.

Sechenov, I. M. *Refleksy golovnogo mozga*. St. Petersburg, 1866.

Shepoval'nikov, N. P. *Fiziologiia kishechnago soka*. Military-Medical Academy Doctoral Dissertation Series. St. Petersburg, 1899.

Smiles, Samuel. *Selections from Lives of the Engineers, with an Account of Their Principal Works*. Ed. and introduction by Thomas Parke Hughes. Cambridge, Mass.: M.I.T. Press, 1966.

————. *Self-Help, with Illustrations of Character, Conduct, and Perseverance.* 1859. Reprint, Chicago: Belford, Clarke and Co., 1884.

Snarskii, A. T. *Analiz normal'nykh uslovii raboty sliunnykh zhelez u sobaki.* Military-Medical Academy Doctoral Dissertation Series. St. Petersburg, 1901.

Soborov, I. K. *Izolirovannyi zheludok pri patologicheskom sostoianii pishchevaritel'nogo kanala.* Military-Medical Academy Doctoral Dissertation Series. St. Petersburg, 1899.

Sokolov, D. A. *25 let bor'by: Vospominaniia vracha.* St. Petersburg, 1910.

Starling, E. H. "The Chemical Correlation of the Functions of the Body." *Lancet,* 5 August 1905, pp. 339–41; 12 August 1905, pp. 423–25; 19 August 1905, pp. 501–3; and 26 August 1905, pp. 579–83.

Svod uzakonenii i rasporiazhenii pravitel'stva po vrachebnoi i sanitarnoi chasti v Imperii, vyp. 1. St. Petersburg, 1895–96.

Tigerstedt, Robert. *A Text-Book of Human Physiology.* New York: D. Appleton and Co., 1906.

Tolochinov, I. F. "K voprosu ob osnovnykh proiavleniiakh uslovnykh i bezuslovnykh refleksov v pervonachal'noi razrabotke ikh metoda. Neprigodnost' terminologii tormazheniia i rastormazhivaniia. Teoriia pobochnykh reflektornykh dug." *Russkii Vrach,* 1913, no. 1: 20–24.

————. "K voprosu ob osnovnykh proiavleniiakh uslovnykh i bezuslovnykh refleksov v pervonachal'noi razrabotke ikh metoda." *Russkii Vrach,* 1913, no. 2: 54–57.

————. *O patologo-anatomicheskikh izmeneniiakh iader cherepnykh nervov i otnosiashchikhsia k nim nervnykh volokon mozgovogo stvola pri narastaiushchem paralichnom slaboumii.* Military-Medical Academy Doctoral Dissertation Series. St. Petersburg, 1900.

————. "Pervonachal'noe primenenie metoda uslovnykh refleksov k izsledovaniiu tsentrov kory bol'shogo mozga u sobak." *Nevrologicheskii vestnik* 19, no. 2 (1912): 410–45.

————. Pervonachal'naia razrabotka sposoba uslovnykh refleksov i obosnovanie termina 'uslovnyi refleks.'" *Russkii Vrach,* 1912, no. 31: 1277–82.

Tolotschinoff, I. "Contribution a l'étude de la physiologie et de la psychologie des glandes salivaries." In *Forhandlingar vid Nordiska Naturforskare och Lakermotet,* 42–46. July 1902.

Trzhetseskii. A. A. "Achylia gastrica." *Russkii arkhiv patologii, klinicheskoi meditsiny i bakteriologii* 9 (1900): 183–205.

Tsion, I. *Kurs fiziologii (Lektsii chitannye v 1872/73 uchebnom godu v Imperatorskoi Mediko-Khirurgicheskoi Akademii).* St. Petersburg, 1873.

————. *Serdtse i mozg.* St. Petersburg, 1873.

Tsitovich, I. S. "Natural'nyi zheludochnyi sok i rezul'taty ego primeneniia u bugorchatkovykh bol'nykh." *Russkii Vrach,* 1907, no. 28: 962.

Ushakov, V. G. *K voprosu o vliianii bluzhdaiushchego nerva na otdelenie zheludochnogo soka u sobaki.* Military-Medical Academy Doctoral Dissertation Series. St. Petersburg, 1894.

Val'ter, A. A. "Excitabilité sécrétoire spécifique de la muqueuse du canal digestif. Cinquième mémoire. Sécrétion pancréatique." *Archives des sciences biologiques* 7 (1899): 1–86.

————. *Otdelitel'naia rabota podzheludochnoi zhelezy.* Military-Medical Academy Doctoral Dissertation Series. St. Petersburg, 1897.

Vasil'ev, V. N. *O vliianii raznogo roda edy na deiatel'nost' podzheludochnoi zhelezy.* Military-Medical Academy Doctoral Dissertation Series. St. Petersburg, 1893.

Verekunov, V., V. Statsenko, and A. Rutkovskii. *Pervaia vserossiiskaia gigienicheskaia vystavka v S. Peterburge: Kratkii ocherk.* St. Petersburg, 1894.

Virshubskii, A. M. "Obobshchenie noveishikh faktov iz fiziologii pishchevareniia s tochki zreniia printsipov: a) spetsificheskoi vozbudimosti i b) vzaimodeistviia funktsii organov." *Praktikcheskaia meditsina* 7 (1900): 129–38.

————. *Rabota zheludochnykh zhelez pri raznykh sortakh zhirnoi pishchi.* Military-Medical Academy Doctoral Dissertation Series. St. Petersburg, 1900.

————. "Staroe i novoe v oblasti sekretornoi funktsii zheludka." *Gazeta Meditsina* 23–24 (1898): 4–14; and 25–26 (1898): 4–8.

Volkovich, A. N. *Fiziologiia i patologiia zheludochnykh zhelez.* Military-Medical Academy Doctoral Dissertation Series. Kronstadt, 1898.

Vremennyi ustav Imperatorskogo Instituta Eksperimental'noi Meditsiny. St. Petersburg, 1894.

Vul'fson, S. G. *Rabota sliunnykh zhelez.* Military-Medical Academy Doctoral Dissertation Series. St. Petersburg, 1898.

Welch, W. H. "The Evolution of Modern Scientific Laboratories." *Nature* 54 (28 May 1896): 87–90.

Woodbury, F. *On Disordered Digestion and Dyspepsia.* Detroit: George S. Davis, 1889.

Zavriev (Abo-Zavaridze), Ia. Kh. *Fiziologiia i patologiia zheludochnykh zhelez sobaki.* Military-Medical Academy Doctoral Dissertation Series. St. Petersburg, 1900.

Zel'geim, A. P. *Rabota sliunnykh zhelez do I posle pererezki nn. Glossopharyngei I linguales.* Military-Medical Academy Doctoral Dissertation Series. St. Petersburg, 1904.

Zhegalov, I. P. *Otdelitel'naia rabota zheludka pri pereviazke protokov podzheludochnoi zhelezy i o belkovom fermente v zhelchi.* Military-Medical Academy Doctoral Dissertation Series. St. Petersburg, 1900.

Zimnitskii, S. S. *Otdelitel'naia rabota zheludochnykh zhelez pri zaderzhke zhelchi v organizme.* Military-Medical Academy Doctoral Dissertation Series. St. Petersburg, 1901

Secondary Sources

I have listed here only those works that were especially useful either for the specific information they provided or for their treatment of general themes addressed in the text.

Adams, Mark. "Science, Ideology, and Structure: The Kol'tsov Institute 1900–1970." In *The Social Context of Soviet Science,* ed. Linda Lubrano and Susan Gross Solomon, 173–204. Boulder, Colo.: Westview Press, 1980.

Alder, Kenneth. "Making Things the Same: Representation, Tolerance and the End of the *Ancien Régime* in France." *Social Studies of Science* 28, no. 4 (August 1998): 499–546.

Andreeva, V. N. "Ivan Petrovich Pavlov kak opponent i retsenzent." *Trudy Instituta Istorii Estestvoznaniia i Tekhniki* 41, no. 10 (1961): 294–323.

Andreeva, V. N., and E. A. Kosmachevskaia. "Nauchnye sviazi shkol I. P. Pavlova i V. M. Bekhtereva." *Zhurnal vysshei nervnoi deiatel'nosti* 42, no. 5 (1992): 1039–45.

Babkin, Boris. *Pavlov: A Biography.* Chicago: University of Chicago Press, 1949.

Bailes, Kendall. *Science and Russian Culture in an Age of Revolution: Vernadsky and His Scientific School, 1863–1945.* Bloomington: Indiana University Press, 1989.

Bickel, Marcel H. *Marceli Nencki.* Bern: Hans Huber, 1972.

Brieger, Gert H. "Dyspepsia: The American Disease? Needs and Opportunities for Research." In *Healing and History: Essays for George Rosen,* ed. Charles E. Rosenberg, 179–90. Folkestone, Kent: Dawson, 1979.

Canguilhem, Georges. *The Normal and Pathological.* New York: Zone, 1991.

———. *A Vital Rationalist: Selected Writings from Georges Canguilhem.* Ed. François Delaporte. New York: Zone, 1994

Chebysheva, N. A. "Nauchno-organizatsionnaia rol' I. P. Pavlova v Institute Eksperimental'noi Meditsiny v 1891–1916 gg. (po materialam arkhiva IEM)." In *Ezhegodnik IEM za 1956,* 637–52. Leningrad, 1957.

Chen, T. S., and P. S. Chen, eds. *The History of Gastroenterology.* New York: Parthenon, 1995.

Clarke, Adele, and Joan Fujimura, eds. *The Right Tools for the Job: At Work in Twentieth-Century Life Sciences.* Princeton: Princeton University Press, 1992.

Clause, Bonnie. "The Wistar Rat as a Right Choice: Establishing Mammalian Standards and the Ideal of a Standardized Mammal." *Journal of the History of Biology* 26 (1993): 329–49.

Coleman, William. "The Cognitive Basis of the Discipline: Claude Bernard on Physiology." *Isis* 76 (1985): 49–70.

———. "Experimental Physiology and Statistical Inference: The Therapeutic Trial in Nineteenth-Century Germany." In *The Probabilistic Revolution.* Vol. 2, ed. Lorenz Krüger, Gerd Gigerenzer, and Mary S. Morgan, 201–26. Cambridge, Mass.: MIT Press, 1987.

Collins, Harry, and Trevor Pinch. *The Golem: What Everyone Should Know about Science.* Cambridge: Cambridge University Press, 1993.

Cunningham, Andrew, and Williams Perry, eds. *The Laboratory Revolution in Medicine.* Cambridge: Cambridge University Press, 1992.

Davenport, Horace W. *A History of Gastric Secretion and Digestion.* New York: Oxford University Press, 1992.

Dror, Otniel E. "The Affect of Experiment: The Turn to Emotions in Anglo-American Physiology, 1900–1940." *Isis* 90 (1999): 205–37.

———. "Creating the Emotional Body: Confusion, Possibilities, and Knowledge." In *An Emotional History of the United States,* ed. Peter N. Stearns and Jan Lewis, 173–94. New York: New York University Press, 1998.

Fearing, Franklin. *Reflex Action: A Study in the History of Physiological Psychology.* Cambridge, Mass.: M.I.T. Press, 1970.

Frank, Robert, Jr. "American Physiologists in German Laboratories, 1865–1914." In *Physiology in the American Context, 1850–1940,* ed. Gerald Geison, 11–46. Bethesda, Md.: American Philosophical Society, 1987.

Frieden, Nancy M. *Russian Physicians in an Era of Reform and Revolution, 1856–1905.* Princeton: Princeton University Press, 1981.

Fruton, Joseph. *Contrasts in Scientific Style: Research Groups in the Chemical and Biochemical Sciences.* Philadelphia: American Philosophical Society, 1990.

———. *Molecules and Life: Historical Essays on the Interplay of Chemistry and Biology.* New York: John Wiley and Sons, 1972.

Fye, W. Bruce. "Carl Ludwig and the Leipzig Physiological Institute: 'A Factory of New Knowledge.'" *Circulation* 74, no. 5 (November 1986): 920–28.

Galison, Peter. "History, Philosophy, and the Central Metaphor." *Science in Context* 2, no. 1 (1988): 59–75.

———. *Image and Logic: A Material Culture of Microphysics.* Chicago: University of Chicago Press, 1997.

Geison, Gerald L. *Michael Foster and the Cambridge School of Physiology: The Scientific Enterprise in Late Victorian Society.* Princeton: Princeton University Press, 1978.

———. *The Private Science of Louis Pasteur.* Princeton: Princeton University Press, 1995.

Golikov, Iu. P., and K. A. Lange. "Stanovlenie pervogo v Rossii issledovatel'skogo uchrezhdeniia v oblasti biologii i meditsiny." In *Pervyi v Rossii issledovatel'skii tsentr v oblasti biologii i meditsiny,* 7–75. Leningrad: Nauka, 1990.

Grebel'skii, P. Kh., and A. B. Mirvis. *Dom Romanovykh.* 2d ed. St. Petersburg, 1992.

Gregory, R. A. "The Gastrointestinal Hormones: An Historical Review." In *The Pursuit of Nature: Informal Essays on the History of Physiology,* ed. A. L. Hodgkin et al., 105–32. Cambridge: Cambridge University Press, 1977.

Gureeva, N. M. "K istorii organizatsii Instituta Eksperimental'noi Meditsiny." In *Ezhegodnik Instituta Eksperimental'noi Meditsiny za 1956,* 628–37. Leningrad, 1957.

———. "Uchastie I. P. Pavlova v deiatel'nosti pedagogicheskogo myzeia voennykh uchebnykh zavedenii." *Fiziologicheskii zhurnal SSSR* 45, no. 9 (1959): 1157–62.

Gureeva, N. M., N. A. Chebysheva, and V. A. Merkulov. *Letopis' zhizni i deiatel'nosti akademika I. P. Pavlova.* Leningrad: Nauka, 1969.

Hacking, Ian. "The Self-Vindication of the Laboratory Sciences." In *Science as Practice and Culture,* ed. Andrew Pickering. Chicago: University of Chicago Press, 1992.

Hannaway, Owen. "Laboratory Design and the Aim of Science: Andreas Libavius versus Tycho Brahe." *Isis* 77 (1986): 585–610.

Harré, Rom. *Great Scientific Experiments.* Oxford: Oxford University Press, 1981.

Holmes, Frederic L. *Between Biology and Medicine: The Formation of Intermediary Metabolism.* Berkeley: University of California Office for History of Science and Technology, 1992.

———. *Claude Bernard and Animal Chemistry: The Emergence of a Scientist.* Cambridge, Mass.: Harvard University Press, 1974.

———. "Early Theories of Protein Metabolism." In *The Origins of Modern Biochemistry: A Retrospect on Proteins,* ed. P. R. Srinivasan, Joseph S. Fruton, and John T. Edsall, 171–87. New York: New York Academy of Sciences, 1979.

———. "The Formation of the Munich School of Metabolism." In *The Investigative Enterprise: Experimental Physiology in Nineteenth-Century Medicine,* ed. William Coleman and Frederic L. Holmes, 179–210. Berkeley: University of California Press, 1988.

————. *Hans Krebs.* 2 vols. New York: Oxford University Press, 1991, 1993.

————. *Lavoisier and the Chemistry of Life: An Exploration of Scientific Creativity.* Madison: University of Wisconsin Press, 1985.

————. "Physiology and Experimental Medicine in the Nineteenth Century" (unpublished manuscript).

Holton, Gerald. "Subelectrons, Presuppositions, and the Millikan-Ehranhaft Dispute." In *The Scientific Imagination: Case Studies,* 25–83. Cambridge: Cambridge University Press, 1978.

Hutchinson, John F. *Champions of Charity: War and the Rise of the Red Cross.* Boulder, Colo.: Westview Press, 1996.

Jacob, François. *The Statue Within: An Autobiography.* New York: Basic Books, 1988.

Johnson, George. *Fire in the Mind: Science, Faith, and the Search for Order.* New York: Vintage Books, 1995.

Joravsky, David. *Russian Psychology: A Critical History.* Oxford: Basil Blackwell, 1989.

Kharauzov, N. A. "K istorii Instituta Eksperimental'noi Meditsiny AMN SSSR." In *Ezhegodnik Instituta Eksperimental'noi Meditsiny za 1956,* 625–28. Leningrad, 1957.

Knorr-Cetina, Karin. *The Manufacture of Knowledge: An Essay on the Constructivist and Contextual Nature of Science.* Oxford: Pergamon Press, 1981.

Kohler, Robert. *Lords of the Fly: Drosophila Genetics and the Experimental Life.* Chicago: University of Chicago Press, 1994.

Kol'tsov, N. K. "Trud zhizni velikogo biologa." *Biologicheskii Zhurnal* 5, no. 3 (1936): 387–402.

Krementsov, Nikolai. *Stalinist Science.* Princeton: Princeton University Press, 1997.

Krohn, Roger. "Why Are Graphs So Central in Science?" *Biology and Philosophy* 6, no. 2 (April 1991): 181–203.

Kusch, Martin. "Recluse, Interlocutor, Interrogator: Natural and Social Order in Turn-of-the-Century Psychological Research Schools." *Isis* 86, no. 3 (1995): 419–39.

Kvasov, D. G., and A. K. Fedorova-Grot. *Fiziologicheskaia Shkola I. P. Pavlova.* Leningrad: Nauka, 1967.

Landes, David S. *The Unbound Prometheus: Technological Change and Industrial Development in Western Europe from 1750 to the Present.* Cambridge: Cambridge University Press, 1969.

Latour, Bruno. "The Costly Ghastly Kitchen." In *The Laboratory Revolution in Medicine,* ed. Andrew Cunningham and Perry Williams, 295–303. Cambridge: Cambridge University Press, 1992.

Latour, Bruno, and Steve Woolgar. *Laboratory Life: The Social Construction of Scientific Facts.* Beverly Hills: Sage, 1979.

Liljestrand, Goran. "The Prize in Physiology or Medicine." In *Nobel: The Man and His Prizes,* ed. H. Schück et al., 131–344. Amsterdam: Elsevier Publishing Co., 1962.

Lindenmeyr, Adele. *Poverty Is Not a Vice: Charity, Society, and the State in Imperial Russia.* Princeton: Princeton University Press, 1996.

Liozner-Kannabikh, S. A. "Iz vospominanii ob Ivane Petroviche Pavlove." *Zhurnal nevropatologii i psikhiatrii imeni S. S. Korsakova* 53, no. 7 (1953): 580–82.

Loir, Adrien. *Le Prince Alexandre D'Oldenbourg, initiator de l'alliance Franco-Russe, et ses relations avec Pasteur.* Le Havre: Havre Eclair, 1933.

Lombard, Warren. "The Life and Work of Carl Ludwig." *Science*, 15 September 1916, 363–75.

Materialy k istorii VIEM. Vol. 1. Moscow, 1941.

Matthews, David M. *Protein Absorption: Development and Present State of the Subject*. New York: Wiley-Liss, 1991.

Merkulov, V. L. "Materialy o druzhbe I. P. Pavlova s R. Tigershtedtom." *Fiziologicheskii zhurnal SSSR* 14, no. 9 (1959): 1162–65.

———. "Rabota nemetskikh i avstro-vengerskikh uchenykh v fiziologicheskom otdele Instituta Eksperimental'noi Meditsiny pod rukovodstvom I. P. Pavlova 1902–1908 gg." *Ezhegodnik Instituta Eksperimental'noi Meditsiny* 4 (1961): 511–20.

Meshkunov, V. S., and A. M. Blokh. "Al'fred Nobel' i Imperatorskii Institut Eksperimental'noi Meditsiny v Sankt-Peterburge." *Voprosy istorii estestvoznaniia i tekhniki* 1 (1994): 121–28.

Orbeli, L. A. *Vospominaniia*. Moscow: Nauka, 1966.

Pashutin, V. I. *Kratkii ocherk Imperatorskoi Voenno-Meditsinskoi Akademii za 100 let eia sushchestvovaniia*. St. Petersburg, 1898.

Pickering, Andrew. *The Mangle of Practice: Time, Agency and Science*. Chicago: University of Chicago Press, 1995.

Porter, Theodore. *Trust in Numbers: The Pursuit of Objectivity in Science and Public Life*. Princeton: Princeton University Press, 1995.

Potter, Brian. "The History of the Disease Called Lupus." *Journal of the History of Medicine and Allied Sciences* 48 (1993): 81–82.

Principe, Lawrence. *The Aspiring Adept: Robert Boyle and His Alchemical Quest*. Princeton: Princeton University Press, 1998.

Rabinbach, Anson. *The Human Motor: Energy, Fatigue, and the Origins of Modernity*. Berkeley: University of California Press, 1992.

Rheinberger, Hans-Jorg. *Toward a History of Epistemic Things: Synthesizing Proteins in the Test Tube*. Stanford: Stanford University Press, 1997.

Salomon, P. "Imperatorskii Institut Eksperimental'noi Meditsiny v S.-Peterburge." *Arkhiv Biologicheskikh Nauk* 1 (1892): 3–22.

Samoilov, V. O., and A. S. Mozzhukhin. *Pavlov v Peterburge-Petrograde-Leningrade*. Leningrad: Lenizdat, 1989.

Samuel, Raphael. "Workshop of the World: Steam Power and Hand Technology in Mid-Victorian Britain." *History Workshop: A Journal of Socialist Historians* 3 (spring 1977): 6–72.

Savich, V. V. "Ivan Petrovich Pavlov: Biograficheskii ocherk." In *Sbornik, posviashchennyi 75-letiiu akademika Ivana Petrovicha Pavlova*, ed. V. L. Omelianskii and L. A. Orbeli, 3–31. Leningrad: Gosudarstvennoe Izdatel'stvo, 1924.

Schaffer, Simon. "Astronomers Mark Time: Discipline and the Personal Equation." *Science in Context* 2, no. 1 (1988): 115–45.

Schröer, Heinz. *Carl Ludwig*. Stuttgart: Wissenschaftliche Verlagsgesellschaft, 1967.

Shapin, Steven. *A Social History of Truth: Civility and Science in Seventeenth-Century England*. Chicago: University of Chicago Press, 1994.

Shapin, Steven, and Simon Schaffer. *Leviathan and the Air-Pump: Hobbes, Boyle, and the Experimental Life*. Princeton: Princeton University Press, 1985.

Silverstein, Arthur. *A History of Immunology.* San Diego: Academic Press, 1989.

Smith, Robert W. "Large-Scale Scientific Enterprise." In *Encyclopedia of the United States in the Twentieth Century,* ed. Stanley Kutler. Vol. 2, 739–65. New York: Charles Scribner's Sons, 1996.

Todes, Daniel P. "Biological Psychology and the Tsarist Censor: The Dilemma of Scientific Development." *Bulletin of the History of Medicine* 58 (1984): 529–44.

———. *Darwin without Malthus: The Struggle for Existence in Russian Evolutionary Thought.* New York: Oxford University Press, 1989.

———. "From Radicalism to Scientific Convention: Biological Psychology in Russia from Sechenov to Pavlov." Doctoral diss., University of Pennsylvania, 1981.

———. "Pavlov and the Bolsheviks." *History and Philosophy of the Life Sciences* 17 (1995): 379–418.

———. "Pavlov's Physiology Factory." *Isis* 88, no. 2 (1997): 205–46.

———. "V. O. Kovalevskii: The Genesis, Content, and Reception of His Paleontological Work." *Studies in History of Biology* 2 (1978): 99–165.

Tugan-Baranovsky, M. I. *The Russian Factory in the Nineteenth Century.* 1907. Trans. Arthur Levin and Clara Levin. Homewood, Ill.: Richard D. Irwin, 1970.

Ushakov, V. G. "Antirabicheskoe otdelenie." In *Materialy k istorii Vserossiiskogo Instituta Eksperimental'noi Meditsiny.* Moscow, 1941.

———. "Iz istorii VIEM." In *Materialy k istorii VIEM.* Moscow, 1941.

Vitte, S. Iu. *Vospominaniia.* Vol. 2. Moscow: Izdatel'stvo Sotsial'no-Ekonomicheskoi Literatury, 1960.

Volodin, B., and V. Demidov. *Zhazhda Istiny.* Moscow: Sovetskaia Rossiia, 1988.

Warner, John Harley. "Ideals of Science and Their Discontents in Late Nineteenth-Century American Medicine." *Isis* 82 (1991): 454–78.

Windholz, George. "Pavlov and the Pavlovians in the Laboratory." *Journal of the History of the Behavioral Sciences* 26 (1990): 64–74.

Windholz, George, and James R. Kuppers. "Pavlov and the Nobel Prize Award." *Pavlovian Journal of Behavioral Science* 25 (1990): 155–62.

Wortman, Richard, S. *Scenarios of Power: Myth and Ceremony in Russian Monarchy.* Vol. 2. Princeton: Princeton University Press, 2000.

SUBJECT INDEX

NAME INDEX